SKELETAL MUSCLE PLASTICITY
AND DISEASE

Advances in Muscle Research

Volume 2

Series Editor

G.J.M. Stienen, *Vrije Universiteit, Amsterdam, The Netherlands*

Skeletal Muscle Plasticity in Health and Disease

From Genes to Whole Muscle

Edited by

Roberto Bottinelli
University of Pavia, Pavia, Italy

and

Carlo Reggiani
University of Padova, Padova, Italy

 Springer

A C.I.P. Catalogue record for this book is available from the Library of Congress.

ISBN-10 1-4020-5176-X (HB)
ISBN-13 978-1-4020-5176-0 (HB)
ISBN-10 1-4020-5177-8 (e-book)
ISBN-13 978-1-4020-5177-7 (e-book)

Published by Springer,
P.O. Box 17, 3300 AA Dordrecht, The Netherlands.

Printed on acid-free paper

CONTENTS

PREFACE

Striated muscles have a striking ability to adapt to variations in activity and working demand. Structure and function change in response to changes in activity in a way that reminds the ability of brain to learn and store memories. The last few years show a rising interest in muscle adaptation not only in basic science but also in applied disciplines. An increasing number of people is interested in training, in improving physical performance or simply in obtaining muscle enlargement in the context of a desire for fitness, health and wellbeing. Closely connected is the increased awareness on the loss of muscle performance that occurs as we get older and relating to disuse, disease and senescence itself.

Plasticity is the most appropriate wording to describe the ability of muscle to adapt. In recent years our understanding of the cellular and molecular mechanisms which contribute to plasticity have steadily mounted. The expression has become popular and widely used since its introduction as a title of a symposium held in Konstanz (Germany) in 1979. The proceedings of the meeting were collected in a book "Plasticity of muscle" edited by Dirk Pette in 1980. The secret wish is that the present book becomes an updating of Pette's important book in this field which, after a quarter of century, has seen an unforeseen accumulation of knowledge. The expansion of our knowledge has been chiefly due to the progress in genetics and molecular biology in eukaryotes and its application to muscle biology.

In our present view the adaptive events involve all muscle fibre structures from myofibrils to mitochondria, membranes, extracellular matrix, as well as the capillaries surrounding muscle fibres. Gene expression studies, particularly after the advent of high-throughput technology, reveal that adaptation is often based at the transcriptional level. Many signalling pathways involving chains of protein kinases and transcription factors mediate transcriptional regulation acting on gene promoter regions. Further regulations occur at translational and post-translational levels that result in rapid re-organization of the muscle fibre proteome when appropriate stimuli are given to the muscle. And contractile and metabolic performance of the muscle change as a direct consequence of the changes in the protein complement.

The present book aims to cover muscle plasticity at the molecular and cellular level as well as at the level of whole muscles in the body. Thus, after an initial overview (Chapter 1) written by Dirk Pette, chapters 2 and 3 (written by Kandarian and Haddad, Pandorf, Giger, Baldwin, respectively) describe the transcriptional regulation and its control mechanisms. Chapter 4 (Schiaffino, Murgia, Sandri) and chapter 5 (Liu, Chen, Randall, Schneider) then provide an overview on intracellular signalling, with particular attention to activity and disuse. The possible contribution of satellite cells as providers of myonuclei during muscle adaptations is discussed in chapter 6 (Gillespie, Holterman, Rudnicki). Chapter 7 (Paine, DelBono) and 8

(Mounier, Bastide, Stevens) describe the structural and functional changes which may occur during muscle adaptations in excitation-contraction coupling and, respectively, in the contractile machinery. Chapter 9 (Narici) enlarges the view to include the changes which occur in muscle architecture during adaptations. Finally, chapter 10 (Harridge) is focussed on muscle plasticity induced by endocrine and paracrine signals whereas chapter 11 (Gea and coworkers) describes how skeletal muscles respond to adapt to systemic diseases.

As suggested by the above short list of the subjects, this book wishes to give a comprehensive and fairly systematic review of the subject. The book is meant to be of help for persons entering in the field of muscle biology, physiology or physiopathology, as PhD students or young physicians specializing in neurology, geriatric medicine, physiotherapy and sport medicine. It is also meant to be a source for updating and refreshing the knowledge of people already working in the field either in basic sciences or in applied and medical disciplines.

The editors wish to thank all authors who have contributed to this book not only with their scientific expertise, but also for their endurance. They also want to thank the team at Springer publishing house, Fabio de Castro, Laura van Mourik and Marieke Mol.

Roberto Bottinelli
University of Pavia, Pavia, Italy

Carlo Reggiani
University of Padova, Padova, Italy

CHAPTER 1

SKELETAL MUSCLE PLASTICITY – HISTORY, FACTS AND CONCEPTS

DIRK PETTE

University of Konstanz, Germany

1.   INTRODUCTION

In 1873, Louis Ranvier characterized "white" and "red" muscles of the rabbit and ray as "tetanic-fast" and "tonic-slow", respectively (Ranvier, 1873). More recently, functional specialization of muscles and muscle fibers has become a major topic of interest. The muscular system is unique in its architectural design and complexity. As such, each skeletal muscle within an organism is distinct and even homologous muscles in different species and strains have been shown to differ. Contributing to this diversity is the design of muscle as a composite tissue with a large variety of fiber phenotypes distributed in varying proportions in each muscle. The phenotypic properties of muscle fibers are determined by endogenous, species-specific programs which can be modulated by a variety of exogenous influences, making muscle a highly adaptive tissue. This adaptive potential marked an important evolutionary achievement because it improves survival under altered environmental conditions.

The change in paradigm from functionally distinct muscles and fiber types that are essentially static structures to that of an ever-changing, dynamic model has taken a long time since Ranvier's characterization of two functionally distinct types of muscle in rabbit and ray (Ranvier, 1873). The initial concept of a few distinct fiber types which emerged from studies combining histochemical with physiological methods, has been increasingly modified by more refined analytical methods, such as immunohistochemistry, microbiochemistry, and molecular biology. Thus, the categorization of fiber types has given way to the notion that terminally differentiated, postmitotic muscle fibers are versatile entities. In this regard, the nerve cross-union experiment performed over four decades ago by Buller, Eccles and Eccles (1960) has become a landmark study for its demonstration of muscle

1

R. Bottinelli and C. Reggiani (eds.), Skeletal Muscle Plasticity in Health and Disease, 1–27.
© 2006 *Springer.*

plasticity. They showed that the acutely denervated slow-twitch soleus muscle of the cat turned faster when it was reinnervated by nerve fibers normally supplying the fast-twitch flexor digitorum longus (FDL) muscle. Conversely, the fast FDL muscle converted into a slower contracting muscle after being reinnervated with the nerve normally supplying the slow soleus muscle. These findings brought to light the phenotypic influence innervation exerts on muscle, and more importantly, demonstrated the functional malleability of fully differentiated mammalian skeletal muscles. The moulding influence of innervation on developing and adult vertebrate muscles was a discovery with an impact similar to the discovery of the specific effects hormones and growth factors exert on cell and tissue differentiation. Since the landmark study of Buller, Eccles and Eccles (1960), numerous studies have increased our understanding of the cellular and molecular mechanisms underlying the processes of myogenesis, muscle fiber differentiation, specialization, and transformation. Although skeletal muscle is a dynamic tissue, the classification of muscle fiber types has played an important role in most of these studies. The development of different types of muscle fibers and their modulation are tightly connected to the evolution of the concept of muscle plasticity and will, therefore, be dealt with in the following paragraphs.

1.1   Muscle Fiber Types

Early attempts at classifying specific fiber types were based solely on light microscopic morphology (Grützner, 1883; Knoll, 1891; Krüger, 1952). A breakthrough, however, came with the introduction of enzyme histochemical methods (Padykula and Herman, 1955; Ogata, 1958a; 1958b; 1958c; Dubowitz and Pearse, 1960; Engel, 1962). Several approaches combining histochemical, biochemical, and physiological methods ultimately led to the delineation of functionally and metabolically distinct fiber types (Guth and Samaha, 1969; Brooke and Kaiser, 1970; Barnard et al., 1971; Burke et al., 1971; Peter et al., 1972; Close, 1972). Based on enzyme histochemistry, two major classifications schemes evolved. One is entirely based on myofibrillar adenosine triphosphatase (mATPase) histochemistry and separates fibers into slow type I and fast type II, including subtypes IIA, IIDX, and IIB (Guth and Samaha, 1969; Brooke and Kaiser, 1970). According to microbiochemical and immunohistochemical analyses, the mATPase profiles of these fiber types correspond to specific myosin isoforms (Staron and Pette, 1986; Pette and Staron, 1990; Gorza, 1990; Hämäläinen and Pette, 1993). The other classification scheme distinguishes between fast-twitch glycolytic (FG), fast-twitch oxidative-glycolytic (FOG), and slow-twitch oxidative (SO) fibers by combining mATPase histochemistry with histochemical assays for enzyme activities specific to aereobic and anaerobic energy metabolism (Barnard et al., 1971).

Although both methods are still in use, their resolution is relatively low in comparison to immunohistochemistry, single fiber biochemistry, and the latest methodological achievements such as real-time RT-PCR and microarray techniques.

According to results obtained using these methods, differences in gene expression patterns between the above-mentioned fiber types can be estimated to be in the range of hundreds. Whole genome gene expression analyses of mouse skeletal muscles predominantly composed of SO + FOG fibers (soleus), and FOG + FG fibers (extensor digitorum longus) provide a rough estimate of the fiber diversity in different muscles. Microarray analysis has identified 70 genes differentially expressed in these two muscles (Wu et al., 2003). A comparison of global mRNA expression patterns of mouse quadriceps and soleus muscles has yielded similar results (Campbell et al., 2001). Still, these results are preliminary and should be viewed with caution, especially since they were obtained from mixed fiber populations (see also (Reggiani and Kronnie, 2004)). It may be speculated, however, that such analyses at the single fiber level will demonstrate an even greater number of genes differentially expressed between fibers within a muscle.

1.2   Muscle Fibers and Myofibrillar Protein Diversity

Protein isoforms represent major elements of functional diversity. Most sarcomeric proteins are expressed in various isoforms. Moreover, many are composed of two or more subunits, with each subunit existing in multiple isoforms (for reviews see (Pette and Staron, 1990; Schiaffino and Reggiani, 1996)). Good examples of protein diversity can be found in the sarcomeric regulatory proteins tropomyosin (TM) and troponin (TN). The dimeric TM molecule exists in up to six different variants composed of homo- or heterodimers of its three major subunit isoforms, whereas the troponin molecule shows even greater diversity. Troponin is composed of three subunits, each existing as two or more isoforms. Of these, TnI and TnC (the inhibitory and Ca^{2+}-binding subunits, respectively) exist as fast and slow isoforms, whereas the tropomyosin-binding TnT subunit exists as at least four fast and three slow isoforms. Four different fast isotroponins may be formed by the assembly of one of the four fast TnT subunits with both the fast TnI and TnC subunits. Assemblies of the slow TnT isoforms with slow TnI, and TnC create a variety of three slow isotroponins. Specific pCa/tension relationships of different fiber types have been attributed to specific distribution patterns of the various isotroponins (Sréter et al., 1975; Schachat et al., 1987; 1990; Galler et al., 1997c; Geiger et al., 1999; Bastide et al., 2002).

1.3   Myosin-Based Fiber Types

Among myofibrillar proteins, myosin displays the greatest diversity. It is composed of two heavy chains (MHC), two essential light chains (ELC), and two regulatory light chains (RLC) according to the formula $[(ELC)_2(RLC)_2(MHC)_2]$. Skeletal muscles of small mammals contain four major MHC isoforms, MHCI in slow type I fibers and three fast isoforms, i.e., MHCIIa, MHCIIdx and MHCIIb, in the fast fiber types IIA, IIDX, and IIB, respectively (Weiss et al., 1999).

Although only one slow fiber type has been characterized to date, evidence exists that type I fibers do not represent a homogeneous population (Galler et al., 1997a; Andruchov et al., 2003). According to combined histochemical and microbiochemical studies, the four MHC-based fiber types can also be delineated by mATPase histochemistry (for reviews see Pette and Staron, 1990; Bottinelli and Reggiani, 2000).

Experiments measuring the maximum unloaded shortening velocity (V_0) of single fibers were a breakthrough in the functional characterization of the four MHC-based fiber types (Bottinelli et al., 1991; Larsson and Moss, 1993; Galler et al., 1994; Bottinelli et al., 1994a; 1994b; 1996; Bottinelli and Reggiani, 2000; Pellegrino et al., 2003)). More recently, a functional characterization of different myosins extracted from single fibers has been performed by measuring the speed of actin translocation *in vitro* (Pellegrino et al., 2003; Toniolo et al., 2004). According to results from both methods, V_0 increases in a sequential order in adult rat muscle fibers such that type I << type IIA < type IIDX < type IIB. Measurements of stretch activation kinetics, relevant to kinetic properties of the myosin head, have demonstrated a similar order (Galler et al., 1997b; Hilber et al., 1999; Andruchov et al., 2004). It should be noted, however, that V_0 values measured on adult muscle fibers have been found to change with age, especially in the case of type I fibers. These apparent age-dependent changes in V_0 may be the result of posttranslational modifications of myosin (e.g., by glycation, oxidative stress, or nitration) (Höök et al., 2001).

Studies on various mammals (mouse, rat, rabbit, horse, rhesus monkey, man) have revealed similar differences in V_0 between MHC-based fiber types in homologous muscles (Rome et al., 1990; Widrick et al., 1997; Canepari et al., 2000; Höök et al., 2001; Pellegrino et al., 2003). Differences, however, exist in the absolute V_0 values of orthologous myosins, i.e., myosins from same fiber types in muscles of different species. As a general rule, V_0 of orthologous myosins decreases with body mass. Consequently, the shortening velocity of orthologous fiber types is higher in small compared to large mammals (Pellegrino et al., 2003). Also the stretch activation kinetics of orthologous myosins have been reported to display a similar relation to body mass (Andruchov et al., 2004). This diversity between orthologous myosins obviously relates to structural differences in the actin-binding surface of the head and also in the S2 subfragment. Its physiological significance has been interpreted as an adjustment of muscle shortening velocity to body size (Pellegrino et al., 2003).

According to single fiber measurements on MHC-based fiber types in muscles of rat and human, the differences in V_0 correspond to similar differences in myofibrillar ATPase activity (Bottinelli et al., 1994c; Stienen et al., 1996; He et al., 2000; Canepari et al., 2000). The correlation between ATPase activity and V_0 is highly relevant with regard to fiber type-specific tension costs (Fig. 1). High contraction velocity is attained at the expense of high ATP consumption, whereas low contraction velocity consumes less ATP and, therefore, appears to be more economical. In this regard, it is of interest to note that measurements

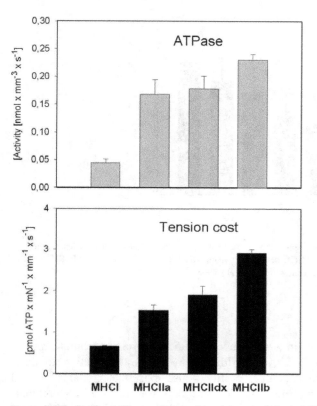

Figure 1. Myofibrillar ATPase activity and tension cost of single MHC-based rat muscle fiber types. Plots were drawn according to the data (means ± SEM) from Bottinelli et al., 1994

of the ATP phosphorylation potential in single fibers from rabbit muscle have revealed gradients that are similar to the differences in ATPase activities and tension costs of rat muscle fibers. The $[ATP]/[ADP_{free}]$ ratio is highest in type IIB fibers, intermediate in types IIDX and IIA, and lowest in type I fibers (Fig. 2) (Conjard et al., 1998).

In addition to these qualitative differences in myofibrillar protein isoforms, MHC-based fiber types also exhibit quantitative differences in specific protein levels. For example, the fast Ca^{2+}-ATPase isoform, SERCA1, is expressed at higher levels in type IIB and IIDX fibers than in type IIA fibers (Maier et al., 1986; Krenács et al., 1989). Also, phospholamban, a regulatory protein of the sarcoplasmic reticulum Ca^{2+}-ATPase, is expressed in type I but not in type II fibers (Jorgensen and Jones, 1986), whereas parvalbumin (a cyotosolic Ca^{2+}-binding protein) is expressed at high levels in type IIB and IIDX fibers, hardly detectable in type IIA, and absent in type I fibers (Heizmann et al., 1982; Schmitt and Pette, 1991).

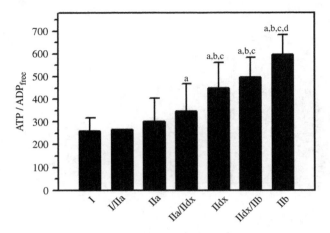

Figure 2. [ATP]/[ADP$_{free}$] ratios in pure and hybrid fiber types from adult rabbit muscles. Values are means ± SD. a. Different from MHCI; b. different from MHCIIa; c. different from MHCIIa/IIdx; d. different from MHCIIdx. Adapted from Conjard et al., 1998

1.4   Hybrid Fibers

Adult muscles normally contain large numbers of fibers expressing a single, unique MHC isoform (MHCI, MHCIIa, MHCIIdx, or MHCIIb). In addition to these "pure" fibers, a considerable number of fibers also exist that contain two and occasionally more MHC isoforms ("hybrid" fibers). These hybrid fibers preferentially display MHC combinations such as (MHCI+MHCIIa), (MHCIIa+MHCIIdx) or (MHCIIdx+MHCIIb) (Pette and Staron, 1997; 2000; Caiozzo et al., 2003). The functionally and energetically determined order of MHC isoforms appears to explain the existence of hybrid fibers with so-called "nearest-neighbor" combinations (Pette and Staron, 1997)not only in transforming muscles but also in muscles under steady state conditions. As such, the coexistence of functionally similar MHC isoforms may serve to optimally adjust muscle fibers to their exact, specific function.

Hybrid fibers obviously bridge the relatively large functional "gaps" between pure fiber types. According to their V_0 values and stretch activation kinetics (Hilber et al., 1999; Bottinelli, 2001; Toniolo et al., 2004), pure and hybrid fibers form a fiber continuum in the order of, type I – type I/IIA – type IIA – type IIA/DX – type IIDX – type IIDX/B – type IIB.

Although the MHC complement of a muscle fiber can be regarded as a major determinant of its contractile speed, modulatory effects of the myosin light chains have been identified (Lowey et al., 1993; Bottinelli, 2001). Taken together, the combinatorial patterns of myosin light and heavy chains form a large spectrum of different isomyosins (Pette and Staron, 1990; Hämäläinen and Pette, 1995) creating a much greater functional diversity at the thick filament level than from MHC diversity alone.

2. PLASTICITY OF MUSCLE

The classic definition of fiber types may appear meaningless in view of the large diversity resulting from myofibrillar protein isoforms and metabolic specialization. In addition, the different fiber type classification schemes (which originally evolved in parallel with methodological developments) have been more recently refined and modified by new insights and methods. Finally, classifying fiber types may be questioned because numerous studies have demonstrated that adult, postmitotic muscle fibers are not fixed cellular modules, but represent versatile entities. Nevertheless, some predominant fiber populations (or "fiber types") can be identified in adult skeletal muscles by distinct protein and metabolic profiles adjusted to specific functional requirements. Persistent changes in functional demand, however, induce adaptive changes and even fiber type transitions (Booth and Baldwin, 1996; Pette and Staron, 1997; Hamilton and Booth, 2000; Flück and Hoppeler, 2003; Rennie et al., 2004). Muscle fibers are capable of adjusting their size and phenotypic properties in response to various endogenous and exogenous influences (such as innervation, total amount and temporal pattern of neuromuscular activity, mechanical loading/unloading, and specific hormones). This adaptive responsiveness has been termed the "plasticity of muscle" (Pette, 1980; 2001).

Nerve cross-union (Buller et al, 1960) provided the first compelling evidence that muscle fiber phenotypes are under neural control. As has been shown in the meantime, the changes in phenotype following cross-reinnervation are primarily due to specific impulse patterns delivered to the muscle (Vrbová, 1963a; Salmons and Vrbová, 1967; Lömo et al., 1974; Hennig and Lömo, 1985). Similar to cross-reinnervation, chronic electrical stimulation of skeletal muscles with slow or fast motoneuron-specific impulse patterns induce profound changes in phenotypic properties (for reviews see (Pette and Vrbová, 1992; Pette and Staron, 1997). Two protocols of electrical stimulation have been derived from motoneuron-specific firing patterns. Chronic low-frequency stimulation (CLFS) mimics the tonic low-frequency impulse pattern delivered to slow-twitch muscle (Salmons and Vrbová, 1969), whereas phasic, high-frequency stimulation mimics the impulse pattern delivered to fast-twitch muscles (Lömo et al., 1974). Electrical stimulation of denervated muscles, denervated/regenerating muscles, or aneural myotube cultures in vitro has also been shown to be a suitable experimental model for studying changes in response to specific impulse patterns (Gundersen et al., 1988; Gorza et al., 1988; Ausoni et al., 1990; Wehrle et al., 1994; Meissner et al., 2001; Liu et al., 2001; Serrano et al., 2001).

Endurance training and chronic low-frequency stimulation are major experimental models for investigating the effects of increased neuromuscular activity (for reviews see (Holloszy and Coyle, 1984; Pette and Vrbová, 1992; Salmons, 1994; Booth et al., 1998). Chronic low-frequency stimulation (CLFS) offers several advantages as compared to exercise training (Pette and Vrbová, 1999), 1) CLFS is a standardized and reproducible regimen, 2) activates all motor units of the stimulated muscle, 3) imposes much higher levels of activity over time than any exercise regimen, 4) the adaptive potential of the target muscle is forced to its maximal limits, 5) secondary systemic effects can be largely excluded because enhanced contractile activity is restricted to the target muscle. Because stimulation-induced changes are reversible, CLFS also provides a suitable experimental model for studying changes in response to decreased neuromuscular activity after cessation of stimulation.

Several experimental protocols are currently in use for studying the effects of mechanical overloading. Mechanical loading can be increased by hypergravity (Tavakol et al., 2002; Bozzo et al., 2004), immobilization in a lengthened position (Goldspink et al., 1992), or by tenotomy, denervation or ablation of synergists (Goldberg, 1967; Roy et al., 1985; Sugiura et al., 1993). The most frequently used models for studying the effects of unloading are microgravity (Desplanches, 1997; Fitts et al., 2001), hind limb suspension (Morey, 1979; Musacchia et al., 1980; Stevens et al., 1999a,b), tenotomy (Vrbová, 1963b; Jamali et al., 2000), immobilization of the muscle in a shortened position (Goldspink, 1977; Loughna et al., 1990), and spinal cord transsection and isolation (Margreth et al., 1980; Roy et al., 1984).

The phenotypic properties of muscle fibers are also under hormonal control. Thyroid hormones appear to have the greatest influence. In general, hypothyroidism causes fast-to-slow transitions with fast-to-slow shifts in MHC isoform expression (Ianuzzo et al., 1977; Fitts et al., 1980; Nwoye and Mommaerts, 1981), while hyperthyroidism causes slow-to-fast shifts (Izumo et al., 1986; Kirschbaum et al., 1990; Fitzsimons et al., 1990; Caiozzo et al., 1992; 1993; d'Albis and Butler-Browne, 1993; Larsson et al., 1994; Li et al., 1996; Canepari et al., 1998). Other hormones have been shown to modify the fiber type composition of specific muscles. For example, testosterone has a significant effect on the fiber type composition of guinea pig temporalis muscle (Gutmann and Hanzlíková, 1970; Lyons et al., 1986), and rabbit masseter muscle (English et al., 1999). An extreme example of sexual dimorphism is the levator ani muscle of the rat. This muscle disappears during development of the female, but can be maintained by testosterone administration (Hanzlíková et al., 1970; Joubert et al., 1994). Hormonal differences, especially testosterone, may also contribute to the gender differences in specific fiber type sizes that ultimately affect the relative concentrations of MHC isoforms (Staron et al., 2000).

2.1   Modes and Grades of Functional Adaptation

The adaptation of muscle fibers to altered functional demands can be interpreted as a process of adjusting energy supply to energy expenditure, and *vice versa*. It consists of quantitative and qualitative changes at both the molecular and cellular levels and affects the elements of energy metabolism and myofibrillar apparatus in a sequential and hierarchical order. Depending on the quality, intensity and duration of the modifying influence, the adaptive process may be restricted to changes in energy metabolism or it may extend to the major systems of energy expenditure, the myofibrillar apparatus and ionic pumps. Quantitative alterations in myofibrillar protein levels ultimately lead to changes in fiber size (atrophy/hypertrophy) (Booth and Criswell, 1997; Fitts et al., 2001; Rennie et al., 2004; Jackman and

Kandarian, 2004). Qualitative changes in the isoform patterns of contractile, regulatory, and functionally related proteins end up in fiber type transitions. Because many of the adaptive changes in response to enhanced neuromuscular activity have been studied with the use of CLFS, the following paragraphs on metabolic adaptations and fiber type transitions will focus on results using this model (Pette and Vrbová, 1992; 1999).

2.2   Metabolic Adaptations

The chemomechanical transformation underlying the contractile process requires a fine tuning between energy supply and energy expenditure. Fuel supply and energy metabolism of a muscle depend on the quality and quantity of active motor units and the duration of their recruitment. For example, the so-called fast-twitch glycolytic fibers are recruited predominantly for short-term, high-intensity performance, because they are rapidly fatigued (Edström and Kugelberg, 1968; Burke et al., 1971). Their ATP supply is based on easily accessible fuels, such as phosphocreatine and glycogen. However, the intracellular stores of these fuels are limited and the ATP yield of anaerobic glycogen catabolism is low. An adaptation of FG fibers to sustained activity, therefore, requires a switch to aerobic-oxidative energy metabolism including the use of additional fuels, such as fatty acids, ketone bodies, and amino acids. The corresponding adaptive changes at the cellular level primarily consist of elevations in mitochondrial content and up-regulation of the corresponding enzyme apparatus.

Skeletal muscles possess a considerable capacity to change their mitochondrial content in response to altered activity levels (Adhihetty et al., 2003; Flück and Hoppeler, 2003). The pioneering studies of Holloszy (1967) and Gollnick (1969) have shown that the activity profiles of enzymes involved in aerobic-oxidative pathways of energy metabolism depend on the training state of the muscle. Due to increases in mitochondrial content and elevations in enzyme activities of terminal substrate oxidation, endurance training enhances fatigue resistance (Saltin and Gollnick, 1983; Holloszy and Coyle, 1984; Booth and Baldwin, 1996; Williams and Neufer, 1996; Turner et al., 1997). The increases in enzyme levels are preceded by alterations in transcriptional rate and elevated levels of specific mRNAs, e.g. (Hood et al., 1989; Puntschart et al., 1995; Zhou et al., 2000; Hood, 2001; Nordsborg et al., 2003; Hildebrandt et al., 2003; Agbulut et al., 2003; Wu et al., 2003). A single bout of exhaustive exercise has been shown to be sufficient to induce several fold increases in transcriptional activities of specific mitochondrial enzymes in human muscle (Pilegaard et al., 2000). However, these increases are transient and cumulative effects of repetitive increases in transcriptional activities seem to be required to elevate protein levels by increases in translational activities.

As shown in studies on rats and rabbits, the metabolic changes in low-frequency stimulated muscles by far exceed those induced by endurance training (Pette and

ıvá, 1992; 1999). The greatly enhanced aerobic-oxidative capacity of low-
ıι juency stimulated rabbit muscles is reflected by several fold increases in
mitochondrial content (Eisenberg and Salmons, 1981; Reichmann et al., 1985) and
enzyme activities of terminal substrate oxidation (transport, activation and oxidation
of fatty acids, ketone body utilization, amino acid oxidation, the citric acid cycle, the
respiratory chain) (Simoneau and Pette, 1988; Hood and Pette, 1989). Conversely,
enzyme activities of glycogen metabolism and glycolysis are markedly reduced.
Improvements in fuel and oxygen supply by increases in capillary density (Skorjanc
et al., 1998; Egginton and Hudlická, 1999; Lloyd et al., 2003; Waters et al., 2004)
and myoglobin content (Holloszy and Coyle, 1984; Michel et al., 1994; Waters
et al., 2004) represent additional adaptive responses contributing to enhanced
fatigue-resistance and endurance.

The metabolic adaptations to reduced neuromuscular activity may be described
as a reversal of the changes induced by increased neuromusclar activity. The major
changes in energy metabolism by reduced neuromuscular activity (e.g., by immobilization, bed rest or after cessation of exercise training) are reflected by decreases in
mitochondrial content and enzyme activities of aerobic-oxidative metabolism. Fast-
twitch and slow-twitch (postural) fibers are differently affected and the metabolic
changes coincide with changes due to fiber atrophy, especially of type I fibers
(Ferretti et al., 1997; Shenkman et al., 2002; Flück and Hoppeler, 2003; Hoppeler
and Flück, 2003).

2.3 Fiber Type Transitions

Persistently altered neuromuscular activity evokes changes beyond the metabolic
adaptations described above. Qualitative and quantitative changes in gene expression
lead to altered isoform profiles of myofibrillar and other sarcomeric proteins
ultimately resulting in fiber type transitions. In general, enhanced neuromuscular
activity (endurance exercise, CLFS) or mechanical load (e.g., stretch) shifts the fiber
population toward slower and metabolically more oxidative phenotypes, whereas
fiber type transitions in the opposite direction result from persistently reduced
neuromuscular activity (Pette and Staron, 1997; Talmadge, 2000; Baldwin and
Haddad, 2001). Fiber type transitions obviously represent an additional adaptive
response to adjust energy expenditure to major and persistent changes in functional
demands.

CLFS, which induces the most extensive fast-to-slow phenotype transitions,
virtually affects all functional elements of the muscle fiber. In addition to the
up-regulation of enzymes gearing aerobic-oxidative metabolic pathways and the
enhanced mitochondrial biogenesis, the changes include fast-to-slow isoform transitions of myosin light and heavy chains, tropomyosin, troponin subunits, α-actinin,
and various proteins of the sarcoplasmic reticulum and cytosolic Ca^{2+}-regulatory
systems. A detailed description of these changes has been given in previous reviews
(Pette and Vrbová, 1992; 1999). A brief summary of the transitions in myosin
isoforms will be given in the following paragraph.

The fast-to-slow transitions in MHC isoform expression follow specific time courses. Transitions between the fast isoforms precede the final fast-to-slow conversion. In rat tibialis anterior and extensor digitorum longus muscles, the conversion is initiated by an exchange of MHCIIb with MHCIIdx, followed by an exchange of MHCIIdx with MHCIIa. The final transition from MHCIIa to MHCI isoform is incomplete and requires very long stimulation periods (> 100 days) (Windisch et al., 1998), perhaps because the slow isoform appears to be expressed only in fibers undergoing degeneration/regeneration (Pette et al., 2002). In the rabbit, where MHCIIdx is the fastest isoform expressed in these muscles, the fast-to-slow conversion proceeds from MHCIIdx to MHCIIa, and finally to MHCI (Leeuw and Pette, 1993).

Sequential fast-to-slow transitions in isomyosin composition have also been observed in nerve cross-union studies on rat muscle. When the fast-twitch extensor digitorum longus muscle of the rat was reinnervated with the nerve normally supplying the slow-twitch soleus muscle, sequential transitions of fast MHC-based isomyosins in the order of MHCIIb $\longrightarrow$ MHCIIdx $\longrightarrow$ MHCIIa preceded the appearance of the slow MHCI-based isomyosin (Mira et al., 1992).

Data from immunohistochemical and single fiber studies have indicated that fiber type transitions occur in a sequential manner (Ausoni et al., 1990; Conjard et al., 1998; Stevens et al., 1999b). For example, chronic low-frequency stimulation of rabbit tibialis anterior muscles for periods of 20 and 30 days result in progressive fast-to-slow shifts of the fiber population (Conjard et al., 1998). At 20 days, type IIDX fibers decreased, while hybrid type IIDX/A fibers increased. At 30 days, the percentage of the type IIDX/A was reduced, while type IIA fibers and hybrid type IIA/I fibers increased. These type IIA/I fibers essentially represented precursors of the ever expanding type I fiber fraction under these chronic stimulation conditions.

The coexistence of different MHC isoforms and isomyosins in single fibers is characteristic of transforming muscle. Increases in the population of hybrid fibers have been observed under various conditions, e.g., in human muscle fibers with endurance training (Klitgaard et al., 1990), strength training (Staron and Hikida, 1992), and in chronically electrostimulated vastus lateralis muscles of spinal cord-injured individuals (Andersen et al., 1996). Increases in the number of hybrid fibers have also been shown to occur under conditions causing slow-to-fast transitions, e.g., unloaded rat soleus muscle (Stevens et al., 1999a), human vastus lateralis muscle after spinal cord injury (Burnham et al., 1997), and soleus muscles of hyperthyroid rats (Li and Larsson, 1997).

A common observation in studies where hybrid fibers have been identified, is the existence of preferential combinations of MHC isoforms, such as MHCI + MHCIIa, MHCIIa + MHCIIdx, and MHCIIdx + MHCIIb. These preferentially-combined isoform pairs can be aligned with the four separate MHC isoforms to form a continuum extending from the slowest to the fastest, MHCI − (MHCI + MHCIIa) − MHCIIa − (MHCIIa/MHCIIdx) − MHCIIdx − (MHCIIdx + MHCIIb) − MHCIIb.

Hybrid fibers containing more than two MHC isoforms (three or more) have also been observed under both normal and transforming conditions. Combinations such as MHCI + MHCIIa + MHCIIdx or MHCIIa + MHCIIdx + MHCIIb in single fibers lend support to the nearest-neighbor rule (Pette and Staron, 1997). Taken together, it appears that the sequence of fast-to-slow MHC isoform transitions follows an order corresponding to gradual differences in contractile speed, stretch activation kinetics, ATPase activity, tension cost, and ATP phosphorylation potential. As a result, MHC isoforms with lower energy cost for force production are up-regulated in response to enhanced neuromuscular activity.

A few single fiber studies have observed exceptions from the above proposed nearest-neighbor rule of MHC isoform combinations. Hybrid or "polymorphic" fibers (Caiozzo et al., 2003) displaying "atypical" MHC combinations have been observed in slow and fast hindlimb rat muscles under various experimental conditions (Caiozzo et al., 1998; Talmadge et al., 1999; Caiozzo et al., 2000) and also in normal rat diaphragm (Caiozzo et al., 2003). These same studies, however, also showed that the vast majority of fibers do exhibit combinations of MHC isoforms which are in agreement with the nearest-neighbor rule .

Time course studies on low-frequency stimulated rabbit muscles have shown that the fast-to-slow transitions in MHC expression do not occur in strict synchrony with changes in the MLC complement (Leeuw and Pette, 1996). Multiple MLC/MHC assemblies, including fast MLC/slow MHC and slow MLC/fast MHC combinations, can be expected to generate a variety of isomyosins resulting in gradual fiber type conversions producing less abrupt changes in contractile properties.

2.4   Fiber Type Specific Transitions

According to the scheme depicted in Fig. 3, the ranges of possible transitions depend on a fiber's position in the fiber type spectrum. The scheme also suggests that, with the exception of those types at the end of the continuum (types I and IIB), muscle fibers are capable of transforming in either direction. Both the direction and extent

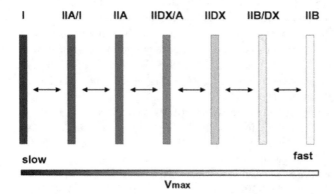

Figure 3. Scheme of a fiber type spectrum with a continuum of pure and hybrid fibers undergoing reversible phenotype transitions according to the nearest neighbor rule

of the transition are determined by the quality, intensity, and duration of the change in functional demand.

Metabolic adaptations appear to depend on similar parameters, i.e., the actual metabolic profile of the fiber, the quality, intensity, and duration of the stimulus. Therefore, enhanced neuromuscular activity should be expected to cause much smaller increases in mitochondrial content and enzymes of terminal substrate oxidation in fibers with intrinsically high aerobic-oxidative potential compared to low aerobic-oxidative potential. Conversely, reduced neuromuscular activity (which elicits metabolic changes in the opposite direction) should be expected to have the greatest impact on "oxidative" fibers compared to "glycolytic" fibers. Indeed, studies on homologous muscles of mouse, rat, guinea pig, and rabbit, exposed to identical protocols of CLFS showed that the induced increases in mitochondrial enzyme activities were inversely related to their basal levels (Simoneau and Pette, 1988). Obviously related to differences in basal metabolism, the baseline levels were highest in mouse, intermediate in guinea pig and rat, and lowest in rabbit.

Fiber type-specific metabolic adaptations have also been observed in a study on CLFS-induced increases in mitochondrial enzyme activities in the fast-twitch rabbit tibialis anterior muscle (Reichmann et al., 1985). The superficial portion of this muscle is characterized by high glycolytic and low mitochondrial enzyme activity levels, whereas the deep portion is more oxidative, containing twofold higher mitochondrial content and enzyme activities compared to the superficial portion. CLFS for 28 days led to an approximately six fold increase in mitochondrial content and enzyme activitiy levels in the superficial portion, compared to an approximately threefold increase in the deep portion. As a result, mitochondrial content and enzyme activitiy levels had become nearly identical in both portions of the muscle.

These and other observations suggest that muscle-specific and also species-specific differences in adaptive responses may, at least partially, be due to differences in fiber type composition. As shown by studies of MHC isoform patterns in muscles of mouse, rat, rabbit, and man, the fiber type composition of homologous muscles differs in an apparently regular manner with regard to body size. It appears as though the larger the animal the greater the shift toward slower fiber types (Hämäläinen and Pette, 1995). This relationship has been interpreted as an additional phenomenon contributing to the scaling of velocity of skeletal muscles with body size *in vivo* (Pellegrino et al., 2003). It remains to be seen in further studies what impact these fiber population shifts have on the adaptive ranges of skeletal muscles. CLFS might be an appropriate experimental model to investigate this problem in more detail under standardized and reproducible conditions.

3.   SIGNALING PATHWAYS INVOLVED IN FIBER TYPE TRANSITIONS

The multiplicity of adaptive changes that take place in adult muscle fibers in response to altered functional demands are based on a molecular reprogramming by orchestrated changes in gene expression coupled with alterations in protein translational activities and/or protein turnover. Considering the general concepts of enhanced neuromuscular activity inducing fast-to-slow and reduced neuromuscular activity inducing slow-to-fast transitions, the question arises as to whether or not changes in both directions could be under control of a common "master" signal. A possible answer emerges from the fact that adaptive responses of muscle fibers primarily encompass the two major systems related to chemomechanical energy transformation, 1) energy metabolism and 2) the myofibrillar apparatus. Obviously, energy supply and energy expenditure are balanced under steady state conditions as reflected by the relationship between ATP phosphorylation potential and myosin complement (Fig. 2). An imbalance of the two systems would not only impair the muscle fiber's performance but might be a major signal which initiates compensatory reactions to counteract the perturbation of the cellular energy state. Observations on the immediate metabolic effects of CLFS on rabbit muscle are in support of this suggestion. The most conspicuous change after the onset of stimulation consists of an immediate and drastic drop in the $[ATP]/[ADP_{free}]$ ratio. According to time course studies and analyses of single fibers, the reduction of the ATP phosphorylation potential persists with ongoing stimulation and equally affects the three fast MHC-based fiber types (Green et al., 1992; Conjard et al., 1998). The changes in the phosphorylation potential of low-frequency stimulated muscle are extreme but not unexpected in view of the extensive fiber type transitions. Smaller increases in work load most likely cause less dramatic shifts of the $[ATP]/[ADP_{free}]$ ratio than CLFS. In any case, disturbances of the ATP phosphorylation potential appears to represent a major signal for immediate and, if necessary, long-term counter reactions in energy supply and energy expenditure. Changes in $[ATP]/[ADP_{free}]$ are transmitted to signaling pathways involved in the control of fiber type-specific gene expression via a network of protein kinases. In addition, energy deprivation reduces the activity of the ATP-driven Ca^{2+}-pump of the sarcoplasmic reticulum (Läuger, 1991). The resulting impaired Ca^{2+} sequestration causes increases in the myoplasmic Ca^{2+} concentration which may then activate multiple Ca^{2+}-regulated pathways involved in the expression of slow fiber type-specific genes. In this regard, changes in the MHC isoform pattern of the gastrocnemius muscles from patients suffering from peripheral arterial occlusive disease are relevant, the MHC profile is significantly shifted toward slower isoforms under conditions of chronically impaired perfusion (Steinacker et al., 2000). Another observation, most likely also related to chronic energy deprivation, concerns the MHC profile of cytochrome oxidase-deficient fibers in muscles of old rats. According to their myosin complement, these fibers have been immunohistochemically identified exclusively as type I fibers (Skorjanc et al., 2001).

A key role in the up-regulation of slow fiber-specific genes has been proposed for a Ca^{2+}-regulated signaling cascade in which calcineurin (CaN), a Ca^{2+}/calmodulin-dependent, serine/threonine protein phosphatase, and Ca^{2+}/calmodulin dependent protein kinase activities (CaMK) play a major role. Ca^{2+}-activated CaN dephosporylates members of the NFAT (nuclear factors of activated T-cells) isoform family in the cytoplasm, thereby unmasking nuclear signals that enhance movement of NFATc proteins into the nucleus. Intranuclear NFAT proteins cooperate with other proteins, specifically members of the MEF2 (myocyte enhancer factor 2) family and activate the transcription of genes preferentially expressed in fatigue-resistant type I and type IIA fibers, such as MHCI, MHCIIa, slow TnI, myoglobin (Chin et al., 1998; Bigard et al., 2000; Olson and Williams, 2000a; Wu et al., 2001; Allen and Leinwand, 2002; Crabtree and Olson, 2002; Bassel-Duby and Olson, 2003; McCullagh et al., 2004; Chin, 2004; Giger et al., 2004; Akimoto et al., 2004; Liu et al., 2005). Inhibition of calcineurin by cyclosporin A suppresses these changes in metabolic and protein profiles. The observation that genetic loss of calcineurin blocks the mechanical overload-induced upregulation of slow fiber type-specific proteins in mouse muscle, convincingly demonstrates the essential role of calcineurin in fiber type transitions (Parsons et al., 2004).

Additional signaling cascades involved in phenotypic expression and adaptive remodeling of muscle fibers include the RAS/MAPK pathway (Murgia et al., 2000) and a protein kinase C-mediated pathway involved in mitochondrial biogenesis (Freyssenet et al., 1999; Freyssenet et al., 2004).

Taken together, Ca^{2+} plays a crucial role in skeletal muscle not only by regulating contraction but also by serving as an important signal for inducing alterations in gene expression. Possibly, this dual role requires the existence of separate Ca^{2+} pools and specific sensors for triggering contraction and Ca^{2+}-regulated gene expression. That changes in intracellular Ca^{2+} homeostasis might be involved in the CLFS-induced fast-to-slow transition process was suggested when Ca^{2+} sequestration by the sarcoplasmic reticulum was found impaired soon after the onset of stimlation (Heilmann and Pette, 1979; Hicks et al., 1997). Significant increases in free myoplasmic Ca^{2+} have been observed in single mouse muscle fibers during fatiguing stimulation (Westerblad and Allen, 1991; Chin and Allen, 1996). Indeed, single fibers from rat extensor digitorum muscles displayed 2.5-fold increases in resting Ca^{2+} concentration following exposure to CLFS (10 Hz) for only 2 hours (Carroll et al., 1999). These elevated Ca^{2+} levels, which were in the range similar to that of slow-twitch fibers from soleus muscle, persisted throughout the stimulation period (10 days). Interestingly, changes in free myoplasmic Ca^{2+} seem to precede and accompany also the slow-to-fast fiber transitions induced by mechanical unloading. Thus, resting Ca^{2+} in rat soleus fibers was found to decrease after 3 days, reaching the low level of the fast-twitch extensor digitorum longus muscle after 14 days of hindlimb suspension. These changes occurred in parallel with a reduced sarcolemmal permeability to cations (Fraysse et al., 2003).

The role of increased Ca^{2+} as a signal for the expression of slow fiber-specific genes has also been investigated in several studies using muscle cell cultures. Significant increases in MHCI mRNA and MHCI protein have been demonstrated in rabbit myotube cultures following an elevation in intracellular Ca^{2+} concentration induced by the Ca^{2+}-ionophore A23187, and these increases were prevented by applying the calcineurin antagonist cyclosporin A (Meissner et al., 2000; Meissner et al., 2001; Kubis et al., 2003). Similar experiments on L6E9 myoblast cultures have shown that cytochrome c gene expression can be induced by A23187 (Freyssenet et al., 1999), as well as by the Ca^{2+} ionophore ionomycin (Ojuka et al., 2003). In addition, stimulation of rabbit myocyte cultures at a low frequency for 24 hours has been shown to fully activate the calcineurin/NFAT pathway and to induce a distinct fast-to-slow transition at the MHC mRNA level (Kubis et al., 2003).

The Ca^{2+}-mediated activation of slow fiber type-specific genes is highly relevant with regard to specific, neurally induced oscillations in free myoplasmic Ca^{2+} concentration. Tonic low-frequency and phasic high-frequency impulse patterns generate different temporal changes in free Ca^{2+} concentration that may exert differential effects on gene expression (Westerblad and Allen, 1991; Chin and Allen, 1996; Olson and Williams, 2000b; Kubis et al., 2003). In this regard, results from a study on the effects of various impulse patterns on the translocation of NFATc in mouse muscle fibers are of special interest. Stimulation of single fibers *in vitro* with impulse patterns of 1 Hz and 50 Hz were inefficient, whereas 10 Hz stimulus trains caused nuclear translocation resulting in characteristic intranuclear NFAT patterns (Liu et al., 2001).

The role of NFAT as a nerve activity sensor in skeletal muscle has been substantiated by studying the transcriptional activity of NFAT in regenerating rat soleus muscle (McCullagh et al., 2004). NFAT activity decreased in denervated soleus muscle but increased by stimulation with a tonic low-frequency impulse pattern. However, a phasic high-frequency impulse pattern typical of fast motoneurons did not increase NFAT activity.

A separate signaling pathway has been implicated in work-induced metabolic adaptations of muscle that relates to the function of 5'AMP-activated protein kinase (AMPK). In addition to phosphorylation by AMPK kinase, AMPK is allosterically activated by 5'AMP and inhibited by phosphocreatine (Winder, 2001; Hardie, 2004). Therefore, AMPK (via adenylate kinase and creatine kinase) is tied directly to the energy state of the ATP/ADP system which makes it a highly responsive sensor of energy deprivation. Activation of AMPK has been suggested to initiate a Ca^{2+}-independent signal chain which triggers exercise-induced metabolic adaptations, such as increases in glucose transporter GLUT4, mitochondrial biogenesis, and elevations in enzymes of aerobic-oxidative metabolism (Winder et al., 2000; Winder, 2001; Sakamoto and Goodyear, 2002), and angiogenesis (Ouchi et al., 2005). Circumstantial evidence exists, however, that the AMPK signaling cascade may also be activated by increases in myoplasmic Ca^{2+}-concentration (Freyssenet et al., 2004). Nevertheless, the contribution of AMPK to the enhanced expression of nuclear and mitochondrial genes related to mitochondrial biogenesis has been convincingly demonstrated in a study on transgenic mice. Contrary to wild type mice, mitochondrial biogenesis was not enhanced in AMPK-deficient mice when intramuscular 5'-AMP levels were increased by feeding the animals the creatine analogue β-guanidinopropionic acid (Zong et al., 2002).

AMPK is a metabolic stress sensor and, therefore, the AMPK-triggered signal chain in muscle seems to be primarily directed toward metabolic adaptations. This has been suggested by studies on rat muscle where chronic activation of AMPK *in vivo* by administration of AICAR (5-aminoimidazole-4-carboxamide-1-beta-D-ribofuranoside) led to increases in mitochondrial enzyme activities but did not lead to changes in MHC isoforms or fiber type transitions (Putman et al., 2003). Early metabolic adaptations in response to enhanced neuromuscular activity may, thus, be mediated by the AMPK signal pathway. More extensive changes in response to stronger and/or longer lasting stimuli might activate Ca^{2+}-dependent signaling cascades involved in the remodeling of energy metabolism and myofibrillar proteins. Although AMPK- and Ca^{2+}-regulated pathways are both connected to the ATP phosphorylation potential of the muscle fiber, differences in threshold might provide the possibility of their separate regulation.

Finally, an independent signaling pathway controlling muscle gene expression and protein turnover has recently been identified (Lange et al., 2005). This pathway relates to the protein kinase domain of titin and its regulation by mechanically induced conformational changes. Titin protein kinase would thus function as a physiological link between sarcomere activity and transcriptional regulation.

4.   CONCLUSIONS

Skeletal muscles of small mammals are composed of pure and hybrid fibers which can be aligned in a fiber type spectrum energetically determined by gradual differences in contractile speed, myofibrillar ATPase activity, tension cost, and ATP phosphorylation potential. Type I fibers represent the slowest and type IIB the fastest members of this spectrum. Similarly, myofibrillar ATPase activity, tension cost, and ATP phosphorylation potential are lowest in type I and highest in type IIB fibers. The fiber populations of different muscles are adjusted to specific functional demands and, therefore, differ in the proportions of the various fiber types. However, the fiber type composition of any given muscle may change because postmitotic, terminally differentiated muscle fibers are versatile entities. They are capable of modifying their metabolic and contractile properties in response to altered functional demands. Qualitative and quantitative changes in gene expression serve to adjust energy supply to energy expenditure. In general, enhanced neuromuscular activity shifts energy metabolism toward aerobic-oxidative pathways and induces fast-to-slow transitions in the isoform spectrum of myofibrillar proteins. Conversely, reduced neuromuscular activity leads to changes in the opposite direction. The adaptive changes in metabolic and contractile properties exhibit characteristics of a dose-response relationship and seem to occur in a sequential order along the energetically determined spectrum of fiber types. While the direction of the transitions toward slower or faster phenotypes depends on the quality, intensity, and duration of the imposed stimulus, the maximum range of transitions depends on the location of the fiber within the spectrum. Type I fibers may convert only in the fast direction and type IIB fibers only in the slow direction, whereas all other fibers of the spectrum are capable of transforming, although with different ranges, in both directions. The metabolic adaptations to sustained alterations in neuromuscular activity seem to obey a similar rule.

The transitions between anaerobic-glycolytic and aerobic-oxidative pathways of energy metabolism depend on the "actual" metabolic profile of the fiber. In other words, "glycolytic" fibers respond to enhanced recruitment with greater increases in aerobic-oxidative potential than "oxidative" fibers. Oxidative fibers are metabolically prepared to meet the energetic demands of sustained activity and, thus, exhibit a relatively small adaptive range for increasing their aerobic-oxidative potential. Conversely, reduced neuromuscular activity hardly affects the metabolic profile of glycolytic fibers, but shifts that of oxidative fibers towards the glycolytic phenotype.

Perturbations of the energy state of the muscle fiber seem to represent major signals for fiber type conversions, especially with regard to transitions in the slow direction. Decreases in the ATP phosphorylation potential, as reflected by reduced $[ATP]/[ADP_{free}]$ ratios, translate into elevated concentrations of two signal molecules, Ca^{2+} and 5'-AMP, triggering several partially overlapping pathways involved in the control of metabolic adaptations and of expression patterns of myofibrillar and other sarcomeric protein isoforms.

REFERENCES

Adhihetty PJ, Irrcher I, Joseph AM, Ljubicic V, Hood DA. (2003). Plasticity of skeletal muscle mitochondria in response to contractile activity. Exp Physiol 88, 99–107.

Agbulut, O., Noirez, P., Beaumont, F., and ButlerBrowne, G.(2003) Myosin heavy chain isoforms in postnatal muscle development of mice. Bio Cell 6, 399–406.

Akimoto T, Ribar TJ, Williams RS, Yan Z. (2004). Skeletal muscle adaptation in response to voluntary running in Ca^{2+}/calmodulin-dependent protein kinase IV-deficient mice. Amer J Physiol 287, C1311–C1319.

Allen DL, Leinwand LA. (2002). Intracellular calcium and myosin isoform transitions. Calcineurin and calcium-calmodulin kinase pathways regulate preferential activation of the IIa myosin heavy chain promoter. J Biol Chem 277, 45323–45330.

Andersen JL, Mohr T, Biering-Sörensen F, Galbo H, Kjaer M. (1996). Myosin heavy chain isoform transformation in single fibres from m. vastus lateralis in spinal cord injured individuals, Effects of long-term functional electrical stimulation (FES). Pflügers Arch 431, 513–518.

Andruchov O, Andruchova O, Wang Y, Galler S. (2003). Functional differences in type-I fibres from two slow skeletal muscles of rabbit. Pflugers Arch Eur J Physiol 446, 752–759.

Andruchov O, Andruchova O, Wang YS, Galler S. (2004). Kinetic properties of myosin heavy chain isoforms in mouse skeletal muscle, comparison with rat, rabbit, and human and correlation with amino acid sequence. Amer J Physiol Cell Physiol 287, C1725–C1732.

Ausoni S, Gorza L, Schiaffino S, Gundersen K, Lömo T. (1990). Expression of myosin heavy chain isoforms in stimulated fast and slow rat muscles. J Neurosci 10, 153–160.

Baldwin KM, Haddad F. (2001). Effects of different activity and inactivity paradigms on myosin heavy chain gene expression in striated muscle. J Appl Physiol 90, 345–357.

Barnard RJ, Edgerton VR, Furukawa T, Peter JB. (1971). Histochemical, biochemical and contractile properties of red, white, and intermediate fibers. Am J Physiol 220, 410–414.

Bassel-Duby R, Olson EN. (2003). Role of calcineurin in striated muscle, development, adaptation, and disease. Biochem Biophys Res Commun 311, 1133–1141.

Bastide B, Kischel P, Puterflam J, Stevens L, Pette D, Jin JP, Mounier Y. (2002). Expression and functional implications of troponin T isoforms in soleus muscle fibers of rat after unloading. Pflügers Arch 444, 345–352.

Bigard X, Sanchez H, Zoll J, Mateo P, Rousseau V, Veksler V, Ventura-Clapier R. (2000). Calcineurin Co-regulates contractile and metabolic components of slow muscle phenotype. J Biol Chem 275, 19653–19660.

Booth FW, Baldwin KM (1996). Muscle plasticity, Energy demanding and supply processes. In, Rowell LB and Shepherd JT (eds) Handbook of Physiology, Section 12, Exercise, Regulation and Integration of Multiple Systems. Oxford University Press, New York, pp 1075–1123

Booth FW, Criswell DS. (1997). Molecular events underlying skeletal muscle atrophy and the development of effective countermeasures. Int J Sports Med 18, S265–S269.

Booth, F. W., Tseng, B. S., Flück, M., and Carson, J. A. (1998). Molecular and cellular adaptation of muscle in response to physical training. Acta Physiol.Scand. 3, 343–350.

Bottinelli R. (2001). Functional heterogeneity of mammalian single muscle fibres: do myosin isoforms tell the whole story? Pflügers Arch 443, 6–17.

Bottinelli R, Betto R, Schiaffino S, Reggiani C. (1994a). Maximum shortening velocity and coexistence of myosin heavy chain isoforms in single skinned fast fibres of rat skeletal muscle. J Muscle Res Cell Motil 15, 413–419.

Bottinelli R, Betto R, Schiaffino S, Reggiani C. (1994b). Unloaded shortening velocity and myosin heavy chain and alkali light chain isoform composition in rat skeletal muscle fibres. J Physiol (Lond) 478, 341–349.

Bottinelli R, Canepari M, Pellegrino MA, Reggiani C. (1996). Force-velocity properties of human skeletal muscle fibres, Myosin heavy chain isoform and temperature dependence. J Physiol (Lond) 495, 573–586.

Bottinelli R, Canepari M, Reggiani C, Stienen GJM. (1994c). Myofibrillar ATPase activity during isometric contraction and isomyosin composition in rat single skinned muscle fibres. J Physiol (Lond) 481, 663–675.

Bottinelli R, Reggiani C. (2000). Human skeletal muscle fibres, molecular and functional diversity. Prog Biophys Mol Biol 73, 195–262.

Bottinelli R, Schiaffino S, Reggiani C. (1991). Force-velocity relationship and myosin heavy chain isoform compositions of skinned fibres from rat skeletal muscle. J Physiol (Lond) 437, 655–672.

Bozzo C, Stevens L, Bouet V, Montel V, Picquet F, Falempin M, Lacour M, Mounier Y. (2004). Hypergravity from conception to adult stage, effects on contractile properties and skeletal muscle phenotype. J Exp Biol 207, 2793–2802.

Brooke MH, Kaiser KK. (1970). Three "myosin adenosine triphosphatase" systems, the nature of their pH lability and sulfhydryl dependence. J Histochem Cytochem 18, 670–672.

Buller AJ, Eccles JC, Eccles RM. (1960). Interactions between motoneurones and muscles in respect of the characteristic speed of their responses. J Physiol (Lond) 150, 417–439.

Burke RE, Levine DN, Zajac FE, Tsairis P, Engel WK. (1971). Mammalian motor units, Physiological-histochemical correlation in three types in cat gastrocnemius. Science 174, 709–712.

Burnham R, Martin T, Stein R, Bell G, Maclean I, Steadward R. (1997). Skeletal muscle fibre type transformation following spinal cord injury. Spinal Cord 35, 86–91.

Caiozzo VJ, Baker MJ, Baldwin KM. (1998). Novel transitions in MHC isoforms, separate and combined effects of thyroid hormone and mechanical unloading. J Appl Physiol 85, 2237–2248.

Caiozzo VJ, Baker MJ, Huang K, Chou H, Wu YZ, Baldwin KM. (2003). Single-fiber myosin heavy chain polymorphism, how many patterns and what proportions? Am J Physiol 285, R570–R580.

Caiozzo VJ, Haddad F, Baker M, McCue S, Baldwin KM. (2000). MHC polymorphism in rodent plantaris muscle, effects of mechanical overload and hypothyroidism. Am J Physiol 278, C709–C717.

Caiozzo VJ, Herrick RE, Baldwin KM. (1992). Response of slow and fast muscle to hypothyroidism – maximal shortening velocity and myosin isoforms. Am J Physiol 263, C86–C94.

Caiozzo VJ, Swoap S, Tao M, Menzel D, Baldwin KM. (1993). Single Fiber Analyses of Type-IIA Myosin Heavy Chain Distribution in Hyperthyroid and Hypothyroid Soleus. Am J Physiol 265, C842–C850.

Campbell WG, Gordon SE, Carlson CJ, Pattison JS, Hamilton MT, Booth FW. (2001). Differential global gene expression in red and white skeletal muscle. Am J Physiol Cell Physiol 280, C763–C768.

Canepari M, Cappelli V, Pellegrino MA, Zanardi MC, Reggiani C. (1998). Thyroid hormone regulation of MHC isoform composition and myofibrillar ATPase activity in rat skeletal muscles. Arch Physiol Biochem 106, 308–315.

Canepari M, Rossi R, Pellegrino MA, Bottinelli R, Schiaffino S, Reggiani C. (2000). Functional diversity between orthologous myosins with minimal sequence diversity. J Muscle Res Cell Motil 21, 375–382.

Carroll S, Nicotera P, Pette D. (1999). Calcium transients in single fibers of low-frequency stimulated fast-twitch muscle of rat. Am J Physiol 277, C1122–C1129.

Chin ER. (2004). The role of calcium and calcium/calmodulin-dependent kinases in skeletal muscle plasticity and mitochondrial biogenesis. Proc Nutr Soc Engl Scot 63, 279–286.

Chin ER, Allen DG. (1996). The role of elevations in intracellular $[Ca^{2+}]$ in the development of low frequency fatigue in mouse single muscle fibres. J Physiol (Lond) 491, 813–824.

Chin ER, Olson EN, Richardson JA, Yano Q, Humphries C, Shelton JM, Wu H, Zhu WG, Basselduby R, Williams RS. (1998). A calcineurin-dependent transcriptional pathway controls skeletal muscle fiber type. Genes Develop 12, 2499–2509.

Close RI. (1972). Dynamic properties of mammalian skeletal muscles. Physiol Rev 52, 129–197.

Conjard A, Peuker H, Pette D. (1998). Energy state and myosin isoforms in single fibers of normal and transforming rabbit muscles. Pflügers Arch 436, 962–969.

Crabtree GR, Olson EN. (2002). NFAT signaling, choreographing the social lives of cells. Cell 109 Suppl S67–79, S67–S79.

d'Albis A, Butler-Browne G. (1993). The hormonal control of myosin isoform expression in skeletal muscle of mammals, a review. Bas Appl Myol 3, 7–16.

Desplanches D. (1997). Structural and functional adaptations of skeletal muscle to weightlessness. Int J Sports Med 18, S259–S264.

Dubowitz V, Pearse AGE. (1960). Reciprocal relationship of phosphorylase and oxidative enzymes in skeletal muscle. Nature 185, 701–702.

Edström L, Kugelberg E. (1968). Histochemical composition, distribution of fibres and fatiguability of single motor units. Anterior tibial muscle of the rat. J Neurol Neurosurg Psychiatry 31, 424–433.

Egginton S, Hudlická O. (1999). Early changes in performance, blood flow and capillary fine structure in rat fast muscles induced by electrical stimulation. J Physiol (Lond) 515, 265–275.

Eisenberg BR, Salmons S. (1981). The reorganization of subcellular structure in muscle undergoing fast-to-slow type transformation. A stereological study. Cell Tissue Res 220, 449–471.

Engel WK. (1962). The essentiality of histo- and cytochemical studies of skeletal muscle in the investigation of neuromuscular disease. Neurology 12, 778–784.

English AW, Eason J, Schwartz G, Shirley A, Carrasco DI. (1999). Sexual dimorphism in the rabbit masseter muscle, Myosin heavy chain composition of neuromuscular compartments. Cells Tissues Organs 164, 179–191.

Ferretti G, Antonutto G, Denis C, Hoppeler H, Minetti AE, Narici MV, Desplanches D. (1997). The interplay of central and peripheral factors in limiting maximal O_2 consumption in man after prolonged bed rest. J Physiol (Lond) 501, 677–686.

Fitts RH, Riley DR, Widrick JJ. (2001). Functional and structural adaptations of skeletal muscle to microgravity. J Exp Biol 204, 3201–3208.

Fitts RH, Winder WW, Brooke MH, Kaiser KK, Holloszy JO. (1980). Contractile, biochemical, and histochemical properties of thyrotoxic rat soleus muscle. Am J Physiol 238, C15–C20.

Fitzsimons DP, Herrick RE, Baldwin KM. (1990). Isomyosin distribution in rodent muscles, effects of altered thyroid state. J Appl Physiol 69, 321–327.

Flück M, Hoppeler H. (2003). Molecular basis of skeletal muscle plasticity–from gene to form and function. Rev Physiol Biochem Pharmacol 146, 159–216.

Fraysse B, Desaphy JF, Pierno S, DeLuca A, Liantonio A, Mitolo CI, Camerino DC. (2003). Decrease in resting calcium and calcium entry associated with slow-to- fast transition in unloaded rat soleus muscle. Faseb J 17, U172–U196.

Freyssenet D, DiCarlo M, Hood DA. (1999). Calcium-dependent regulation of cytochrome c gene expression in skeletal muscle cells – Identification of a protein kinase C-dependent pathway. J Biol Chem 274, 9305–9311.

Freyssenet D, Irrcher I, Connor MK, DiCarlo M, Hood DA. (2004). Calcium-regulated changes in mitochondrial phenotype in skeletal muscle cells. Amer J Physiol Cell Physiol 286, C1053–C1061.

Galler S, Hilber K, Gohlsch B, Pette D. (1997a). Two functionally distinct myosin heavy chain isoforms in slow skeletal muscle fibres. FEBS Lett 410, 150–152.

Galler S, Hilber K, Pette D. (1997b). Stretch activation and myosin heavy chain isoforms of rat, rabbit and human skeletal muscle fibres. J Muscle Res Cell Motil 18, 441–448.

Galler S, Schmitt T, Pette D. (1994). Stretch activation, unloaded shortening velocity, and myosin heavy chain isoforms of rat skeletal muscle fibres. J Physiol (Lond) 478, 523–531.

Galler S, Schmitt TL, Hilber K, Pette D. (1997c). Stretch activation and isoforms of myosin heavy chain and troponin-T of rat skeletal muscle fibres. J Muscle Res Cell Motil 18, 555–561.

Geiger PC, Cody MJ, Sieck GC. (1999). Force-calcium relationship depends on myosin heavy chain and troponin isoforms in rat diaphragm muscle fibers. J Appl Physiol 87, 1894–1900.

Giger JM, Haddad F, Qin AX, Baldwin KM. (2004). Effect of cyclosporin A treatment on the in vivo regulation of type I MHC gene expression. J Appl Physiol 97, 475–483.

Goldberg AL. (1967). Work-induced growth of skeletal muscle in normal and hypophysectomized rats. Am J Physiol 213, 1193–1198.

Goldspink DF. (1977). The influence of immobilization and stretch on protein turnover of rat skeletal muscle. J Physiol (Lond) 264, 267–282.

Goldspink G, Scutt A, Loughna PT, Wells DJ, Jaenicke T, Gerlach GF. (1992). Gene expression in skeletal muscle in response to stretch and force generation. Am J Physiol 262, R356–R363.

Gollnick PD, King DW. (1969). Effect of exercise and training on mitochondria of rat skeletal muscle. Am J Physiol 216, 1502–1509.

Gorza L. (1990). Identification of a novel type 2 fiber population in mammalian skeletal muscle by combined use of histochemical myosin ATPase and anti-myosin monoclonal antibodies. J Histochem Cytochem 38, 257–265.

Gorza L, Gundersen K, Lömo T, Schiaffino S, Westgaard RH. (1988). Slow-to-fast transformation of denervated soleus muscles by chronic high-frequency stimulation in the rat. J Physiol (Lond) 402, 627–649.

Green HJ, Düsterhöft S, Dux L, Pette D. (1992). Metabolite patterns related to exhaustion, recovery, and transformation of chronically stimulated rabbit fast-twitch muscle. Pflügers Arch 420, 359–366.

Grützner P. (1883). Zur Physiologie und Histologie der Skelettmuskeln. Breslauer Ärztl Z 5, 257–258.

Gundersen K, Leberer E, Lömo T, Pette D, Staron RS. (1988). Fibre types, calcium-sequestering proteins and metabolic enzymes in denervated and chronically stimulated muscles of the rat. J Physiol (Lond) 398, 177–189.

Guth L, Samaha FJ. (1969). Qualitative differences between actomyosin ATPase of slow and fast mammalian muscle. Exp Neurol 25, 138–152.

Gutmann E, Hanzlíková V. (1970). Effect of androgens on histochemical fibre type. Differentiation in the temporal muscle of the guinea pig. Histochemie 24, 287–291.

Hamilton MT, Booth FW. (2000). Skeletal muscle adaptation to exercise, a century of progress. J Appl Physiol 88, 327–331.

Hanzlíková V, Schiaffino S, Settembrini P. (1970). Histochemical fiber types characteristics in the normal and the persistent levator ani muscle of the rat. Histochemie 22, 45–50.

Hardie DG. (2004). AMP-activated protein kinase, A key system mediating metabolic responses to exercise. Med Sci Sport Exercise 36, 28–34.

Hämäläinen N, Pette D. (1993). The histochemical profiles of fast fiber types IIB, IID and IIA in skeletal muscles of mouse, rat and rabbit. J Histochem Cytochem 41, 733–743.

Hämäläinen N, Pette D. (1995). Patterns of myosin isoforms in mammalian skeletal muscle fibres. Microsc Res Tech 30, 381–389.

He ZH, Bottinelli R, Pellegrino MA, Ferenczi MA, Reggiani C. (2000). ATP consumption and efficiency of human single muscle fibers with different myosin isoform composition. Biophys J 79, 945–961.

Heilmann C, Pette D. (1979). Molecular transformations in sarcoplasmic reticulum of fast-twitch muscle by electro-stimulation. Eur J Biochem 93, 437–446.

Heizmann CW, Berchtold MW, Rowlerson AM. (1982). Correlation of parvalbumin concentration with relaxation speed in mammalian muscles. Proc Natl Acad Sci USA 79, 7243–7247.

Hennig R, Lömo T. (1985). Firing patterns of motor units in normal rats. Nature 314, 164–166.

Hicks A, Ohlendieck K, Göpel SO, Pette D. 1997. Early functional and biochemical adaptations to low-frequency stimulation of rabbit fast-twitch muscle. Am J Physiol 273, C297–C305.

Hilber K, Galler S, Gohlsch B, Pette D. (1999). Kinetic properties of myosin heavy chain isoforms in single fibers from human skeletal muscle. FEBS Lett 455, 267–270.

Hildebrandt AL, Pilegaard H, Neufer PD. (2003). Differential transcriptional activation of select metabolic genes in response to variations in exercise intensity and duration. Amer J Physiol 285, E1021–E1027.

Holloszy JO. (1967). Biochemical adaptations in muscle. Effects of exercise on mitochondrial oxygen uptake and respiratory enzyme activity in skeletal muscle. J Biol Chem 242, 2278–2282.

Holloszy JO, Coyle EF. (1984). Adaptations of skeletal muscle to endurance exercise and their metabolic consequences. J Appl Physiol 56, 831–838.

Hood DA. (2001). Invited Review, contractile activity-induced mitochondrial biogenesis in skeletal muscle. J Appl Physiol 90, 1137–1157.

Hood DA, Pette D. (1989). Chronic long-term stimulation creates a unique metabolic enzyme profile in rabbit fast-twitch muscle. FEBS Lett 247, 471–474.

Hood DA, Zak R, Pette D. (1989). Chronic stimulation of rat skeletal muscle induces coordinate increases in mitochondrial and nuclear mRNAs of cytochrome c oxidase subunits. Eur J Biochem 179, 275–280.

Hoppeler H, Flück M. (2003). Plasticity of skeletal muscle mitochondria, structure and function. Med Sci Sports Exerc 35, 95–104.

Höök P, Sriramoju V, Larsson L. (2001). Effects of aging on actin sliding speed on myosin from single skeletal muscle cells of mice, rats, and humans. Am J Physiol 280, C782–C788.

Ianuzzo D, Patel P, Chen V, O'Brien P, Williams C. (1977). Thyroidal trophic influence on skeletal muscle myosin. Nature 270, 74–76.

Izumo S, Nadal-Ginard B, Mahdavi V. (1986). All members of the MHC multigene family respond to thyroid hormone in a highly tissue-specific manner. Science 231, 597–600.

Jackman RW, Kandarian SC. (2004). The molecular basis of skeletal muscle atrophy. Amer J Physiol 287, C834–C843.

Jamali AA, Afshar P, Abrams RA, Lieber RL. (2000). Invited review, Skeletal muscle response to tenotomy. Muscle Nerve 23, 851–862.

Jorgensen AO, Jones LR. (1986). Localization of phospholamban in slow but not fast canine skeletal muscle fibers. An immunocytochemical and biochemical study. J Biol Chem 261, 3775–3781.

Joubert Y, Tobin C, Lebart MC. (1994). Testosterone-induced masculinization of the rat levator ani muscle during puberty. Dev Biol 162, 104–110.

Kirschbaum BJ, Kucher H-B, Termin A, Kelly AM, Pette D. (1990). Antagonistic effects of chronic low frequency stimulation and thyroid hormone on myosin expression in rat fast-twitch muscle. J Biol Chem 265, 13974–13980.

Klitgaard H, Bergman O, Betto R, Salviati G, Schiaffino S, Clausen T, Saltin B. (1990). Co-Existence of myosin heavy chain I and IIA isoforms in human skeletal muscle fibres with endurance training. Pflügers Arch 416, 470–472.

Knoll P. (1891). Über protoplasmaarme und protoplasmareiche Musculatur. Denkschr Kais Akad Wiss Wien, math -naturwiss Cl 58, 633–700.

Krenács T, Molnar E, Dobo E, Dux L. (1989). Fibre typing using sarcoplasmic reticulum Ca^{2+}-ATPase and myoglobin immunohistochemistry in rat gastrocnemius muscle. Histochem J 21, 145–155.

Krüger P (1952) Tetanus und Tonus der quergestreiften Skelettmuskeln der Wirbeltiere und des Menschen. Akademische Verlagsgesellschaft Geest & Portig K.-G., Leipzig

Kubis HP, Hanke N, Scheibe RJ, Meissner JD, Gros G. (2003). Ca^{2+} transients activate calcineurin/NFATc1 and initiate fast-to-slow transformation in a primary skeletal muscle culture. Am J Physiol 285, C56–C63.

Lange S, Xiang F, Yakovenko A, Vihola A, Hackman P, Rostkova E, Kristensen J, Brandmeier B, Franzen G, Hedberg B, Gunnarsson LG, Hughes SM, Marchand S, Sejersen T, Richard I, Edstrom

L, Ehler E, Udd B & Gautel M. (2005). The kinase domain of titin controls muscle gene expression and protein turnover. Science 308, 1599–1603.

Larsson L, Li XP, Teresi A, Salviati G. (1994). Effects of thyroid hormone on fast- and slow-twitch skeletal muscles in young and old rats. J Physiol (Lond) 481, 149–161.

Larsson L, Moss RL. (1993). Maximum velocity of shortening in relation to myosin isoform composition in single fibres from human skeletal muscles. J Physiol (Lond) 472, 595–614.

Läuger P (1991) Electrogenic ion pumps. Sinauer Assoc., Sunderland, MA, USA

Leeuw T, Pette D. (1993). Coordinate changes in the expression of troponin subunit and myosin heavy chain isoforms during fast-to-slow transition of low- frequency stimulated rabbit muscle. Eur J Biochem 213, 1039–1046.

Leeuw T, Pette D. (1996). Coordinate changes of myosin light and heavy chain isoforms during forced fiber type transitions in rabbit muscle. Dev Genet 19, 163–168.

Li WP, Hughes SM, Salviati G, Teresi A, Larsson L. (1996). Thyroid hormone effects on contractility and myosin composition of soleus muscle and single fibres from young and old rats. J Physiol (Lond) 494, 555–567.

Li XP, Larsson L. (1997). Contractility and myosin isoform compositions of skeletal muscles and muscle cells from rats treated with thyroid hormone for 0, 4 and 8 weeks. J Muscle Res Cell Motil 18, 335–344.

Liu Y, Cseresnyés Z, Randall WR, Schneider MF. (2001). Activity-dependent nuclear translocation and intranuclear distribution of NFATc in adult skeletal muscle fibers. J Cell Biol 155, 27–39.

Liu Y, Shen T, Randall WR & Schneider MF. (2005). Signaling pathways in activity-dependent fiber type plasticity in adult skeletal muscle. J Muscle Res Cell Motil 26, 13–21.

Lloyd PG, Prior BM, Yang HT, Terjung RL. (2003). Angiogenic growth factor expression in rat skeletal muscle in response to exercise training. Am J Physiol 284, H1668–H1678.

Loughna PT, Izumo S, Goldspink G, Nadal-Ginard B. (1990). Disuse and passive stretch cause rapid alterations in expression of developmental and adults contractile protein genes in skeletal muscle. Development 109, 217–223.

Lowey S, Waller GS, Trybus KM. (1993). Skeletal muscle myosin light chains are essential for physiological speeds of shortening. Nature 365, 454–456.

Lömo T, Westgaard RH, Dahl HA. (1974). Contractile properties of muscle, control by pattern of muscle activity in the rat. Proc R Soc Lond B 187, 99–103.

Lyons GE, Kelly AM, Rubinstein NA. (1986). Testosterone-induced changes in contractile protein isoforms in the sexually dimorphic temporalis muscle of the guinea pig. J Biol Chem 261, 13278–13284.

Maier A, Leberer E, Pette D. (1986). Distribution of sarcoplasmic reticulum Ca-ATPase and of calsequestrin in rabbit and rat skeletal muscle fibers. Histochemistry 86, 63–69.

Margreth A, Dalla Libera L, Salviati G, Ischia N. (1980). Spinal transection and the postnatal differentiation of slow myosin isoenzymes. Muscle Nerve 3, 483–486.

McCullagh KJA, Calabria E, Pallafacchina G, Ciciliot S, Serrano AL, Argentini C, Kalhovde JM, Lomo T, Schiaffino S. (2004). NFAT is a nerve activity sensor in skeletal muscle and controls activity-dependent myosin switching. Proc Nat Acad Sci Usa 101, 10590–10595.

Meissner JD, Gros G, Scheibe RJ, Scholz M, Kubis HP. (2001). Calcineurin regulates slow myosin, but not fast myosin or metabolic enzymes, during fast-to-slow transformation in rabbit skeletal muscle cell culture. J Physiol (Lond) 533, 215–226.

Meissner JD, Kubis HP, Scheibe RJ, Gros G. (2000). Reversible Ca^{2+}-induced fast-to-slow transition in primary skeletal muscle culture cells at the mRNA level. J Physiol (Lond) 523, 19–28.

Michel JB, Ordway GA, Richardson JA, Williams RS. (1994). Biphasic induction of immediate early gene expression accompanies activity-dependent angiogenesis and myofiber remodeling of rabbit skeletal muscle. J Clin Invest 94, 277–285.

Mira J-C, Janmot C, Couteaux R, d'Albis A. (1992). Reinnervation of denervated extensor digitorum longus of the rat by the nerve of the soleus does not induce the type I myosin synthesis directly but through a sequential transition of type II myosin isoforms. Neurosci Lett 141, 223–226.

Morey ER. (1979). Spaceflight and bone turnover, correlation with a new rat model of weightlessness. Bioscience 29, 168–172.

Murgia M, Serrano AL, Calabria E, Pallafacchina G, Lomo T, Schiaffino S. (2000). Ras is involved in nerve-activity-dependent regulation of muscle genes. Nat Cell Biol 2, 142–147.

Musacchia XJ, Deavers DR, Meininger GA, Davis TP. (1980). A model for hypokinesia, effects on muscle atrophy in the rat. J Appl Physiol 48, 479–486.

Nordsborg N, Bangsbo J, Pilegaard H. (2003). Effect of high-intensity training on exercise-induced gene expression specific to ion homeostasis and metabolism. J Appl Physiol 95, 1201–1206.

Nwoye L, Mommaerts WFHM. (1981). The effects of thyroid status on some properties of rat fast-twitch muscle. J Muscle Res Cell Motil 2, 307–320.

Ogata T. (1958a). A histochemical study of the red and white muscle fibers II Activity of the cytochrome oxidase in muscle fibers. Acta Med Okayama 12, 228–232.

Ogata T. (1958b). A histochemical study of the red and white muscle fibers III Activity of the diphos-phopyridine nucleotide diaphorase and triphosphopyridine nucleotide diaphorase in muscle fibers. Acta Med Okayama 12, 233–240.

Ogata T. (1958c). A histochemical study of the red and white muscle fibers Part I Activity of the succinoxydase system in muscle fibers. Acta Med Okayama 12, 216–227.

Ojuka EO, Jones TE, Han DH, Chen M, Holloszy JO. (2003). Raising Ca2+ in L6 myotubes mimics effects of exercise on mitochondrial biogenesis in muscle. FASEB J 17, 675–681.

Olson EN, Williams RS. (2000a). Calcineurin signaling and muscle remodeling. Cell 101, 689–692.

Olson EN, Williams RS. (2000b). Remodeling muscles with calcineurin. BioEssays 22, 510–519.

Ouchi N, Shibata R & Walsh K. (2005). AMP-activated protein kinase signaling stimulates VEGF expression and angiogenesis in skeletal muscle. Circ Res 96, 838–846.

Padykula HA, Herman E. (1955). Factors affecting the activity of adenosine triphosphatase and other phosphatases as measured by histochemical techniques. J Histochem Cytochem 3, 161–167.

Parsons SA, Millay DP, Wilkins BJ, Bueno OF, Tsika GL, Neilson JR, Liberatore CM, Yutzey KE, Crabtree GR, Tsika RW, Molkentin JD. (2004). Genetic loss of calcineurin blocks mechanical overload-induced skeletal muscle fiber type switching but not hypertrophy. J Biol Chem 279, 26192–26200.

Pellegrino MA, Canepari M, Rossi R, DAntona G, Reggiani C, Bottinelli R. (2003). Orthologous myosin isoforms and scaling of shortening velocity with body size in mouse, rat, rabbit and human muscles. J Physiol (Lond) 546, 677–689.

Peter JB, Barnard RJ, Edgerton VR, Gillespie CA, Stempel KE. (1972). Metabolic profiles of three fiber types of skeletal muscle in guinea pigs and rabbits. Biochemistry 11, 2627–2633.

Pette D (1980) Plasticity of Muscle. de Gruyter, Berlin New York

Pette D. (2001). Historical Perspectives, Plasticity of mammalian skeletal muscle. J Appl Physiol 90, 1119–1124.

Pette D, Sketelj J, Skorjanc D, Leisner E, Traub I, Bajrovic F. (2002). Partial fast-to-slow conversion of regenerating rat fast-twitch muscle by chronic low-frequency stimulation. J Muscle Res Cell Motil 23, 215–221.

Pette D, Staron RS. (1990). Cellular and molecular diversities of mammalian skeletal muscle fibers. Rev Physiol Biochem Pharmacol 116, 1-76.

Pette D, Staron RS. (1997). Mammalian skeletal muscle fiber type transitions. Int Rev Cytol 170, 143–223.

Pette D, Staron RS. (2000). Myosin isoforms, muscle fiber types, and transitions. Microsc Res Tech 50, 500–509.

Pette D, Vrbová G. (1992). Adaptation of mammalian skeletal muscle fibers to chronic electrical stimulation. Rev Physiol Biochem Pharmacol 120, 116–202.

Pette D, Vrbová G. (1999). Invited review, What does chronic electrical stimulation teach us about muscle plasticity? Muscle Nerve 22, 666–677.

Pilegaard H, Ordway GA, Saltin B, Neufer PD. (2000). Transcriptional regulation of gene expression in human skeletal muscle during recovery from exercise. Am J Physiol 279, E806–E814.

Puntschart A, Claassen H, Jostarndt K, Hoppeler H, Billeter R. (1995). mRNAs of enzymes involved in energy metabolism and mtDNA are increased in endurance-trained athletes. Am J Physiol 38, C619–C625.

Putman CT, Kiricsi M, Pearcey J, O'Brian C, Maclean I, Murdoch G, Pette D. (2003). AMPK activation increases UCP-3 and enzyme activities in rat muscle without fiber type transitions. J Physiol (Lond) 551, 169–178.

Ranvier L. (1873). Proprietés et structures différentes des muscles rouges et des muscles blancs chez les lapins et chez les raies. C r Acad Sci Paris 77, 1030–1034.

Reggiani C, Kronnie GT. (2004). Muscle plasticity and high throughput gene expression studies. J Muscle Res Cell Motil 25, 231–234.

Reichmann H, Hoppeler H, Mathieu-Costello O, von Bergen F, Pette D. (1985). Biochemical and ultrastructural changes of skeletal muscle mitochondria after chronic electrical stimulation in rabbits. Pflügers Arch 404, 1–9.

Rennie MJ, Wackerhage H, Spangenburg EE, Booth FW. (2004). Control of the size of the human muscle mass. Annu Rev Physiol 66, 799–828.

Rome LC, Sosnicki AA, Goble DO. (1990). Maximum velocity of shortening of three fibre types from horse soleus muscle, implications for scaling with body size. J Physiol (Lond) 431, 173–185.

Roy RR, Baldwin KM, Martin TP, Chimarusti SP, Edgerton VR. (1985). Biochemical and physiological changes in overloaded rat fast- and slow-twitch ankle extensors. J Appl Physiol 59, 639–646.

Roy RR, Sacks RD, Baldwin KM, Short M, Edgerton VR. (1984). Interrelationships of contraction time, V_{max}, and myosin ATPase after spinal transection. J Appl Physiol 56, 1594–1601.

Sakamoto K, Goodyear LJ. (2002). Invited review, intracellular signaling in contracting skeletal muscle. J Appl Physiol 93, 369–383.

Salmons S. (1994). Exercise, stimulation and type transformation of skeletal muscle. Int J Sports Med 15, 136–141.

Salmons S, Vrbová G. (1967). Changes in the speed of mammalian fast muscle following long- term stimulation. J Physiol (Lond) 192, 39–40P.

Salmons S, Vrbová G. (1969). The influence of activity on some contractile characteristics of mammalian fast and slow muscles. J Physiol (Lond) 201, 535–549.

Saltin B, Gollnick PD (1983). Skeletal muscle adaptability, significance for metabolism and performance. In, Peachey LD, Adrian RH, and Geiger SR (eds) Handbook of Physiology, Sect. 10, Skeletal Muscle. Williams & Wilkins, Baltimore MD, pp 555–631

Schachat F, Briggs MM, Williamson EK, McGinnis H (1990). Expression of fast thin filament proteins. Defining fiber archetypes in a molecular continuum. In, Pette D (ed) The Dynamic State of Muscle Fibers. de Gruyter, Berlin New York, pp 279–291

Schachat FH, Diamond MS, Brandt PW. (1987). Effect of different troponin T-tropomyosin combinations on thin filament activation. J Mol Biol 198, 551–554.

Schiaffino S, Reggiani C. (1996). Molecular diversity of myofibrillar proteins, Gene regulation and functional significance. Physiol Rev 76,371–423.

Schmitt T, Pette D. (1991). Fiber type-specific distribution of parvalbumin in rabbit skeletal muscle – a quantitative immunohistochemical and microbiochemical study. Histochemistry 96, 459–465.

Serrano AL, Murgia M, Pallafacchina G, Calabria E, Coniglio P, Lomo T, Schiaffino S. (2001). Calcineurin controls nerve activity-dependent specification of slow skeletal muscle fibers but not muscle growth. Proc Natl Acad Sci USA 98, 13108–13113.

Shenkman BS, Nemirovskaya TL, Belozerova IN, Mazin MG, Matveeva OA. (2002). Mitochondrial adaptations in skeletal muscle cells in mammals exposed to gravitational unloading. J Gravit Physiol 9, 159–162.

Simoneau J-A, Pette D. (1988). Species-specific effects of chronic nerve stimulation upon tibialis anterior muscle in mouse, rat, guinea pig, and rabbit. Pflügers Arch 412, 86–92.

Skorjanc D, Dünstl G, Pette D. (2001). Mitochondrial enzyme defects in normal and low-frequency-stimulated muscles of young and aging rats. J Gerontol A Biol Sci Med Sci 56, B503–B509.

Skorjanc D, Jaschinski F, Heine G, Pette D. (1998). Sequential increases in capillarization and mitochondrial enzymes in low-frequency stimulated rabbit muscle. Am J Physiol 274, C810–C818.

Sréter FA, Elzinga M, Mabuchi K. (1975). The N-methylhistidine content of myosin in stimulated and cross- reinnervated skeletal muscles of the rabbit. FEBS Lett 57, 107–111.

Staron RS, Hagerman FC, Hikida RS, Murray TF, Hostler DP, Crill MT, Ragg KE, Toma K. (2000). Fiber type composition of the vastus lateralis muscle of young men and women. J Histochem Cytochem 48, 623–629.

Staron RS, Hikida RS. (1992). Histochemical, biochemical, and ultrastructural analyses of single human muscle fibers, with special reference to the C-fiber population. J Histochem Cytochem 40, 563–568.

Staron RS, Pette D. (1986). Correlation between myofibrillar ATPase activity and myosin heavy chain composition in rabbit muscle fibers. Histochemistry 86, 19–23.

Steinacker JM, Opitz-Gress A, Baur S, Lormes W, Bolkart K, Sunder-Plassmann L, Liewald F, Lehmann M, Liu YF. (2000). Expression of myosin heavy chain isoforms in skeletal muscle of patients with peripheral arterial occlusive disease. J Vasc Surg 31, 443–449.

Stevens L, Gohlsch B, Mounier Y, Pette D. (1999a). Changes in myosin heavy chain mRNA and protein isoforms in single fibers of unloaded rat soleus muscle. FEBS Lett 463, 15–18.

Stevens L, Sultan KR, Peuker H, Gohlsch B, Mounier Y, and Pette D. (1999b). Time-dependent changes in myosin heavy chain mRNA and protein isoforms in unloaded soleus muscle of rat. Am J Physiol 277, C1044–1049

Stienen GJM, Kiers JL, Bottinelli R, Reggiani C. (1996). Myofibrillar ATPase activity in skinned human skeletal muscle fibres, Fibre type and temperature dependence. J Physiol (Lond) 493, 299–307.

Sugiura T, Miyata H, Kawai Y, Matoba H, Murakami N. (1993). Changes in myosin heavy chain isoform expression of overloaded rat skeletal muscles. Int J Biochem 25, 1609–1613.

Talmadge RJ. (2000). Myosin heavy chain isoform expression following reduced neuromuscular activity, Potential regulatory mechanisms. Muscle Nerve 23, 661–679.

Talmadge RJ, Roy RR, Edgerton VR. (1999). Persistence of hybrid fibers in rat soleus after spinal cord transection. Anat Rec 255, 188–201.

Tavakol M, Roy RR, Kim JA, Zhong H, Hodgson JA, Hoban-Higgins TM, Fuller CA, Edgerton VR. (2002). Fiber size, type, and myosin heavy chain content in rhesus hindlimb muscles after 2 weeks at 2 G. Aviat Space Environ Med 73, 551–557.

Toniolo L, Patruno M, Maccatrozzo L, Pellegrino MA, Canepari M, Rossi R, D'Antona G, Bottinelli R, Reggiani C, Mascarello F. (2004). Fast fibres in a large animal, fibre types, contractile properties and myosin expression in pig skeletal muscles. J Exp Biol 207, 1875–1886.

Turner DL, Hoppeler H, Claassen H, Vock P, Kayser B, Schena F, Ferretti G. (1997). Effects of endurance training on oxidative capacity and structural composition of human arm and leg muscles. Acta Physiol Scand 161,459–464.

Vrbová G. (1963a). The effect of motoneurone activity on the speed of contraction of striated muscle. J Physiol (Lond) 169, 513–526.

Vrbová G. (1963b). The effect of tenotomy on the speed of contraction of fast and slow mammalian muscles. J Physiol (Lond) 166, 241–250.

Waters RE, Rotevatn S, Li P, Annex BH, Yan Z. (2004). Voluntary running induces fiber type-specific angiogenesis in mouse skeletal muscle. Amer J Physiol 287, C1342–C1348.

Wehrle U, Düsterhöft S, Pette D. (1994). Effects of chronic electrical stimulation on myosin heavy chain expression in satellite cell cultures derived from rat muscles of different fiber-type composition. Differentiation 58, 37–46.

Weiss A, Schiaffino S, Leinwand LA. (1999). Comparative sequence analysis of the complete human sarcomeric myosin heavy chain family, Implications for functional diversity. J Mol Biol 290, 61–75.

Westerblad H, Allen DG. (1991). Changes of myoplasmic calcium concentration during fatigue in single mouse muscle fibers. J Gen Physiol 98, 615–635.

Widrick JJ, Romatowski JG, Karhanek M, Fitts RH. (1997). Contractile properties of rat, rhesus monkey, and human type I muscle fibers. Am J Physiol 272, R34–R42.

Williams RS, Neufer PD (1996). Regulation of gene expression in skeletal muscle by contractile activity. In, Rowell LB and Shepherd JT (eds) The Handbook of Physiology, Section 12, Exercise, Regulation and Integration of Multiple Systems. Oxford University Press, New York, pp 1124–1150

Winder WW. (2001). Invited review, Energy-sensing and signaling by AMP-activated protein kinase in skeletal muscle. J Appl Physiol 91, 1017–1028.

Winder WW, Holmes BF, Rubink DS, Jensen EB, Chen M, Holloszy JO. (2000). Activation of AMP-activated protein kinase increases mitochondrial enzymes in skeletal muscle. J Appl Physiol 88, 2219–2226.

Windisch A, Gundersen K, Szabolcs MJ, Gruber H, Lömo T. (1998). Fast to slow transformation of denervated and electrically stimulated rat muscle. J Physiol (Lond) 510, 623–632.

Wu H, Gallardo T, Olson EN, Williams RS, Shohet RV. (2003). Transcriptional analysis of mouse skeletal myofiber diversity and adaptation to endurance exercise. J Muscle Res Cell Motil 24, 587–592.

Wu H, Rothermel B, Kanatous S, Rosenberg P, Naya FJ, Shelton JM, Hutcheson KA, DiMaio JM, Olson EN, Bassel-Duby R, Williams RS. (2001). Activation of MEF2 by muscle activity is mediated through a calcineurin-dependent pathway. EMBO J 20, 6414–6423.

Zhong H, Roy RR, Siengthai B & Edgerton VR. (2005). Effects of inactivity on fiber size and myonuclear number in rat soleus muscle. J Appl Physiol 99, 1494–1499.

Zhou M, Lin BZ, Coughlin S, Vallega G, Pilch PF. (2000). UCP-3 expression in skeletal muscle, effects of exercise, hypoxia, and AMP-activated protein kinase. Am J Physiol 279, E622–E629.

Zong H, Ren JM, Young LH, Pypaert M, Mu J, Birnbaum MJ, Shulman GI. (2002). AMP kinase is required for mitochondrial biogenesis in skeletal muscle in response to chronic energy deprivation. Proc Natl Acad Sci USA 99, 15983–15987.

CHAPTER 2

LARGE SCALE GENE EXPRESSION PROFILES AS TOOLS TO STUDY SKELETAL MUSCLE ADAPTATION

SUSAN C. KANDARIAN

Boston University, Boston, MA 02215, USA

1.   INTRODUCTION

The quest to identify the entire collection of expressed mRNAs in a given cell or tissue has been ongoing since the development of technologies that allowed for the measurement of large-scale gene expression. Of equal or greater interest is discovering the total number and identities of mRNAs that are differentially expressed in tissues in response to a multitude of conditions. That is, learning about the differences in gene expression between control and experimental conditions, especially normal and pathological states. There are several terms used to denote the measurement of large-scale differential gene expression. These include global gene expression, massively parallel gene expression, transcriptional profiling, and gene expression profiling. The information sought is the most complete mRNA phenotype of the cell or tissue under study, a kind of physiology-specific functional genomics. But of what usefulness is this in understanding muscle adaptation? What do we hope to gain by carrying out expression profiling experiments on muscle? Since muscle remodeling relies to a significant extent on the transcriptional regulation of structural and regulatory genes, in most studies investigators are trying to obtain a window into the molecular processes underlying muscle adaptation. With more sophisticated designs and data analysis, expression profiling experiments can reveal quite specific and useful information as will be discussed in this review.

The technology leading to tools that allow for gene expression profiling was initiated with the partial sequencing of thousands of cDNAs made from mRNAs expressed in a given tissue (via cDNA libraries). These sequences, expressed sequence tags (ESTs), are collections that gave a rough idea of the types and

29

R. Bottinelli and C. Reggiani (eds.), Skeletal Muscle Plasticity in Health and Disease, 29–54.
© 2006 *Springer.*

numbers of mRNAs expressed in a tissue (Ewing and Claverie, 2000). At about the same time, RNA subtraction libraries, serial analysis of gene expression (SAGE) and differential display were used by some laboratories to measure large scale gene expression (e.g. (Chu and Paul, 1998; Liotta and Petricoin, 2000; Velculescu et al., 1995). These methods for measuring differential expression of many mRNAs in a tissue are labor intensive and somewhat difficult. Many more labs began to use expression profiling as cDNA and oligonucleotide microarrays became available because of the advances in technology and the fact that core facilities could carry out the bulk of the technical procedures after a total RNA sample was provided. With time, it seems that the most popular and easiest way to obtain a global inventory of differentially expressed mRNAs in a tissue is by using high-density synthetic oligonucleotide microarrays, particularly the Affymetrix GeneChips. The reason for this is that the photolithographic nanotechnology used affords more and more gene products from the genome of a given organism to be assayed on a very small platform (Lipshutz et al., 1999). Also, they are commercially available, they have high specificity, and the annotation of the gene products represented on the microarrays continues to be improved. The probe sets applied to these arrays are mostly obtained from NCBI Unigene databases, which are repositories for clustered EST data, so that even unknown gene products can be assayed. For certain labs, producing cDNA microarrays is relatively easy and less expensive than using oligonucleotide microarrays, but the specificity and the number of mRNAs that are typically measured is lower.

1.1   The Evolving Utility of Gene Expression Profiling

The number of mRNAs that can be assayed on existing microarrays and the design of new ones are dependent on the continued worldwide efforts on genome analysis, including the identification of all the expressed gene products and their functions. From these efforts come improvements in annotation of genes in public databases and the use of ontology in an attempt to unify biological function (Ashburner et al., 2000). Clarification of gene names and functions alone, will greatly increase the amount of knowledge that can be gleaned from even the existing expression profiling papers, since typically, about a third of the differen-tially expressed genes found in most datasets have unknown names or functions. Sometimes investigators pursue the study of these gene products by identifying the complete sequence, determining any homologies or motifs, and assessing gene function. Several important discoveries have been made using this approach in muscle as discussed in Section 2.

One of the best gene expression inventories of skeletal muscle was published from human vastus lateralis muscle using SAGE; 12,000 unique tags (mRNAs) were found (Welle et al., 1999). In 2003, Welle et al. reported approximately 13,000 unique mRNAs were expressed in the vastus muscle of humans using GeneChips that assayed for 33,000 unique gene products. In a study on normal and dystrophic muscle in humans where 65,000 oligonucleotide probes were used, approximately 30% of these mRNAs were present in muscle (Bakay et al., 2002). However, we do not know how many of these, almost 20,000 probe sets represent the same gene product because of the redundancy built into the GeneChip microarrays. In this study the human U95A through U95E GeneChips were used in addition to a custom MuscleChip containing 3600 probes. Since there is about 10–20% probe redundancy on GeneChips, this suggests at least 13,000 unique gene products were expressed in skeletal muscle, similar to what Welle et al. (2003) found using a similar GeneChip platform. Thus, these studies give us a reasonable estimate of the number of different gene products expressed in skeletal muscle. We do not yet know the total number of genes in any mammalian genome, but the current best estimate for mice and humans is 20,000–25,000 (Stein, 2004). In mammals, the number of genes is much smaller than the number of gene products because of the possibility of multiple alternatively spliced mRNAs from a given gene. This number, the total number of expressed gene products, is not yet known for any mammal. When the total number of unique gene products that can be expressed by a given genome is determined then we will be able to truly measure "genome-wide" gene expression. In the meantime, many labs have performed gene expression profiling studies where hundreds, and usually thousands of mRNAs are measured at the same time. NCBI's gene expression omnibus (GEO) is a repository of gene expression profiling experiments. It is therefore becoming a valuable database from which to extract new relationships from many different tissues and cells under a variety of conditions.

1.2   Computational Tools and Statistics

Expression profiling can create long gene lists of genes that are differentially expressed in two or more muscle conditions. Ideally, in order to assess differences in expressed genes from multiple samples, appropriate statistics should be applied to determine meaningful differences. Many labs have begun to employ multiple test corrections (Benjamini and Hochberg, 1995) which reduce the number of type I errors (false positives). These types of statistics are usually an important first step when the comparisons of gene expression differences are made between control and experimental groups. Expression profiling datasets, even when solid statistics are performed and when genes are well annotated, can be difficult to interpret. This is because of the complexity of making sense out of a list of differentially expressed genes that have little apparent relationship to each other. This is in turn, a reflection of our limitation in understanding the biology we seek to study.

The method for data analysis of differential gene expression most often used is by functional category. In some papers hierarchical clustering, k-means clustering or self-organizing maps have also been used and can help to understand differentially expressed mRNA patterns, particularly when a time course design is used. Briefly, hierarchical clustering is used to identify genes whose expression is changed to a similar extent in response to an intervention. This method was borrowed from phylogenetic and sequence analysis (Eisen et al., 1998). The idea here is that genes whose expression change is similar may be co-regulated by the same upstream genes or signaling pathways, and may even share regulatory elements in promoter regions (sometimes called "guilt by association"). K-means clustering is used to identify (i.e., cluster) genes with similar patterns of change over time, or in multiple conditions, without regard for the magnitude of change. A self-organizing map also clusters genes on the basis of the similarity of expression over time except that genes with similar patterns but different magnitudes of change are clustered separately. These tools are useful when trying to create order in a mass of many differentially expressed genes. They determine how many different types of transcription profiles define the response to a stimulus. For instance they reveal whether a cluster of genes with a given pattern of expression has an early or late activation, and whether the activation is transient or sustained. Also, this type of clustering identifies the most common expression pattern in response to a stimulus. When a molecular signature is being developed from tissues representing two or more conditions, different types of computational tools are used (Golub et al., 1999), and these also have the effect of reducing type I errors. There are multiple reviews that detail the use of these data analysis methods including their limitations (e.g. (Gilbert et al., 2000; Quackenbush, 2001; Shannon et al., 2003). Examples of some of these methods for reducing and extracting meaning from gene expression datasets will be presented, using the primary literature as examples. Finally, gene expression datasets can be used to help identify target genes of specific transcription factors that are activated in muscle in response to a specific stimulus.

1.3   Considerations About Design and Data Interpretation

Although some studies have used multiple test corrections to reduce false positive findings, some investigators have felt the need to reconfirm a number of the gene expression findings of a microarray study. Unless one is embarking on further study of single genes revealed from a microarray study, reconfirmation is not necessary, especially if the appropriate statistics are being used. Expression profiling is often used to get a general picture of changes, and patterns of change, so that if a few of the reported significant observations are in fact false-positives, the paper need not be retracted. Also, if the GeneChip microarray platform is being used, these microarrays are highly redundant for the genes they assay. This is one test for how well a microarray is quantifying a specific gene product. One can also use the existing literature to re-confirm a number of genes on the microarray that have already been measured by other labs using microarrays or traditional techniques for assessing relative mRNA levels. A number of studies have used the pooling of samples because of limited sample size and subject number, and to reduce variability of the expression for a given gene. If possible, this approach is not recommended because it does not allow for statistical testing or for the identification of the extent of variability of gene expression - this information is important to data interpretation.

Another issue to be addressed when interpreting microarray data when a tissue such as skeletal muscle is being studied is the fact that there is more than one cell type that comprises a tissue. Therefore one does not know which differentially expressed mRNAs in a given list or cluster come from muscle cells versus other cell types such as endothelial cells, fibroblasts, infiltrating immune cells, and so on. In cases where conclusions are being made on the basis of one or several genes with ambiguous localization, the best course is to carry out experiments on whole muscle to determine the location of either the mRNA (in situ hybridization) or the protein product (immunohistochemistry) of the gene.

As mentioned, going back over a dataset with updated annotations and functional information may reveal new ideas and improved biological understanding. Another way that a dataset becomes more useful over time is with additional datasets on the same, related, or contrasting experimental interventions. For instance, if a gene is found to be upregulated in many types of atrophy, but then is found to also be upregulated with exercise training, it may be a gene that is a general responder to new stimuli and not specific to any one process under study. Thus knowledge of many datasets taken together provides new rather than redundant information. Examples of this are discussed.

A question sometimes posed is: are we really improving our understanding of muscle biology by using these somewhat involved and expensive tools? I believe the answer is yes, but perhaps more slowly than we had hoped. Another question is: are the computational methods and tools, and our ability to apply them to gene expression datasets lagging behind the many gene lists generated by laboratories? Again I believe the answer is yes, although the level of sophistication of data analysis continues to improve. In this chapter I will try to give examples of how knowledge about mRNA expression profiling in a number of different conditions has contributed to our understanding of muscle adaptation.

2. MUSCLE ADAPTATION AND GENE EXPRESSION PROFILING

One of the first papers where gene expression profiling was used to understand muscle plasticity was the comparison of young and old skeletal muscle of mice, with and without caloric restriction; the latter was used to determine if age-associated changes could be reversed since general markers of age are decreased with regulated feeding (Lee et al., 1999). Many papers on aging using parallel gene expression methods followed, in an attempt to understand the marked decline in muscle mass and function that is associated with age (sarcopenia). Muscle atrophy studies are perhaps the second most abundant in the use of expression profiling in an attempt to obtain a window into the molecular underpinnings of this process. Papers where hypertrophic stimuli are being studied are used to contrast those on atrophy. Studies on muscle regeneration, muscle development, and circadian rhythms have also used microarrays to better understand muscle adaptation. In addition, microarrays have been used to identify gene targets of transcription factors and signaling proteins. In this review, I will focus on areas of muscle adaptation that have benefited most from analysis by microarrays. These studies should represent a reasonable sampling of the work on muscle adaptation using expression profiling.

2.1   Aging

The first study on aging skeletal muscle using expression profiling was on mouse gastrocnemius muscle (Lee et al., 1999). In this study, the age-related gene expression changes were tested for whether caloric restriction, known to retard the aging process in general, could reverse these changes. One of the earliest versions of high density oligonucletide microarrays was used in this study, assaying 6347 unique mRNAs. Although the statistics were not near as rigorous as those currently used, this study revealed for the first time that only a small number of genes were differentially expressed by 2-fold or greater suggesting that sarcopenia was a long-term subtle process not associated with widespread or large changes in gene expression. The lack of many large magnitude changes was again found in aging rat soleus muscles (Pattison et al., 2003b), and this was even more evident in aging human vastus lateralis muscles (Giresi et al., 2005; Welle et al., 2003). Several changes in the mouse dataset were consistent with known physiological changes in aged muscle such as a reduced expression of genes involved in energy metabolism and an induction in a few genes involved in neuronal growth. Aging is associated not only with increased fatigability and reduced capacity for oxidative metabolism (Lindstrom et al., 1997) but there is also a loss of motor innervation of specific muscle fibers followed by reinnervation by other motor units (Larsson, 1995; Lexell, 1997). There was some evidence for stress gene induction and oxidative stress markers. Some of the heat shock genes such as HSP27 and HSP71 were induced while HSP70 was down regulated. GADD45 was also upregulated and thought to be associated with stress and DNA damage, but it is now known that many skeletal muscle adaptations are associated with an upregulation of GADD45 (see below), so its specificity, at the mRNA level, as a marker of DNA damage is unclear. There were decreases in several genes encoding proteins involved in proteasome-mediated proteolysis, although the opposite was true for the rat (Pattison et al., 2003b) and human (Giresi et al., 2005; Welle et al., 2003). Part of the reason that some of the findings of this initial mouse study were not seen in subsequent studies may have to do with the small number of samples used and limited statistical analysis. The fact that different species and muscles were studied could also contribute to some of the differences.

At least 6 genes were differentially expressed (greater than 2-fold) in one of the first human sarcopenia papers to use microarrays (containing 588 cDNAs) (Jozsi et al., 2000). While there was only one microarray per group (samples were pooled), most results were re-confirmed by RT-PCR. Like the mouse dataset, these data also suggested a stress/damage response in the vastus muscle of older vs. younger men. An interesting aspect of this study was the much smaller change in gene expression in response to one bout of resistance exercise in the old subjects compared with the younger subjects. This is reminiscent of another more recent paper on the reduced ability of aged rat soleus muscle to recover from immobilization atrophy compared to younger muscle (Pattison et al., 2003a).

A comprehensive study of aging soleus muscles in 3.5 month and 30.5 month old rats showed that using 20–25 samples per group greatly increased the power of the microarray analysis (Pattison et al., 2003b). This is important because many individual statistical tests are performed with microarray analysis, and so multiple test correction is needed. Of the 24,000 probes contained on the microarrays in this study, 682 were differentially expressed based on t-tests with a Bonferroni multiple test correction, and of these, 413 had functions documented in Internet searchable databases. Consistent with other aging studies, there were decreases in energy metabolism genes. A few upregulated genes were involved with oxidative stress/anti-oxidant responses and complement activation. Most of the decreases in expression with age were associated with extracellular matrix genes, also seen with cachexia (Lecker et al., 2004) and disuse atrophy (Stevenson et al., 2003). Follistatin was increased 3-fold, similar to that found in the vastus muscle of old vs younger men (Welle et al., 2003) and also in unloaded rat soleus muscles (Stevenson et al., 2003). Follistatin physically interacts with myostatin to inhibit its binding and growth preventing activity (Hill et al., 2002). Thus the role of increased follistatin mRNA is unclear. Perhaps this increase is a compensatory effect to counter myostatin activity, or there may be a paradoxical function of this protein. As mentioned, both the aging rat study and an aging human study (Welle et al., 2003) genes involved with proteasome-mediated proteolysis were increased. Consistent with this, the rat study showed a marked increase in the ubiquitin ligase atrogin (Pattison et al., 2003b), and in two human studies there was an increase in another F-box gene called F-box and leucine rich repeat protein 11 (Giresi et al., 2005; Welle et al., 2003). Other similarities between these two human aging papers where oligonucleotide microarrays were used to study the vastus muscle were a 3-fold increase in the cell cycle inhibitor p21, increases in several genes involved in RNA binding or splicing, increases in several immune genes, decreases in ion/amino acid transporter genes and energy metabolism genes. Metallothionin mRNAs were increased and these genes are normally upregulated in response to trace heavy metal accumulation particularly copper and iron (Coyle et al., 2002). Iron accumulation has been shown to play an important role in both inactivity and age related atrophy processes (Cook and Yu, 1998; Kondo et al., 1992). Metallothionein is also upregulated in skeletal muscle in response to starvation-and cachexia-induced atrophy (Lecker et al., 2004). While the role of ROS and trace metal accumulation during sarcopenia or other atrophies is not well understood, some believe that ROS production in muscle wasting conditions is a trigger for the activation of proteolytic systems that promote muscle protein degradation (Reid and Li, 2001).

One of the few sarcopenia studies in which women were studied showed similar age related changes as men including increased p21, hnRNA binding proteins, and decreases in genes involved in energy metabolism (Welle et al., 2004), although another study on aging muscle showed that gender had a greater effect on the number of differentially expressed genes than did age (Roth et al., 2002).

Lastly, there was a distinct increase in Foxo1A in the Welle et al. (2003) human dataset and an increase in Foxo3A in our dataset (Giresi et al., 2005). This is relevant, as Foxo proteins have been identified as having a role in several atrophy models in vivo and in cell culture (Sandri et al., 2004; Stitt et al., 2004), and with aging (Machida and Booth, 2004). Overall, from the aging microarray studies, genes that may be good candidates for further study include the Foxo proteins, follistatin, the RNA splicing genes, and selected immune and oxidative stress genes (e.g., metallothionein). Also, the glucose, amino acid, and ion transporter genes which are downregulated (Giresi et al., 2005) may be a key to the lower adaptive capacity of aged muscle, and perhaps less than optimal cellular homeostasis.

A different approach to understanding the mRNA phenotype of aged muscle was recently shown by the computation of a molecular signature of sarcopenia (Giresi et al., 2005). Molecular signatures were originally developed by Golub et al. (1999) to distinguish different cancer subtypes in human tissue based on "unsupervised" computational analysis of the gene expression profiles from these samples. The aim of computing a molecular signature is to identify the most economical number of genes that can best identify or predict the class of a new sample. In the case of aging muscle, 45 genes were found to best identify young vs. older muscle, and together these genes were considered to be a signature of aged muscle. The entire 45 genes taken as a whole possesses the predictive power of the signature, rather than individual genes representing absolute markers of age alone. Age related changes in the expression of genes that are not part of the aging signature do not indicate that they are unimportant in the progression of sarcopenia. The purpose of the signature is to distinguish an aging sample based on a set of the best predictive genes, and to determine if pharmacologic or exercise interventions can alter the signature. Many of the genes mentioned in this section as being differentially expressed with age are contained in this 45-gene sarcopenia signature.

2.2   Muscle Atrophy

Besides sarcopenia, there is disease-associated muscle atrophy (including fasting) and disuse atrophy. It is well known that increased proteolysis plays an important role in the protein loss with these types of atrophy. Further, the ubiquitin-proteasome system is involved in the bulk of protein degradation with atrophy (Jagoe and Goldberg, 2001; Lecker, 2003). Consistent with this, in disuse atrophies, one group of investigators found two ubiquitin ligase genes markedly upregulated with limb immobilization, denervation and muscle unloading, and these mRNAs were identified using differential display (Bodine et al., 2001a). These two genes, now familiar in the muscle field, were both found to be ubiquitin protein ligases important in regulating protein degradation during atrophy. One was named Muscle Atrophy F-box (MAFbx) containing ubiquitin ligase and the other was designated Muscle Ring Finger 1 (MuRF1) ubiquitin ligase. Knockout of either of these genes inhibited atrophy due to denervation by about one-half. At the same time another lab found, using a microarray platform, that with several different types of cachexia and with fasting there was a marked increase in a gene they named atrogin-1 (Gomes et al., 2001), which was found to be the same gene as MAFbx. Shortly there-after, several groups using microarrays or RT-PCR also found these two ubiquitin ligases were upregulated in atrophy due to spinal isolation or unloading, respectively (Haddad et al., 2003; Stevenson et al., 2003). An important aspect of these protein ligases is that they are muscle specific and, in addition, ubiquitin ligases are known to have significant substrate specificity. Therefore, much work has been stimulated on muscle atrophy due to the discovery of these ubiquitin ligases (rev. by (Sartorelli and Fulco, 2004). These ubiquitin ligases are thought to be key in initiating a majority of the protein degradation in both cachexia atrophy and disuse atrophy (Bodine et al., 2001a; Gomes et al., 2001). Thus they may be important in developing therapeutic targets to mitigate protein loss. Some of the ubiquitins, and one or two of the ubiquitin conjugating enzymes were also upregulated in disease or disuse atrophy (Batt et al., 2006; Lecker et al., 2004; Pattison et al., 2003c; St-Amand et al., 2001; Stevenson et al., 2003; Wittwer et al., 2002).

Further evidence that components of the ubiquitin-proteasome pathway were upregulated at the mRNA level was identified in the unloading time course study using microarrays (Stevenson et al., 2003). Besides upregulation of MuRF1 and atrogin, Nedd4 another ubiquitin ligase was upregulated to a similar extent as atrogin1. Denervated rat gastrocnemius muscle (1 month) showed the same results for these 3 genes (Batt et al., 2006). Nedd4 has been implicated in targeting membrane proteins for degradation by the proteasome (Snyder et al., 2001) but has not previously been shown to be upregulated during atrophy. A reason all atrophy studies have not shown upregulation of ubiquitin ligases may be because they seem to have a transient upregulation, at least with disuse protocols (Fig. 1). None of these three ubiquitin ligases were shown to be upregulated in several microarray studies with either early (Bey et al., 2003) or later time points being studied (Batt et al., 2006; Pattison et al., 2003c; St-Amand et al., 2001; Wittwer et al., 2002). One of the other interesting findings in the time course study was the striking co-activation of many different gene products comprising structural and regulatory components of the 26S proteasome (Fig. 1). The increase peaks at day 4 of unloading and is less elevated at days 7 and 14. Previous work showed the activation of several proteasome subunits with atrophy (Bey et al., 2003; Ikemoto et al., 2001; Taillandier et al., 1996a), but the co-regulation of many proteasome genes was a novel finding, a demonstration of the advantage of using a temporal study design. This pattern of co-regulation suggests that each subunit may share common *cis*-regulatory elements. Muscle from diabetic, uremic, tumor bearing and fasting mice also show upregulation of multiple proteasomal subunits at a single but relatively early time point of muscle wasting (Lecker et al., 2004). In comparison, long term denervation (2–3 months) studies do not show upregulation of proteasomal proteins (Batt et al., 2006; Kostrominova et al., 2005) and this is likely due to the transient upregulation (Batt et al., 2006).

Other proteolytic systems that contribute to muscle proteolysis during disuse or disease atrophy are the lysosomal cathepsins, Ca^{2+}-dependent calpains, and caspase-3 (Du et al., 2004; Ikemoto et al., 2001; Taillandier et al., 1996b; Tidball and Spencer, 2002; Tischler et al., 1990; Williams et al., 1999). Focused traditional studies on calpains and caspases have important roles in atrophy that involve initial

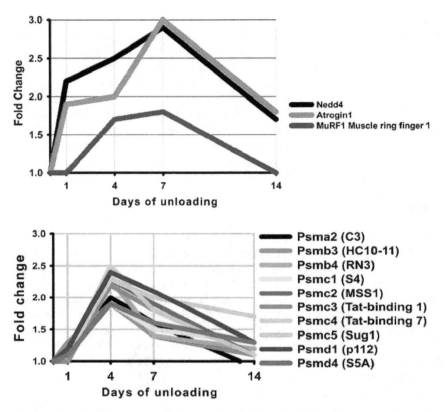

Figure 1. Top: Fold change in mRNA expression of three ubiquitin ligases in rat soleus muscle over a time course of hindlimb unloading. The fold change is relative to weightbearing control muscles. Bottom: Fold change in mRNA expression of proteasome subunits in the same study. The Affymetrix rat U34A GeneChip was used. Modified from Stevenson et al J. Physiol. 551.1, 33-48, 2003. A color figure is freely accessible via the website of the book: http://www.springer.com/1-4020-5176-x

cleaving of specific substrates, before proteasomeal degradation (Du et al., 2004; Tidball and Spencer, 2002; Tischler et al., 1990; Williams et al., 1999). The differential expression of calpain or caspase mRNAs in microarray studies are varied however. The mRNA of calpain 2 was increased (1.8 fold) at 9 days of unloading, but our microarray data showed no change from 1-14 days and calpain 3 decreased moderately. This decrease in calpain 3 was also seen in 1 month denervated muscle. In another microarray study (1,176 cDNAs interrogated) 2 month denervated EDL muscles showed an increase in calpain 2 (Kostrominova et al., 2005). In sepsis, there is an increase in the expression of calpains at the mRNA level using standard techniques and, dantrolene reverses the sepsis effect on myofibrillar disassembly (Williams et al., 1999). Taken together, the role of calpains in protein degradation appears to be regulated differently in different types of atrophy, and the use of microarrays to study calpains has not illuminated any consistent patterns.

In contrast, expression profiling has shown that the cathepsins are consistently and markedly upregulated at the mRNA level with unloading or immobilization (Deval et al., 2001; Lecker et al., 2004; Pattison et al., 2003c; Stevenson et al., 2003), denervation (Batt et al., 2006), and disease wasting (Deval et al., 2001; Lecker et al., 2004; Pattison et al., 2003c; Stevenson et al., 2003). Cellular proteins targeted by lysosomal cathepsins are often membrane proteins (Mayer, 2000). The activation of cathepsins B, C and L has been shown with atrophy (Batt et al., 2006; Bey et al., 2003; Ikemoto et al.,2001; Taillandier et al., 1996b). The time course of this activation with respect to other proteolytic genes showed that some of the cathepsins (cathepsins C, D and L) are upregulated to similar levels as components of the ubiquitin-proteasome system at early time points, but they are increased to even greater levels at later time points (7 and 14 days) (Pattison et al., 2003c; Stevenson et al., 2003). Unlike the ubiquitin ligases, the sustained increase in cathepsin mRNAs may be the reason that the latter are found in most papers where only one time point was studied. Thus expression profiling has shown that cathepsins likely have an important role in both disuse (Batt et al., 2006; Pattison et al., 2003c; Stevenson et al., 2003) and disease (Deval et al., 2001; Lecker et al., 2004) atrophy but the exact role and whether they work independently or in conjunction with the ubiquitin proteasome system, is an important avenue for study.

Other proteases that are differentially expressed with disuse as discovered by microarray studies are several serine proteases and serine protease inhibitors (serpins). In the unloading time course study Spin2B is upregulated 4-fold at day 1, reaching its peak at 10-fold by days 4 and 7 (Stevenson et al., 2003). Several serine proteases are gradually upregulated over the time course of unloading (kallikrein 1, subtilisin-like Ca^{2+}-dependent serine protease and mannose-binding protein associated serine protease). Kallikrein genes are also upregulated in 2-month denervated EDL muscle (Kostrominova et al., 2005). The coordinated actions of serine proteases and their inhibitors are involved in a variety of cellular processes and have been well studied in other cell types (Sangorrin et al., 2002). Recent work has identified a serine protease and its endogenous inhibitor that play a role in the disassembly of normal myofibrillar protein turnover in mice (Sangorrin et al., 2002). Many of these serine proteases have been characterized based on their involvement in complement activation, or have been localized to the ECM of cells (Kaplan et al., 1999; Sim and Laich, 2000). Therefore, it is also possible that they exist in the extracellular space and mediate matrix remodeling of muscle or endothelial cells.

The composition of the ECM in soleus muscle is altered after unloading (Miller et al., 2001). In both disease and disuse models of atrophy, significant downregulation of ECM genes such as collagens and fibronectin have been identified (Cros et al., 2001; Lecker et al., 2004; Stevenson et al., 2003). Thus significant remodeling of the extracellular matrix may be a necessary aspect of all types of muscle atrophy and it requires further study. It seems logical that a smaller muscle fiber requires less elastic structure and extracellular matrix.

Consistent with the idea that decreased protein synthesis is involved in the loss of muscle protein, eukaryotic translation initiation factor 4E binding protein 1 (4EBP-1) is upregulated by 1 day and is maintained at 14 days of unloading (Stevenson et al., 2003). When unphosphorylated, 4EBP-1 is a translational repressor that inhibits eukaryotic translation initiation factor 4E (eIF4E)-dependent mRNA translation. An early (12 h) increase in 4EBP-1 was also seen with muscle unloading (Bey et al., 2003). 4EBP-1/eIF4E complexes are enriched in rat medial gastrocnemius extracts after 14 days of hindlimb unloading suggesting a role in decreased protein synthesis (Bodine et al., 2001b). 4EBP-1 is also markedly upregulated in fasting and disease-associated atrophy (Lecker et al., 2004). These data support the idea that translational inhibition is a hallmark of disuse and disease types of atrophy. Papers focused on the role of deactivated Akt playing an important role in disease and disuse atrophy are consistent with this finding (Bodine et al., 2001b; Latres et al., 2005; Sandri et al., 2004; Stitt et al., 2004) and they have opened an important new avenue of study regarding the signaling pathways involved in triggering the atrophy process.

The transcription factor Foxo1 (a phosphorylation target of Akt) was upregulated in atrophied muscle due to fasting (Kamei et al., 2003; Lecker et al., 2004), cancer, uremia, or diabetes (Lecker et al., 2004). We have also shown an increase in Foxo3 mRNA in unloaded muscle (unpublished data). This observation, originally identified in cachectic muscle may have spawned the more detailed experiments showing involvement of Foxo1 (and Foxo3) in atrophy (Kamei et al., 2004; Sandri et al., 2004; Stitt et al., 2004) and in sarcopenia (Machida and Booth, 2004).

A difference in cachexia vs disuse atrophy is that in the former, there is no slow-to-fast phenotype transition. In fact, glycolytic genes are downregulated in cachexia (Lecker et al., 2004) while they are upregulated in disuse atrophy (Stevenson et al., 2003). Disuse due to muscle unloading leads to marked changes in expression of genes encoding myosin isoforms and components of excitation-contraction coupling consistent with a faster phenotype (Cros et al., 2001; Stevenson et al., 2003) but this is not the case in cachexia. Markers of oxidative stress such as glutathione-S-transferase isoforms and the selenoproteins are upregulated with unloading, immobilization and denervation (Kostrominova et al., 2005; Pattison et al., 2003c; St-Amand et al., 2001; Stevenson et al., 2003), while chaparones are downregulated (Pattison et al., 2003c; Stevenson et al., 2003) but these changes have not been reported for cachexia or fasting atrophy (Lecker et al., 2004).

TNF-alpha has been shown to be a major component of muscle wasting due to disease states such as cancer, diabetes, end-stage heart failure, etc. As a model of these conditions, in one study mice were treated with daily injections of TNF for 1, 3 or 5 days and the effects on skeletal muscle were analyzed using oligonucleotide microarrays containing 11,000 probes (Alon et al., 2003). A k-means clustering algorithm was used to identify 28 different temporal gene expression patterns due to TNF treatment. This algorithm distinguishes temporal patterns of gene expression based on the shape of the curve without consideration to magnitude of change. Of the 1037 genes that were differentially expressed (at least 3-fold) at one or more time points of TNF treatment, some were similar to changes seen in cachectic muscle due to fasting, uremia, diabetes, or cancer. These mRNAs included upregulated metallothionein II, several proteasome subunits, and downregulation of some energy metabolism genes. Changes seen with TNF treatment that were not shown in the cachexia microarray study (Lecker et al., 2004) were activation of stress response genes, acute phase genes, and inflammatory markers. Changes in transcription factor expression and structural genes differed between cachexia and TNF treatment of mice. This study was helpful in determining that cachexia has effects on skeletal muscle that are more than the effects of TNF alone.

When the disuse atrophy studies are taken together, they suggest that atrophy is associated with changes in mRNA levels of genes involved in protein synthesis, proteolysis, oxidative stress, chaperones, growth and cell cycle regulation, structural and regulatory genes of the extracellular matrix and cytoskeleton. However, the use of a time course study of disuse atrophy expanded upon these findings (Stevenson et al., 2003). What is often seen in time course studies is that during early periods after the stimulus induction, there are more genes induced than at later time points when a steady state phenotype is attained (Stevenson et al., 2003; Zhao et al., 2002). This is consistent with the idea that initially, there is an induction of more cellular processes, and this is reflected by a greater number of mRNA changes. The application of k-means clustering allowed a timeline of the atrophy process with respect to the behavior of genes in multiple functional categories (Fig. 2). Regulatory genes (transcription factors, signaling, and growth/proliferation genes) were often upregulated early, in either a transient or sustained manner. Thus, regulatory genes may be involved in the initial switching mechanism that sets in motion atrophic changes. But regulatory genes also populated clusters with later patterns of activation, which is consistent with the idea that disuse is marked by several phases in which regulatory factors are sequentially activated or deactivated. Also, regulatory genes involved in compensatory or secondary processes would be expected to change at later time points. On the other hand, structural genes tended to have less varied patterns of activation. Differential expression of mRNAs that are maintained with atrophy after 2 weeks were often genes associated with structural or phenotype changes including ECM genes, cytoskeletal genes, myosin and other fast vs. slow phenotype determining genes. Now that many differentially expressed genes and gene expression patterns with various types of atrophy have been revealed, ongoing work is aimed at more in-depth studies on specific genes to identify their role in atrophy as has been done with genes such as FOXO and atrogin1.

2.3 Atrophy Studies Using Genetic Models

Another useful way that microarrays have been used is to study the global phenotypic effects of gene overexpression or gene knockout in skeletal muscle. When Foxo1 was overexpressed in skeletal muscle, muscles were significantly atrophied, there were fewer type I fibers, and there was impaired glucose tolerance

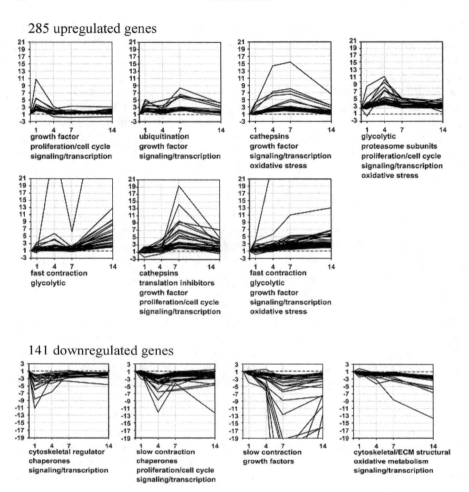

Figure 2. Use of k-means clustering to determine whether there is order in a temporal expression profiling experiment; in this case, numbers along the x-axis indicate days of rat hindlimb unloading (taken from Stevenson et al J. Physiol. 551.1, 33-48, 2003). y-axis is fold change in gene expression in unloaded vs. weightbearing soleus muscle. Broad functional categories represented in each cluster are indicated below each graph. Data indicate that there is not simply a global decrease in gene expression with atrophy, it is a highly regulated process, and there are a distinct number of gene expression patterns. The complexity of upregulated patterns is greater than downregulated patterns

(Kamei et al., 2004). Microarray and RT-PCR analysis of quadriceps muscle showed decreased expression of genes encoding the slow isoforms of myofibrillar proteins, decreased transferrin receptor, increased Gadd45, atrogin, and cathepsin L. All of these changes were similar to those seen in the atrophied soleus after unloading (Stevenson et al., 2003) and transferrin receptor, cathepsin L and atrogin changes were similar to that seen with cachexia (Lecker et al., 2004). However, there were also many genes expressed differently in unloaded and cachectic compared to overexpressing Foxo1 muscle. This type of study points to candidate proteins for further study, proteins that may regulate atrophy induced by Foxo1 or that may be downstream Foxo1 target genes.

In another study on atrophy, muscle specific expression of a constitutively active form of IκB kinase beta (IKKb) showed a distinctive muscle atrophy phenotype suggesting NF-kB activation leads to marked atrophy (Cai et al., 2004). While not using microarrays, these investigators used real-time RT-PCR to measure the expression of 36 genes in response to the IKKb activation. They were able to show that there was no immune activation based on assessment of 20 immune ligands and receptor mRNAs. They also showed that IKKb overexpression was associated with atrophy markers such as two upregulated proteasome subunits as well as MuRF1, but not atrogin1. NF-kB family members known to be self regulated by NF-kB were all upregulated. Thus this atrophy is not associated with inflammation as is seen with true cachexia, but it is associated with some classic mRNA markers of atrophy. Taken together, these studies show that Foxo1- and IKKb-induced atrophy are not phenotypically identical, at the mRNA level. Both genes, or their family members, have been shown to play a role in atrophy in vivo (Cai et al., 2004; Hunter and Kandarian, 2004; Sandri et al., 2004).

2.4   Atrophy in Muscle Cell Culture

In order to dissect the details of regulatory processes underlying atrophy of muscle, many investigators have used cell culture models that are easier to manipulate and assay compared with whole muscle. For the most part these have involved the treatment of cultured myotubes with exogenous cytokines (e.g., TNF-alpha) or glucocorticoids (i.e. dexamethasone). These compounds have been used because they are thought to be involved in triggering atrophy in various types of cachexia, and in fact, they do induce increased protein degradation in cultured myotubes similar to that seen during cachexia and fasting in vivo (Li et al., 2003; Li and Reid, 2000; Li et al., 1998; Sacheck et al., 2004; Shah et al., 2000; Thompson et al., 1999). We do not know however, the extent to which the mRNA profile of muscle cells treated with Dex or TNF resembles that of muscle atrophy due to disease or disuse, although there are a significant number of similarities in the mRNA expression profile of disease-induced and disuse-induced atrophy as discussed above. In a recent study, starvation of mature myotubes by failure to replenish media every 48 hrs leads to a marked atrophy (Stevenson et al., 2005). While morphologically atrophied, the transcription profile of these cells was quite distinct from any expression profile published for whole muscle due to either reduced use or to disease. The only similarity was an increase in the translation initiation factor 4EBP-1 and a decrease in ECM genes. Other hallmark mRNA changes associated with atrophy were not seen such as the increase in cathepsin or ubiquitin-proteasome genes. These data show that while the gross morphology of atrophied muscle fibers may be similar in whole muscle vs. myotube culture, the processes by which this phenotype is achieved differs markedly. Further, knowledge of the transcriptional profile of TNF or Dex treated cells would prove useful for comparison to disease induced wasting in vivo.

2.5   High Intensity Contractions and Hypertrophy

There have been efforts to identify the expression profile of muscle after high intensity muscle contractions (Chen et al., 2002), repeated bouts of which lead to hypertrophy, and of muscle hypertrophy after the induction of "overload" due to synergist ablation (Carson et al., 2002). The purpose of these studies was to identify key mRNAs or patterns of mRNA changes that would provide insight into the regulation of the development of muscle hypertrophy. Thus while these studies use different models of increased contraction, both provide an expression profile at an early time point after the induction of the exercise; in the case of the high intensity muscle contractions, 1 and 6 hr time points were studied in the tibialis muscle. In the synergist ablation study the soleus was studied at 3 days after the increased loaded contraction. This is considered to be a time in which true growth is beginning and inflammation due to the surgery is waning. Six hours after a bout of high intensity contractions, induction of 60 genes was found, many of which are involved in growth and proliferation (e.g., bFGF, c-fos, Egr-1, Egr-2, CARP), but there was also upregulation of growth inhibiting genes such as p53 and GADD45. Three days after synergist ablation, the soleus also showed increased CARP, GADD45, Egr-2, and c-fos expression. A study in humans, 6 hrs after a bout of resistance exercise showed increased CARP expression (Zambon et al., 2003). These results not only suggest interesting molecules for further study, but they take on further meaning when compared to expression profiles from atrophying muscles. For instance, the increased contraction studies show marked increases in the muscle-specific growth factor CARP, while the time course of unloading shows a marked and sustained decrease in the expression of CARP (Stevenson et al., 2003). There are other genes responding in opposite fashions to atrophy vs hypertrophy inducing stimuli. Unloading shows decreases in expression of collagen genes, cytoskeletal genes, and tissue inhibitor of metalloproteinase 1, while hypertrophy shows increases in these genes. The transcription factor Nur77 was upregulated with soleus atrophy but markedly downregulated with hypertrophy (Carson et al., 2002). IL-15 was upregulated with atrophy but downregulated 6 hours after high intensity contractions in the tibialis. Homer2, a protein known to form complexes with membrane channels such as the glutamate receptor, IP3 receptor, and ryanodine receptor (Ward et al., 2004) was upregulated due to high intensity muscle contractions but downregulated in a sustained manner with unloading or with immobilization (Pattison et al., 2003c; Stevenson et al., 2003). On the other hand, there are genes that respond similarly to either hypertrophy or atrophy stimuli. For instance, GADD45 is also increased with unloading, thus, this may be a gene that is induced with many types of perturbations, and appears not to be specific to hypertrophy or atrophy. The growth inhibitor p21 and the signaling gene JAK2, were both induced in both the 3-day overloaded soleus and with unloading, showing that these genes respond similarly to opposite types of stimuli. Further evidence that p21 is a ubiquitous responder in muscle is that it is also upregulated with fasting (Jagoe et al., 2002), aging (Giresi et al., 2005; Welle et al., 2003), and with one bout of resistance exercise in humans (Zambon et al., 2003). Based on the expression profiling of these studies, some genes that show promise for further study on muscle hypertrophy are CARP, Egr-1, Egr-2, IL-15, Homer2 (also known as Vesl), and Nur77. A partial list of candidate genes that may be interesting for further study during several different types of muscle adaptation processes is presented in Table 1.

There is one comprehensive study on acute and long term adaptive changes of mRNA expression to voluntary wheel running in mice (Choi et al., 2005). On a high density cDNA array containing almost 16,000 probes (10,615 unique genes), 900 showed a 2-fold change or greater at 3, 6, 12, or 24 hrs after a single 12 hr bout of intermittent voluntary running. Thus the authors were able to create self-organizing maps to identify the patterns of gene expression change after recovery from running.

Table 1. Selected genes, or families of genes, from expression profiling studies that may be interesting candidates for further study. Myofibrillar and extracellular matrix genes, while changed in most plasticity conditions, are not included in table. List is partial, and is meant to stimulate discussion

Aging	Disuse	Cachexia	Contraction/ Hypertrophy	Regeneration
		mRNAs with increased expression		
Foxo	GST	Foxo genes	homer2	calpain 6
ubiquitin ligases	ubiquitin ligases	ubiquitin ligases	CARP	Targets of MyoD
hnRNA splicing	cathepsins	cathepsins	Vegf	Egf
hnRNA binding	gutamine synthatase	glutamine synthatase	Nur77	TGFb
follistatin	follistatin	Slc7a8 (a.a. transporter)	Egr-1, Egr-2	Igf-1
C/EBPbeta	4EBP1	4EBP1	Igf-1	Chemokines (C-C)
metallothioneins	serine protease inhibitors	metallothioneins		biglycan
period2	IGFBP5			fasciclin I-like
complement component 1	Cited2			ATF3
	Selenoproteins			MRFs
	Activin IIB receptor			
	IL-15			
		mRNAs with decreased expression		
transporter genes (i.e. solute carriers)	homer 2	transferrin receptor	IL-15	cell cycle genes
	Vegf	IGFBP5		
	Annexin 5			
TBZF	Timp1			
DAAM2				

With a few exceptions, most clusters represented genes that had transient changes in expression. Some examples were Fos, Vegf, and Pgc1. The greatest number of differentially expressed genes was found at 24 hrs suggesting that this is a time of high biological adaptation in muscle, at least to a single bout of exercise. Several of the changes in gene expression were similar to that seen with eccentric exercise such as Fos, Gadd45a, cardiac alpha-actin, and Carp. In contrast, mice that had performed nightly voluntary wheel running for 4 weeks showed the least number of differentially expressed genes; ostensibly because a steady state adaptation had occurred; there were only $\sim$50 genes differentially regulated in a sustained manner that characterized the new more oxidative phenotype.

2.6   Muscle Regeneration and Development

Skeletal muscle has the unique ability to regenerate after injury. In rodents, small muscles can regenerate entirely after exposure to cardiotoxin for example, or in response to cutting the blood vessels. Muscle regeneration is thus an adaptation to injury and interestingly, many features of the process, except the initial degeneration phase, recapitulate myogenic development. Given the complex but staged nature of this process, it is ideal to study using gene expression profiling over the time course of the process, in order to identify transcriptional cascades. One experiment was done, using oligonucleotide microarrays containing 10,000 probes, over the course of regenerating mouse gastrocnemius muscle (0, 12hr, 1 day, 2 days, 4 days, 10 days) after injection of cardiotoxin (Zhao et al., 2002). The focus was on identifying target genes of the well known myogenic transcription factor MyoD, which has a central role in both myogensis and in regenerating muscle. To do this, the temporal expression pattern of known downstream MyoD target genes, such as Ulip, were identified. Other genes that had an expression pattern similar to Ulip were assigned to the "Ulip cluster" of genes (n=47). Promoter databases and sequence analysis tools were then used to show that a subset of these genes had potential MyoD binding sites in their promoters. In one case from this cluster, gel-shift assays and chromatin immunoprecipitation were used to discover a functional binding site in the promoter of the gene "Slug", a member of the snail/slug family of transcriptional repressors. The functionality of this relationship was demonstrated *in vivo* using slug-dependent reporter constructs and through the demonstration that Slug null mice show impaired ability to regenerate after injury. The use of transcription profiling in this way is very useful for identifying unknown target genes of transcription factors thought to be involved in regeneration. This public database could be used to discover other targets of either MyoD or other transcription factors during regeneration.

In another study the gastrocnemius was again injected with cardiotoxin to study expression profiling during regeneration at 0.25, 0.5, 1, 2, 5, 7, and 14 days (Goetsch et al., 2003). This experiment was performed using the same high-density oligonucleotide microarray as the study above. However in this work, the differentially expressed mRNAs were presented in functional clusters and in clusters created using self organizing maps. This work showed coordinated expression of specific growth factors, myogenic genes, inflammatory genes, and genes for proteins that comprise the extracellular matrix. This work supports other physiological and morphological studies by clearly presenting the temporal changes in expression of those genes involved with modulating the changes in physiology during regeneration. Further experiments on two extracellular matrix genes (biglycan and periostin) that were differentially expressed, during regeneration, showed their specificity as mesenchymal derivatives during development.

In related work, temporal expression profiling was used to investigate how MyoD orchestrates the process of differentiation (Bergstrom et al., 2002). MyoD overexpression was used to induce differentiation in MyoD-/- / Myf5-/- mouse fibroblasts, and expression was measured at several time points using microarrays. By using cyclohexamide to inhibit synthesis of other activated regulatory factors the authors were able to strictly identify targets of MyoD. Using k-means cluster analysis they were able to show that MyoD can initiate several distinct subprograms of gene expression through promoter-specific recognition rather than global activation of all MyoD regulated genes.

Other studies on gene expression profiling in muscle during differentiation have identified new genes that may be involved in this process. One recent study used the differentiation of the C2C12 cell line as a model of development to identify the time course of gene induction during differentiation (Tomczak et al., 2004). Many genes were differentially regulated over 12 days of differentiation and during cell cycle withdrawal there were marked increases in Vcam1, Itgb3, Vcl, and Ptger4, the latter has not previously been associated with muscle development. This type of study using high-density microarrays and others like it (Sterrenburg et al., 2004) present a treasure trove of candidate genes for further analysis in myogenesis.

The molecules necessary for differentiation in cell culture were also studied in a cell line lacking muscle derived IGF-II (C2AS12). When treated with differentiation media, these cells normally die, but the addition of IGF rescues them (Kuninger et al., 2004). Thus, by comparing cells with and without IGF treatment in the differentiation media, the mRNAs that are "recovered" by IGF treatment were identified by microarray analysis (36,000 genes probed). Theoretically, the differentially expressed genes would be involved in IGF-dependent survival and differentiation in this cell line. Of the 90 genes induced, 28 were muscle specific, others were regulatory genes, and 33 were as yet unidentified. One interesting gene that was identified was a muscle specific gene, hemojuvelin, which is know to be expressed in somites during development and is thought to be a member of a family of genes involved in cell migration guidance.

2.7 Muscular Dystrophies

While not central to a chapter on adaptation, there is a solid literature on discoveries about dystrophic muscle using microarrays. Thus, only an overview will be given here despite the 25 or so papers now published on expression profiling of dystrophic muscle. Typically, the differentially expressed genes in dystrophic muscle are consistent with the histological pathology (Bakay et al., 2002; Chen et al., 2000; Haslett et al., 2002; Porter et al., 2003). The lack of dystrophin, the hallmark of most muscular dystrophies, leads to a dissociation of the mechanical linkage of dystrophin, which normally connects the extracellular matrix to the cytoskeleton. Thus there is thought to be a loss in the stabilization of the sarcolemma during contractile activity and this triggers necrosis (Campbell, 1995; Petrof et al., 1993). Secondary effects of dystrophin deficiency are fibrosis, inflammation, and the failure for normal muscle regeneration (Blake et al., 2002). Duchenne muscular dystrophy (DMD) in humans is lethal by the second or third decade of life whereas the genetic mouse model of dystrophin deficiency (the mdx mouse) is less severe. The extraocular muscle is unaffected in dystrophin deficiency and the diaphragm is more heavily affected than limb muscles leading investigators to the conclusion that more than the genetic defect is involved in the pathogenesis of affected muscle (Porter et al., 2003). An inflammatory response, likely to muscle damage, is a common response in both mdx and DMD and the expression profiles of both reflect activation of immune and inflammation genes (Bakay et al., 2002; Chen et al., 2000; Haslett et al., 2002; Porter et al., 2002; Porter et al., 2003; Tseng et al., 2002). The increase in extracellular matrix genes is a common finding in DMD and mdx mice but only the fibrosis genes are seen in DMD (Porter et al., 2002). Another interesting difference between DMD and the mdx mouse is that DMD shows marked reduction in genes involved in energy metabolism (Chen et al., 2000) while mdx muscle does not (Tseng et al., 2002). Increased expression of neonatal and perinatal myosin isoforms is prominent in DMD (Chen et al., 2000; Haslett et al., 2002) but not in mdx muscle (Porter et al., 2002; Tseng et al., 2002) suggesting an abnormal developmental program in the human form of the disease. Many of the cathepsins are activated in mdx (Porter et al., 2002; Tseng et al., 2002), but not in DMD (Bakay et al., 2002). Time course studies during the development of the disease in mdx mice have also been useful to understand the underlying molecular events associated with the pathogenesis (Porter et al., 2003). Taken together, these studies have provided scientists with focused avenues for study of muscular dystrophies as well as the types of genes that may determine the severity and progression of the disease.

2.8   Circadian Rhythms and Exercise

The existence of circadian clocks in peripheral tissues, including skeletal muscle, has been established in the past several years (Yamazaki et al., 2002). The existence of a clock in skeletal muscle was defined by expression of the core circadian rhythm regulatory genes such as clock, Bmal, period (Per) and cryptochromes (Cry) (Zambon et al., 2003). In addition to the core clock genes several other genes exhibit circadian expression in skeletal muscle and these include, Nfil3, C/EBPbeta, Myf6, Hat, and Gadd45. It was also shown that one bout of resistance exercise in humans can regulate clock genes (Per2, Cry1, Bmal) in the exercised leg but not the control leg. This suggests that the muscle clock is affected by exercise independent of signals from the central clock (the suprachiasmatic nucleus), and, that exercise may directly affect circadian regulated genes. Exercise also upregulated 12 diurnal-regulated genes that are normally repressed in the morning, and downregulated 29 diurnal genes normally induced in the morning. Thus, not only do both exercise and circadian rhythms rely on the transcriptional regulation of genes, but these data also support the idea that exercise may reset circadian rhythms by affecting gene expression. Further work in this interesting area will be forthcoming.

3.   CONCLUSIONS

A significant literature has accumulated in the past six years in which gene expression profiling has been used to better understand muscle plasticity. This review summarizes a considerable sampling of these studies. Depending in part on study design, expression profiling experiments can be difficult to interpret. However, careful analysis of quantitative expression data and especially the addition of time course analyses can insure a high probability of useful molecule identification. The most useful approaches have been: 1. the identification of interesting or unknown differentially expressed genes and their subsequent characterization; 2. time course studies using cluster analysis and functional categorization to identify co-regulated genes (functional genomics); 3. comparative studies using multiple conditions; 4. time course studies used for identification of downstream target genes of activated transcription factors; and 5. comparison of expression profiles from wild-type and transgenic (or knockout) animals. The use of computationally derived molecular signatures from expression profiling has also proven useful. It addition, it has become evident that comparison of expression profiles from multiple studies can be helpful in deciphering genes that may have important regulatory function in muscle adaptation. The most significant advances in knowledge using microarrays so far have come from studies on atrophy, aging, regeneration, and muscle dystrophy. In some cases this is due to the shear number of papers on the topic. New computational programs for analyzing gene expression datasets, including those that make associations among known genes, and the improvement in annotation and function of all the expressed genes in mammalian genomes will greatly increase the power of studies to be done, and will allow for further interpretation of those already performed.

ACKNOWLEDGEMENTS

The author is grateful to Dr. Robert Jackman for editorial assistance. This review was written during support from NIH (R01 AR41705) and NASA (NNA04CD02G).

REFERENCES

Alon, T., Friedman, J. M., and Socci, N. D. (2003). Cytokine-induced patterns of gene expression in skeletal muscle tissue. *J Biol Chem* 278, 32324–32334.

Ashburner, M., Ball, C. A., Blake, J. A., Botstein, D., Butler, H., Cherry, J. M., Davis, A. P., Dolinski, K., Dwight, S. S., Eppig, J. T., Harris, M. A., Hill, D. P., Issel-Tarver, L., Kasarskis, A., Lewis, S., Matese, J. C., Richardson, J. E., Ringwald, M., Rubin, G. M., and Sherlock, G. (2000). Gene ontology: tool for the unification of biology. The Gene Ontology Consortium. *Nat Genet* 25, 25–29.

Bakay, M., Zhao, P., Chen, J., and Hoffman, E. P. (2002). A web-accessible complete transcriptome of normal human and DMD muscle. *Neuromuscul Disord* 12 Suppl 1, S125–141.

Batt, J., Bain, J., Goncalves, J., Michalski, B., Plant, P., Fahnestock, M., and Woodgett, J. (2006). Differential gene expression profiling of short and long term denervated muscle. *Faseb J* 20, 115–117.

Benjamini, Y., and Hochberg, Y. (1995). Controlling the false discovery rate: a practical and powerful approach to multiple testing. *J Roy Stat Soc B* 57, 289–300.

Bergstrom, D. A., Penn, B. H., Strand, A., Perry, R. L., Rudnicki, M. A., and Tapscott, S. J. (2002). Promoter-specific regulation of MyoD binding and signal transduction cooperate to pattern gene expression. *Mol Cell* 9, 587–600.

Bey, L., Akunuri, N., Zhao, P., Hoffman, E. P., Hamilton, D. G., and Hamilton, M. T. (2003). Patterns of global gene expression in rat skeletal muscle during unloading and low-intensity ambulatory activity. *Physiol Genomics*.

Blake, D. J., Weir, A., Newey, S. E., and Davies, K. E. (2002). Function and genetics of dystrophin and dystrophin-related proteins in muscle. *Physiol Rev* 82, 291–329.

Bodine, S. C., Latres, E., Baumhueter, S., Lai, V. K., Nunez, L., Clarke, B. A., Poueymirou, W. T., Panaro, F. J., Na, E., Dharmarajan, K., Pan, Z. Q., Valenzuela, D. M., DeChiara, T. M., Stitt, T. N., Yancopoulos, G. D., and Glass, D. J. (2001a). Identification of ubiquitin ligases required for skeletal muscle atrophy. *Science* 294, 1704–1708.

Bodine, S. C., Stitt, T. N., Gonzalez, M., Kline, W. O., Stover, G. L., Bauerlein, R., Zlotchenko, E., Scrimgeour, A., Lawrence, J. C., Glass, D. J., and Yancopoulos, G. D. (2001b). Akt/mTOR pathway is a crucial regulator of skeletal muscle hypertrophy and can prevent muscle atrophy in vivo. *Nat Cell Biol* 3, 1014–1019.

Cai, D., Frantz, J. D., Tawa, N. E., Jr., Melendez, P. A., Oh, B. C., Lidov, H. G., Hasselgren, P. O., Frontera, W. R., Lee, J., Glass, D. J., and Shoelson, S. E. (2004). IKKbeta/NF-kappaB activation causes severe muscle wasting in mice. *Cell* 119, 285–298.

Campbell, K. P. (1995). Three muscular dystrophies: loss of cytoskeleton-extracellular matrix linkage. *Cell* 80, 675–679.

Carson, J. A., Nettleton, D., and Reecy, J. M. (2002). Differential gene expression in the rat soleus muscle during early work overload-induced hypertrophy. *Faseb J* 16, 207–209.

Chen, Y. W., Nader, G. A., Baar, K. R., Fedele, M. J., Hoffman, E. P., and Esser, K. A. (2002). Response of rat muscle to acute resistance exercise defined by transcriptional and translational profiling. *J Physiol* 545, 27–41.

Chen, Y. W., Zhao, P., Borup, R., and Hoffman, E. P. (2000). Expression profiling in the muscular dystrophies: identification of novel aspects of molecular pathophysiology. *J Cell Biol* 151, 1321–1336.

Choi, S., Liu, X., Li, P., Akimoto, T., Lee, S. Y., Zhang, M., and Yan, Z. (2005). Transcriptional profiling in mouse skeletal muscle following a single bout of voluntary running: evidence of increased cell proliferation. *J Appl Physiol* 99, 2406–2415.

Chu, C. C., and Paul, W. E. (1998). Expressed genes in interleukin-4 treated B cells identified by cDNA representational difference analysis. *Mol Immunol* 35, 487–502.

Cook, C. I., and Yu, B. P. (1998). Iron accumulation in aging: modulation by dietary restriction. *Mech Ageing Dev* 102, 1–13.

Coyle, P., Philcox, J. C., Carey, L. C., and Rofe, A. M. (2002). Metallothionein: the multipurpose protein. *Cell Mol Life Sci* 59, 627–647.

Cros, N., Tkatchenko, A. V., Pisani, D. F., Leclerc, L., Leger, J. J., Marini, J. F., and Dechesne, C. A. (2001). Analysis of altered gene expression in rat soleus muscle atrophied by disuse. *J Cell Biochem* 83, 508–519.

Deval, C., Mordier, S., Obled, C., Bechet, D., Combaret, L., Attaix, D., and Ferrara, M. (2001). Identification of cathepsin L as a differentially expressed message associated with skeletal muscle wasting. *Biochem J* 360, 143–150.

Du, J., Wang, X., Miereles, C., Bailey, J. L., Debigare, R., Zheng, B., Price, S. R., and Mitch, W. E. (2004). Activation of caspase-3 is an initial step triggering accelerated muscle proteolysis in catabolic conditions. *J Clin Invest* 113, 115–123.

Eisen, M. B., Spellman, P. T., Brown, P. O., and Botstein, D. (1998). Cluster analysis and display of genome-wide expression patterns. *Proc Natl Acad Sci U S A* 95, 14863–14868.

Ewing, R. M., and Claverie, J. M. (2000). EST databases as multi-conditional gene expression datasets. *Pac Symp Biocomput*, 430–442.

Gilbert, D. R., Schroeder, M., and van Helden, J. (2000). Interactive visualization and exploration of relationships between biological objects. *Trends Biotechnol* 18, 487–494.

Giresi, P. G., Stevenson, E. J., Theilhaber, J., Koncarevic, A., Parkington, J., Fielding, R. A., and Kandarian, S. C. (2005). Identification of a molecular signature of sarcopenia. *Physiol Genomics*.

Goetsch, S. C., Hawke, T. J., Gallardo, T. D., Richardson, J. A., and Garry, D. J. (2003). Transcriptional profiling and regulation of the extracellular matrix during muscle regeneration. *Physiol Genomics* 14, 261–271.

Golub, T. R., Slonim, D. K., Tamayo, P., Huard, C., Gaasenbeek, M., Mesirov, J. P., Coller, H., Loh, M. L., Downing, J. R., Caligiuri, M. A., Bloomfield, C. D., and Lander, E. S. (1999). Molecular classification of cancer: class discovery and class prediction by gene expression monitoring. *Science* 286, 531–537.

Gomes, M. D., Lecker, S. H., Jagoe, R. T., Navon, A., and Goldberg, A. L. (2001). Atrogin-1, a muscle-specific F-box protein highly expressed during muscle atrophy. *Proc Natl Acad Sci U S A* 98, 14440–14445.

Haddad, F., Roy, R. R., Zhong, H., Edgerton, V. R., and Baldwin, K. M. (2003). Atrophy responses to muscle inactivity. II. Molecular markers of protein deficits. *J Appl Physiol* 95, 791–802.

Haslett, J. N., Sanoudou, D., Kho, A. T., Bennett, R. R., Greenberg, S. A., Kohane, I. S., Beggs, A. H., and Kunkel, L. M. (2002). Gene expression comparison of biopsies from Duchenne muscular dystrophy (DMD) and normal skeletal muscle. *Proc Natl Acad Sci U S A* 99, 15000–15005.

Hill, J. J., Davies, M. V., Pearson, A. A., Wang, J. H., Hewick, R. M., Wolfman, N. M., and Qiu, Y. (2002). The myostatin propeptide and the follistatin-related gene are inhibitory binding proteins of myostatin in normal serum. *J Biol Chem* 277, 40735–40741.

Hunter, R. B., and Kandarian, S. C. (2004). Disruption of either the Nfkb1 or the Bcl3 gene inhibits skeletal muscle atrophy. *J Clin Invest* 114, 1504–1511.

Ikemoto, M., Nikawa, T., Takeda, S., Watanabe, C., Kitano, T., Baldwin, K. M., Izumi, R., Nonaka, I., Towatari, T., Teshima, S., Rokutan, K., and Kishi, K. (2001). Space shuttle flight (STS-90) enhances degradation of rat myosin heavy chain in association with activation of ubiquitin-proteasome pathway. *Faseb J* 15, 1279–1281.

Jagoe, R. T., and Goldberg, A. L. (2001). What do we really know about the ubiquitin-proteasome pathway in muscle atrophy? *Curr Opin Clin Nutr Metab Care* 4, 183–190.

Jagoe, R. T., Lecker, S. H., Gomes, M., and Goldberg, A. L. (2002). Patterns of gene expression in atrophying skeletal muscles: response to food deprivation. *Faseb J* 16, 1697–1712.

Jozsi, A. C., Dupont-Versteegden, E. E., Taylor-Jones, J. M., Evans, W. J., Trappe, T. A., Campbell, W. W., and Peterson, C. A. (2000). Aged human muscle demonstrates an altered gene expression profile consistent with an impaired response to exercise. *Mech Ageing Dev* 120, 45–56.

Kamei, Y., Miura, S., Suzuki, M., Kai, Y., Mizukami, J., Taniguchi, T., Mochida, K., Hata, T., Matsuda, J., Aburatani, H., Nishino, I., and Ezaki, O. (2004). Skeletal muscle FOXO1 (FKHR) transgenic mice have less skeletal muscle mass, down-regulated Type I (slow twitch/red muscle) fiber genes, and impaired glycemic control. *J Biol Chem* 279, 41114–41123.

Kamei, Y., Mizukami, J., Miura, S., Suzuki, M., Takahashi, N., Kawada, T., Taniguchi, T., and Ezaki, O. (2003). A forkhead transcription factor FKHR up-regulates lipoprotein lipase expression in skeletal muscle. *FEBS Lett* 536, 232–236.

Kaplan, F., Ledoux, P., Kassamali, F. Q., Gagnon, S., Post, M., Koehler, D., Deimling, J., and Sweezey, N. B. (1999). A novel developmentally regulated gene in lung mesenchyme: homology to a tumor-derived trypsin inhibitor. *Am J Physiol* 276, L1027–1036.

Kondo, H., Miura, M., Kodama, J., Ahmed, S. M., and Itokawa, Y. (1992). Role of iron in oxidative stress in skeletal muscle atrophied by immobilization. *Pflugers Arch* 421, 295–297.

Kostrominova, T. Y., Dow, D. E., Dennis, R. G., Miller, R. A., and Faulkner, J. A. (2005). Comparison of gene expression of 2-mo denervated, 2-mo stimulated-denervated, and control rat skeletal muscles. *Physiol Genomics* 22, 227–243.

Kuninger, D., Kuzmickas, R., Peng, B., Pintar, J. E., and Rotwein, P. (2004). Gene discovery by microarray: identification of novel genes induced during growth factor-mediated muscle cell survival and differentiation. *Genomics* 84, 876–889.

Larsson, L. (1995). Motor units: remodeling in aged animals. *J Gerontol A Biol Sci Med Sci* 50 Spec No, 91–95.

Latres, E., Amini, A. R., Amini, A. A., Griffiths, J., Martin, F. J., Wei, Y., Lin, H. C., Yancopoulos, G. D., and Glass, D. J. (2005). Insulin-like Growth Factor-1 (IGF-1) Inversely Regulates Atrophy-induced Genes via the Phosphatidylinositol 3-Kinase/Akt/Mammalian Target of Rapamycin (PI3K/Akt/mTOR) Pathway. *J Biol Chem* 280, 2737–2744.

Lecker, S. H. (2003). Ubiquitin-protein ligases in muscle wasting: multiple parallel pathways? *Curr Opin Clin Nutr Metab Care* 6, 271–275.

Lecker, S. H., Jagoe, R. T., Gilbert, A., Gomes, M., Baracos, V., Bailey, J., Price, S. R., Mitch, W. E., and Goldberg, A. L. (2004). Multiple types of skeletal muscle atrophy involve a common program of changes in gene expression. *Faseb J* 18, 39–51.

Lee, C. K., Klopp, R. G., Weindruch, R., and Prolla, T. A. (1999). Gene expression profile of aging and its retardation by caloric restriction [see comments]. *Science* 285, 1390–1393.

Lexell, J. (1997). Muscle capillarization: morphological and morphometrical analyses of biopsy samples. *Muscle Nerve Suppl* 5, S110–112.

Li, Y. P., Lecker, S. H., Chen, Y., Waddell, I. D., Goldberg, A. L., and Reid, M. B. (2003). TNF-alpha increases ubiquitin-conjugating activity in skeletal muscle by up-regulating UbcH2/E220k. *Faseb J* 17, 1048–1057.

Li, Y. P., and Reid, M. B. (2000). NF-kappaB mediates the protein loss induced by TNF-alpha in differentiated skeletal muscle myotubes. *Am J Physiol Regul Integr Comp Physiol* 279, R1165–R1170.

Li, Y. P., Schwartz, R. J., Waddell, I. D., Holloway, B. R., and Reid, M. B. (1998). Skeletal muscle myocytes undergo protein loss and reactive oxygen-mediated NF-kappaB activation in response to tumor necrosis factor alpha. *Faseb J* 12, 871–880.

Lindstrom, B., Lexell, J., Gerdle, B., and Downham, D. (1997). Skeletal muscle fatigue and endurance in young and old men and women. *J Gerontol A Biol Sci Med Sci* 52, B59–66.

Liotta, L., and Petricoin, E. (2000). Molecular profiling of human cancer. *Nat Rev Genet* 1, 48–56.

Lipshutz, R. J., Fodor, S. P., Gingeras, T. R., and Lockhart, D. J. (1999). High density synthetic oligonucleotide arrays. *Nat Genet* 21, 20–24.

Machida, S., and Booth, F. W. (2004). Increased nuclear proteins in muscle satellite cells in aged animals as compared to young growing animals. *Exp Gerontol* 39, 1521–1525.

Mayer, R. J. (2000). The meteoric rise of regulated intracellular proteolysis. *Nat Rev Mol Cell Biol* 1, 145–148.

Miller, T. A., Lesniewski, L. A., Muller-Delp, J. M., Majors, A. K., Scalise, D., and Delp, M. D. (2001). Hindlimb unloading induces a collagen isoform shift in the soleus muscle of the rat. *Am J Physiol Regul Integr Comp Physiol* 281, R1710–1717.

Pattison, J. S., Folk, L. C., Madsen, R. W., and Booth, F. W. (2003a). Selected Contribution: Identification of differentially expressed genes between young and old rat soleus muscle during recovery from immobilization-induced atrophy. *J Appl Physiol* 95, 2171–2179. Epub 2003 Aug 2171.

Pattison, J. S., Folk, L. C., Madsen, R. W., Childs, T. E., and Booth, F. W. (2003b). Transcriptional profiling identifies extensive downregulation of extracellular matrix gene expression in sarcopenic rat soleus muscle. *Physiol Genomics* 15, 34–43.

Pattison, J. S., Folk, L. C., Madsen, R. W., Childs, T. E., Spangenburg, E. E., and Booth, F. W. (2003c). Expression profiling identifies dysregulation of myosin heavy chains IIb and IIx during limb immobilization in the soleus muscles of old rats. *J Physiol* 553, 357–368.

Petrof, B. J., Shrager, J. B., Stedman, H. H., Kelly, A. M., and Sweeney, H. L. (1993). Dystrophin protects the sarcolemma from stresses developed during muscle contraction. *Proc Natl Acad Sci U S A* 90, 3710–3714.

Porter, J. D., Khanna, S., Kaminski, H. J., Rao, J. S., Merriam, A. P., Richmonds, C. R., Leahy, P., Li, J., Guo, W., and Andrade, F. H. (2002). A chronic inflammatory response dominates the skeletal muscle molecular signature in dystrophin-deficient mdx mice. *Hum Mol Genet* 11, 263–272.

Porter, J. D., Merriam, A. P., Leahy, P., Gong, B., and Khanna, S. (2003). Dissection of temporal gene expression signatures of affected and spared muscle groups in dystrophin-deficient (mdx) mice. *Hum Mol Genet* 12, 1813–1821.

Quackenbush, J. (2001). Computational analysis of microarray data. *Nat Rev Genet* 2, 418–427.

Reid, M. B., and Li, Y. P. (2001). Tumor necrosis factor-alpha and muscle wasting: a cellular perspective. *Respir Res* 2, 269–272.

Roth, S. M., Ferrell, R. E., Peters, D. G., Metter, E. J., Hurley, B. F., and Rogers, M. A. (2002). Influence of age, sex, and strength training on human muscle gene expression determined by microarray. *Physiol Genomics* 10, 181–190.

Sacheck, J. M., Ohtsuka, A., McLary, S. C., and Goldberg, A. L. (2004). IGF-I stimulates muscle growth by suppressing protein breakdown and expression of atrophy-related ubiquitin ligases, atrogin-1 and MuRF1. *Am J Physiol Endocrinol Metab* 287, E591–601.

Sandri, M., Sandri, C., Gilbert, A., Skurk, C., Calabria, E., Picard, A., Walsh, K., Schiaffino, S., Lecker, S. H., and Goldberg, A. L. (2004). Foxo transcription factors induce the atrophy-related ubiquitin ligase atrogin-1 and cause skeletal muscle atrophy. *Cell* 117, 399–412.

Sangorrin, M. P., Martone, C. B., and Sanchez, J. J. (2002). Myofibril-bound serine protease and its endogenous inhibitor in mouse: extraction, partial characterization and effect on myofibrils. *Comp Biochem Physiol B Biochem Mol Biol* 131, 713–723.

Sartorelli, V., and Fulco, M. (2004). Molecular and cellular determinants of skeletal muscle atrophy and hypertrophy. *Sci STKE* 2004, re11.

Shah, O. J., Anthony, J. C., Kimball, S. R., and Jefferson, L. S. (2000). 4E-BP1 and S6K1: translational integration sites for nutritional and hormonal information in muscle. *Am J Physiol Endocrinol Metab* 279, E715–729.

Shannon, W., Culverhouse, R., and Duncan, J. (2003). Analyzing microarray data using cluster analysis. *Pharmacogenomics* 4, 41–52.

Sim, R. B., and Laich, A. (2000). Serine proteases of the complement system. *Biochem Soc Trans* 28, 545–550.

Snyder, P. M., Olson, D. R., McDonald, F. J., and Bucher, D. B. (2001). Multiple WW domains, but not the C2 domain, are required for inhibition of the epithelial Na+ channel by human Nedd4. *J Biol Chem* 276, 28321–28326.

St-Amand, J., Okamura, K., Matsumoto, K., Shimizu, S., and Sogawa, Y. (2001). Characterization of control and immobilized skeletal muscle: an overview from genetic engineering. *Faseb J* 15, 684–692.

Stein, L. D. (2004). Human genome: end of the beginning. *Nature* 431, 915–916.

Sterrenburg, E., Turk, R., t Hoen, P. A., van Deutekom, J. C., Boer, J. M., van Ommen, G. J., and den Dunnen, J. T. (2004). Large-scale gene expression analysis of human skeletal myoblast differentiation. *Neuromuscul Disord* 14, 507–518.

Stevenson, E. J., Giresi, P. G., Koncarevic, A., and Kandarian, S. C. (2003). Global analysis of gene expression patterns during disuse atrophy in rat skeletal muscle. *J Physiol* 551, 33–48.

Stevenson, E. J., Koncarevic, A., Giresi, P. G., Jackman, R. W., and Kandarian, S. C. (2005). The transcriptional profile of a myotube starvation model of atrophy. *J Appl Physiol* 98, 1396–1406.

Stitt, T. N., Drujan, D., Clarke, B. A., Panaro, F., Timofeyva, Y., Kline, W. O., Gonzalez, M., Yancopoulos, G. D., and Glass, D. J. (2004). The IGF-1/PI3K/Akt pathway prevents expression of muscle atrophy-induced ubiquitin ligases by inhibiting FOXO transcription factors. *Mol Cell* 14, 395–403.

Taillandier, D., Aurousseau, E., Meynial-Denis, D., Bechet, D., Ferrara, M., Cottin, P., Ducastaing, A., Bigard, X., Guezennec, C. Y., Schmid, H. P., and et, a. l. (1996a). Coordinate activation of lysosomal, Ca 2+-activated and ATP-ubiquitin-dependent proteinases in the unweighted rat soleus muscle. *Biochem J* 316, 65–72.

Taillandier, D., Aurousseau, E., Meynial-Denis, D., Bechet, D., Ferrara, M., Cottin, P., Ducastaing, A., Bigard, X., Guezennec, C. Y., Schmid, H. P., and et al. (1996b). Coordinate activation of lysosomal, Ca 2+-activated and ATP-ubiquitin-dependent proteinases in the unweighted rat soleus muscle. *Biochem J* 316, 65–72.

Thompson, M. G., Thom, A., Partridge, K., Garden, K., Campbell, G. P., Calder, G., and Palmer, R. M. (1999). Stimulation of myofibrillar protein degradation and expression of mRNA encoding the ubiquitin-proteasome system in C(2)C(12) myotubes by dexamethasone: effect of the proteasome inhibitor MG-132. *J Cell Physiol* 181, 455–461.

Tidball, J. G., and Spencer, M. J. (2002). Expression of a calpastatin transgene slows muscle wasting and obviates changes in myosin isoform expression during murine muscle disuse. *J Physiol* 545, 819–828.

Tischler, M. E., Rosenberg, S., Satarug, S., Henriksen, E. J., Kirby, C. R., Tome, M., and Chase, P. (1990). Different mechanisms of increased proteolysis in atrophy induced by denervation or unweighting of rat soleus muscle. *Metabolism* 39, 756–763.

Tomczak, K. K., Marinescu, V. D., Ramoni, M. F., Sanoudou, D., Montanaro, F., Han, M., Kunkel, L. M., Kohane, I. S., and Beggs, A. H. (2004). Expression profiling and identification of novel genes involved in myogenic differentiation. *Faseb J* 18, 403–405.

Tseng, B. S., Zhao, P., Pattison, J. S., Gordon, S. E., Granchelli, J. A., Madsen, R. W., Folk, L. C., Hoffman, E. P., and Booth, F. W. (2002). Regenerated mdx mouse skeletal muscle shows differential mRNA expression. *J Appl Physiol* 93, 537–545.

Velculescu, V. E., Zhang, L., Vogelstein, B., and Kinzler, K. W. (1995). Serial analysis of gene expression. *Science* 270, 484–487.

Ward, C. W., Feng, W., Tu, J., Pessah, I. N., Worley, P. K., and Schneider, M. F. (2004). Homer protein increases activation of Ca2+ sparks in permeabilized skeletal muscle. *J Biol Chem* 279, 5781–5787.

Welle, S., Bhatt, K., and Thornton, C. A. (1999). Inventory of high-abundance mRNAs in skeletal muscle of normal men. *Genome Res* 9, 506–513.

Welle, S., Brooks, A. I., Delehanty, J. M., Needler, N., Bhatt, K., Shah, B., and Thornton, C. A. (2004). Skeletal muscle gene expression profiles in 20–29 year old and 65–71 year old women. *Exp Gerontol* 39, 369–377.

Welle, S., Brooks, A. I., Delehanty, J. M., Needler, N., and Thornton, C. A. (2003). Gene expression profile of aging in human muscle. *Physiol Genomics* 14, 149–159.

Williams, A. B., Decourten-Myers, G. M., Fischer, J. E., Luo, G., Sun, X., and Hasselgren, P. O. (1999). Sepsis stimulates release of myofilaments in skeletal muscle by a calcium-dependent mechanism. *Faseb J* 13, 1435–1443.

Wittwer, M., Fluck, M., Hoppeler, H., Muller, S., Desplanches, D., and Billeter, R. (2002). Prolonged unloading of rat soleus muscle causes distinct adaptations of the gene profile. *Faseb J* 16, 884–886.

Yamazaki, S., Straume, M., Tei, H., Sakaki, Y., Menaker, M., and Block, G. D. (2002). Effects of aging on central and peripheral mammalian clocks. *Proc Natl Acad Sci U S A* 99, 10801–10806.

Zambon, A. C., McDearmon, E. L., Salomonis, N., Vranizan, K. M., Johansen, K. L., Adey, D., Takahashi, J. S., Schambelan, M., and Conklin, B. R. (2003). Time- and exercise-dependent gene regulation in human skeletal muscle. *Genome Biol* 4, R61.

Zhao, P., Iezzi, S., Carver, E., Dressman, D., Gridley, T., Sartorelli, V., and Hoffman, E. P. (2002). Slug is a novel downstream target of MyoD. Temporal profiling in muscle regeneration. *J Biol Chem* 277, 30091–30101.

CHAPTER 3

STRIATED MUSCLE PLASTICITY: REGULATION OF THE MYOSIN HEAVY CHAIN GENES

FADIA HADDAD, CLAY E. PANDORF, JULIA M. GIGER
AND KENNETH M. BALDWIN
University of California Irvine, Irvine, CA 92697, USA

1.   INTRODUCTION AND OVERVIEW

All living organisms possess the inherent capacity to alter the structural and functional properties of their organ systems in accordance with the environmental conditions imposed on a particular system. These changes are largely the manifestation of altered protein expression in which either the amount and/or type of protein are altered in order to meet the imposed functional demands. This adaptive plasticity of protein expression involves a complex process centered on the general theme of altered gene expression. The goal of this chapter is to examine our current understanding of the chain of events known to be involved in the adaptive process of striated muscles (i.e., cardiac and skeletal) whereby specific genes and their protein products undergo altered expression in response to various mechanical and hormonal stimuli. In order to focus the discussion, we will examine primarily the regulation of expression of the contractile protein, myosin heavy chain (MHC). This protein, which is an important structural and regulatory protein component of the contractile apparatus, can be expressed as different isoforms, thereby impacting the functional diversity of striated muscle. Since the MHC gene family is keenly sensitive to mechanical and hormonal stimuli, it will serve as a cellular "marker" for muscle plasticity and adaptive responses.

55

R. Bottinelli and C. Reggiani (eds.), Skeletal Muscle Plasticity in Health and Disease, 55–89.
© 2006 *Springer*.

2.   FUNDAMENTAL CONCEPTS

2.1   Organelle Plasticity

In this section we will be dealing with the concept of "muscle plasticity," which is the ability of a given muscle cell to alter either 1) the quantity (amount) of protein, and/or 2) the type of protein (i.e., phenotype or isoform) comprising its different subcellular components in response to any stimulus that disrupts its normal homeostasis. For example, a given muscle cell may respond to chronic increases in mechanical stress (physical activity) by increasing its cross-sectional area or cell volume such that all of the subcellular components remain in normal proportion to one another. In this case the muscle expands both its protein mass and mechanical strength without qualitatively changing any other inherent functional property such as endurance and/or contractile speed. On the other hand, a given myocyte may respond to the same perturbation by both increasing its mass and altering the type of myosin heavy chain (MHC) isoform that it expresses in the myofilaments. In this situation, while the muscle becomes both larger and stronger due to the increase in contractile protein accumulation, its intrinsic contractile properties also become transformed due to the altered myosin phenotype that is expressed. Thus, the muscle's plasticity potential may involve 1) a change in the amount of protein, 2) the type of protein isoform it expresses, and 3) a combination of the two. The derived functional consequences of such a transformation will depend on both the magnitude and extent of the quantitative and qualitative alteration in protein expression.

2.2   Protein Isoforms

Based on the above, it is apparent that a muscle cell's capacity for adaptation is related to its genetic capability of expressing different isoforms (molecular species) of a given protein. By definition, isoforms of a particular protein are molecules with slight variations in amino acid composition, thereby altering either the structural, functional, and/or enzymatic properties of that protein. Isoform species have been identified for a large number of muscle proteins involved in processes governing ion transport, contraction/relaxation, and energy metabolism. Within this chapter, considerable emphasis will be devoted to the regulation of the MHC gene family of proteins.

2.3   Protein Turnover

All proteins comprising mammalian cells are not stable. That is, each protein is in a continuous process of being synthesized (i.e., protein synthesis) and subsequently degraded (i.e., protein degradation). The balance between protein synthesis and degradation determines whether there is a stable amount, or a net gain or loss of that protein in the striated muscle cell depending on the physiological situation. This fundamental process of protein turnover also allows for qualitative remodeling of the muscle to occur so that one isoform of a given protein can be eventually replaced by another one that is better suited for a specific physio-logical/pathological state. Further, the inherent turnover rate of a given protein is dictated by its biological half-life $(T_{1/2})$, i.e., the time in which 50% of it's content is degraded. Under normal steady state conditions, muscle myofibrillar (contractile apparatus) proteins have relatively long half-lives, on the order of days (seven to 10 days for MHC). Interestingly, the half-life of the myofibrillar proteins is significantly reduced if these proteins become disassembled, as seen during low levels of contractile activity (Samarel et al., 1992). Furthermore, under conditions of muscle unloading, such as during exposure to microgravity or chronic bedrest, it has been shown that protein synthesis in general is reduced while the protein degradation process is increased (Booth and Criswell, 1997). Consequently, there is a rapid and marked degree of atrophy in those muscles normally used to support the body against the force of gravity. Therefore, the process of protein turnover provides a mechanism in which both the type and amount of protein comprising cellular systems can be altered quickly in accordance with the internal and external environmental conditions imposed.

2.4   The Nature of the Adaptive Stimulus: Intensity, Frequency, and Duration

In order for any gene/protein to undergo altered expression as outlined in the above section on protein turnover, a highly specific adaptive stimulus must be applied to the cell system of sufficient intensity, duration, and frequency. In terms of physical stress or the lack thereof, the altered stimulus can be continuous, such as chronic exposure to neural inactivation, or to states of unloading as occurs during spaceflight. Also, the stimulus (or lack thereof) can be of constant magnitude, i.e., the total elimination of gravity acting on the body over a period of many days or the chronic absence of a specific hormone such as thyroxin. Under these situations, the imposition of a stimulus is associated with altered cellular processes that lead to altered gene expression in favor of the new physiological/pathological state. Most of the time, however, the stimulus is intermittent. For example, typical heavy resistance training paradigms are usually performed $\sim$ one hour per day in the form of a series of individual contractions rather than as a continuous overload stress being applied to a given muscle group. Often, the stimulus is gradually increased over several days/weeks to reach some designated peak level, i.e., building up running duration to one hour per day at a specified speed; or the slow elevation of blood pressure affecting the after-load of the heart. In these last examples, due to the intermittent nature of applying the stimulus, and the fact that an optimal stimulus intensity and duration are only gradually reached over time in order to induce a particular adaptation, the resulting adaptive response follows a complex kinetic pattern, which involves a cumulative process of stimulation and recovery phases that eventually lead to altered gene expression.

For most interventions, there is undoubtedly some critical threshold stimulus needed to induce an adaptation in gene expression, as exemplified by the adaptive process of cardiac or skeletal muscle enlargement. This can be defined as the minimal level of mechanical force necessary to affect some step(s) in the flow of gene information (see below) to eventually result in muscle fiber hypertrophy and/or a change in MHC phenotype. For example, recent studies on animal models suggest that approximately 50 near-maximal high resistance contractions, performed in a single training session, are sufficient to transiently induce both the level of mRNA and the rate of protein synthesis for marker contractile proteins (actin and myosin). While this stimulus is sufficient to transiently perturb the processes necessary for enhancing contractile protein expression, the stimulus must be repeated a sufficient number of times (12 to15 training sessions) at sufficiently close intervals, i.e., a minimum of every other day, for the collective process to induce a net expansion of the contractile protein pool to result in a measurable degree of muscle fiber enlargement. Further, merely contracting the muscle without an accompanied high level of sustained force production appears to be insufficient to induce the hypertrophy response (unpublished observations). These examples illustrate the overall complexity of imposing an appropriate stimulus paradigm on a given cell system in order to achieve a specific adaptive response.

2.5   The Flow of Genetic Information

In order to understand the nature of protein expression in adapting mammalian cell systems, including striated muscle, it is important to define the cascade of events in which the expression of a given protein is regulated (Fig. 1). Before a gene is expressed into protein, it is manifest into different forms by undergoing a series of processes. The first process occurs in the nucleus and involves gene transcription by RNA polymerase to produce a primary transcript (pre-mRNA), which is the unprocessed, nascent transcript. The next process is pre-mRNA processing which consists of three steps: a) capping at the 5' end which involves a unique 5' $\rightarrow$ 5' linkage to a 7-methylguanylate, b) polyadenylation at the 3' end which involves addition of a poly(A) tail consisting of 150 or more adenosine residues, and c) splicing which consists of removing the introns and joining the exons.

This processing results in the mature mRNA that is exported from the nucleus and serves as the template for protein translation. The third process is mRNA translation into polypeptide chains on the ribosomes and their assembly into functional proteins. The steady-state levels of gene products (pre-mRNA, mRNA, and protein) depend on the balance between their synthesis and their degradation. Thus the regulation can occur at several levels, i.e., transcriptional; post-transcriptional/pre-translational; translational; and post-translational (figure 1). Translational and post-translational events include the processes of protein synthesis and degradation which will not be discussed in this chapter. Instead, we will focus the discussion on transcriptional, post-transcriptional/pre-translational events involved in the regulation of MHC gene expression.

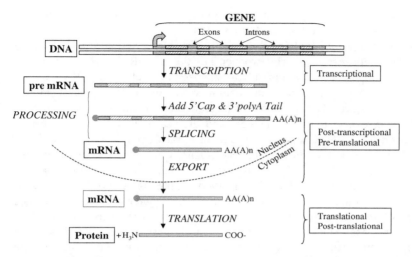

Figure 1. Flow of Genetic Information and Key Steps in the Regulation of Gene Expression. The level of protein expressed in the cell results from the net balance between protein synthesis and protein degradation. Protein synthesis can be regulated via several processes including those operating at the transcriptional, post-transcriptional, pre-translational, translational, and post-translational levels. The product of each step is subjected to degradation control

3.   STUDYING GENE TRANSCRIPTION AND ITS REGULATION

3.1   Approaches to Determine Transcription of Endogenous MHC Genes

In the majority of the studies reported to date involving changes in MHC isoform expression and hence phenotypic transformation in response to various stimuli, most investigations have been focused on either mRNA or protein analyses, which are markers of pre-translational or translational events that do not confer insight on transcriptional regulatory events. In order to study transcriptional control of any endogenous gene (e.g., its inherent regulation in its normal physiological setting) it is necessary to examine the nascent product of the transcriptional event. Two approaches have been used: 1) nuclear run-on assays, and 2) determinations of pre-mRNA products. The first operates by examining radio-labeled nucleic acid incorporation into transcriptional products (RNAs) as the transcriptional process is being completed in isolated nuclei obtained from tissues representing a particular physiological state. These transcriptional products are subsequently screened via hybridization to probes representing specific genes of interest. This assay is tedious, requires a large amount of starting material, is time consuming (3-4 days), and requires the use of large amounts of radio-labeled nucleotides. It also is necessary to use relatively pure nuclei, which are difficult to obtain from the highly structured striated muscles. Due to these limitations, very few studies have used nuclear run on assays to study muscle gene transcription (Boheler et al., 1992; Cox and Buckingham, 1992; Hildebrandt and Neufer, 2000).

The second approach of studying endogenous gene transcription is to analyze the expression of the primary transcript product, i.e., the pre-mRNA. With techniques using RT-PCR approaches, it is possible to analyze the level of pre-mRNA of any specific gene as long as sequence information is available on the gene of interest and its mRNA product. The success of this method relies on the use of gene-specific primers for the reverse transcription step, and the use of PCR primers targeting intronic regions of the gene sequence. The reverse transcription (RT) step makes a cDNA of RNA products specific to the target gene, while the PCR step amplifies this product taking advantage of targeting intronic sequences that are not present in the more abundant mRNA product. This method requires the use of DNase treated RNA in order to avoid amplification of genomic DNA which has the same sequence as the pre-mRNA of the specific gene. This method is simple, versatile, and does not require a large amount of tissue and thus is ideal to use to study muscle specific gene transcription when sample amounts are scarce such as in human studies. We have recently explored this approach to gain insight on the mode of regulation of MHC genes in cardiac and skeletal muscle under conditions of altered MHC expression. Results of such analyses will be reviewed in a later section of this chapter.

3.2   Experimental Approaches to Study Regulation of Gene Transcription

In eukaryotes, major control of transcription occurs at the initiation step. Transcription of specific genes is initiated via a basal promoter located upstream of the Transcription Start Site (TSS). This structure and its location allows the RNA polymerase II and other members of the basal transcription machinery to assemble and transcribe the downstream gene. The basal promoter activity is affected by surrounding regulatory transcription factors which bind to specific DNA sequences (cis elements) normally located 5' upstream of the basal promoter TSS, although some elements occur downstream, and collectively function to either enhance or inhibit transcription. A central question in cell biology is how cis-acting DNA sequences and trans-acting factors interact to precisely regulate eukaryotic gene expression. One common experimental approach to study promoter regulation is via reporter gene sytems which were developed based on recombinant DNA technology (Alam and Cook, 1990). The development of the reporter gene system has been an invaluable tool to achieve great advancement in the study of transcriptional regulation of many genes including the MHC genes. The system operates on the principle that the transcriptional activity of a promoter of interest can be monitored via measurement of expression of a linked reporter gene not normally expressed in the cell type of interest. This is accomplished by appropri- ately fusing the DNA promoter fragment directly upstream of a reporter gene in a vector lacking promoter activity. This fusion forms a heterologous expression vector that can be transfected into living cells. Once introduced into the cell the vector enters the nucleus, where it is exposed to the same nuclear milieu of transcription factors as the endogenous genes (Fig. 2). Changes in reporter gene expression reflect altered activity of the upstream promoter. This allows characterizing of a given promoter specificity (e.g. muscle-type specificity) and responsiveness to external stimuli (e.g. hormone, activity). Furthermore, cis-regulatory elements can be functionally characterized by constructing successive deletions, by altering sequence orientation, or by site-directed mutagenesis. Several reporter genes are used in mammalian promoter studies including chloramphenicol acetyltransferase (CAT), firefly luciferase (F-Luc), renilla luciferase (R-Luc), and β galactosidase from which expression can be assayed in cell and tissue lysates using simple procedures.

At least four different transfection techniques have been employed in order to introduce chimeric promoter-reporter constructs into muscle cells. These approaches include 1) transgenic animals, 2) direct gene transfer, 3) retroviral vectors, and 4) adenoviral vectors. These approaches have been applied to both *in vitro* systems such as myoblasts and myocytes in culture (Allen et al., 2001b; Takeda, 1992),

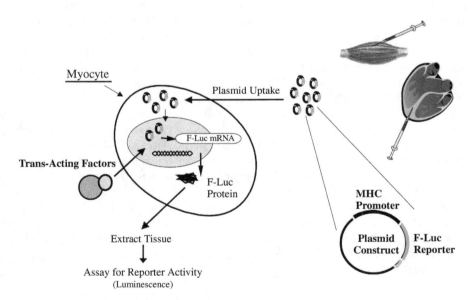

Figure 2. Direct Gene Transfer Approach to Study Specific Promoter Activity. Schematic diagram of the direct gene transfer approach. Plasmid DNA containing the promoter reporter construct is injected into striated muscle (skeletal muscle or myocardium) and is taken up into the nucleus of the cells. The plasmid DNA remains extrachromosomal (episomal) but it is exposed to the same milieu of transcription factors/cofactors as the genome. Promoter activity is translated into reporter expression, which can be determined via a chemiluminescent assay of the tissue extract. A color figure is freely accessible via de website of the book: A color figure is freely accessible via the website of the book: http://www.springer.com/1-4020-5176-x

and *in vivo* systems such as transgenics (Karasseva et al., 2003; Knotts et al., 1994; Knotts et al., 1995; Knotts et al., 1996; McCarthy et al., 1997; Rindt et al., 1995; Subramaniam et al., 1991; Tsika et al., 1996), or direct gene transfer (Corin et al., 1995; Giger et al., 2000; Lupa-Kimball and Esser, 1998; Ojamaa et al., 1995; Swoap, 1998; Wright et al., 1999). These promoter studies demonstrated that transcriptional control of the MHC gene promoters is multifactorial and it involves a complex interaction between cis-acting DNA sequences, their cognate transacting protein factors, and the basic transcription machinery. Transgenic animals and direct gene transfer are the most commonly used of the four approaches and will be the focus of this section.

The transgenic mouse approach involves the insertion of a promoter-reporter construct, referred to as the transgene, into the genome at an early embryonic stage of the animal. Every cell in the mouse will eventually contain the genomically inserted chimeric transgene as the animal develops. Transcription of the gene can then be monitored by simply assaying for reporter activity in the various tissues of interest. In contrast to the transgenic approach, the direct gene transfer approach involves the ligation of a specific gene promoter into a DNA plasmid vector containing a reporter gene. The chimeric plasmid construct is then directly injected into the skeletal muscle of interest or into the apex of the myocardium where it is taken up by the nuclei of the surrounding muscle fibers (Fig. 2). The mechanism by which this occurs is not yet understood. The plasmid constructs remain episomal within the nuclei and are exposed to the same milieu of transcription factors as the endogenous genes within the genome. These plasmid constructs are detectable within the nuclei for a long period of time after their injection because differentiated skeletal muscle cells remain in a post-mitotic stage and the turnover rate of their nuclei is relatively slow (Wolff et al., 1992). The direct gene transfer technique is cost and time effective, is not limited to the species of animal that can be used, and is ideal for the study of transcriptional regulation (promoter activity) in striated muscles because of this tissue's unique DNA uptake properties. Although both methods have their limitations, they have been invaluable in providing insight into how the MHC gene promoters are regulated *in vivo* (Giger et al., 2000; Giger et al., 2002; Giger et al., 2004a; Giger et al., 2004b; Huey et al., 2003; Huey et al., 2002; Knotts et al., 1994; Knotts et al., 1996; Molkentin et al., 1996; Molkentin et al., 1994; Ojamaa et al., 1995; Rindt et al., 1995; Subramaniam et al., 1993; Swoap, 1998; Tsika et al., 2002; Vyas et al., 2001; Wright et al., 2001).

3.3 Phylogenetic Analyses in Studying Transcriptional Regulation

Another approach of studying sites of regulation in gene promoters is via phylogenetic footprinting which involves identification of conserved regulatory sequences in the 5' flanking regions as well as intronic regions of genes. Phylogenetic footprinting is a comparative genomic technique based on the observation that important regulatory sequences are conserved across species (Wasserman and

Fickett, 1998; Wasserman et al., 2000). With the rat genomic resources available, as well as the mouse and human genome sequence, comparative sequence analyses across species can give insight into potential regulatory sites. This in-silico approach has been utilized recently by Konig et al in an effort to understand developmental regulation of the MHC gene family (Konig et al., 2002). Furthermore, in order to gain insight on regulatory regions of cardiac MHC intergenic sequence, the entire intergenic sequence among 5 species including the rat, mouse, and human was subjected to phylogenetic footprinting analysis (Haddad et al., 2003a). These analyses show that conserved regulatory elements are clustered mainly in two domains, a proximal module consisting of the already characterized α promoter (Molkentin et al., 1996), and a distal module located between 1.2 and 1.6 kb relative to the α transcription start site (TSS) (Haddad et al., 2003a). This distal module, due to its strategic location, was implicated in regulating both the α MHC as well as the antisense β RNA transcription (Haddad et al., 2003a).

4.   THE MYOSIN HEAVY GENE FAMILY

4.1   MHC Gene Organization and Expression

The myosin heavy chain isoforms are encoded by a highly conserved multigene family and are highly regulated by developmental factors and exogenous stimuli in a muscle type specific fashion (for review, see Schiaffino and Reggiani, 1996). At least ten MHC isoforms have been identified in mammalian striated muscles (defined herein as heart and skeletal muscle), and they have been designated as: 1) embryonic; 2) neonatal; 3) cardiac alpha (α); 4) cardiac beta (β) or slow, type I (as expressed in skeletal muscle); 5) fast IIa; 6) fast IIx/IId; 7) fast IIb; 8) extraocular; 9) mandibular or masticatory (m-MHC); and 10) the slow tonic MHC. All these isoforms are the products of distinct genes. Cardiac β and skeletal slow, type I are identical products of the same gene. The MHC gene family is highly conserved through millions of years of evolution in terms of genomic organization, head to tail orientation, and the order of the MHC genes on the chromosomes (Gulick et al., 1991; Mahdavi, 1986; Saez et al., 1987; Weiss et al., 1999; Weydert et al., 1985). The cardiac β and α genes are located on chromosomes 15 (rat) and 14 (mouse and human), respectively, with the β tandemly located upstream of the α with ~4.5 kb separating the two. The rat skeletal muscle MHCs are clustered on chromosome 10 (17 in human, and 11 in mouse) in the order of embryonic, IIa, IIx, IIb, neonatal, and extraocular spanning approximately 400 kb with the IIa, IIx, and IIb MHC existing in a relatively tight cluster spanning ~90kb (Fig. 3). The masticatory MHC (mMHC, MYH16) is located on chromosome 7 in human (Hoh, 2002), while the slow tonic MHC is located on chromosome 20 (Berg et al., 2001). This chapter will exclude discussion on the extraocular, m-MHC, and the slow tonic MHC because they are not widely expressed and little is known about their regulation. The embryonic and neonatal isoforms are expressed predominantly in developing skeletal muscles, and they can also be detected in the adult muscle

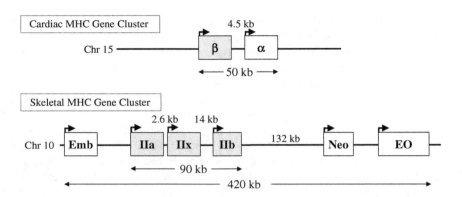

Figure 3. Organization of the MHC Genes on the Chromosomes. Cardiac MHC α and β genes are clustered on chromosome 15 in the rat and they span a distance of ~50 kb with a 4.5 kb of intergenic DNA between β and α. The Skeletal MHC genes are clustered on chromosome 10 of the rat and the fast MHC span a distance of ~90kb. The IIa and IIx MHCs are separated by only 2.6 kb of intergenic DNA. This MHC gene organization, order, head to tail orientation, and spacing has been conserved through millions of years of evolution and could be of great significance to the way these genes are regulated in response to various stimuli. Human and mouse cardiac MHCs are found on chromosome 14; whereas human skeletal MHCs are found on chromosome 17, and the mouse skeletal MHC are found on chromosome 11. Distance between genes may vary among species

in regenerating fibers (d'Albis et al., 1988), as well as in specialized adult muscles such as the masseter and extraocular muscles (Schiaffino and Reggiani, 1996). The α and β MHCs are expressed chiefly in cardiac muscle. The α MHC isoform was thought to be expressed exclusively in cardiac muscle; however, it was reported to be expressed in the extraocular and masseter muscles (Schiaffino and Reggiani, 1996). The β (type I) MHC expression is not confined only to cardiac muscle, but also is found in embryonic skeletal muscle, and is the major isoform expressed in adult slow skeletal muscle (Lompre et al., 1984; Schiaffino and Reggiani, 1996). In addition to type I MHC, adult skeletal muscles express various proportions of type IIa, IIx and IIb MHC, known as the fast isoforms. Of interest to note is that the IIb MHC protein is not detected in human muscle although its transcript can be detected (Weiss et al., 1999); and recent reports demonstrate the absence of functional IIb MHC isoform in equine skeletal muscles and in other large size animals (Chikuni et al., 2004). The clustered organization and proximity of the MHC genes (Fig. 3) is intriguing and is thought to have occurred via MHC gene duplication (Goodson, 1993; Hoh, 2002; Mahdavi et al., 1984); and this clustering may have important implications for the regulation of these genes as a group. This observation brings up the question of the MHC genomic organization as being essential for their mode of regulation. Our data support the notion that this alignment is of regulatory significance and is essential for coordinated shifts in MHC expression between adjacent MHC genes ((Haddad et al., 2003a) and unpublished observations).

4.2   MHC Regulation as a Model For Studying Muscle Plasticity

It is widely accepted that the contractile properties of striated muscles are deter-mined mainly by the pattern of myosin heavy chain (MHC) isoforms that is expressed in the muscle fibers. For almost 50 years it has been appreciated that skeletal muscle fibers are clustered into relatively similar groups of fibers with common biochemical, metabolic and physiological properties, with each fiber of the group innervated by the same motor neuron. These clusters have been termed motor units with each having a unique "fiber typing" profile. Current fiber typing methodology is based on either biochemical properties, histochemical staining for ATPase, or myosin heavy chain identification. Based on various biochemical properties, muscle fibers have been classified as fast oxidative, slow oxidative, or fast glycolytic. Myosin ATPase staining methods use differences in pH sensitivity of the different myosins, and fibers are classified based on the intensity of staining. Using this method, muscle fibers were initially identified as type I, IIA, and IIB (Pette et al., 1999; Staron, 1997). Improvement on this technique lead to the identification of more fiber types of intermediate staining intensity and were termed as IC, IIC, IIAC, and IIAB (Staron, 1997). Using immuno-histochemical analyses, muscle fibers can be classified based on the MHC isoform being expressesd using isoform-specific antibodies. Alternatively, using gel electrophoresis to separate the MHC isoforms, MHC composition of single individual fibers can be accurately quantified and this method proved that a single fiber can express more than one single MHC isoform. Considerable data generated over the past two decades by MHC-based immuno-histochemical and/or SDS-PAGE methods demonstrated that normal muscles from mammals contain a sizeable proportion of MHC hybrids, which coexpress, at the protein level, two, three and even four MHC isoforms (Caiozzo et al., 2003). The complexity of MHC isoform expression in single fibers is increased in transforming muscles, because a higher percentage of the fiber population express multiple combina-tions (Caiozzo et al., 1998; Caiozzo et al., 2003; Pette and Staron, 2001). The significance of MHC polymorphism in single myofibers is not clear; it could be a characteristic of fibers with high adaptive potential, i.e., hybrid fibers are more suitable to switch phenotype to meet new functional demands. For the most part, the functional properties, e.g., contractile speed or rate of force devel-opment of these fiber-types, are regulated primarily by the pattern of MHC isoform expression (Caiozzo et al., 2003). Thus, it becomes apparent that a great deal of the functional diversity, as well as the adaptive plasticity of the muscle fiber, and hence the individual motor units are regulated by the pattern of MHC isoforms that is expressed in individual fibers making up the spectrum of motor units. Consequently, if one gains an understanding of the plasticity of expression of the MHC gene family and the mechanisms of such regulation, considerable insight will be gained in understanding the adaptive nature of the skeletal muscle system.

5.   STIMULI SPECIFICITY IMPACTING MUSCLE PLASTICITY

It has been long appreciated that the protein expression of the MHC genes is highly plastic in that it can be modulated by a variety of factors, including (but not limited to) fetal/embryonic developmental programs (Adams et al., 1999a; Condon, 1990; Cox and Buckingham, 1992; Mahdavi et al., 1987; McKoy, 1998), innervation pattern and associated neuronal firing patterns (Pette and Staron, 1990; Schiaffino and Reggiani, 1996), hormonal (Caiozzo and Haddad, 1996), and mechanical activity/inactivity factors (Caiozzo and Haddad, 1996; Talmadge, 2000). A primary goal of this chapter is to review MHC gene expression in different models of muscle plasticity as well as to delineate what is known about the mechanism underlying this plasticity. An additional goal is to highlight some novel findings regarding the mechanism of MHC gene switching via coordinated regulation of linked MHC genes. In the following sections we will first focus on cardiac MHC gene switching between alpha and beta in response to hypothyroidism, diabetes, and pressure overload. Then, we will review some findings on skeletal muscle MHC gene switching in fast muscle between IIb and IIx in response to intermittent resistance training, and in slow muscle between IIa and IIx in response to chronic inactivity (spinal isolation model).

6.   MHC PLASTICITY IN CARDIAC MUSCLE

6.1   Developmental Aspect

During the fetal developmental period of the heart in both humans and small animals, the β MHC is the principal isoform that is initially expressed in the ventricles (Chizzonite and Zak, 1984; Lompre et al., 1984), whereas the α MHC is the principal isoform expressed in the atria of small and large neonatal and adult animals (Lompre et al., 1981). However, in rodents, shortly after birth, when thyroid hormone (T3) titers begin to increase progressively, expression of the β MHC becomes rapidly decreased and is replaced by the α MHC such that by $\sim$ three weeks of age the α MHC is exclusively expressed in the ventricles of small mammals (but not in large mammals such as humans). This early postnatal MHC shift from β to α is thyroid state-sensitive and is much delayed if the animals were made hypothyroid via pharmacological treatments (Morkin, 2000). As the rodent heart continues to mature to adulthood, the β MHC becomes re-expressed, accounting for $\sim$10-15% of the total MHC pool. This regulation of the β MHC is thought to be due to an increase in the mechanical stress gradient that occurs across the wall of the myocardium as the chamber enlarges during maturation to adulthood (Swynghedauw, 1999; Yazaki et al., 1989). In humans, throughout the growth and development phase of the heart, it appears that the β MHC remains as the dominant isoform that is expressed in the ventricle under normal conditions. In the normal human heart, α MHC mRNA is detected up to 30% while its corresponding protein comprises only 7% of the total MHC expressed (Lowes et al., 1997; Miyata et al., 2000; Nakao et al., 1997). Not much is known about the developmental regulation of cardiac MHC in the human heart; however, in rodents, the developmental cardiac MHC alterations occur at both the protein and mRNA levels and thus are regulated at the pre-translational level, which could be either transcriptional or post-transcriptional. Although few studies have attempted to deal with this question directly, the developmental cardiac MHC regulation is thought to occur at the transcriptional level. Several transcription factors are thought to be involved in the developmental control of cardiac MHC and these include the thyroid receptors (TR), retinoid receptors (RXR), GATA4, NFK2.5, MEF2c, and others (Morkin, 2000).

6.2   MHC Plasticity in the Adult Heart

As described above, the adult heart of small animals such as rodents and mice chiefly expresses the α MHC throughout the adult state. This MHC pattern can change toward increased β MHC expression under several patho-physiological conditions including a) chronic pressure overload (induced hypertension) (Haddad et al., 1995; Lompre et al., 1979; Mercadier et al., 1981; Morkin, 1993; Morkin, 2000), b) hypothyroidism/thyroid deficiency (TD) (Caiozzo and Haddad, 1996; Dillmann, 1984; Gustafson, 1985), c) diabetes (Dillmann, 1989); (Dhalla et al., 1998; Haddad et al., 1997a; Haddad et al., 1997b), and d) chronic energy (caloric) deprivation (Haddad, 1993; Swoap et al., 1994). In the models of hypertension and caloric restriction, the heart can increase the expression of the β MHC ~ two- to four-fold above the control levels, whereas β MHC can become exclusively expressed in the case of TD and of diabetes (Haddad et al., 1997b).

The remodeling of the MHC motor proteins that occurs in the rodent heart such as the increase in the β MHC expression in response to pressure overload is of physiological importance. This increased relative β MHC expression contributes to increasing the economy of force production at the cross bridge level (Alpert and Mulieri, 1986; Swynghedauw, 1999). On the other hand, during the conditions of diabetes, thyroid deficiency and energy deprivation, the strong bias to greater expression of the β MHC has been correlated with an intrinsic cardiac functional state characterized by a low intrinsic contractility state, heart rate, and cardiac output, e.g., conditions consistent with both the animal's global and the heart's substrate energy utilization. Consequently, the heart of small animals can be remodeled both biochemically and functionally in accordance with the chronic level of energy demand placed on either the heart itself or the organism in general. There is evidence that similar functional adaptations can occur in the cardiovascular system of humans, and these adaptations appear to be associated with some degree of plasticity involving the MHC isoforms (Miyata et al., 2000; Morkin, 2000; Reiser et al., 2001). For example, human heart failure is associated with low contractility state, and this is attributed, in part, to a decreased α MHC protein expression in the failing heart. It was observed that in human hearts, an increased expression of α MHC correlates well with improvements in left ventricular function (Abraham et al., 2002).

Thus, although α MHC is expressed at a low proportion in the human heart, its expression is important to maintain a normal level of contractility state.

6.3   Mechanisms Underlying Cardiac MHC Plasticity

In all the above models of cardiac MHC gene plasticity, the change in MHC gene expression is detected at both the protein and mRNA levels; and thus the regulation occurs at the pre-translational level based on the scheme outlined in figure 1. Much information has been accumulated on transcriptional regulation of cardiac MHC genes and several transcription factors have been identified in the regulation of cardiac MHCs such as TEF-1, MEF-2, USF1, GATA-4, SRF, and TR (thyroid receptors). It is well agreed upon that there is no single factor responsible for this regulation, rather it appears that it is multifactorial involving interaction among several transacting factors (Mably and Liew, 1996; Morkin, 2000; Samuel et al., 1995). Of all the regulatory stimuli impacting cardiac MHC plasticity, thyroid hormone is the most potent. T3 regulates α and β MHC mRNA expression in an antithetical fashion, increasing the expression of α MHC while simultaneously decreasing that of the β MHC (Rottman et al., 1990). While T3 regulation of the α MHC gene has been well linked to thyroid response elements on the α MHC gene promoter, thus inferring regulation to be at the transcriptional level, the regulation of the β MHC gene transcription by T3 has been more difficult to elucidate. A recent study by our group utilized pre-mRNA analyses in order to estimate gene transcription (Haddad et al., 2003a) and its regulation under a hypothyroid state and in response to streptozotocin-induced diabetes. Our results (Haddad et al., 2003a) can be summarized as follows. The α MHC gene regulation is consistent with classical transcriptional regulation in these models. There was a strong correspondence between the α MHC pre-mRNA product and the corresponding mRNA product among the two interventions, i.e., they both decreased to the same level (Fig. 4A and B). However, in analysis of the β MHC pre-mRNA we observed that while there was increased expression in pre-mRNA levels under conditions of thyroid deficiency and diabetes, these increases were far below the level of increase seen for the mRNA product, suggesting that transcriptional activity by the β MHC promoter cannot account for all of the changes observed for mRNA expression (figure 4C and D). In that study we discovered the existence of an antisense β MHC transcript that appears to play a role in post-transcriptional regulation of the β MHC gene expression. Apparently, the antisense β-RNA can align with and interfere with processing of any sense pre-mRNA that is generated by the β 5' promoter such that there is little or no β mature mRNA product detected under normal physiological conditions. However, both hypothyroid and diabetic states were shown to involve a down-regulation of the antisense β MHC RNA, the absence of which enables the β sense pre-mRNA to be processed into mature β mRNA (Fig. 4 and 5). Interestingly, there was a strong direct correlation between the α MHC pre-mRNA and the antisense β RNA, suggesting that the sense α and the antisense β transcriptions are regulated via a common mechanism or a common

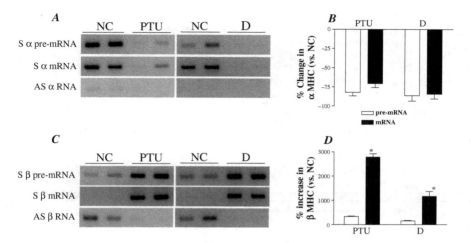

Figure 4. Cardiac MHC RNA Expression in Ventricles from Normal Control (NC), Hypothyroid (PTU) and Diabetic (D) Rats. A, α MHC RNA (pre-mRNA, mRNA, and antisense) analyses using RT-PCR (see (Haddad et al., 2003a) for details). B, Percent change in α pre-mRNA and α mRNA in both PTU and D hearts. *p<0.05 pre-mRNA vs mRNA. n=6 rats/group. C, β MHC RNA (pre-mRNA, mRNA, and antisense) analyses using RT-PCR (see (Haddad et al., 2003a) for details. D, Percent increase in β pre-mRNA and β mRNA in both PTU and D hearts. *p<0.05 pre-mRNA vs mRNA. n=6 rats/group

promoter region (Figure 5). In more recent studies we followed this same approach to examine the regulation of cardiac MHC genes in response to pressure overload. Our results (manuscript in preparation) provide evidence that pressure overload involves the same type of regulation concerning both transcriptional regulation of the β MHC through the upstream promoter as well as post-transcriptional regulation via an antisense RNA that interferes with the mRNA processing. These studies provide the first set of observations to indicate that cardiac MHC genes are linked via a common regulatory mechanism whereby α MHC gene transcription can simultaneously impact the β MHC gene's post-transcriptional processing. Collectively, these observations provide fundamental insight as to the antithetical regulation of cardiac α and β MHC regulation.

7. MHC PLASTICITY IN LIMB SKELETAL MUSCLE

7.1 MHC Expression and Muscle Fiber Specificity

Mammalian skeletal muscles can be generally classified into two major groups, slow-twitch and fast-twitch based on their intrinsic contractile properties which are, in part, determined by their MHC expression profiles (Pette and Staron, 1990; Schiaffino and Reggiani, 1996). Slow muscles of mammals express a predominance of the slow type I MHC isoform with some varied proportion of type IIa, the slowest of the fast MHCs (Baldwin, 1996; Fitts et al., 1991; Fitts and Widrick, 1996; Rivero

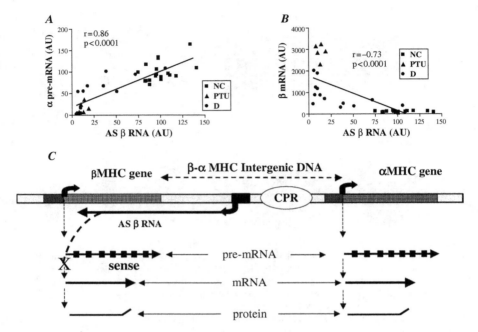

Figure 5. Coordinated Regulation of Cardiac MHC via Expression of an Antisense RNA Transcript. A, Relationship between AS β RNA and α pre-mRNA. Strong positive correlation between α pre-mRNA and antisense (AS) β RNA expression suggesting a common regulation. Data points are from individual samples representing NC, PTU, or D groups. The line was generated by linear regression analysis; r designates the correlation coefficient. (See (Haddad et al., 2003a) for details). B, Relationship between β mRNA and AS β RNA. B mRNA expression indirectly correlates with the antisense β RNA. Data points are from individual samples representing NC, PTU, or D groups. The line was generated by linear regression analysis; r designates the correlation coefficient. C, Schematic of cardiac MHC gene expression. Cardiac MHC genes are located on the same chromosome. Each gene is ~25kb, and they are separated by ~4.5kb. Both β MHC and α MHC genes are transcribed into their corresponding pre-mRNA. On the opposite strand, the β antisense RNA is transcribed and is originated from the intergenic region between α and β genes. The antisense β RNA is proposed to inhibit the processing of the sense β pre-mRNA into mature β mRNA. We propose that both sense α and antisense β transcriptions are regulated by a common promoter region (CPR)

et al., 1999; Schiaffino and Reggiani, 1996; Talmadge, 2000). This profile is exemplified in muscles such as the soleus and vastus intermedius, which play a strong role in antigravity function. Fast muscles of small mammals such as the gastrocnemius-plantaris complex, the vastus lateralis, the vastus medialis, the extensor digitorum longus, and the tibialis anterior express a predominance of the two fast isoforms, IIx and IIb with variable proportions of the two isoforms depending on the muscle, the region of the muscle, or the specific animal model (Baldwin, 1996; Baldwin and Haddad, 2001; Demirel et al., 1999; Fitts et al., 1991; Fitts and Widrick, 1996). In human muscle, there appear to be only two fast isoforms that are expressed (IIa and IIx) in addition to that of the slow, type I MHC. MHC analyses of human soleus muscle show that it expresses an approximately equal proportion of type I and IIa isoforms (Harridge et al., 1998). In contrast, human fast muscles such as the vastus lateralis express all three types of MHC isoforms in variable proportions (Harridge et al., 1998; Larsson, 1993) depending on the physical fitness and activity profiles of the subject. MHC expression is highly dynamic and undergoes major changes during development as well as in the adult stage in order to meet new functional demands. Besides development, MHC gene expression is impacted by hormonal state and by mechanical activity, with the latter depending on both the magnitude of the loading state and neural input.

7.2   Developmental Aspect

While the formation of the skeletal muscle system begins during the latter stages of fetal development, it is important to recognize that in terms of MHC gene expression, both slow and fast muscles are still in an undifferentiated state at the time of birth. For example, in the rodent soleus muscle, which is destined to be comprised chiefly of slow, type I fibers, both the embryonic and neonatal isoforms are abundantly expressed shortly after birth, accounting for ~50% of the total MHC pool with the remaining protein comprised of the slow type I MHC (Adams et al., 1999b; Butler-Browne, 1984; Condon, 1990; d'Albis, 1989). In fast muscles, the degree of undifferentiation is even greater, as the embryonic and neonatal MHC isoforms account for ~90% of the total MHC pool shortly after birth (Adams et al., 1999b; Di Maso et al., 2000). During the first three to four weeks of neonatal development, both slow and fast muscles rapidly grow and differentiate into their adult MHC phenotype (Adams et al., 1999b; d'Albis, 1989; Di Maso et al., 2000). While this process requires an intact nerve (Adams et al., 1999b), the two muscle types appear to differ substantially in the primary factors that drive this transformation process. Based on recent studies comparing muscle development of neonatal rats raised in a microgravity environment of spaceflight versus those rodents that have developed on Earth, it appears that slow muscle types are heavily dependant on weight bearing activity (which begins during the second week following birth) for both normal growth and the optimal expression of type I MHC (Adams et al., 2000). In the absence of the weight bearing state, muscle growth is dramatically reduced, and the MHC pool in the soleus muscle becomes biased to fast MHC expression with IIa and IIx MHCs combined to comprise up to 60–70% of the total MHC expressed as determined at the end of a 16 day space flight that started when the animals were either 7 days or 14 days old (Adams et al., 2000). In contrast, it appears that weight bearing activity is not essential for the fast-type muscles to achieve the adult fast MHC phenotype, which as mentioned above, normally consists chiefly of the fast IIx and IIb isoforms (Adams et al., 2000). Instead, an intact thyroid state (e.g., circulating T3 level) appears to be necessary to down regulate the neonatal isoform and replace its expression with the IIb isoform (Adams et al., 2000; Di Maso et al., 2000). If insufficient T3 is available, the fast muscle remains undifferentiated for a longer duration; whether this undifferentiated state remains permanent with severe thyroid deficiency remains to be determined (Di Maso et al., 2000). Interestingly, TD in the neonatal state actually induces greater expression of the slow type I MHC in both slow and fast skeletal muscles and this occurs independently of gravity (Adams et al., 2000). Thus, it appears that circulating T3 is necessary for the normal growth axis of the body as well as the skeletal muscle system to occur. However, in the euthyroid animal, weight bearing activity is necessary for skeletal muscle to fully express the type I (β) MHC gene in antigravity muscles. These studies have focused on the protein and mRNA expression of the MHC, and they proved that the regulation occurs mainly via pre-translational processes. However, the developmental regulation of skeletal MHC gene expression is thought to occur via transcriptional and post-transcriptional processes (Cox and Buckingham, 1992; Cox et al., 1991) and it involves interaction between several factors. Myogenic factors such as the MyoD/MRF and the MEF2 families play a pivotal role in the activation and maintenance of the myogenic differentiation pathway (Black, 1998; Molkentin and Olson, 1996). In addition, several ubiquitous and muscle-specific factors play important roles in muscle gene activation during terminal differentiation; these include AP-1, Oct-1, SRF, SP1, TEF, GATA, and thyroid receptors. Computational sequence analyses show that the promoter region for all skeletal MHC isoforms have potential binding sites to the above transcription factors. The interplay among these factors in orchestrating muscle-type specific regulation MHC isoform expression is still obscure at this stage.

7.3   Altered MHC Gene Expression in Adult Skeletal Muscle

After the differentiation stage, the MHC profile in any particular muscle is plastic and can change depending on the hormonal (chiefly T3) and activity profile imposed on the target muscle. An increased T3 titer generally reduces slow MHC expression while increasing that of the faster MHCs, while a reduced T3 titer causes the reverse alterations (see Caiozzo and Haddad, 1996, for review). Unloading a muscle from weight bearing activity such as during hindlimb suspension, limb immobilization, bed rest, or exposure to microgravity or making it inactive by inhibiting activation of the motor neurons such as by denervation, or following spinal cord injury, MHC gene expression becomes biased to faster, type II isoform profiles. On the other hand, chronic overloading or intermittent resistance loading the muscle will create shifts to the slower MHCs (see Baldwin and Haddad, 2001). For detailed descriptions the reader also is referred to several excellent review articles on this subject (Baldwin and Haddad, 2002; Pette and Staron, 2000; Schiaffino and Reggiani, 1996; Talmadge, 2000).

In studies involving changes in MHC expression and gene switching in response to various stimuli, the majority of the data have been focused on either mRNA or protein analyses, which are markers of pre-translational and translational events, respectively. Based on these studies, most of the reported MHC plasticity has occurred at the pre-translational level, as there is a tight correlation between the protein and mRNA changes (Caiozzo et al., 2003; Caiozzo et al., 2000; Haddad et al., 1993; Haddad et al., 1998; Huey et al., 2001). However, these levels of analyses do not confer insight on transcriptional regulatory events. There is a strong consensus that MHC gene plasticity is occurring at the transcriptional level. Several studies have focused on characterizing transcription factors involved in MHC regulation in the normal state, those that confer muscle type diversity, as well as during the dynamic states under altered mechanical activity, neural input, and loading states (Allen et al., 2001b; Giger et al., 2000; Giger, 2002; Giger et al., 2004a; Huey et al., 2003; Huey et al., 2002; Karasseva et al., 2003; McCarthy et al., 1997; McCarthy et al., 1999; Olson and Williams, 2000; Schiaffino et al., 1999; Swoap, 1998; Tsika et al., 2004; Vyas et al., 2001; Wheeler, 1999).

7.4 Transcription Regulation of Skeletal MHC Genes

Based on pre-mRNA analyses, fiber type specific expression of MHC genes appears to be regulated at the transcriptional level. For example, RT-PCR analyses of type I and IIb MHC RNA expression in soleus and white medial gastrocnemius (WMG) show that the pre-mRNA expression as a marker of transcription corresponds well to that of the MHC mRNA. Both type I MHC pre-mRNA and mRNA are strongly expressed in the soleus but not the WMG; whereas, IIB pre-mRNA and mRNA are strongly expressed in the WMG but not in the soleus (Fig. 6). The exact regulatory mechanism for muscle fiber-type specific expression of MHC genes is not quite clear; however, several studies propose interactions among several transcription factors as being responsible for this specificity. For example, early

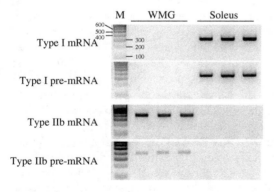

Figure 6. MHC RNA Expression in Soleus vs. WMG Muscles. Representative MHC mRNA and pre-mRNA depicting expression of type I and IIb MHC in soleus and WMG muscles. PCR was carried out following selective RT reactions for either type I or IIB MHC. For mRNA amplification, cDNA was diluted 40x and PCR was carried out for 25 cycles. For pre-mRNA amplifications, cDNA was diluted 5x and the PCR was carried out for 30 cycles. M: DNA molecular weight marker, 100 bp-ladder

studies on troponin I promoters identified two clusters of regulatory elements, the SURE and FIRE that consist of several cis-regulatory elements such as CCAC box, MEF2, E box, and NFAT binding sites, whereby specific transcription factors interact together to enhance gene expression in a fiber-type specific fashion (Calvo et al., 2001; Nakayama, 1996). Whether the same type of clusters are on the MHC gene promoters has not yet been defined, although these cis elements can be identified on each MHC promoter, but in a more dispersed distribution.

With regard to muscle type specific gene expression, recently the calcineurin-NFAT signaling pathway was implicated in the regulation of slow fiber-type specific gene expression (Chin, 1998; Olson and Williams, 2000). Due to their motoneurons firing pattern as well as their total activity, slow muscle fibers maintain higher levels of intracellular free calcium than fast fibers, and this has been implicated in the regulation of fiber type specific gene expression. This chronic increase in intracellular calcium activates calcineurin and other calcium sensitive kinases thereby activating NFAT and MEF-2 transcription factors, which in turn, function to activate transcription of slow muscle-specific genes (Olson and Williams, 2000; Wu et al., 2000). While some reports support this notion (Kubis et al., 2003; Serrano et al., 2001), others have failed to show any muscle type specificity linked to the calcineurin-NFAT signaling pathway. For example, Swoap and coworkers, have shown that overexpression of activated calcineurin or NFAT2 in both cultured C2C12 muscle cells and in muscles enhanced the transcription of muscle promoters but this effect was not specific to slow or fast fibers (Swoap et al., 2000). Also, although an NFAT binding site was identified on the SURE element of the troponin I promoter, mutation of this element did not inhibit its activity in slow muscle fibers (Calvo et al., 1999). Furthermore, type I MHC promoter activity was reduced in the soleus of cyclosporin A (Calcineurin NFAT inhibitor) treated rats; however, this inhibitory effect could not be linked to any of the putative NFAT binding sites of the promoter (Giger et al., 2004a). Thus, the role of calcineurin-NFAT signaling pathway remains controversial concerning its involvement in regulating muscle type specific genes. In addition to NFAT, the peroxisome proliferator-activated receptor gamma coactivator (PGC-1α) has been implicated in slow phenotype determination (Lin et al., 2002). It was shown that PGC1α increases the expression of type I fibrillar proteins by co-activating MEF2 transcription factors, thereby coordinating the expression of both metabolic and contractile properties of type I myofibers; however, the exact involvement of PGC1 α in slow MHC gene regulation is largely unknown.

While the NFAT calcineurin pathway and the PGC1α transcription factors are believed to control slow fiber gene expression, their involvement with fast gene expression is not clear. Recent studies have implicated transcription factors Six1 and Eya1 in fast muscle gene transcription (Grifone et al., 2004), since overexpression of Six-1 and Eya-1 in soleus muscle induced fiber transition from type I/IIa to IIx/IIb similar to that observed in unloading.

Collectively, a complex transcription control mechanism is thought to regulate muscle fiber type-specific expression of the MHC genes under normal conditions and in response to a variety of stimuli. Analyses of type I and IIb MHC pre-mRNA in models showing slow to fast transition and vice-versa confirm that these shifts are primarily driven via transcriptional processes. Type I MHC pre-mRNA expression is significantly decreased in the soleus in response to the models of spinal isolation (SI, Haddad et al., 2003b), denervation (Huey et al., 2003), and hindlimb suspension (Giger et al., 2004b), while IIb MHC pre-mRNA is increased in soleus in SI and hindlimb suspension, and decreased in the overloaded plantaris and in the intermittently trained WMG (data not shown). Analyses of IIa and IIx pre-mRNA expression in SI-treated soleus, SI-treated VI, and trained WMG show that these muscles are subjected to a more complex type of regulation involving transcriptional as well as post-transcriptional processes which appear to control mRNA processing via expression of antisense transcripts. This novel phenomena will be covered in a later section. As to the transcription factors involved in MHC gene regulation associated with phenotype shifts, most of the information has been obtained from promoter studies *in vivo* or *in vitro* as delineated in the next section.

7.4.1   Type I MHC promoter regulation in skeletal muscle

Considerable amount of information exists in the literature on the regulation of the slow type I MHC promoter in vivo. This promoter was examined for fiber type specific expression as well as under unloading and overloading conditions. Based on promoter studies, fiber type specificity is retained in the proximal promoter within 408 bp of the transcription start site (Giger et al., 2000). The proximal 400 bp of the type I MHC promoter was shown to contain the responsiveness to altered loading state. This 400 bp region of the promoter contains the basal promoter (TATA box) and harbors a collection of regulatory cis elements that serve as the binding sites of specific transcription factors such as the TEF-1/MCAT, A/T rich GATA, GC rich SP1/3, E-Box bHLH related factors. Interaction of these TFs with the basal promoter regulate MHC transcription leading to altered MHC expression. For example, the TEF-1 transcription factor was implicated in conferring responsiveness to unloading in the spinal isolation model (SI, which keeps the intact nerve-muscle connection electrically silent and renders the muscle unloaded, Huey et al., 2002), denervation (Huey et al., 2003), and in hindlimb suspension (Giger et al., 2004b). Interestingly, TEF-1 was also implicated in the regulation of type I MHC in response to muscle overload (Giger, 2002; Karasseva, 2003). Recently, the Sp1/Sp3 transcription factor has been shown to confer responsiveness of the type I MHC promoter in the unloading/overloading states. (Tsika et al., 2004). Additional studies show that factors binding to E-box and A/T rich regions of the proximal type I promoter are also involved in its regulation in the non weight bearing state (Giger et al., 2004b).

7.4.2   Regulation of the fast MHC promoters in skeletal muscle cells

With the exception of the IIb MHC, little is known about the regulation of the fast MHC promoters in skeletal muscle cells under *in vivo* conditions in intact muscles. However, Allen and coworkers (Allen et al., 2001b) have characterized and compared 1000 bp sequences 5' to the TSS for the IIa, IIx, and IIb promoters in C2C12 cells. Their findings suggest that these 1kb promoters conferred muscle specificity and that greater expression was seen in differentiated myotubes compared to myoblasts. Moreover, it appears that all three promoters bind to and are activated by MEF-2 factors (over-expressed) to a similar extent, but that the IIa promoter is more responsive to over-expressed activated calcineurin and NFAT factors compared to the IIx and IIb promoters. On the other hand, the IIb promoter was more responsive to MyoD factors even though all three promoters have MyoD binding sites. Interestingly, it was reported that whereas all three promoters contained conserved CArG–like elements, only the IIb promoter had a perfect consensus for binding the serum response factor (SRF). However, the IIb promoter was the least responsive to the deletion of this element. In contrast, the IIx promoter was activated and the IIa promoter was inhibited upon deletion of the CArG element, suggesting that the CArG/SRF have differential effects on fast MHC regulation (Allen et al., 2001b). Early studies on the IIb MHC promoter came from studies by Whalen and associates (Takeda, 1992). Several muscle-specific regulatory elements were identified on the proximal IIb promoter including potential sites of binding of myogenic regulatory factors (MRF's), serum response factor (SRF), and myocyte enhancer factor (MEF-2). Furthermore, the IIb promoter activity as studied in muscle cell culture has been shown to be induced by MRF over-expression independently of the E-Box (MRF binding site) (Takeda, 1995). Also the IIb promoter activity in myocyte cell culture was reduced upon mutation of either the proximal AT1 or the AT2 elements, abolishing the binding to MEF2 and Oct1 transcription factors, respectively (Lakich et al., 1998). Other *in vivo* studies demonstrated that the proximal promoter of IIb between -295 and $+13$ bp relative to TSS is sufficient to achieve a fiber type specific expression as well as induction during slow to fast MHC transition in response to unloading of soleus muscle (Swoap, 1998). Also, we found that intronic sequences within 2kb from the TSS on the 3' flanking region of the IIb promoter were necessary to confer the responsiveness of the promoter activity to overload when transfected into the plantaris muscle (manuscript in preparation). Furthermore, a proximal E box was found necessary for high levels of IIb promoter activity in fast twitch muscle and was implicated in the responsiveness to unloading. (Swoap, 1998; Wheeler, 1999). Other studies showed that MyoD affects IIb expression in a muscle type specific fashion. MyoD-/- transgenic mice were associated with blunting IIb MHC expression only in the diaphragm and in unloaded soleus (Seward et al., 2001).

7.5   Post-Transcriptional Regulation of Skeletal MHC Gene Expression

7.5.1   *IIa-IIx interplay in electrically silenced vastus intermedius and soleus muscles*

In the spinal isolation (SI) model, the hindlimb muscles are electrically silenced while maintaining the motoneuron-muscle connectivity (Grossman et al., 1998). When subjected to this procedure, slow muscles undergo a rapid transformation that results in a shift of myosin heavy chain (MHC) isoforms from slow ATPase isoforms to faster ones (i.e. Type I/IIa to Type IIx/IIb). While we have previously shown that type I MHC is downregulated at the transcriptional level, based on type I pre-mRNA and mRNA analyses (Haddad et al., 2003b), it is not clear at what level the regulation of IIa, IIx, and IIb occurs. We have performed extensive analyses on the vastus intermedius (VI) RNA, and our data suggest that while type I, IIx, and IIb regulation occurs at the transcriptional level, type IIa regulation occurs via a post-transcriptional process. We have found that type IIa pre-mRNA does not significantly change while the mRNA expression decreases by ~60% in the SI muscle compared to the control state (Fig. 7). Analyses of RNA products of the gene locus revealed the existence of an antisense RNA that was detected in the intergenic region between IIa and IIx genes. This antisense RNA extended into the 3' end of the IIa gene region to form an antisense transcript, complementary to the IIa sense strand. The expression of this antisense IIa increased by ~10 fold in the SI-treated VI muscle as compared to the control (see Figure 7).

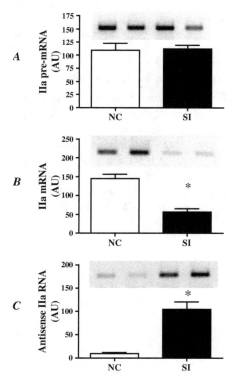

Figure 7. IIa MHC RNA Expression in the VI Muscle in Response to SI Model of Complete Muscle Inactivation. A, IIa MHC pre-mRNA expression in NC vs. SI soleus as determined by RT PCR. AU; arbitrary unit, N=6/group. B, IIa MHC mRNA expression in NC vs. SI soleus. *$P<0.05$ vs NC. C, antisense IIa RNA expression in NC vs. SI soleus as determined by RT PCR targeting the antisense in the IIa-IIx intergenic region

This naturally occurring IIa antisense RNA transcript was detected using RT-PCR primers targeting the intergenic region between the IIa and IIx genes. The discrepancy between the IIa pre-mRNA (whose expression does not change) vs. IIa mRNA (whose expression significantly decrease) in the SI muscle is due to post-transcriptional gene silencing processes, where there is likely formation of double stranded RNA that inhibits processing of pre-mRNA. This IIa antisense RNA was strongly correlated ($r=0.92$, $p<0.0001$) with IIx pre-mRNA (Fig. 8). Transcription of the natural antisense RNA from the IIa-IIx intergenic MHC gene locus appears to be coordinated with the downstream gene (i.e. IIx), and results in attenuation of IIa pre-mRNA processing into mRNA, with a concomitant increase in IIx mRNA. RNA analyses indicate that the same type of regulation occurs in the soleus muscle following SI, i.e., type IIa expression is regulated via post-transcriptional events involving an antisense IIa RNA. This coordinated RNA shift between two adjacent MHC genes (IIa, IIx) appears to underlie the molecular process observed in the classically defined transitions of one MHC isoform to another in response to various stimuli, referred to as muscle plasticity. Interstingly, this coordinated regulation is somehow similar to that observed for the cardiac α and β MHC genes, suggesting that a common scheme of MHC gene regulation occurs in both cardiac and skeletal muscle.

7.5.2   IIx-IIb interplay in resistance trained white medial gastrocnemius

The same post-transcriptional regulation of fast MHC genes appears to occur in the white medial gastrocnemius (WMG). The medial gastrocnemius muscle in rodents is a heterogeneous muscle which can be dissected into a superficial fast white region and a red mixed fiber region. While the mixed MG region expresses all 4 adult MHC isoforms, with type IIx and IIb being the most abundant, the WMG expresses only the IIx and IIb, with IIb making up >70% of the total MHC pool under normal conditions. However, intermittent resistance training in the isometric mode via electrical stimulation causes a rapid shift in MHC expression in this

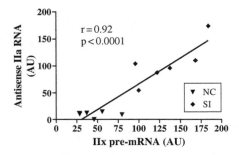

Figure 8. Relationship Between Antisense IIa RNA and IIx pre-mRNA. Antisense IIa RNA expression directly correlates with IIx pre mRNA expression. Data points are from individual samples representing NC and SI groups. The plot was generated by linear regression analysis; r designates the correlation coefficient. AU: arbitrary unit

muscle. Caiozzo et al have shown that after only 2 training sessions, one can detect a significant increase in IIx mRNA expression and a corresponding decrease in IIb MHC mRNA expression (Caiozzo et al., 1996). In order to determine if these rapid shifts in MHCs are mediated through transcriptional processes, we have analyzed the MHC pre-mRNA and mRNA expression in the WMG after undergoing 4 training sessions. MHC mRNA composition analyses confirmed that we have a phenotype shift between IIx and IIb MHC isoforms. The MHC mRNA composition of control WMG consisted of 26% IIx and 74% IIb on average, while that of trained muscle consisted of 45% IIx, 51% IIb, including a small proportion (<2% ea) of IIa, embryonic and neonatal MHC mRNA. Analyses of the MHC pre-mRNA shows that the down regulation of IIb expression is determined at the transcriptional level, since both the pre-mRNA and mRNA decreased to the same extent (63% decrease in pre-mRNA and 68% decrease in mRNA). IIx MHC RNA analyses show that while both the pre-mRNA and IIx mRNA were increased, the increase was greater at the mRNA level. Further analyses enabled us to detect an antisense IIx RNA species whose expression decreases in the trained WMG (Fig. 9). Therefore, it appears that under control conditions, this antisense IIx RNA inhibits full expression of the IIx mRNA, but this inhibition is partially removed in the trained muscle, which allows for a rapid shift from IIb to IIx. This antisense IIx RNA is detected in

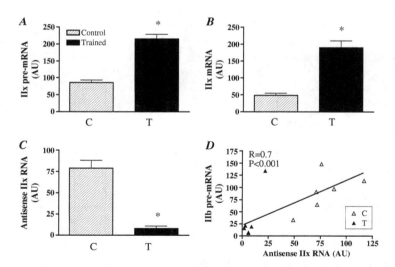

Figure 9. IIx MHC RNA Expression in the WMG Muscle Following Short-Term Resistance Training (T). A, IIx MHC pre-mRNA expression in control (C) vs. T WMG as determined by RT PCR. AU; arbitrary unit, N=6/group, *P<0.05 vs C. B, IIx MHC mRNA expression in C vs. T WMG. C, Antisense IIx RNA expression in C vs. T WMG as determined by RT PCR targeting the antisense in the intergenic region between IIx and IIb. D, Relationship between antisense IIx RNA and IIb pre-mRNA. The antisense IIx RNA expression directly correlates with IIb pre mRNA expression. Data points are from individual samples representing Control and Trained groups. The plot was generated by linear regression analysis; r is correlation coefficient

the intergenic region between IIx and IIb genes and extends towards the IIx gene to overlap and complement the sense IIx transcript. Of interest to note is that the IIx antisense expression also strongly correlated with that of the IIb pre-mRNA (Fig. 10) suggesting that a tight coordinated regulation occurs between these two physically linked genes similar to what we observed in the heart involving alpha and beta MHC regulation, and IIa-IIx in the VI muscle.

7.6   Do MHC Isoform Shifts Occur in a Specific Order?

Adaptive changes in the expression of MHC protein isoforms result in fiber type transitions encompassing a spectrum of pure and hybrid fibers. It has been proposed that MHC isoform shifts follow a general scheme of sequential and reversible transitions from slow to fast and vice versa in the order of I ↔ IIa ↔ IIx ↔ IIb (Pette and Staron, 2000; Termin, 1989). This theory proposes that the MHC shifts occur in a graded manner involving sequential inactivation and activation of genes in the above delineated order. Although this proposed sequence is manifest in some models of MHC transitions, it fails to explain all the observed MHC shifts. For example, in soleus muscle undergoing slow to fast transition in response to inactivity or unloading, there is a high number of fibers co-expressing type I and IIx MHCs without the expression of the IIa MHC (so called "jump fibers"). These patterns of change do not correspond to the above sequential order. Given that the type I gene is physically separated from the fast MHCs on a different chromosome, it appears more likely that this gene is regulated independently of the fast MHC cluster. Our observations suggest that the slow MHC type I gene and the IIa MHC, the slower isoform of the fast type II MHC genes, are regulated in the same direction

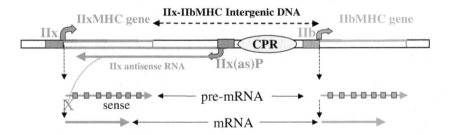

Figure 10. Coordinated Regulation Between IIx and IIb MHC Genes in the WMG. Schematic of IIx and IIb MHC gene expression in the WMG. IIx and IIb MHC genes are positioned in tandem on the same chromosome. Each gene is ~25kb, and they are separated by ~13-14kb on rat chromosome 11. Both IIx MHC and IIb MHC genes are transcribed into their corresponding pre-mRNA. On the opposite strand, the IIx antisense RNA is transcribed and is originated from the intergenic region between the two genes. The antisense IIx RNA is proposed to inhibit the processing of the sense IIx pre-mRNA into the mature IIx mRNA. We propose that both sense IIb and antisense IIx transcriptions are regulated by a common promoter region (CPR). A color figure is freely accessible via the website of the book: http://www.springer.com/1-4020-5176-x

under certain conditions of altered activity. Furthermore, the IIa, IIx, and IIb MHC gene expression appears to be tightly coordinated based on their physical linkage that involves regulation by an antisense RNA as shown in the above section. For example, during unloading conditions such as in the hindlimb suspension and SI models, both the type I and IIa genes become down regulated, while the IIx and IIb genes are upregulated. When IIx expression increases, also an antisense IIa RNA increases, which contributes to a decreased IIa mRNA expression (see above section 7.5.1). These transformation are compatible with the existence of the "jump fibers" co-expressing type I and IIx MHC in these models (Talmadge, 2000).

8.   ARE MHC ISOFORMS FUNCTIONALLY DISTINCT?

The adult fast MHC isoforms IIa, IIb, and IIx are approximately 93% identical at the amino acid level (Weiss and Leinwand, 1996; Weiss et al., 1999), suggesting that they share close functional properties. Information on the question of functional identity of these genes has been generated from studies involving two lines of MHC null mice, the IIx null and the IIb null (Acakpo-Satchivi et al., 1997). These mice exhibited some distinct growth and muscle defects suggesting that IIx and IIb isoforms are functionally distinct and both are required for normal muscle function in mice (Allen et al., 2001a; Allen and Leinwand, 2001; Sartorius, 1998). In the IIx null mice, IIa MHC expression appeared to increase in certain fibers to compensate for the loss of IIx expression, but apparently this compensation is not complete and results in an abnormal phenotype (Allen et al., 2000). Similarly, in the IIb null mice, IIx expression increases in muscle fibers but the response does not result in complete compensation, which is evidenced by exaggerated muscle atrophy and weakness. In terms of analyzing cardiac MHC isoform function, an α null mice was generated (Acakpo-Satchivi et al., 1997). The α MHC gene null mutation was lethal in the fetal stage for homozygotes, while in the heterozygous mice α null resulted in severe cardiac malformities. These observations further suggest that despite cardiac MHC genes being more than 90% similar in their amino acid sequence and sharing similar structure, these genes are not redundant and indicate that the β MHC gene could not compensate for the α deletion. Furthermore, a forced over expression of the β MHC isoform in the mouse heart caused a gradual replacement of the α MHC gene, and these mice exhibited a normal apparent phenotype except that the heart was more energetically efficient and has a lower intrinsic contractility (Krenz et al., 2003), which is compatible with the different physiological and biochemical properties of these two MHC proteins.

9.   FUTURE DIRECTIONS

Collectively, this chapter describes striated muscle plasticity with a focus on the regulation of adult MHC gene expression via transcriptional and post-transcriptional processes. We have presented information favoring the idea that the organization of the MHC genes is of functional significance and that such an organization plays a role in the regulation of expression of these genes under a variety of experimental perturbations. Future research should focus on identifying transcription factors involved in MHC gene fiber-type specificity and plasticity in response to various stimuli. Further, it will be of critical interest to identify how various mechanical alterations impact signaling pathways and how these could influence the expression and function of various transcription factors. Clearly, further research is needed in order to fully understand the regulation of the MHC gene family and hence the fundamental properties of muscle plasticity.

REFERENCES

Abraham, W. T., Gilbert, E. M., Lowes, B. D., Minobe, W. A., Larrabee, P., Roden, R. L., Dutcher, D., Sederberg, J., Lindenfeld, J. A., Wolfel, E. E. et al. (2002). Coordinate changes in Myosin heavy chain isoform gene expression are selectively associated with alterations in dilated cardiomyopathy phenotype. *Molecular Medicine* 8, 750–60.

Acakpo-Satchivi, L. J. R., Edelmann, W., Sartorius, C., Lu, B. D., Wahr, P. A., Watkins, S. C., Metzger, J. M., Leinwand, L. and Kucherlapati, R. (1997). Growth and Muscle Defects in Mice Lacking Adult Myosin Heavy Chain Genes. *J. Cell Biol.* 139, 1219–1229.

Adams, G. R., Haddad, F. and Baldwin, K. M. (1999a). Time course of changes in markers of myogenesis in overloaded rat skeletal muscles. *Journal of Applied Physiology* 87, 1705–12.

Adams, G. R., Haddad, F., McCue, S. A., Bodell, P. W., Zeng, M., Qin, L., Qin, A. X. and Baldwin, K. M. (2000). Effects of spaceflight and thyroid deficiency on rat hindlimb development. II. Expression of MHC isoforms. *Journal of Applied Physiology* 88, 904–16.

Adams, G. R., McCue, S. A., Zeng, M. and Baldwin, K. M. (1999b). Time course of myosin heavy chain transitions in neonatal rats: importance of innervation and thyroid state. *American Journal of Physiology* 276, R954–61.

Alam, J. and Cook, J. L. (1990). Reporter genes: application to the study of mammalian gene transcription. *Analytical Biochemistry* 188, 245–54.

Allen, D. L., Harrison, B. C. and Leinwand, L. A. (2000). Inactivation of myosin heavy chain genes in the mouse: diverse and unexpected phenotypes. *Microscopy Research and Technique* 50, 492–9.

Allen, D. L., Harrison, B. C., Sartorius, C., Byrnes, W. C. and Leinwand, L. A. (2001a). Mutation of the IIB myosin heavy chain gene results in muscle fiber loss and compensatory hypertrophy. *Am J Physiol Cell Physiol* 280, C637–645.

Allen, D. L. and Leinwand, L. A. (2001). Postnatal Myosin Heavy Chain Isoform Expression in Normal Mice and Mice Null for IIb or IId Myosin Heavy Chains. *Developmental Biology* 229, 383–395.

Allen, D. L., Sartorius, C. A., Sycuro, L. K. and Leinwand, L. A. (2001b). Different Pathways Regulate Expression of the Skeletal Myosin Heavy Chain Genes. *J. Biol. Chem.* 276, 43524–43533.

Alpert, N. R. and Mulieri, L. A. (1986). Functional consequences of altered cardiac myosin isoenzymes. *Medicine and Science in Sports and Exercise* 18, 309–13.

Baldwin, K. M. (1996). Effects of altered loading states on muscle plasticity: what have we learned from rodents? *Medicine and Science in Sports and Exercise* 28, S101–6.

Baldwin, K. M. and Haddad, F. (2001). Effects of different activity and inactivity paradigms on myosin heavy chain gene expression in striated muscle. *Journal of Applied Physiology* 90, 345–57.

Baldwin, K. M. and Haddad, F. (2002). Skeletal muscle plasticity: cellular and molecular responses to altered physical activity paradigms. *Am J Phys Med Rehabil* 81, S40–51.

Berg, J. S., Powell, B. C. and Cheney, R. E. (2001). A Millennial Myosin Census. *Mol. Biol. Cell* 12, 780–794.

Black, B. L., and E.N. Olson. (1998). Transcriptional control of muscle development by myocyte enhancer factor-2 (MEF2) proteins. *Annu. Rev. Cell Dev. Biol.* 14, 167–196.

Boheler, K. R., Chassagne, C., Martin, X., Wisnewsky, C. and Schwartz, K. (1992). Cardiac expressions of alpha- and beta-myosin heavy chains and sarcomeric alpha-actins are regulated through transcriptional mechanisms. Results from nuclear run-on assays in isolated rat cardiac nuclei. *Journal of Biological Chemistry* 267, 12979–85.

Booth, F. W. and Criswell, D. S. (1997). Molecular events underlying skeletal muscle atrophy and the development of effective countermeasures. *International Journal of Sports Medicine* 18 Suppl 4, S265–9.

Butler-Browne, G. S., and R.G. Whalen. (1984). Myosin isozyme transitions occurring during the postnatal development of the rat soleus muscle. *Developmental Biology* 102, 324–334.

Caiozzo, V. J., Baker, M. J. and Baldwin, K. M. (1998). Novel transitions in MHC isoforms: separate and combined effects of thyroid hormone and mechanical unloading. *Journal of Applied Physiology* 85, 2237–48.

Caiozzo, V. J., Baker, M. J., Huang, K., Chou, H., Wu, Y. Z. and Baldwin, K. M. (2003). Single-fiber myosin heavy chain polymorphism: how many patterns and what proportions? *Am J Physiol Regul Integr Comp Physiol* 285, R570–580.

Caiozzo, V. J., F. Haddad, M. Baker, S. McCue, and K. M. Baldwin. (1999). Mechanical overload exerts stronger control than hypothyroidism on MHC expression in rodent fast twitch muscle. *American Journal of Physiology* (submitted).

Caiozzo, V. J., Haddad, F., Baker, M., MuCue, S. and Baldwin, K. M. (2000). MHC polymorphism in rodent plantaris muscle: effects of mechanical overload and hypothyroidism. *Am J Physiol Cell Physiol* 278, C709–17.

Caiozzo, V. J. and Haddad, F. (1996). Thyroid hormone: modulation of muscle structure, function, and adaptive responses to mechanical loading. *Exercise and Sport Sciences Reviews* 24, 321–61.

Caiozzo, V. J., Haddad, F., Baker, M. J. and Baldwin, K. M. (1996). Influence of mechanical loading on myosin heavy-chain protein and mRNA isoform expression. *Journal of Applied Physiology* 80, 1503–12.

Calvo, S., Venepally, P., Cheng, J. and Buonanno, A. (1999). Fiber-type-specific transcription of the troponin I slow gene is regulated by multiple elements. *Molecular and Cellular Biology* 19, 515–25.

Calvo, S., Vullhorst, D., Venepally, P., Cheng, J., Karavanova, I. and Buonanno, A. (2001). Molecular Dissection of DNA Sequences and Factors Involved in Slow Muscle-Specific Transcription. *Mol. Cell. Biol.* 21, 8490–8503.

Chikuni, K., Muroya, S. and Nakajima, I. (2004). Absence of the functional Myosin heavy chain 2b isoform in equine skeletal muscles. *Zoolog Sci* 21, 589–96.

Chin, E. R., E.N. Olson, J.A. Richardson, Q. Yang, C. Humphries, J. M. Shelton, H. Wu, W. Zhu, R. Bassel-Duby, and R. Sanders Williams. (1998). A calcineurin-dependent transcriptionial pathway controls skeletal muscle fiber type. *Genes and Development* 12, 1–11.

Chizzonite, R. A. and Zak, R. (1984). Regulation of myosin isoenzyme composition in fetal and neonatal rat ventricle by endogenous thyroid hormones. *Journal of Biological Chemistry* 259, 12628–32.

Condon, K., L. Silberstein, H. M. Blau, and W.J. Thompson. (1990). Develpment of muscle fiber types in the prenatal rat hindlimb. *Develpmental Biology* 138, 256–274.

Corin, S., Levitt, L., O'Mahoney, J., Joya, J., Hardeman, E. and Wade, R. (1995). Delineation of a Slow-Twitch-Myofiber-Specific Transcriptional Element by Using in vivo Somatic Gene Transfer. *PNAS* 92, 6185–6189.

Cox, R. D. and Buckingham, M. E. (1992). Actin and myosin genes are transcriptionally regulated during mouse skeletal muscle development. *Developmental Biology* 149, 228–34.

Cox, R. D., Weydert, A., Barlow, D. and Buckingham, M. E. (1991). Three linked myosin heavy chain genes clustered within 370 kb of each other show independent transcriptional and post-transcriptional regulation during differentiation of a mouse muscle cell line. *Developmental Biology* 143, 36–43.

d'Albis, A., Couteaux, R., Janmot, C., Roulet, A. and Mira, J. C. (1988). Regeneration after cardiotoxin injury of innervated and denervated slow and fast muscles of mammals. Myosin isoform analysis. *European Journal of Biochemistry* 174, 103–10.

d'Albis, A., R. Couteaux, C. Janmot, and A. Roulet. (1989). Specific programs of myosin expression in the postnatal development of rat muscles. *Eur. J. Biochem.* 183, 583–590.

Demirel, H. A., Powers, S. K., Naito, H., Hughes, M. and Coombes, J. S. (1999). Exercise-induced alterations in skeletal muscle myosin heavy chain phenotype: dose-response relationship. *Journal of Applied Physiology* 86, 1002–8.

Dhalla, N. S., Liu, X., Panagia, V. and Takeda, N. (1998). Subcellular remodeling and heart dysfunction in chronic diabetes [editorial]. *Cardiovascular Research* 40, 239–47.

Di Maso, N. A., Caiozzo, V. J. and Baldwin, K. M. (2000). Single-fiber myosin heavy chain polymorphism during postnatal development: modulation by hypothyroidism. *Am J Physiol Regul Integr Comp Physiol* 278, R1099–106.

Dillmann, W. H. (1984). Hormonal influences on cardiac myosin ATPase activity and myosin isoenzyme distribution. *Molecular and Cellular Endocrinology* 34, 169–81.

Dillmann, W. H. (1989). Diabetes and thyroid-hormone-induced changes in cardiac function and their molecular basis. *Annual Review of Medicine* 40, 373–94.

Fitts, R. H., McDonald, K. S. and Schluter, J. M. (1991). The determinants of skeletal muscle force and power: their adaptability with changes in activity pattern. *Journal of Biomechanics* 24 Suppl 1, 111–22.

Fitts, R. H. and Widrick, J. J. (1996). Muscle mechanics: adaptations with exercise-training. *Exercise and Sport Sciences Reviews* 24, 427–73.

Giger, J. M., Haddad, F., Qin, A. X. and Baldwin, K. M. (2000). In vivo regulation of the beta-myosin heavy chain gene in soleus muscle of suspended and weight-bearing rats. *Am J Physiol Cell Physiol* 278, C1153–61.

Giger, J. M., Haddad, F., Qin, A. X. and Baldwin, K. M. (2002). Functional overload increases [beta]-MHC promoter activity in rodent fast muscle via the proximal MCAT ([beta]e3) site. *Am J Physiol Cell Physiol* 282, C518-C527.

Giger, J. M., Haddad, F., Qin, A. X. and Baldwin, K. M. (2004a). Effect of cyclosporin A treatment on the in vivo regulation of type I MHC gene expression. *Journal of Applied Physiology* 97, 475–483.

Giger, J. M., Haddad, F., Qin, A. X., Zeng, M. and Baldwin, K. M. (2004b). The Effect of Unloading on Type I Myosin Heavy Chain Gene Regulation in Rat Soleus Muscle. *Journal of Applied Physiology*, 01099.2004.

Goodson, H. V., and J.A. Spudich. (1993). Molecular evolution of the myosin family: Relationships derived from comparisons of amino acid sequences. *Proc. Natl. Acad. Sci. USA* 90, 659–663.

Grifone, R., Laclef, C., Spitz, F., Lopez, S., Demignon, J., Guidotti, J.-E., Kawakami, K., Xu, P.-X., Kelly, R., Petrof, B. J. et al. (2004). Six1 and Eya1 Expression Can Reprogram Adult Muscle from the Slow-Twitch Phenotype into the Fast-Twitch Phenotype. *Mol. Cell. Biol.* 24, 6253–6267.

Grossman, E. J., Roy, R. R., Talmadge, R. J., Zhong, H. and Edgerton, V. R. (1998). Effects of inactivity on myosin heavy chain composition and size of rat soleus fibers. *Muscle and Nerve* 21, 375–89.

Gulick, J., Subramaniam, A., Neumann, J. and Robbins, J. (1991). Isolation and characterization of the mouse cardiac myosin heavy chain genes. *Journal of Biological Chemistry* 266, 9180–5.

Gustafson, T. A., B. E. Markham, and E. Morkin. (1985). Analysis of thyroid hormone effects on myosin heavy chain gene expression in cardiac and soleus muscles using a novel dot-blot mRNA assay. *Biochem. Biophys. Res. Commun.* 130, 1161–1167.

Haddad, F., Bodell, P. W. and Baldwin, K. M. (1995). Pressure-induced regulation of myosin expression in rodent heart. *Journal of Applied Physiology* 78, 1489–95.

Haddad, F., Bodell, P. W., McCue, S. A. and Baldwin, K. M. (1997a). Effects of diabetes on rodent cardiac thyroid hormone receptor and isomyosin expression. *American Journal of Physiology* 272, E856–63.

Haddad, F., Bodell, P. W., Qin, A. X., Giger, J. M. and Baldwin, K. M. (2003a). Role of antisense RNA in coordinating cardiac myosin heavy chain gene switching. *J. Biol. Chem.* 278, 37132–37138.

Haddad, F., Herrick, R. E., Adams, G. R. and Baldwin, K. M. (1993). Myosin heavy chain expression in rodent skeletal muscle: effects of exposure to zero gravity. *Journal of Applied Physiology* 75, 2471–7.

Haddad, F., Masatsugu, M., Bodell, P. W., Qin, A., McCue, S. A. and Baldwin, K. M. (1997b). Role of thyroid hormone and insulin in control of cardiac isomyosin expression. *Journal of Molecular and Cellular Cardiology* 29, 559–69.

Haddad, F., P.W. Bodell, S.A. McCue, R.E. Herrick, and K.M. Baldwin. (1993). Food restriction-induced transformations in cardiac functional and biochemical properties in rats. *J. Appl. Physiol.* 74, 606–612.

Haddad, F., Qin, A. X., Zeng, M., McCue, S. A. and Baldwin, K. M. (1998). Interaction of hyperthyroidism and hindlimb suspension on skeletal myosin heavy chain expression. *Journal of Applied Physiology* 85, 2227–36.

Haddad, F., Roy, R. R., Zhong, H., Edgerton, V. R. and Baldwin, K. M. (2003b). Atrophy responses to muscle inactivity:II. molecular markers of protein deficit. *Journal of Applied Physiology* 95, 791–802.

Harridge, S. D., Bottinelli, R., Canepari, M., Pellegrino, M., Reggiani, C., Esbjornsson, M., Balsom, P. D. and Saltin, B. (1998). Sprint training, in vitro and in vivo muscle function, and myosin heavy chain expression. *Journal of Applied Physiology* 84, 442–9.

Hildebrandt, A. L. and Neufer, P. D. (2000). Exercise attenuates the fasting-induced transcriptional activation of metabolic genes in skeletal muscle. *Am J Physiol Endocrinol Metab* 278, E1078–86.

Hoh, J. F. Y. (2002). 'Superfast' or masticatory myosin and the evolution of jaw-closing muscles of vertebrates. *J Exp Biol* 205, 2203–2210.

Huey, K. A., Haddad, F., Qin, A. X. and Baldwin, K. M. (2003). Transcriptional regulation of the type I myosin heavy chain gene in denervated rat soleus. *Am J Physiol Cell Physiol* 284, C738–748.

Huey, K. A., Roy, R. R., Baldwin, K. M. and Edgerton, V. R. (2001). Temporal effects of inactivty on myosin heavy chain gene expression in rat slow muscle. *Muscle and Nerve* 24, 517–26.

Huey, K. A., Roy, R. R., Haddad, F., Edgerton, V. R. and Baldwin, K. M. (2002). Transcriptional regulation of the type I myosin heavy chain promoter in inactive rat soleus. *Am J Physiol Cell Physiol* 282, C528-C537.

Karasseva, N., Tsika, G., Ji, J., Zhang, A., Mao, X. and Tsika, R. (2003). Transcription Enhancer Factor 1 Binds Multiple Muscle MEF2 and A/T-Rich Elements during Fast-to-Slow Skeletal Muscle Fiber Type Transitions. *Mol. Cell. Biol.* 23, 5143–5164.

Knotts, S., Rindt, H., Neumann, J. and Robbins, J. (1994). In vivo regulation of the mouse beta myosin heavy chain gene. *Journal of Biological Chemistry* 269, 31275–82.

Knotts, S., Rindt, H. and Robbins, J. (1995). Position independent expression and developmental regulation is directed by the beta myosin heavy chain gene's 5' upstream region in transgenic mice. *Nucleic Acids Research* 23, 3301–9.

Knotts, S., Saanchez, A., Rindt, H. and Robbins, J. (1996). Developmental Modulation of a beta myosin heavy chain promoter-driven transgene. *Developmental Dynamics* 206, 182–92.

Konig, S., Burkman, J., Fitzgerald, J., Mitchell, M., Su, L. and Stedman, H. (2002). Modular Organization of Phylogenetically Conserved Domains Controlling Developmental Regulation of the Human Skeletal Myosin Heavy Chain Gene Family. *J. Biol. Chem.* 277, 27593–27605.

Krenz, M., Sanbe, A., Bouyer-Dalloz, F., Gulick, J., Klevitsky, R., Hewett, T. E., Osinska, H. E., Lorenz, J. N., Brosseau, C., Federico, A. et al. (2003). Analysis of Myosin Heavy Chain Functionality in the Heart. *J. Biol. Chem.* 278, 17466–17474.

Kubis, H.-P., Hanke, N., Scheibe, R. J., Meissner, J. D. and Gros, G. (2003). Ca2+ transients activate calcineurin/NFATc1 and initiate fast-to-slow transformation in a primary skeletal muscle culture. *Am J Physiol Cell Physiol* 285, C56–63.

Lakich, M. M., Diagana, T. T., North, D. L. and Whalen, R. G. (1998). MEF-2 and Oct-1 Bind to Two Homologous Promoter Sequence Elements and Participate in the Expression of a Skeletal Muscle-specific Gene. *J. Biol. Chem.* 273, 15217–15226.

Larsson, L., and R.L. Moss. (1993). Maximum velocity of shortening in relatioin to myosin isoform composition in single fibres from human skeletal muscles. *J. Physiol.* 472, 595–614.

Lin, J., Wu, H., Tarr, P. T., Zhang, C.-Y., Wu, Z., Boss, O., Michael, L. F., Puigserver, P., Isotani, E., Olson, E. N. et al. (2002). Transcriptional co-activator PGC-1[alpha] drives the formation of slow-twitch muscle fibres. *Nature* 418, 797–801.

Lompre, A. M., Mercadier, J. J., Wisnewsky, C., Bouveret, P., Pantaloni, C., D'Albis, A. and Schwartz, K. (1981). Species- and age-dependent changes in the relative amounts of cardiac myosin isoenzymes in mammals. *Developmental Biology* 84, 286–290.

Lompre, A. M., Nadal-Ginard, B. and Mahdavi, V. (1984). Expression of the cardiac ventricular alpha- and beta-myosin heavy chain genes is developmentally and hormonally regulated. *Journal of Biological Chemistry* 259, 6437–46.

Lompre, A. M., Schwartz, K., d'Albis, A., Lacombe, G., Van Thiem, N. and Swynghedauw, B. (1979). Myosin isoenzyme redistribution in chronic heart overload. *Nature* 282, 105–7.

Lowes, B. D., Minobe, W., Abraham, W. T., Rizeq, M. N., Bohlmeyer, T. J., Quaife, R. A., Roden, R. L., Dutcher, D. L., Robertson, A. D., Voelkel, N. F. et al. (1997). Changes in Gene Expression in the Intact Human Heart . Downregulation of alpha -Myosin Heavy Chain in Hypertrophied, Failing Ventricular Myocardium. *J. Clin. Invest.* 100, 2315–2324.

Lupa-Kimball, V. A. and Esser, K. A. (1998). Use of DNA injection for identification of slow nerve-dependent regions of the MLC2s gene. *American Journal of Physiology* 274, C229–35.

Mably, J. D. and Liew, C. C. (1996). Factors involved in cardiogenesis and the regulation of cardiac-specific gene expression. *Circulation Research* 79, 4–13.

Mahdavi, V., Chambers, A. P. and Nadal-Ginard, B. (1984). Cardiac alpha- and beta-myosin heavy chain genes are organized in tandem. *Proceedings of the National Academy of Sciences of the United States of America* 81, 2626–30.

Mahdavi, V., E. Strehler, M. Periasamy, D.F. Wieczorek, S. Izumo and B. Nadal-Ginard. (1986). Sacromeric myosin heavy chain gene family: Organisation and pattern of expression. *Med. Sci. Sports Exerc.* 18, 229–308.

Mahdavi, V., Izumo, S. and Nadal-Ginard, B. (1987). Developmental and hormonal regulation of sarcomeric myosin heavy chain gene family. *Circulation Research* 60, 804–14.

McCarthy, J. J., Fox, A. M., Tsika, G. L., Gao, L. and Tsika, R. W. (1997). beta-MHC transgene expression in suspended and mechanically overloaded/suspended soleus muscle of transgenic mice. *Amer J Physiol* 272, R1552–61.

McCarthy, J. J., Vyas, D. R., Tsika, G. L. and Tsika, R. W. (1999). Segregated regulatory elements direct beta-myosin heavy chain expression in response to altered muscle activity. *Journal of Biological Chemistry* 274, 14270–9.

McKoy, G., M.E. Leger, F. Bacou, and G. Goldspink. (1998). Differential expression of myosin heavy chain mRNA and protein isoforms in four functionally diverse rabbit skeletal muscles during pre- and postnatal development. *Developmental Dynamics* 211, 193–203.

Mercadier, J. J., Lompre, A. M., Wisnewsky, C., Samuel, J. L., Bercovici, J., Swynghedauw, B. and Schwartz, K. (1981). Myosin isoenzyme changes in several models of rat cardiac hypertrophy. *Circulation Research* 49, 525–32.

Miyata, S., Minobe, W., Bristow, M. R. and Leinwand, L. A. (2000). Myosin Heavy Chain Isoform Expression in the Failing and Nonfailing Human Heart. *Circ Res* 86, 386–390.

Molkentin, J. D., Jobe, S. M. and Markham, B. E. (1996). Alpha-myosin heavy chain gene regulation: delineation and characterization of the cardiac muscle-specific enhancer and muscle-specific promoter. *Journal of Molecular and Cellular Cardiology* 28, 1211–25.

Molkentin, J. D., Kalvakolanu, D. V. and Markham, B. E. (1994). Transcription factor GATA-4 regulates cardiac muscle-specific expression of the alpha-myosin heavy-chain gene. *Molecular and Cellular Biology* 14, 4947–57.

Molkentin, J. D. and Olson, E. N. (1996). Combinatorial control of muscle development by basic helix-loop-helix and MADS-box transcription factors. *Proceedings of the National Academy of Sciences of the United States of America* 93, 9366–73.

Morkin, E. (1993). Regulation of myosin heavy chain genes in the heart. *Circulation* 87, 1451–60.

Morkin, E. (2000). Control of cardiac myosin heavy chain gene expression. *Microscopy Research and Technique* 50, 522–31.

Nakao, K., Minobe, W., Roden, R., Bristow, M. R. and Leinwand, L. A. (1997). Myosin Heavy Chain Gene Expression in Human Heart Failure. *J. Clin. Invest.* 100, 2362–2370.

Nakayama, M., J. Stauffer, J. Cheng, S.Banerjee-Basu, E. Wawrousek, and A. Buonanno. (1996). Common core sequences are found in skeletal muscle Slow- and Fast-Fiber-Type-Specific regulatory elements. *Mol. Cell Biol.* 16, 2408–2417.

Ojamaa, K., Samarel, A. M. and Klein, I. (1995). Identification of a contractile-responsive element in the cardiac alpha-myosin heavy chain gene. *Journal of Biological Chemistry* 270, 31276–81.

Olson, E. N. and Williams, R. S. (2000). Remodeling muscles with calcineurin. *Bioessays* 22, 510–9.

Pette, D., Peuker, H. and Staron, R. S. (1999). The impact of biochemical methods for single muscle fibre analysis. *Acta Physiologica Scandinavica* 166, 261–77.

Pette, D. and Staron, R. S. (1990). Cellular and molecular diversities of mammalian skeletal muscle fibers. *Reviews of Physiology Biochemistry and Pharmacology* 116, 1–76.

Pette, D. and Staron, R. S. (2000). Myosin isoforms, muscle fiber types, and transitions. *Microscopy Research and Technique* 50, 500–9.

Pette, D. and Staron, R. S. (2001). Transitions of muscle fiber phenotypic profiles Myosin isoforms, muscle fiber types, and transitions. *Histochemistry and Cell Biology* 115, 359–72.

Reiser, P. J., Portman, M. A., Ning, X.-H. and Moravec, C. S. (2001). Human cardiac myosin heavy chain isoforms in fetal and failing adult atria and ventricles. *Am J Physiol Heart Circ Physiol* 280, H1814–1820.

Rindt, H., Knotts, S. and Robbins, J. (1995). Segregation of cardiac and skeletal muscle-specific regulatory elements of the beta-myosin heavy chain gene. *Proceedings of the National Academy of Sciences of the United States of America* 92, 1540–4.

Rivero, J. L., Talmadge, R. J. and Edgerton, V. R. (1999). Interrelationships of myofibrillar ATPase activity and metabolic properties of myosin heavy chain-based fibre types in rat skeletal muscle. *Histochemistry and Cell Biology* 111, 277–87.

Rottman, J. N., Thompson, W. R., Nadal-Ginard, B. and Mahdavi, V. (1990). Myosin heavy chain gene expression: interplay of cis and trans factors determines hormonal and tissue specificity. New York: de Gruyter.

Saez, L. J., Gianola, K. M., McNally, E. M., Feghali, R., Eddy, R., Shows, T. B. and Leinwand, L. A. (1987). Human cardiac myosin heavy chain genes and their linkage in the genome. *Nucleic Acids Research* 15, 5443–59.

Samarel, A. M., Spragia, M. L., Maloney, V., Kamal, S. A. and Engelmann, G. L. (1992). Contractile arrest accelerates myosin heavy chain degradation in neonatal rat heart cells. *American Journal of Physiology* 263, C642–52.

Samuel, J. L., Dubus, I., Farhadian, F., Marotte, F., Oliviero, P., Mercadier, A., Contard, F., Barrieux, A. and Rappaport, L. (1995). Multifactorial regulation of cardiac gene expression: an in vivo and in vitro analysis. *Annals of the New York Academy of Sciences* 752, 370–86.

Sartorius, C. A., B. D. Lu, L. Acakpo-Satchivi, R. P. Jacobsen, W.C. Byrnes, and L.A. Leinwand. (1998). Myosin heavy chains IIa and IId are functionally distinct in the mouse. *Journal of Cell Biology* 141, 943–953.

Schiaffino, S., Murgia, M., Serrano, A. L., Calabria, E. and Pallafacchina, G. (1999). How is muscle phenotype controlled by nerve activity? *Ital J Neurol Sci* 20, 409–12.

Schiaffino, S. and Reggiani, C. (1996). Molecular diversity of myofibrillar proteins: gene regulation and functional significance. *Physiological Reviews* 76, 371–423.

Serrano, A. L., Murgia, M., Pallafacchina, G., Calabria, E., Coniglio, P., Lomo, T. and Schiaffino, S. (2001). Calcineurin controls nerve activity-dependent specification of slow skeletal muscle fibers but not muscle growth. *PNAS* 98, 13108–13113.

Seward, D. J., Haney, J. C., Rudnicki, M. A. and Swoap, S. J. (2001). bHLH transcription factor MyoD affects myosin heavy chain expression pattern in a muscle-specific fashion. *Am J Physiol Cell Physiol* 280, C408–413.

Staron, R. S. (1997). Human skeletal muscle fiber types: delineation, development, and distribution. *Can. J. Appl. Physiol.* 22, 307–327.

Subramaniam, A., Gulick, J., Neumann, J., Knotts, S. and Robbins, J. (1993). Transgenic analysis of the thyroid-responsive elements in the alpha-cardiac myosin heavy chain gene promoter. *Journal of Biological Chemistry* 268, 4331–6.

Subramaniam, A., Jones, W. K., Gulick, J., Wert, S., Neumann, J. and Robbins, J. (1991). Tissue-specific regulation of the alpha-myosin heavy chain gene promoter in transgenic mice. *Journal of Biological Chemistry* 266, 24613–20.

Swoap, S. J. (1998). In vivo analysis of the myosin heavy chain IIB promoter region. *American Journal of Physiology* 274, C681–7.

Swoap, S. J., Haddad, F., Bodell, P. and Baldwin, K. M. (1994). Effect of chronic energy deprivation on cardiac thyroid hormone receptor and myosin isoform expression. *American Journal of Physiology* 266, E254–60.

Swoap, S. J., Hunter, R. B., Stevenson, E. J., Felton, H. M., Kansagra, N. V., Lang, J. M., Esser, K. A. and Kandarian, S. C. (2000). The calcineurin-NFAT pathway and muscle fiber-type gene expression. *Am J Physiol Cell Physiol* 279, C915–24.

Swynghedauw, B. (1999). Molecular mechanisms of myocardial remodeling. *Physiological Reviews* 79, 215–62.

Takeda, S., D.L. North, M.M. Lakich, S.D. Russell, and R.G. Whalen. (1992). A possible regulatory role for conserved promotor motifs in an adult-specific muscle myosin gene from mouse. *J. Biol. Chem.* 267, 16957–16967.

Takeda, S., D.L. North, T. Diagana, Y. Miyagoe, M.M. Lakich, and R.G. Whalen. (1995). Myogenic regulatory factors can activate TATA-containing promotor elements via an E-box independent mechanism. *J. Biol. Chem.* 270, 15664–15670.

Talmadge, R. J. (2000). Myosin heavy chain-isoform expression following reduced neuromuscular activity: potential regulatory mechanisms. *Muscle and Nerve* 23, 661–679.

Termin, A., R.S. Staron, and D. Pette. (1989). Myosin heavy chain isoforms in histochemically defined fiber types of rat muscle. *Histochemistry* 92, 453–457.

Tsika, G., Ji, J. and Tsika, R. (2004). Sp3 Proteins Negatively Regulate {beta} Myosin Heavy Chain Gene Expression during Skeletal Muscle Inactivity. *Mol. Cell. Biol.* 24, 10777–10791.

Tsika, G. L., Wiedenman, J. L., Gao, L., McCarthy, J. J., Sheriff-Carter, K., Rivera-Rivera, I. D. and Tsika, R. W. (1996). Induction of beta-MHC transgene in overloaded skeletal muscle is not eliminated by mutation of conserved elements. *American Journal of Physiology* 271, C690–9.

Tsika, R. W., McCarthy, J., Karasseva, N., Ou, Y. and Tsika, G. L. (2002). Divergence in species and regulatory role of beta -myosin heavy chain proximal promoter muscle-CAT elements. *Am J Physiol Cell Physiol* 283, C1761–1775.

Vyas, D. R., McCarthy, J. J., Tsika, G. L. and Tsika, R. W. (2001). Multiprotein complex formation at the beta myosin heavy chain distal muscle CAT element correlates with slow muscle expression but not mechanical overload responsiveness. *Journal of Biological Chemistry* 276, 1173–84.

Wasserman, W. W. and Fickett, J. W. (1998). Identification of regulatory regions which confer muscle-specific gene expression. *Journal of Molecular Biology* 278, 167–181.

Wasserman, W. W., Palumbo, M., Thompson, W., Fickett, J. W. and Lawrence, C. E. (2000). Human-mouse genome comparisons to locate regulatory sites. *Nat Genet* 26, 225–8.

Weiss, A. and Leinwand, L. A. (1996). The mammalian myosin heavy chain gene family. *Annual Review of Cell and Developmental Biology* 12, 417–39.

Weiss, A., McDonough, D., Wertman, B., Acakpo-Satchivi, L., Montgomery, K., Kucherlapati, R., Leinwand, L. and Krauter, K. (1999). Organization of human and mouse skeletal myosin heavy chain gene clusters is highly conserved. *Proceedings of the National Academy of Sciences of the United States of America* 96, 2958–63.

Weydert, A., Daubas, P., Lazaridis, I., Barton, P., Garner, I., Leader, D. P., Bonhomme, F., Catalan, J., Simon, D., Guenet, J. L. et al. (1985). Genes for skeletal muscle myosin heavy chains are clustered and are not located on the same mouse chromosome as a cardiac myosin heavy chain gene. *Proceedings of the National Academy of Sciences of the United States of America* 82, 7183–7.

Wheeler, M. T., E.C. Snyder, M.N. Patterson, and S.J. Swoap. (1999). An E-box within the MHC IIB gene is bound by MyoD and is required for gene expressioin in fast muscle. *Am. J. Physiol.* 276, C1069-C1078.

Wolff, J. A., Ludtke, J. J., Acsadi, G., Williams, P. and Jani, A. (1992). Long-term persistence of plasmid DNA and foreign gene expression in mouse muscle. *Human Molecular Genetics* 1, 363–9.

Wright, C. E., Bodell, P. W., Haddad, F., Qin, A. X. and Baldwin, K. M. (2001). In vivo regulation of the beta-myosin heavy chain gene in hypertensive rodent heart. *Am J Physiol Cell Physiol* 280, C1262–76.

Wright, C. E., Haddad, F., Qin, A. X., Bodell, P. W. and Baldwin, K. M. (1999). In vivo regulation of beta-MHC gene in rodent heart: role of T3 and evidence for an upstream enhancer. *American Journal of Physiology* 276, C883–91.

Wu, H., Naya, F. J., McKinsey, T. A., Mercer, B., Shelton, J. M., Chin, E. R., Simard, A. R., Michel, R. N., Bassel-Duby, R., Olson, E. N. et al. (2000). MEF2 responds to multiple calcium-regulated signals in the control of skeletal muscle fiber type. *Embo Journal* 19, 1963–73.

Yazaki, Y., Tsuchimochi, H., Kurabayashi, M. and Komuro, I. (1989). Molecular adaptation to pressure overload in human and rat hearts. *Journal of Molecular and Cellular Cardiology* 21 Suppl 5, 91–101.

CHAPTER 4

SIGNALING PATHWAYS CONTROLLING MUSCLE FIBER SIZE AND TYPE IN RESPONSE TO NERVE ACTIVITY

STEFANO SCHIAFFINO, MARCO SANDRI AND MARTA MURGIA
Department of Biomedical Sciences, University of Padova, CNR Institute of Neurosciences, Venetian Institute of Molecular Medicine and Dulbecco Telethon Institute, Italy

The plasticity of skeletal muscle, i.e. the ability to change the size of its constituent fibers (muscle hypertrophy or atrophy) or their type (switching from a fast-glycolytic to a slow-oxidative phenotype or vice versa), is mainly dictated by use or disuse, namely by the amount and pattern of muscle activity that results from i) motor neuron firing patterns and ii) mechanical loading conditions. The effect of activity is clearly illustrated by the dramatic changes in muscle fiber size and type induced by specific training protocols in athletes or by forced inactivity due to bone fractures or by muscle unloading in a microgravity environment as in space flights. Changes in muscle fiber size and type affect muscle functional properties, including i) the amount of force generated during contraction, which is mainly determined by muscle size, ii) the speed of shortening, that reflects myosin isoform composition, and iii) the resistance to fatigue, that depends on the energetic potential primarily determined by the mitochondrial oxidative enzymes.

In this chapter, we will discuss the mechanisms that mediate the effect of nerve activity on the skeletal muscle phenotype, including the intracellular signaling pathways activated by specific patterns of action potential at the plasma membrane and the transcription factors and transcriptional co-activators/co-repressors that are responsible for the final control of muscle gene expression. This process can be called excitation-transcription coupling by analogy to excitation-contraction coupling and likely involves a variety of pathways triggered by various steps along the excitation-contraction coupling itself as well as by post-contraction events (Fig. 1). One can thus envisage activity-dependent signaling pathways triggered by changes in intracellular Ca^{2+}, such as calcineurin and calcium/calmodulin-dependent protein kinases (CaMKs), pathways triggered by mechanical effects of muscle contraction, such as titin kinase and possibly protein kinase B (PKB)/Akt,

91

R. Bottinelli and C. Reggiani (eds.), Skeletal Muscle Plasticity in Health and Disease, 91–119.
© 2006 *Springer.*

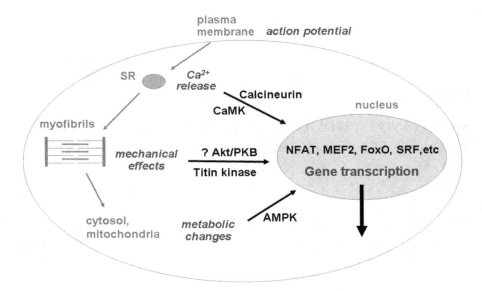

Figure 1. The scheme illustrates some of the pathways that might mediate the effect of nerve activity on muscle gene regulation. These include pathways triggered by changes in i) intracellular Ca^{2+} concentration, such as calcineurin and calcium/calmodulin-dependent protein kinase (CaMK), ii) mechanical effects of muscle contraction, titin kinase and possibly Akt /PKB, and iii) metabolic post-contraction events, such as the increase in intracellular AMP, caused by ATP consumption and adenylate kinase activity ($2ADP \rightarrow ATP + AMP$), and consequent activation of AMP-activated protein kinase (AMPK). Different transcription factors, including the nuclear factor of activated T cells (NFAT), MEF2, FoxO and the serum response factor (SRF) represent the final effectors controlling gene transcription

and pathways triggered by metabolic changes induced by contraction, such as AMP kinase (AMPK). In this chapter we will mostly focus on those activity-dependent pathways that have been clearly shown to affect muscle fiber size and type, with particular reference to the calcineurin and Akt pathways.

1.   EXPERIMENTAL APPROACHES TO IDENTIFY ACTIVITY-DEPENDENT SIGNALING PATHWAYS IN SKELETAL MUSCLE

The signaling pathways that mediate the effect of nerve activity on muscle fiber size and type have been intensively investigated during the last few years. Rapid progress in this field is mainly due to the introduction of genetic approaches to either block or stimulate specific signaling pathways both *in vitro* and *in vivo*. As usual in biological problems, many variables, including unpredictable variables, can affect the result of any single experimental approach, therefore multiple approaches are required to conclusively demonstrate the role of a given signaling pathway. Let us consider a simple situation in which a specific kinase, say kinase X, is postulated to affect nerve activity-dependent gene expression in skeletal muscle by inducing the translocation into the nucleus of a transcription factor Y, that binds to the promoter of an activity-dependent target muscle gene Z, and stimulates its transcription. To validate the role of the X-Y pathway in regulating gene Z, one would like to collect the following evidence:

a) The level of activity of kinase X, as determined by enzymatic assays or by immunoblotting with antibodies specific for the active form of X, should be compatible with the proposed role, namely should be increased by muscle activity, when the transcription of the target gene Z is stimulated, and decreased by inactivity.

b) Gain-of-function approaches should show that overexpression of a constitutively active mutant of kinase X causes activation of transcription factor Y (see *d* below) and up-regulation of target gene Z. A caveat here is that non-physiological super-activation of kinase X can stimulate or repress downstream effectors that are not affected under physiological conditions, thus leading to false conclusions.

c) Loss-of-function approaches should be used to block kinase X activity: the block can be induced at the DNA level, as in knockout mice, or at the RNA level, using RNA interference (RNAi) techniques, or at the protein level, by overexpressing dominant negative mutants or inhibitory peptides. These procedures, leading to kinase X inhibition, should block the activation of transcription factor Y and down-regulation of target gene Z.

d) The response of transcription factor Y should be monitored by showing that in response to muscle activity i) Y translocates from the cytoplasm to the nucleus, as determined by specific antibodies or by transfection with a Green Fluorescent Protein (GFP) fusion protein, and ii) Y becomes transcriptionally active, as determined by co-transfection with appropriate reporter genes, e. g. luciferase driven by the promoter of the target gene Z or by an artificial promoter consisting of specific DNA binding sites for Y. Chromatin immunoprecipitation (ChIP) assays can also be used to demonstrate that muscle activity leads to *in vivo* binding of transcription factor Y to the promoter of target gene Z.

e) Gain-of-function and loss-of-function approaches, similar to those described above, should be used to validate the role of transcription factor Y.

These various approaches can be applied to cultured muscle cells or to *in vivo* systems. Studies in cultured muscle cells allow to analyze pure muscle cell populations, as the myogenic cell line C2C12, to obtain highly efficient gene transfer using adenoviral vectors and to complement genetic with pharmacological approaches to block the various pathways. However, a serious limitation of *in vitro* studies is that the electrical and mechanical conditions that characterize muscle activity *in vivo* cannot be reproduced in culture. A powerful *in vivo* approach is the generation of knockout mice, in which specific genes are selectively inactivated, and transgenic mice that overexpress a component of a signaling pathway or a constitutively active mutant or an inhibitory factor. The role of specific genes in mediating the effect of muscle activity can thus be studied in a physiological *in vivo* context. But a problem with such mouse models is that germline mutations might have effects other than somatic mutations, owing to developmental compensation (Sage et al., 2003). In addition, some mutations can lead to early embryonic lethality, preventing studies in adult animals. These problems can be circumvented by using conditional knockout mice in which genes can be inactivated in a spatially and temporally controlled manner, however this is a complex and time consuming method.

An alternative approach is the generation of "transgenic muscles" through the intramuscular injection of plasmids coding for constitutively active mutants of specific signal transducers or dominant negative mutants or small interfering RNAs (siRNAs) to knockdown the expression of specific genes. Intramuscular DNA injection is sufficient to induce gene transfer in regenerating muscle (Vitadello et al., 1994), whereas DNA injection followed by electroporation is required to obtain efficient delivery of plasmids in adult skeletal muscle. When electroporation is applied, analyses are usually performed at least one week after transfection, to avoid changes caused by local damage and inflammation (Bertrand et al., 2003). The advantage of this approach is that signaling pathways are selectively perturbed in adult muscle, thus reducing the possibility of compensatory changes. A limitation of this method is that only a proportion of the fibers present in the injected muscle are transfected, thus precluding biochemical studies of transfected muscles. However, transfected fibers can be identified using tagged constructs or by co-transfecting plasmids coding for GFP: the effect of activity changes in transfected *versus* untransfected fibers can thus be compared within the same muscle. In addition, it is possible to determine the effect of the induced perturbations on specific transcription factors by co-transfecting plasmids coding for appropriate reporters. This "transgenic muscle" approach has been applied in our laboratory to identify signal transduction pathways and transcription factors responsible for activity-dependent changes in the muscle phenotype. A variant of this approach is the transfection of the regenerating rat soleus, a typical slow-twitch muscle, either in the presence or in the absence of innervation (Fig. 2). In the absence of the nerve the muscle remains atrophic, the slow myosin is not expressed and the soleus muscle undergoes a "default" differentiation to a fast-like phenotype, while in the presence of the nerve the regenerating soleus muscle grows rapidly and acquires in a few days a predominantly slow profile (Jerkovic et al., 1997). The effect of innervation can be reproduced by electrical stimulation of the denervated muscle with an impulse pattern mimicking the low frequency trains typical of the slow motor neurons (Kalhovde et al., 2005). The regenerating soleus thus provides a good model to examine the effect of selective perturbations of signaling pathways on muscle fiber size and type in response to nerve activity. Using this model, as well as transfection experiments in adult muscles, we have been able to demonstrate the role of the Ras-ERK and calcineurin-NFAT pathways in the induction and maintenance of the slow phenotype (Murgia et al., 2000; Serrano et al., 2001; McCullagh et al., 2004), and the role of the Akt-mTOR and Akt-FoxO pathways in muscle atrophy and hypertrophy (Pallafacchina et al., 2002; Sandri et al., 2004).

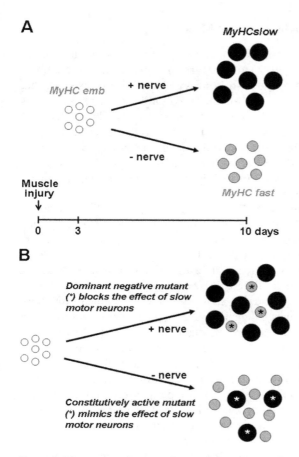

Figure 2. The regenerating rat soleus muscle as a model to identify activity-dependent signaling pathways in skeletal muscle. A: Bupivacaine (marcaine) injection causes necrosis of muscle fibers and consequent activation of satellite cells leading to muscle regeneration. At 3 days after injury the regenerating soleus consists of small myotubes expressing embryonic myosin heavy chain (MyHC). During subsequent days in response to slow motor neuron activity the regenerating fibers grow in size and start to express MyHC-slow but in the absence of innervation muscle growth is blocked and fibers undergo a default differentiation leading to MyHC-fast gene expression. B: Transfection at day 3 with a dominant negative mutant of a component of a crucial signaling pathway blocks the effect of slow motor neuron activity in transfected fibers, labeled with asterisks, of the innervated regenerating soleus. On the other hand transfection of the regenerating denervated soleus with a constitutively active mutant induces muscle growth and upregulation of the MyHC-slow gene in transfected fibers

2.   SIGNALING PATHWAYS INVOLVED
IN ACTIVITY-DEPENDENT CHANGES IN MUSCLE
FIBER TYPE

2.1   Fiber Types in Mammalian Skeletal Muscle

Mammalian skeletal muscles are composed of two major fiber types with different physiological properties, slow (type 1) and fast (type 2) fibers. Type 1 and 2 fibers are usually identified according to the myosin heavy chain (MyHC) isoform that they express, but differ also with respect to other myofibrillar proteins, as well as sarcoplasmic reticulum, mitochondrial and cytosolic components. The existence of distinct fast and slow gene programs is supported by whole-genome gene expression analysis: a recent study identified 70 genes differentially expressed by 3-fold or greater in slow soleus vs. fast EDL muscle (Wu et al., 2003). Type 1 fibers have a slow velocity of shortening but are resistant to fatigue. Type 2 fibers contract rapidly and therefore generate more power, but have low resistance to fatigue. MyHC isoforms are the main determinants of the contractile properties. The metabolic profile of the fibers ultimately determines their fatigability: slow fibers are rich in mitochondria and have a predominantly oxidative metabolism, while fast fibers rely on glycolysis for the production of ATP. Type 1 fibers also appear red in a freshly cut muscle, due to their enrichment in myoglobin, a heme containing protein that serves as oxygen storage. Type 2 fibers can be further divided into three subtypes, 2A, 2X (also called 2D) and 2B, showing graded functional properties, with 2B being the fastest and most fatigable and 2A being relatively slower but with an oxidative metabolism and rich in mitochondria (Fig. 3) (Schiaffino and Reggiani, 1996). Furthermore, intermediate fiber types containing more than one MyHC isoform (e.g. slow and 2A, or 2A and 2X, or 2X and 2B) are normally present in adult muscles. The relative distribution of the different fiber types varies between mammalian species and only two fast fiber types, corresponding to the 2A and 2X types, are found in most human skeletal muscles (Smerdu et al., 1994).

The composition in slow and fast fibers of a given muscle is the result of developmental instructions and activity-dependent plasticity. The identity of muscle fibers is specified through two waves of myogenic differentiation during embryonic development (Schiaffino and Reggiani, 1996). In the mouse embryo primary myofibers are formed around embryonic day 12 (E12), prior to the onset of innervation, which occurs around E14. Primary myotubes can mature into either slow or fast muscle fibers, independent of innervation but depending on the specific muscle context. The mechanisms responsible for this early fiber type diversification in mammalian skeletal muscle have not yet been identified. After E14, secondary myofibers are formed which colonize the muscles and progressively acquire the identity of mature fibers. Most adult muscle fibers derive from secondary fibers. Two embryo-specific MyHC isoforms, MyHC-embryonic and MyHC-neonatal/perinatal are progressively replaced by adult 2A, 2X and 2B isoforms during the first weeks after birth in rat skeletal muscle (DeNardi et al., 1993). At this point, polyneuronal innervation disappears and, in mature motor units, nerve-muscle interactions

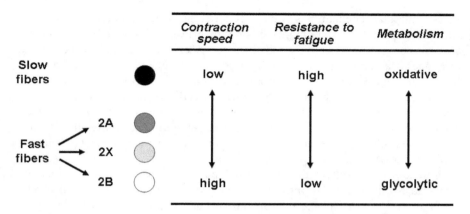

Figure 3. The spectrum of fiber types in mammalian skeletal muscle with the corresponding major functional and metabolic properties

become instructive (Lømo, 2003). The pattern of nerve activity acting at the plasma membrane is precisely translated into the coordinated regulation of complex fiber type-specific gene programs in the nucleus. The role of the nerve on muscle properties was first demonstrated by nerve cross-union studies (Buller et al., 1960) and the role of electrical patterns of impulses by chronic electrical stimulation experiments (Salmons and Vrbova, 1967). A large number of stimulation experiments, using impulse patterns mimicking the firing pattern of slow and fast motor neurons and applied either onto the nerve or directly onto denervated muscles, have shown that the muscle fiber type composition and physiological properties can be partly changed by electrical activity (Pette, 2001). The changes in MyHC gene expression generally follow the sequence MyHC-slow ↔ MyHC-2A ↔ MyHC-2X ↔ MyHC2B (Schiaffino and Reggiani, 1994; Pette and Staron, 1997). However, the transformation of the muscle fiber type induced by electrostimulation is usually incomplete due to intrinsic differences between muscles.

A network of intracellular signals, that are beginning to be understood, is involved in mediating the effect of nerve activity on muscle gene regulation. Most published evidence point to one nerve activity-dependent signaling pathway as the main control of MyHC switching and fiber type, the calcium-calcineurin-NFAT pathway. In addition, some major determinants of the metabolic oxidative profile of muscle fibers, the peroxisome proliferator activated receptor-γ coactivator-1α (PGC-1α), the peroxisome proliferator activated receptor δ (PPARδ) and the transcription factor myogenin, have been identified. These mechanisms of fiber type specification and their potential connections will be discussed separately.

2.2   The Calcineurin Pathway

The firing pattern of slow and fast motor neurons differs in frequency and duration. Accordingly, the frequency of the action potentials and of calcium release from the sarcoplasmic reticulum is significantly different in slow and fast fibers. Mean cytosolic calcium levels are higher in slow than in fast fibers and, although this issue has not been systematically addressed, it is likely that this fundamental second messenger is a major trigger of activity-dependent fiber type specification. Indeed, the signaling mediated by calcineurin, a serine/threonine phosphatase activated by calcium-calmodulin, has been shown to be a key activator of the slow muscle gene program. As required for an intracellular mediator of slow motor neuron activity, calcineurin can be activated by long-lasting low-amplitude increases in cytosolic calcium, like those predicted for slow but not fast nerve stimulation (Olson and Williams, 2000).

Calcineurin is a heterodimer of a catalytic (CnA) and a regulatory subunit (CnB), each with various isoforms transcribed from different genes or produced by alternative splicing. Calcineurin is ubiquitous and highly conserved in evolution. A fundamental regulatory cue of calcineurin is the presence of a calmodulin binding and a pseudo-substrate autoinhibitory domain, which allows exposure of the catalytic domain upon calmodulin binding. Calcineurin is the molecular target of the immunosuppressive drugs cyclosporin A (CsA) and FK506. The first observation that calcineurin plays a role in the control of slow skeletal muscle genes was obtained by overexpressing active calcineurin in cultured muscle cells (Chin et al., 1998). Activated calcineurin caused activation of slow fiber-specific gene program, while the calcineurin inhibitor CsA caused slow to fast fiber transition. More detailed analysis of this phenomenon in adult rat muscle has revealed that calcineurin inhibition causes decrease of MyHC slow and 2A with an enrichment in fibers containing MyHC 2X and 2B, i.e. a net switch towards a fast fatigable phenotype. Several studies based on transgenic mice overexpressing calcineurin and on calcineurin knockout mice, as well as on transfection experiments in adult muscles, have also reinforced the notion that calcineurin activation triggers a slow muscle transcriptional program while repressing the fast muscle gene program (Fig. 4) (Schiaffino and Serrano, 2002).

Calcineurin activity is modulated not only by calmodulin but also by a number of physiological inhibitory proteins, that control the intensity and duration of the signal. Among such endogenous inhibitory proteins are cain (*alias* cabin-1), calsarcins and RCAN1 (regulator of calcineurin 1, *alias* DSCR1/MCIP1/calcipressin). Interestingly, calcineurin controls transcription of its inhibitor MCIP1, which gives rise to a stimulation-dependent feedback loop. In a recent study, MCIP1 has been overexpressed specifically in the skeletal muscle of a transgenic mouse line under the control of the myogenin promoter, which is activated from embryonic day 8 throughout adult life (Oh et al., 2005). In these mice, therefore, calcineurin activity was inhibited from the very beginning of muscle differentiation. Transgenic mice show completely normal muscles at birth, with a fast/slow fiber composition and MyHC content indistinguishable from that of controls. Surprisingly however, transgenic mice begin loosing slow fibers at post-natal day 7 and by day 14 all slow fibers have switched to 2A. These data suggest that calcineurin signaling is not required for the initial diversification of

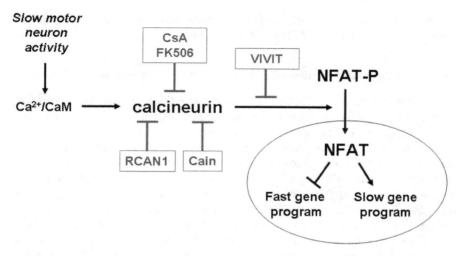

Figure 4. The scheme illustrates the role of the calcineurin-NFAT pathway in activity-dependent muscle fiber type specification. Calcineurin is activated by calcium/calmodulin in response to slow motor neuron activity and dephosphorylates NFAT inducing its translocation to the nucleus, where NFAT activates the slow gene program and blocks the fast gene program. Various inhibitors of this pathway are also indicated, including the drugs cyclosporine A (CsA) and FK506, the endogenous protein inhibitors Cain (*alias* Cabin1) and RCAN1 (*alias* DSCR1/ MCIP1/calcipressin) and the NFAT peptide inhibitor, VIVIT

fast and slow fibers in the embryo but is necessary for the nerve activity-dependent maintenance of slow fibers in the adult life.

The calcineurin-interacting proteins calsarcins are also relevant for skeletal muscle fiber type diversification. These proteins have been shown to co-localize with calcineurin at the Z-disc of the myofibrils and their expression pattern is fiber type-specific, with calsarcin-1 being expressed in slow fibers and heart and calsarcin-2 and −3 in fast fibers. The role of these proteins as negative modulators of calcineurin activity is demonstrated by the finding that calsarcin-1 deficient mice show an increase in calcineurin activity and in the number of slow fibers (Frey et al., 2004).

Several studies using gain-of-function and loss-of-function approaches have also contributed to define the role of calcineurin in slow muscle fibers. Mice overexpressing calcineurin show an increased number of slow fibers (Naya et al., 2000). On the other hand, CnAα or CnAβ null mice show a down-regulation of the slow gene program in skeletal muscles (Parsons et al., 2003). Likewise, mice with muscle-specific deletion of the calcineurin regulatory subunit CnB show markedly impaired fast-to-slow fiber type switching following functional overload (Parsons et al., 2004). While the role of calcineurin in nerve activity-dependent fiber type specification is well established, there are controversies concerning a role of calcineurin in muscle hypertrophy (Schiaffino and Serrano, 2002). Recent studies show a reduced growth of the slow soleus but not of fast muscles in calcineurin Aβ

null mice (Parsons et al., 2003) and in mice with muscle-specific overexpression
of the calcineurin inhibitor MCIP1 (Oh et al., 2005).

2.3   Downstream Effectors of Calcineurin: NFAT Transcription Factors

Calcineurin influences gene expression in the myofibers mainly by dephospho-
rylating a family of transcription factors denominated nuclear factor of activated
T cells (NFAT), which then translocate to the nucleus and bind to specific sequences
on the promoters of target genes (Crabtree and Olson, 2002; Hogan et al., 2003).
This gene family is composed of five members, four of which activated by
calcineurin, namely NFATc1, -c2, -c3 and -c4 (also called, respectively, NFAT2/c,
-1/p, -4/x and -3). The fifth member, NFAT5, is not calcium-calcineurin regulated
and responds to osmotic stress. NFATc1, -c2 and -c3 also display splice variants and
multiple mRNAs transcribed from alternative promoters. The interaction of these
transcription factors with calcineurin depends on a specific consensus sequence at
the N-terminus of NFAT, PxIxIT, where x can be any amino acid. This sequence
has been used to generate a competitive peptide inhibitor, VIVIT, which blocks
the interaction of all NFATs with calcineurin (Aramburu et al., 1999). The 16mer
VIVIT peptide, fused to GFP, has been successfully used as an inhibitor of NFAT
signaling in many cell systems.

The defining structural motif of this family is the regulatory domain, NFAT
homology region, which contains multiple regulatory serine residues. It has been
show that 13 of these serines are dephosphorylated by calcineurin and some of
them unmask a nuclear localization signal causing nuclear import. Without sufficient calcineurin activation, the critical serine residues are kept phosphorylated
in the cytosol by so-called maintenance kinases, two of which are CK1 and
glycogen-synthase kinase 3 or GSK3 (Okamura et al., 2004). Once dephosphorylated and imported into the nucleus, NFATs are eventually rephosphorylated by
export kinases, including GSK3, JNK and p38, which lead to nuclear export. An
important aspect of this complex activation mechanism is that partial dephosphorylation of NFAT can be sufficient for nuclear translocation but not for full transcriptional activity. The level of NFAT-regulated transcription, therefore, depends on the
nuclear translocation but is finely tuned by the ratio between the main input signal,
i.e. calcium-calcineurin, and other signaling pathways modulating the activity of
regulatory kinases. Once in the nucleus, NFAT proteins bind to DNA through the
so called REL homology region and cooperate with a number of transcriptional
partners. AP1 (activator protein 1), a dimer of Fos and Jun, is the best characterized
transcriptional interactor of NFAT in T cell activation; a number of other positive
and negative regulators of NFAT have been identified to date. Myocyte enhancer
factor-2 (MEF2) has also been implicated as a partner of NFAT in muscle gene
regulation (Wu et al., 2001).

NFATs have been characterized in T lymphocytes, where most seminal studies
have been conducted. These transcription factors are, however, ubiquitous and
increasing evidence shows their involvement in gene regulation in many tissues. The

analysis of the functional role of NFATs is complex, due to the existence of various genes and alternatively spliced gene products. Since all NFATs bind the same DNA consensus sequence, the question arises whether there can be functional redundancy among the members of this gene family. This issue is still open, although many regulatory events have been described in which a single NFAT isoform is necessary and sufficient. Since the relative expression of the various NFATs varies in different cell types, it is likely that some mechanisms of NFAT-dependent gene regulation are highly tissue-specific and rely on specific combinations of NFATs. Various NFAT isoforms are necessary for the development of the cardiovascular system and over-expression of either calcineurin or NFATc4 in the heart of transgenic mice leads to cardiac hypertrophy. NFATc3/NFATc4 double knockouts, on the other hand, have generalized defects in vascular patterning and angiogenesis that lead to death in early embryonic development (Crabtree and Olson, 2002). The cardiomyocytes of these mice also exhibit mitochondrial defects that impair cardiac morphogenesis and function.

The notion that NFATs may play a role in skeletal muscle plasticity is comparatively recent but growing evidence indicates that these transcription factors are involved in virtually all stages of the life cycle of a skeletal muscle fiber. Interestingly, all four calcium-dependent NFATs are expressed in adult rat skeletal muscle (Hoey et al., 1995). Studies in cultured muscle cells have shown that calcineurin promotes myoblast differentiation and that NFATc3 is activated in myoblasts before their fusion to myotubes. NFATc1 and -c2, on the other hand, can be found in the nuclei of myotubes in culture (Abbott et al., 1998).

The main evidence that NFATs play important roles in muscle development *in vivo* comes from the phenotypic analysis of knockout mice. NFATc2-null and NFATc3-null mice are viable and fertile, although their immune response is impaired. Interestingly, both mice have reduced skeletal muscle mass but this is due to completely different cellular defects. The muscles of NFATc2 null mice show smaller fiber cross-sectional areas than those of controls but the total number of fibers per muscle is not altered (Horsley et al., 2001). In contrast, fibers from NFATc3 null mice are normal in size but reduced in number with respect to controls (Kegley et al., 2001). A very small reduction in the percentage of myofibers expressing MyHC-slow was also observed in NFATc2-null mice, whereas NFATc3-null mice had largely normal fiber type composition. Taken together, these results indicate that NFATc2 and -c3 play fundamental roles in establishing the muscle architecture during development. NFAT has also been shown to play a major role in nerve-dependent fiber type specification in adult muscle. NFAT transcriptional activity is higher in slow than in fast muscles (Parsons et al., 2003; McCullagh et al., 2004) and is activated by tonic low frequency stimulation that mimics the electrical activity of slow motor neurons (McCullagh et al., 2004). Furthermore, NFAT inhibition with VIVIT, that blocks all NFAT isoforms, downregulates MyHC-slow gene expression and upregulates the fast MyHC-2X and -2B isoforms in adult rat soleus. These results support the notion that calcineurin-NFAT signaling acts as a nerve activity sensor in skeletal muscle and controls nerve activity-dependent myosin switching. Interestingly, this parallels the observation that calcineurin inhibition with MCIP1 in transgenic mice affects the maintenance of slow fibers in the adult (see above). The finding that no significant effect on fiber type composition was found in NFATc2- and NFATc3-null mice may be due to redundancy in the role of different NFAT isoforms.

Altogether, NFAT signaling seems to play a role in all phases of muscle development, from early differentiation and myotube formation to activity-dependent plasticity in the adult. A recent study has elegantly shown that a very early role of NFAT in muscle cell fusion consists in the control of interleukin-4 (IL4) secretion by newly formed myotubes (Horsley et al., 2003). NFATc2 regulates IL4 expression, which promotes myoblast fusion and leads to muscle growth. Transfection experiments with cain/cabin-1 over-expression in adult muscle indicate however that calcineurin-NFAT signaling is still necessary for the maintenance of MyHC slow post-natally, i.e. at a time where myoblast fusion is no longer occurring under normal conditions (Serrano et al., 2001; McCullagh et al., 2004). Mechanistically therefore, NFAT transcription factors could control different events at different stages of muscle differentiation. Furthermore, there are indications that these diverse stage-specific functions are controlled by different NFATs, but the attribution of a specific role to each NFAT isoform requires further investigation.

2.4   Other Signaling Pathways Affecting Muscle Fiber Type

Other signal transduction pathways have been implicated in the control of activity-dependent muscle plasticity. Their physiological significance and possible cross-talk with the calcineurin-NFAT pathway will be discussed.

Ras-MAPK. The small GTPase Ras has been shown to induce the expression of a slow muscle gene program in regenerating rat soleus muscle, thus mimicking the effect of slow motor neuron activity. When soleus, a typical slow muscle, is allowed to regenerate in the absence of the nerve, i.e. denervated when the myotoxic damage is induced, it expresses by default only the fast isoforms of MyHC (Fig. 2). When such a denervated muscle is transfected with a constitutively active Ras mutant (RasV12) during the regeneration process, however, MyHC slow is upregulated and the fast isoforms are downregulated (Murgia et al., 2000). The same effect can be observed with a double mutant of Ras (RasV12S35), which activates only the ERK-MAPK pathway, or with a constitutively active form of the MAPK kinase. Consistently, an increase in MAPK-ERK activity can be observed in regenerating innervated muscle when MyHC-slow begins to be expressed, i.e. presumably when the neuromuscular junctions of the newly formed fibers are correctly re-assembled and functional. Although these studies have not been systematically extended to adult muscle, there are indications that the Ras-MAPK pathway also plays a role in adult fiber plasticity. When adult muscles are denervated and electrostimulated for 24 hours with a pattern of activity (20 Hz) that mimics the firing pattern of slow motor neurons, a sixfold increase in MAPK-ERK activity can be observed with respect to denervated unstimulated muscles (Murgia et al., 2000). In contrast a fast nerve-like stimulation (150 Hz) did not have any effect.

The notion that both the calcineurin-NFAT and Ras-MAPK pathways regulate the expression of slow muscle genes is reminiscent of a number of gene regulation events in other cell contexts. The Ras-MAPK pathway ultimately controls the activity of AP1 (Fos/Jun dimers), which binds cooperatively with NFAT to composite DNA sites. This signaling paradigm controls fundamental physiological responses such as IL-2 production by T lymphocytes, which only occurs upon concomitant activation of both signals by independent receptors. In the heart, both signals are necessary for cell growth and, additionally, there is extensive cross talk since the MAPK-ERK pathway enhances the transcriptional activity of NFAT. The nature of the putative interaction between calcineurin and Ras signaling in skeletal muscle remains to be determined.

CaMK-MEF2. Calcium/calmodulin-dependent protein kinase IV (CaMKIV) has been shown to synergize with calcineurin in the differentiation of cultured myocytes, acting on the transcription factor MEF2 (Wu et al., 2002). However, CaMKIV is not expressed in adult skeletal muscle, thus suggesting that another member of the CaMK family may control MEF2 activity in this system. The MEF2 transcription factor family comprises four members and various splice variants, which are normally present in the nucleus but are inhibited by the interaction with class II histone deacetylases (HDACs). This enzyme activity contributes to keep chromatin in a "closed state" that inhibits transcription. CaMK stimulates myogenesis by disrupting MEF2-HDAC complexes via phosphorylation and nuclear export of HDAC (McKinsey et al., 2000a; McKinsey et al., 2000b). An indication that MEF2 might also be involved in muscle fiber plasticity comes from the observation that some transgenic lines expressing a MEF2-dependent transgene show increased expression after exercise or muscle electrical stimulation. It has been suggested, therefore, that MEF2 could be another effector of calcineurin signaling in the control of slow muscle genes (Wu et al., 2000).

Protein Kinase C (PKC). In skeletal muscle, the calcium and phospholipid-dependent PKCα and the calcium-independent PKCθ are the most abundant isoforms of this family of serine/threonine kinases (Osada et al., 1992). The activity of PKCθ has also been shown to be higher in fast than in slow muscle, thus suggesting that this signaling pathway might take part in fiber type specification (Donnelly et al., 1994). This has been confirmed in muscle fiber cultures from avian muscles using the PKC inhibitor staurosporin and over-expressing either wild type or dominant negative PKCα and PKCθ (DiMario, 2001). The expression of the avian slow isoform of MyHC, MyHC2, is blocked by over-expression of wild type PKCθ but is induced by inhibition of PKCθ. Thus, not only is the distri-bution of PKCθ fiber type-specific, but it also has a causative role in the control of myosin heavy chain isoforms. However, these findings are based on studies on cultured avian muscle cells. It remains to be determined whether this is true also in a mammalian system and whether and how muscle activity is coupled to PKCθ stimulation in adult skeletal muscle.

PPARs. The tonic activity of slow muscles is supported by a continuous energy provision. Beta-oxidation of fatty acids fuels the contraction of slow/type I fibers, which are accordingly rich in mitochondria. The metabolic phenotype of muscle fibers is also specified by developmental instructions but it can be modified in adult life, to a certain extent, by activity. Endurance training causes a shift towards a more aerobic metabolism, with an increase in mitochondrial content and in the activity of oxidative enzymes. In the last few years, some intracellular mediators have been identified that control the metabolic profile of slow fibers, namely the peroxisome proliferator-activated receptor delta (PPARδ), the peroxisome proliferator-activated receptor-gamma coactivator 1alpha (PGC-1α) and the basic helix-loop-helix (bHLH) transcription factor myogenin.

PPARs are nuclear receptors, transcription factors activated by lipids that play an important role in lipid metabolism in many tissues. PPARδ knockout mice show extensive defects in the proliferation and differentiation of many cell types, which result in reduced body size and adipose tissue mass, together with neural and skin defects. Muscle-specific over-expression of PPARδ in transgenic mice, on the other hand, causes a large increase in the number of fibers with an oxidative metabolism (Luquet et al., 2003; Wang et al., 2004). These mice also display a net reduction in body fat, caused by a reduction in the size but not in the number of adipocytes with respect to control mice. Both an increase in the oxidative capability of muscle fibers and a decrease in the fat mass are a characteristic of muscle remodelling that follows endurance training. Interestingly, the expression of PPARδ is increased, both at the RNA and protein level, after three weeks of aerobic exercise; the molecular mechanisms underlying this increase have not been elucidated. These results suggest that PPARδ could be one of the mediators of the activity-dependent increase in oxidative capability.

PGC-1α. PGC-1α is a transcriptional co-activator, i.e. a regulator of transcription that binds transcription factors but not DNA directly, which was identified and cloned as an interactor of PPARγ in brown adipose tissue (Puigserver and Spiegelman, 2003; Gabellini et al., 2005). PGC-1α is expressed in slow but not in fast muscles and is readily induced by exercise in both rodents and humans. The transcription of PGC-1α is regulated by Ca^{2+}-dependent pathways, such as CaMK and calcineurin (Handschin et al., 2003), and also by energy deprivation via AMPK activity (Zong et al., 2002). When ectopically expressed in fast fibers, PGC-1α activates mitochondrial biogenesis and the synthesis of oxidative enzymes and makes the muscles more resistant to fatigue (Gabellini et al., 2005). Conversely, PGC-1α-deficient mice are exercise intolerant and slow muscles from these mice have diminished mitochondrial number and respiratory capacity (Leone et al., 2005). The promoter region of many mitochondrial genes encoded in the nucleus contains consensus sequences for the transcription factors nuclear respiratory factor (NRF)-1 and 2, which are controlled by PGC-1α at two levels: i) PGC1α increases the expression of both NRF-1 and -2; ii) PGC-1α co-activates the transcriptional activity of NRF-1 by physically interacting with it. NRF-1 and -2 are in turn able to stimulate the expression of the mitochondrial transcription factor A (Tfam), a mitochondrial matrix protein essential for the replication and transcription of mitochondrial DNA. A homologue of PGC-1α, named PGC-1β, is expressed at high levels in skeletal and cardiac muscle and has been shown to exert similar functions.

AMPK. The AMP-activated protein kinase (AMPK) is an evolutionarily conserved sensor of cellular energy status, which is activated by conditions leading to ATP consumption and consequent AMP increase (Kahn et al., 2005). AMPK, which consists of three subunits α, β and γ, is activated by contractile activity in skeletal muscle and appears to mediate some of the adaptive changes induced by exercise. Experiments with transgenic mice expressing a dominant negative mutant of AMPK showed that AMPK is required for mitochondrial biogenesis in skeletal muscle in response to chronic energy deprivation (Zong et al., 2002). Chronic treatment with a drug called AICAR, that is converted within the cells to an AMP analogue and thus leads to AMPK activation, was likewise found to induce increased activities of mitochondrial oxidative enzymes but no change in MyHC composition (Putman et al., 2003). The up-regulation of mitochondrial enzymes by AMPK activation might be mediated by PGC1α, whose expression is also up-regulated by AMPK activation (Kahn et al., 2005), however the exercise-induced increase in transcription of PGC1α was not affected by knockout of $\tilde{\alpha}$AMPK (Jorgensen et al., 2005).

Myogenin. The four members of the MyoD family of bHLH transcription factors, MyoD, Myf-5, myogenin and MRF4, play a fundamental role in myogenesis but their role in fiber-type-specific gene regulation in the adult remains to be established. Interestingly, when over-expressed in glycolytic fibers of adult muscles myogenin was shown to cause an increase in the content of oxidative enzymes and a shift towards a slow fiber metabolism (Hughes et al., 1999). Myogenin was unable to change the expression pattern of MyHC, thus confirming that the contractile and metabolic phenotypes tend to be regulated by different signals rather than being part of the same signaling pathway.

3. SIGNALING PATHWAYS INVOLVED IN ACTIVITY-DEPENDENT CHANGES IN MUSCLE FIBER SIZE

3.1 Determinants of Muscle Atrophy and Hypertrophy

The maintenance of skeletal muscle mass, like the mass of all tissues, depends on protein turnover and cell turnover (Sartorelli and Fulco, 2004). Protein turnover reflects the balance between dynamic anabolic reactions, namely new protein formation (protein synthesis) and catabolic reactions, namely degradation of existing proteins (protein breakdown), determining the level of muscle proteins within each muscle fiber (Glass, 2003). The physiological conditions promoting muscle growth, therefore, do so by increasing protein synthesis and/or decreasing protein degradation. Cell turnover, on the other hand, has in fact two meanings in the context of skeletal muscle tissue. One type of cell turnover, that could more appropriately be defined fiber turnover, is the loss of muscle fibers and the formation of new muscle fibers through muscle regeneration. Another type of cell turnover, that could more appropriately be defined nuclear turnover and takes place within each muscle fiber, reflects the balance between increase in muscle fiber nuclei due to proliferation of satellite cells followed by their incorporation into the fibers and loss of muscle fiber nuclei due to focal apoptosis. Protein turnover and nuclear turnover are usually co-ordinated processes so that the size of each nuclear domain (quantity of cytoplasm/number of nuclei within that cytoplasm) tends to remain constant. For example, satellite cell incorporation into the fibers takes place during post-natal muscle growth (Moss and Leblond, 1971) and compensatory hypertrophy (Schiaffino et al., 1976), concomitantly with increased protein synthesis. A discussion of satellite cells and their contribution to nuclear turnover in muscle fibers is outside the scope of this chapter (see in this book the chapter written by Gillespie, Holterman, and Rudnicki). We will discuss here the signaling pathways controlling activity-dependent protein turnover in skeletal muscle fibers and will especially focus on the kinase Akt, also called protein kinase B (PKB), which has emerged as a major evolutionarily conserved signaling pathway in the control of cell size.

3.2   The Akt/PKB Pathway

The role of Akt in muscle growth was first suggested by the finding that an active Ras double mutant (RasV12C40), that stimulates selectively the Akt pathway acting through phosphatidylinositol 3 kinase (PI3K), promotes muscle growth, thus opening new perspectives in the signaling of fiber size (Murgia et al., 2000). This observation was subsequently confirmed by the finding that muscle fiber hypertrophy is induced by over-expression of a constitutive active form of Akt after plasmid DNA injection in adult skeletal muscle (Bodine et al., 2001b; Pallafacchina et al., 2002). Interestingly, denervation atrophy is completely rescued by active Akt. Similar results were obtained with the generation of a conditional transgenic mouse in which Akt is expressed in skeletal muscle only after tamoxifen treatment (Lai et al., 2004).

In mammals there are three Akt genes, Akt1 (PKBα), Akt2 (PKBβ) and Akt3 (PKBγ), which appear to have distinct functions. In skeletal muscle Akt1 and Akt2 are expressed at higher levels compared to Akt3, which is mainly expressed in the brain. Targeted deletion experiments have shown that Akt1 null mice display growth retardation and muscle atrophy, while Akt2 null mice suffer from a type 2 diabetes-like syndrome and Akt3 null mice have impaired brain development (Yang et al., 2004). Taken together with other observations, these results suggest that Akt1 is a major mediator of skeletal muscle hypertrophy.

Once established that Akt plays a crucial role in muscle growth, it remains to be defined which are the upstream signals triggering Akt activation and the downstream targets involved in muscle hypertrophy. Exercise in vivo is associated with activation of Akt1 but not Akt2 and Akt3 in contracting muscles (Turinsky and

Damrau-Abney, 1999). Akt activity was increased in the rat plantaris after functional overload induced by elimination of synergistic muscles (Bodine et al., 2001b). Other studies in rats and humans support the notion that Akt activity is increased in response to muscle contractile activity (Nader and Esser, 2001; Sakamoto et al., 2002; Sakamoto et al., 2003; Sakamoto et al., 2004). Surprisingly, this effect was observed only in the fast EDL not in the slow soleus muscle (Sakamoto et al., 2003; Sakamoto et al., 2004). The finding that passive stretch of the fast rat EDL muscle can also induce Akt activation has suggested that mechanical tension may be a part of the mechanism by which contraction activates Akt in fast-twitch muscles (Sakamoto et al., 2003). However, it remains to be established how mechanical stress is converted to Akt activation.

Akt activity is also increased in response to hormonal and growth factor stimulation, in particular insulin is known to activate Akt2 while insulin-like growth factor 1 (IGF-1) activates Akt1. IGF-1 is among the best characterized muscle growth-promoting factor. In addition to circulating IGF-1, mainly synthesized by the liver, local production by skeletal muscle of distinct IGF-1 splicing products has recently raised considerable interest. A specific IGF-1 splicing product is postulated to mediate load- and stretch-induced adaptations in skeletal muscle (Goldspink, 1999). Increased IGF-1 gene expression has been demonstrated following functional overload induced by elimination of synergistic muscles (McCall et al., 2003). Muscle-specific over-expression in transgenic mice of an IGF1 isoform locally expressed in skeletal muscle results in muscle hypertrophy (Musaro et al., 2001). Although these results are suggestive for an autocrine/paracrine role of local IGF-1 in activity-dependent muscle plasticity, direct evidence for such a role through loss-of-function approaches, such as knockout or knockdown experiments, has not yet been reported. Akt activation is induced by growth factors, such as IGF-1, through the generation of phosphatidylinositol-3,4,5-triphosphates produced by PI3K, which is opposed by the activity of the phosphatase PTEN. Phosphatidylinositol-3,4,5-triphosphates recruit Akt to the plasma membrane by binding to its N-terminal pleckstrin homology domain. At the membrane, Akt is phosphorylated on separate residues by two distinct kinases, PDK1 and the mTOR-rictor complex.

Two major downstream branches of the Akt pathway, that are relevant to muscle plasticity, are the mTOR pathway, which is activated by Akt and controls protein synthesis, and the FoxO pathway, which is inhibited by Akt and controls protein degradation (Fig. 5). The major effectors of mTOR and FoxO involved in muscle atrophy and hypertrophy will be discussed below. A third downstream effector of Akt, glycogen synthase kinase 3β (GSK3β), has been shown to modulate hypertrophy in muscle cell culture. GSK3β is inhibited by Akt and in turn blocks the eukaryotic initiation factor 2B (eIF2B), which is involved in protein synthesis. Expression of a dominant negative, kinase inactive form of GSK3β induces hypertrophy of cultured myotubes (Rommel et al., 2001). However, it remains to be proven *in vivo* whether inhibiting the negative action of GSK3β on eIF2B is sufficient to promote muscle growth.

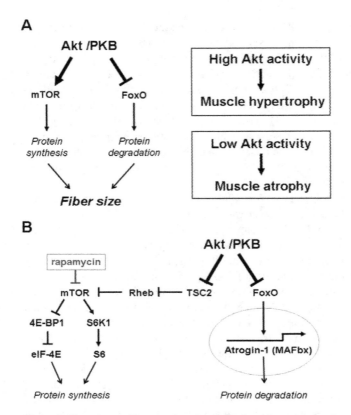

Figure 5. The schemes illustrate the role of Akt-dependent pathways in muscle fiber size regulation. A, a simplified scheme showing that, in response to muscle activity, and possibly muscle loading, Akt/PKB is activated and controls two major downstream pathways: the kinase mammalian target of rapamycin (mTOR), that regulates protein synthesis, and the transcription factor FoxO, that regulates protein degradation. Since Akt exerts a positive regulation on mTOR and a negative regulation on FoxO, increased Akt activity causes muscle hypertrophy, while decreased Akt activity causes muscle atrophy. B, a more detailed scheme showing that Akt activates mTOR, a target of the inhibitory drug rapamycin, indirectly by inhibiting tuberous sclerosis complex 2 (TSC2), a GTPase-activating protein (GAP) that inactivates the small G protein Rheb. Akt phosphorylation of TSC2 inhibits GAP activity, allowing the formation of Rheb-GTP that activates mTOR. mTOR stimulates translation initiation through the activation of the S6K1, which phosphorylates the ribosomal protein S6, and the inactivation of the 4E-BP1 protein, an inhibitor of the translation initiation factor eIF4E. Akt also inhibits FoxO, which controls protein degradation by up-regulating the muscle-specific ubiquitin ligase atrogin 1, *alias* MAFbx, a major component of the ubiquitin-proteasome system in skeletal muscle

3.3   Downstream Effectors of Akt: mTOR and Muscle Hypertrophy

The kinase mTOR (mammalian target of rapamycin) has recently emerged as a key regulator of cell growth that integrates signals from growth factors, nutrients and energy status to control protein synthesis and other cell functions

(Hay and Sonenberg, 2004). As the name implies, mTOR is selectively inhibited by rapamycin, a drug used as immunosuppressant in organ transplantation: rapamycin binds to members of the FK binding protein (FKBP) family and the complex rapamycin/FKBP binds to mTOR and blocks its activity.

The role of mTOR in muscle growth was demonstrated by *in vivo* studies showing that rapamycin blocks overload hypertrophy of adult muscle and growth of regenerating muscle (Bodine et al., 2001b; Pallafacchina et al., 2002). The activation of mTOR by Akt is indirect and involves the phosphorylation and inhibition by Akt of tuberous sclerosis complex 2 (TSC2). TSC2 is a GTPase activating protein (GAP) that functions together with TSC1 to inactivate the small G protein Rheb that in turn activates mTOR in complex with the raptor adapter protein (raptor-mTOR complex). As indicated above, mTOR can also form a complex with another adapter protein, called rictor: the rictor-mTOR complex, which has an essential role in Akt phosphorylation and activation, is rapamycin-insensitive.

The effect of mTOR via TORc1 on the translation machinery and protein synthesis is mediated by phosphorylation of the ribosomal protein S6 kinases (S6K1 and 2) and of 4E-BP1, a repressor of the cap-binding protein eIF4E. S6K1 appears to be an important effector of the Akt pathway, as muscle fibers are smaller in S6K1 null mice and their hypertrophic response to IGF-1 and to activated Akt is blunted (Ohanna et al., 2005). mTOR is also involved in transcriptional regulation, as the global gene expression profiles of cultured muscle cells treated with IGF-1 showed that most of the genes induced by IGF-1 were blocked by rapamycin treatment (Latres et al., 2005). However, the transcription factors induced by mTOR have not been identified.

3.4   Downstream Effectors of Akt: FoxO and Muscle Atrophy

A major progress in the dissection of the molecular mechanism of muscle atrophy has been the identification of two muscle-specific ubiquitin ligases, atrogin-1/MAFbx and MuRF1, that are up-regulated in different models of muscle atrophy and are responsible for the increased protein degradation through the ubiquitin-proteasome system (Bodine et al., 2001a; Gomes et al., 2001). In fact, knockout mice for either atrogin-1 or MuRF1 are resistant to denervation atrophy (Bodine et al., 2001a). A subsequent crucial step was the identification of the signaling pathway which regulates the expression of the two ubiquitin ligases. The up-regulation of atrogin-1/MAFbx and MuRF1 is blocked by Akt that functions through negative regulation of the FoxO family of transcription factors (Sandri et al., 2004; Stitt et al., 2004). FoxOs comprise three isoforms, called FoxO1, FoxO3 and FoxO4. FoxO phosphorylation by Akt promotes the export of FoxOs from the nucleus to the cytoplasm. As predicted, the reduction in the activity of the Akt pathway observed in *in vitro* models of muscle atrophy resulted in decreased levels of phosphorylated FoxO in the cytoplasm and marked increase of nuclear FoxO protein. The translocation and activity of FoxO is required for the up-regulation of atrogin-1/MAFbx and MuRF1, and FoxO3 was found to be sufficient to promote atrogin-1 expression and muscle atrophy also when transfected in skeletal muscles *in vivo* (Sandri et al., 2004). Accordingly, FoxO1 transgenic mice showed markedly reduced muscle mass and fiber atrophy, further supporting the notion that FoxO is sufficient to promote loss of muscle mass (Kamei et al., 2004). On the other hand, the knockdown of FoxO expression by RNAi is able to block the up-regulation of atrogin-1 expression during atrophy (Sandri et al., 2004). Together these findings indicate that muscle atrophy is an active process controlled by a specific signaling pathway and transcriptional program.

3.5   The Titin Kinase-MURF2-SRF Pathway: A Mechanical Sensor?

An emerging novel concept in muscle biology is that signals dependent on muscle activity and specifically on mechanical load may arise in the sarcomere, the basic unit of the contractile machinery of striated muscles, and from the sarcomere can be transmitted to the nucleus and affect gene expression (Lange et al., 2005a). The giant elastic protein titin, that spans half sarcomere extending from Z-disk to M-band and interacts with a number of muscle proteins, provides an exciting example of a sarcomeric activity-dependent signaling complex (signalosome). A unique property of titin is the presence in the M-band region of this protein of a serine/threonine kinase domain that can be induced to acquire an open active conformation by stretch and contraction (Grater et al., 2005). In active muscle cells, the titin kinase domain (TK) is linked through two zinc-finger scaffolding proteins, nbr1 and p62, to a member of the muscle-specific RING-finger proteins, MURF2 (Lange et al., 2005b). In the absence of mechanical activity, the signalosome is dissociated and MURF2 translocates to the nucleus where it can interact with the serum response transcription factor SRF, leading to nuclear export of SRF and loss of SRF-dependent gene expression. This pathway may thus control muscle growth because SRF is known to regulate muscle gene expression and conditional deletion of the SRF gene causes severe skeletal muscle hypoplasia during the perinatal period (Li et al., 2005). SRF regulates muscle gene expression by binding serum response elements (SRE) in target genes and seems to integrate different growth promoting pathways: for example, SRF is a target of Akt signaling (Wang et al., 1998) and can recruit the androgen receptor to muscle gene promoters (Vlahopoulos et al., 2005). The titin kinase-SRF pathway described above is probably just one of several links between the sarcomere and the nucleus that are beginning to emerge (Lange et al., 2005a).

3.6   Other Signaling Pathways Affecting Muscle Fiber Size

Two other signaling pathway, which are known to control muscle fiber size but whose role as activity-dependent pathways remains to be established, are the NF-κB and myostatin pathways.

NF-κB. The NF-κB transcription factors, which play a major role as mediators of immunity and inflammation, are also expressed in skeletal muscle and appear to mediate the effect of several cytokines, in particular TNFα, on muscle wasting and cachexia. In the inactive state NF-κB is sequestered in the cytoplasm by a family of inhibitory proteins called IκB. In response to TNFα, the IκB kinase (IKK) complex phosphorylates IκB resulting in its ubiquitination and proteasomal degradation; this leads to nuclear translocation of NF-κB and activation of NF-κB-mediated gene transcription. Muscle-specific over-expression of IKKβ in transgenic mice leads to severe muscle wasting, that is mediated, at least in part, by the ubiquitin-ligase MuRF1 but not by atrogin-1/MAFbx (Cai et al., 2004). On the other hand, muscle-specific inhibition of NF-κB by transgenic expression of a constitutively active IκB mutant leads to no overt phenotype, but denervation atrophy is substantially reduced. Muscle atrophy induced by hindlimb unloading is likewise abolished in mice with a knockout of p105/p50 NF-κB1 gene (Hunter and Kandarian, 2004). NF-κB is transiently activated after an acute bout of physical exercise, possibly due to increased oxidant production following muscle contraction, however it is not known whether this has any effect on activity-dependent gene regulation (Ji et al., 2004; Ho et al., 2005). Cytokine genes represent a major target of NF-κB in many cell types, and physical exercise is accompanied by increased expression and secretion of various cytokines (Ho et al., 2005). However, persistent activation of NF-κB alone in transgenic mice is not sufficient to induce the expression of these genes in skeletal muscle, suggesting that other signaling input may be required (Ho et al., 2005).

Myostatin. Myostatin, a member of the TGFβ family, is expressed and secreted predominantly by skeletal muscle and functions as a negative regulator of muscle growth. Mutations of the myostatin gene lead to muscle hypertrophy in mice and cattle, and a loss-of-function mutation in the human myostatin gene was also found to induce increased muscle mass (Lee, 2004). Muscle hypertrophy was induced in transgenic mice expressing a truncated and inactive activin receptor (ActRIIB), a type II TGFbeta receptor to which myostatin binds, or by over-expression of follistatin, an inhibitor of myostatin. It is not yet clear whether myostatin regulates fiber size by acting only on satellite cells or also on the fiber themselves. There are few studies and conflicting results on the effect of exercise on myostatin gene expression in skeletal muscle (see Coffey et al., 2006).

4.   CLINICAL IMPLICATIONS: SIGNALING PATHWAYS AS POTENTIAL THERAPEUTIC TARGETS

An understanding of the signaling pathways that control muscle atrophy/hypertrophy and fast/slow fiber type composition may have a significant impact on human health and quality of life because it can provide novel therapeutic targets for the treatment of a variety of conditions ranging from normal aging to neuromuscular disorders to chronic degenerative diseases. Indeed, insufficient physical activity is a major risk factor, together with tobacco use and poor diet, for cardiovascular disease, cancer and diabetes, the major chronic diseases that account for about two thirds of all deaths in Europe and United States (Eyre et al., 2004). Although daily physical activity is known to confer health benefits in these chronic disorders, exercise is not possible for many patients, or may even be deleterious for others. For example, in

Duchenne Muscular Dystrophy (DMD), characterized by cell membrane instability caused by the lack of dystrophin, exercise can precipitate the rupture of muscle fibers. Therefore it will be important to identify drugs that can mimic the effect of exercise by targeting muscle signaling pathways to prevent muscle atrophy, induce hypertrophy or promote appropriate muscle fiber type shifts.

Skeletal muscle atrophy and consequent loss of muscle strength, a condition referred to as sarcopenia, represents a serious problem in normal aging, especially when additional diseases or bone fractures lead to forced bed rest as in hospitalized geriatric patients. Muscle weakness has itself a deleterious influence on hip fracture incidence because of its effect on the risk of falls. To break this vicious circle that leads to physical frailty often associated with aging it is necessary to find novel pharmacological treatments to increase muscle mass and strength.

Recent studies have shown that muscle degeneration can be reduced in muscular dystrophies such as the dystrophin-deficient mdx mouse, an animal model of DMD, either by over-expressing positive regulators of muscle growth, such as IGF-1 (Barton et al., 2002), or by removing negative regulators of muscle growth, such as myostatin (Bogdanovich et al., 2002). A calcineurin-induced shift from the glycolytic/fast to the oxidative/slow fiber type profile may also be relevant to the progression of dystrophy, because in the mdx mouse slow muscle fibers are less susceptible to damage induced by eccentric contractions than fast fibers (Moens et al., 1993). It was recently reported that mdx mice crossed to mice expressing an activated calcineurin mutant show attenuated dystrophic pathology with significant reductions in the extent of central nucleation and fiber size variability (Chakkalakal et al., 2004). This effect was interpreted as the consequence of increased utrophin A expression that accompanies the transition to a more oxidative muscle fiber phenotype induced by calcineurin activity (Chakkalakal et al., 2003). It is known that utrophin can compensate for the lack of dystrophin and up-regulation of utrophin might become a major therapeutic strategy for DMD (Khurana and Davies, 2003). Clearly, to implement these promising approaches for the clinical treatment of muscular dystrophies we need a precise understanding of the pathways that control muscle growth and fiber type specification.

The beneficial role of physical exercise in preventing insulin resistance and type 2 diabetes is now well established by a large number of epidemiological studies. Endurance exercise promotes a shift from the fast/glycolytic to the slow/oxidative phenotype, thus promoting insulin sensitivity because glucose transport is greater in slow/oxidative fibers. On the other hand, insulin resistance, a major risk factor for type 2 diabetes, is increased in healthy offspring of patients with type 2 diabetes mellitus, due to a reduction in mitochondrial content, which in turn might be attributable to a reduced ratio of slow/oxidative to fast/glycolytic muscle fibers (Petersen et al., 2004; Morino et al., 2005). Experimental studies in animals support the concept that skeletal muscle reprogramming to a slow/oxidative phenotype can protect against the development of insulin resistance and type 2 diabetes. For example, transgenic mice over-expressing activated calcineurin in skeletal muscles show reduced dietary-induced insulin resistance (Ryder et al., 2003). These results suggest that increasing the proportion of slow/oxidative muscle fibers by stimulating the calcineurin, PGC1α and the PPARδ pathways, may improve insulin-stimulated glucose transport and help to overcome the mitochondrial dysfunction and metabolic defects associated with insulin-resistant states.

5.   PERSPECTIVES AND OPEN ISSUES

Activity-dependent signaling has become a central issue in muscle biology and the growing appreciation of the importance of exercise for human health will certainly boost research in this field in the coming years. There are many unanswered questions to be addressed, including both the role of specific signaling pathways and the cross talk between different pathways. In this chapter the regulation of fiber type and fiber size has been presented in distinct sections and shown to involve distinct pathways, however there are connections between these pathways and the same pathway may affect both fiber type and size. For example, GSK3 is involved in the export of NFAT from the nucleus (see section 2.3), but is also a target of Akt (see section 3.2). The role of FoxO has been discussed in the context of fiber size regulation, however transgenic mice overexpressing FoxO1 in skeletal muscle show a decreased number of slow type 1 fibers, in addition to muscle atrophy (Kamei et al., 2004).

Finally, we wish to point out one area of research that has not yet been sufficiently explored and will probably attract more interest in the future. Muscle activity is accompanied by systemic and local responses mediated both by circulating hormones acting on skeletal muscle and by factors released by muscle and acting either in a autocrine-paracrine manner onto the muscle fibers themselves or on other tissues. Among the hormonal responses increased by exercise, the acute elevations in catecholamines are especially interesting with respect to changes in muscle phenotype, because beta-agonists such as clenbuterol, acting through β2 adrenoceptors, are known to cause muscle hypertrophy and a slow-to-fast fiber type switch. Activation of beta-adrenoceptors is know to increase intracellular cAMP levels and activate protein kinase A (PKA) and the transcription factor CREB, however this pathway has not been explored in contracting muscle and its targets are unknown. Interestingly, some effects of catecholamines could be mediated by local production of IGF-I and IGF-II by skeletal muscle (Sneddon et al., 2001; Awede et al., 2002). The activity-dependent release of "myokines" from skeletal muscle is the object of growing interest. The notion that skeletal muscle, like adipose tissue, may be considered an endocrine organ has emerged during the last years (Pedersen et al., 2003) and a first attempt at defining a skeletal muscle secretome based on bioinformatics analyses has been recently reported (Bortoluzzi et al., 2006). A total of 319 proteins, including 78 still uncharacterized proteins, were identified as potentially secreted proteins based on the presence of a signal peptide required for entry into the secretory pathway and lack of transmembrane domains or intracellular localization signals. One would predict that further studies based on a variety of approaches, including microarray data and proteomics, will soon follow. It will be especially important to extend the analysis from normal muscle to exercised muscle, because it is clear that some proteins are only produced and secreted in response to muscle activity or other forms of stress. For example, IL-6 gene expression is negligible in normal muscle but is increased >100-fold during intense exercise (Pedersen et al., 2003). It will thus be interesting to compare secretomes from exercised, hypoxic or regenerating muscle to identify specific signals released by muscle cells in different conditions.

ACKNOWLEDGEMENTS

Our research was supported by grants from the European Commission (Network of Excellence MYORES, contract LSHG-CT-2004-511978, and Integrated Project EXGENESIS, contract LSHM-CT-2004-005272), Ministero dell'Università e della Ricerca Scientifica e Tecnologica of Italy (PRIN 2004 and FIRB 2001 RBNE015AX4), Telethon (grant n. GGP04227) and Agenzia Spaziale Italiana (ASI).

REFERENCES

Abbott KL, Friday BB, Thaloor D, Murphy TJ & Pavlath GK. (1998). Activation and cellular localization of the cyclosporine A-sensitive transcription factor NF-AT in skeletal muscle cells. Mol Biol Cell 9, 2905–2916.
Aramburu J, Yaffe MB, Lopez-Rodriguez C, Cantley LC, Hogan PG & Rao A. (1999). Affinity-driven peptide selection of an NFAT inhibitor more selective than cyclosporin A. Science 285, 2129–2133.
Awede BL, Thissen JP & Lebacq J. (2002). Role of IGF-I and IGFBPs in the changes of mass and phenotype induced in rat soleus muscle by clenbuterol. Am J Physiol Endocrinol Metab 282, E31–37.
Barton ER, Morris L, Musaro A, Rosenthal N & Sweeney HL. (2002). Muscle-specific expression of insulin-like growth factor I counters muscle decline in mdx mice. J Cell Biol 157, 137–148.
Bertrand A, Ngo-Muller V, Hentzen D, Concordet JP, Daegelen D & Tuil D. (2003). Muscle electro-transfer as a tool for studying muscle fiber-specific and nerve-dependent activity of promoters. Am J Physiol Cell Physiol 285, C1071–1081.
Bodine SC, Latres E, Baumhueter S, Lai VK, Nunez L, Clarke BA, Poueymirou WT, Panaro FJ, Na E, Dharmarajan K, Pan ZQ, Valenzuela DM, DeChiara TM, Stitt TN, Yancopoulos GD & Glass DJ. (2001a). Identification of ubiquitin ligases required for skeletal muscle atrophy. Science 294, 1704–1708.
Bodine SC, Stitt TN, Gonzalez M, Kline WO, Stover GL, Bauerlein R, Zlotchenko E, Scrimgeour A, Lawrence JC, Glass DJ & Yancopoulos GD. (2001b). Akt/mTOR pathway is a crucial regulator of skeletal muscle hypertrophy and can prevent muscle atrophy in vivo. Nat Cell Biol 3, 1014–1019.
Bogdanovich S, Krag TO, Barton ER, Morris LD, Whittemore LA, Ahima RS & Khurana TS. (2002). Functional improvement of dystrophic muscle by myostatin blockade. Nature 420, 418–421.
Bortoluzzi S, Scannapieco P, Cestaro A, Danieli GA & Schiaffino S. (2006). Computational reconstruction of the human skeletal muscle secretome. PROTEINS: Structure, Function, and Bioinformatics, 62, 776–792.
Buller AJ, Eccles JC & Eccles RM. (1960). Interactions between motoneurones and muscles in respect of the characteristic speeds of their responses. J Physiol 150, 417–439.
Cai D, Frantz JD, Tawa NE, Jr., Melendez PA, Oh BC, Lidov HG, Hasselgren PO, Frontera WR, Lee J, Glass DJ & Shoelson SE. (2004). IKKbeta/NF-kappaB activation causes severe muscle wasting in mice. Cell 119, 285–298.

Chakkalakal JV, Harrison MA, Carbonetto S, Chin E, Michel RN & Jasmin BJ. (2004). Stimulation of calcineurin signaling attenuates the dystrophic pathology in mdx mice. Hum Mol Genet. 13, 379–388.

Chakkalakal JV, Stocksley MA, Harrison MA, Angus LM, Deschenes-Furry J, St-Pierre S, Megeney LA, Chin ER, Michel RN & Jasmin BJ. (2003). Expression of utrophin A mRNA correlates with the oxidative capacity of skeletal muscle fiber types and is regulated by calcineurin/NFAT signaling. Proc Natl Acad Sci U S A 100, 7791–7796.

Chin ER, Olson EN, Richardson JA, Yang Q, Humphries C, Shelton JM, Wu H, Zhu W, Bassel-Duby R & Williams RS. (1998). A calcineurin-dependent transcriptional pathway controls skeletal muscle fiber type. Genes Dev 12, 2499–2509.

Coffey VG, Shield A, Canny BJ, Carey KA, Cameron-Smith D & Hawley JA. (2006). Interaction of contractile activity and training history on mRNA abundance in skeletal muscle from trained athletes. Am J Physiol Endocrinol Metab 290, E849–855.

Crabtree GR & Olson EN. (2002). NFAT signaling. Choreographing the social lives of cells. Cell 109 Suppl, S67–79.

DeNardi C, Ausoni S, Moretti P, Gorza L, Velleca M, Buckingham M & Schiaffino S. (1993). Type 2X-myosin heavy chain is coded by a muscle fiber type-specific and developmentally regulated gene. J Cell Biol 123, 823–835.

DiMario JX. (2001). Protein kinase C signaling controls skeletal muscle fiber types. Exp Cell Res 263, 23–32.

Donnelly R, Reed MJ, Azhar S & Reaven GM. (1994). Expression of the major isoenzyme of protein kinase-C in skeletal muscle, nPKC theta, varies with muscle type and in response to fructose-induced insulin resistance. Endocrinology 135, 2369–2374.

Eyre H, Kahn R, Robertson RM, Clark NG, Doyle C, Hong Y, Gansler T, Glynn T, Smith RA, Taubert K & Thun MJ. (2004). Preventing cancer, cardiovascular disease, and diabetes: a common agenda for the American Cancer Society, the American Diabetes Association, and the American Heart Association. Circulation 109, 3244–3255.

Frey N, Barrientos T, Shelton JM, Frank D, Rutten H, Gehring D, Kuhn C, Lutz M, Rothermel B, Bassel-Duby R, Richardson JA, Katus HA, Hill JA & Olson EN. (2004). Mice lacking calsarcin-1 are sensitized to calcineurin signaling and show accelerated cardiomyopathy in response to pathological biomechanical stress. Nat Med 10, 1336–1343.

Gabellini D, D'Antona G, Moggio M, Prelle A, Zecca C, Adami R, Angeletti B, Ciscato P, Pellegrino MA, Bottinelli R, Green MR & Tupler R. (2005). Facioscapulohumeral muscular dystrophy in mice overexpressing FRG1. Nature.

Glass DJ. (2003). Signalling pathways that mediate skeletal muscle hypertrophy and atrophy. Nat Cell Biol 5, 87–90.

Goldspink G. (1999). Changes in muscle mass and phenotype and the expression of autocrine and systemic growth factors by muscle in response to stretch and overload. J Anat 194, 323–334.

Gomes MD, Lecker SH, Jagoe RT, Navon A & Goldberg AL. (2001). Atrogin-1, a muscle-specific F-box protein highly expressed during muscle atrophy. Proc Natl Acad Sci U S A 98, 14440–14445.

Grater F, Shen J, Jiang H, Gautel M & Grubmuller H. (2005). Mechanically induced titin kinase activation studied by force-probe molecular dynamics simulations. Biophys J 88, 790–804.

Handschin C, Rhee J, Lin J, Tarr PT & Spiegelman BM. (2003). An autoregulatory loop controls peroxisome proliferator-activated receptor gamma coactivator 1alpha expression in muscle. Proc Natl Acad Sci U S A 100, 7111–7116.

Hay N & Sonenberg N. (2004). Upstream and downstream of mTOR. Genes Dev 18, 1926–1945.

Ho RC, Hirshman MF, Li Y, Cai D, Farmer JR, Aschenbach WG, Witczak CA, Shoelson SE & Goodyear LJ. (2005). Regulation of IkappaB kinase and NF-kappaB in contracting adult rat skeletal muscle. Am J Physiol Cell Physiol 289, C794–801.

Hoey T, Sun YL, Williamson K & Xu X. (1995). Isolation of two new members of the NF-AT gene family and functional characterization of the NF-AT proteins. Immunity 2, 461–472.

Hogan PG, Chen L, Nardone J & Rao A. (2003). Transcriptional regulation by calcium, calcineurin, and NFAT. Genes Dev 17, 2205–2232.

Horsley V, Friday BB, Matteson S, Kegley KM, Gephart J & Pavlath GK. (2001). Regulation of the growth of multinucleated muscle cells by an NFATC2- dependent pathway. J Cell Biol 153, 329–338.

Horsley V, Jansen KM, Mills ST & Pavlath GK. (2003). IL-4 Acts as a Myoblast Recruitment Factor during Mammalian Muscle Growth. Cell 113, 483–494.

Hughes SM, Chi MM, Lowry OH & Gundersen K. (1999). Myogenin induces a shift of enzyme activity from glycolytic to oxidative metabolism in muscles of transgenic mice. J Cell Biol 145, 633–642.

Hunter RB & Kandarian SC. (2004). Disruption of either the Nfkb1 or the Bcl3 gene inhibits skeletal muscle atrophy. J Clin Invest 114, 1504–1511.

Jerkovic R, Argentini C, Serrano-Sanchez A, Cordonnier C & Schiaffino S. (1997). Early myosin switching induced by nerve activity in regenerating slow skeletal muscle. Cell Struct Funct 22, 147–153.

Ji LL, Gomez-Cabrera MC, Steinhafel N & Vina J. (2004). Acute exercise activates nuclear factor (NF)-kappaB signaling pathway in rat skeletal muscle. Faseb J 18, 1499–1506.

Jorgensen SB, Wojtaszewski JF, Viollet B, Andreelli F, Birk JB, Hellsten Y, Schjerling P, Vaulont S, Neufer PD, Richter EA & Pilegaard H. (2005). Effects of alpha-AMPK knockout on exercise-induced gene activation in mouse skeletal muscle. Faseb J 19, 1146–1148.

Kahn BB, Alquier T, Carling D & Hardie DG. (2005). AMP-activated protein kinase: ancient energy gauge provides clues to modern understanding of metabolism. Cell Metab 1, 15–25.

Kalhovde JM, Jerkovic R, Sefland I, Cordonnier C, Calabria E, Schiaffino S & Lomo T. (2005). "Fast" and "slow" muscle fibres in hindlimb muscles of adult rats regenerate from intrinsically different satellite cells. J Physiol 562, 847–857.

Kamei Y, Miura S, Suzuki M, Kai Y, Mizukami J, Taniguchi T, Mochida K, Hata T, Matsuda J, Aburatani H, Nishino I & Ezaki O. (2004). Skeletal muscle FOXO1 (FKHR) transgenic mice have less skeletal muscle mass, down-regulated Type I (slow twitch/red muscle) fiber genes, and impaired glycemic control. J Biol Chem 279, 41114–41123.

Kegley KM, Gephart J, Warren GL & Pavlath GK. (2001). Altered primary myogenesis in NFATC3(-/-) mice leads to decreased muscle size in the adult. Dev Biol 232, 115–126.

Khurana TS & Davies KE. (2003). Pharmacological strategies for muscular dystrophy. Nat Rev Drug Discov 2, 379–390.

Lai KM, Gonzalez M, Poueymirou WT, Kline WO, Na E, Zlotchenko E, Stitt TN, Economides AN, Yancopoulos GD & Glass DJ. (2004). Conditional activation of akt in adult skeletal muscle induces rapid hypertrophy. Mol Cell Biol 24, 9295–9304.

Lange S, Ehler E & Gautel M. (2005a). From A to Z and back? Multicompartment proteins in the sarcomere. Trends Cell Biol. 16, 11–18.

Lange S, Xiang F, Yakovenko A, Vihola A, Hackman P, Rostkova E, Kristensen J, Brandmeier B, Franzen G, Hedberg B, Gunnarsson LG, Hughes SM, Marchand S, Sejersen T, Richard I, Edstrom L, Ehler E, Udd B & Gautel M. (2005b). The kinase domain of titin controls muscle gene expression and protein turnover. Science 308, 1599–1603.

Latres E, Amini AR, Amini AA, Griffiths J, Martin FJ, Wei Y, Lin HC, Yancopoulos GD & Glass DJ. (2005). Insulin-like growth factor-1 (IGF-1) inversely regulates atrophy-induced genes via the phosphatidylinositol 3-kinase/Akt/mammalian target of rapamycin (PI3K/Akt/mTOR) pathway. J Biol Chem 280, 2737–2744.

Lee SJ. (2004). Regulation of muscle mass by myostatin. Annu Rev Cell Dev Biol 20, 61–86.

Leone TC, Lehman JJ, Finck BN, Schaeffer PJ, Wende AR, Boudina S, Courtois M, Wozniak DF, Sambandam N, Bernal-Mizrachi C, Chen Z, Holloszy JO, Medeiros DM, Schmidt RE, Saffitz JE, Abel ED, Semenkovich CF & Kelly DP. (2005). PGC-1alpha deficiency causes multi-system energy metabolic derangements: muscle dysfunction, abnormal weight control and hepatic steatosis. PLoS Biol 3, e101.

Li S, Czubryt MP, McAnally J, Bassel-Duby R, Richardson JA, Wiebel FF, Nordheim A & Olson EN. (2005). Requirement for serum response factor for skeletal muscle growth and maturation revealed by tissue-specific gene deletion in mice. Proc Natl Acad Sci U S A 102, 1082–1087.

Lømo T. (2003). Nerve-muscle interactions. In *Clinical neurophysiology of disorders of muscle and the neuromuscular junction in adults and children IFSCN Handbook of Clinical Neurophysiology*, ed. Stålberg E, pp. 47–65. Elsevier, Amstredam.

Luquet S, Lopez-Soriano J, Holst D, Fredenrich A, Melki J, Rassoulzadegan M & Grimaldi PA. (2003). Peroxisome proliferator-activated receptor delta controls muscle development and oxidative capability. Faseb J 17, 2299–2301.

McCall GE, Allen DL, Haddad F & Baldwin KM. (2003). Transcriptional regulation of IGF-I expression in skeletal muscle. Am J Physiol Cell Physiol 285, C831–839.

McCullagh KJ, Calabria E, Pallafacchina G, Ciciliot S, Serrano AL, Argentini C, Kalhovde JM, Lomo T & Schiaffino S. (2004). NFAT is a nerve activity sensor in skeletal muscle and controls activity-dependent myosin switching. Proc Natl Acad Sci U S A.

McKinsey TA, Zhang CL, Lu J & Olson EN. (2000a). Signal-dependent nuclear export of a histone deacetylase regulates muscle differentiation. Nature 408, 106–111.

McKinsey TA, Zhang CL & Olson EN. (2000b). Activation of the myocyte enhancer factor-2 transcription factor by calcium/calmodulin-dependent protein kinase-stimulated binding of 14-3-3 to histone deacetylase 5. Proc Natl Acad Sci U S A 97, 14400–14405.

Moens P, Baatsen PH & Marechal G. (1993). Increased susceptibility of EDL muscles from mdx mice to damage induced by contractions with stretch. J Muscle Res Cell Motil 14, 446–451.

Morino K, Petersen KF, Dufour S, Befroy D, Frattini J, Shatzkes N, Neschen S, White MF, Bilz S, Sono S, Pypaert M & Shulman GI. (2005). Reduced mitochondrial density and increased IRS-1 serine phosphorylation in muscle of insulin-resistant offspring of type 2 diabetic parents. J Clin Invest 115, 3587–3593.

Moss FP & Leblond CP. (1971). Satellite cells as the source of nuclei in muscles of growing rats. Anat Rec 170, 421–435.

Murgia M, Serrano AL, Calabria E, Pallafacchina G, Lømo T & Schiaffino S. (2000). Ras is involved in nerve-activity-dependent regulation of muscle genes. Nat Cell Biol 2, 142–147.

Musaro A, McCullagh K, Paul A, Houghton L, Dobrowolny G, Molinaro M, Barton ER, H LS & Rosenthal N. (2001). Localized Igf-1 transgene expression sustains hypertrophy and regeneration in senescent skeletal muscle. Nat Genet 27, 195–200.

Nader GA & Esser KA. (2001). Intracellular signaling specificity in skeletal muscle in response to different modes of exercise. J Appl Physiol 90, 1936–1942.

Naya FJ, Mercer B, Shelton J, Richardson JA, Williams RS & Olson EN. (2000). Stimulation of slow skeletal muscle fiber gene expression by calcineurin in vivo. J Biol Chem 275, 4545–4548.

Oh M, Rybkin, II, Copeland V, Czubryt MP, Shelton JM, van Rooij E, Richardson JA, Hill JA, De Windt LJ, Bassel-Duby R, Olson EN & Rothermel BA. (2005). Calcineurin is necessary for the maintenance but not embryonic development of slow muscle fibers. Mol Cell Biol 25, 6629–6638.

Ohanna M, Sobering AK, Lapointe T, Lorenzo L, Praud C, Petroulakis E, Sonenberg N, Kelly PA, Sotiropoulos A & Pende M. (2005). Atrophy of S6K1(-/-) skeletal muscle cells reveals distinct mTOR effectors for cell cycle and size control. Nat Cell Biol 7, 286–294.

Okamura H, Garcia-Rodriguez C, Martinson H, Qin J, Virshup DM & Rao A. (2004). A conserved docking motif for CK1 binding controls the nuclear localization of NFAT1. Mol Cell Biol 24, 4184–4195.

Olson EN & Williams RS. (2000). Calcineurin signaling and muscle remodeling. Cell 101, 689–692.

Osada S, Mizuno K, Saido TC, Suzuki K, Kuroki T & Ohno S. (1992). A new member of the protein kinase C family, nPKC theta, predominantly expressed in skeletal muscle. Mol Cell Biol 12, 3930–3938.

Pallafacchina G, Calabria E, Serrano AL, Kalhovde JM & Schiaffino S. (2002). A protein kinase B-dependent and rapamycin-sensitive pathway controls skeletal muscle growth but not fiber type specification. Proc Natl Acad Sci U S A 99, 9213–9218.

Parsons SA, Millay DP, Wilkins BJ, Bueno OF, Tsika GL, Neilson JR, Liberatore CM, Yutzey KE, Crabtree GR, Tsika RW & Molkentin JD. (2004). Genetic loss of calcineurin blocks mechanical overload-induced skeletal muscle fiber type switching but not hypertrophy. J Biol Chem 279, 26192–26200.

Parsons SA, Wilkins BJ, Bueno OF & Molkentin JD. (2003). Altered skeletal muscle phenotypes in calcineurin Aalpha and Abeta gene-targeted mice. Mol Cell Biol 23, 4331–4343.

Pedersen BK, Steensberg A, Fischer C, Keller C, Keller P, Plomgaard P, Febbraio M & Saltin B. (2003). Searching for the exercise factor: is IL-6 a candidate? J Muscle Res Cell Motil 24, 113–119.

Petersen KF, Dufour S, Befroy D, Garcia R & Shulman GI. (2004). Impaired mitochondrial activity in the insulin-resistant offspring of patients with type 2 diabetes. N Engl J Med 350, 664–671.

Pette D. (2001). Historical Perspectives: plasticity of mammalian skeletal muscle. J Appl Physiol 90, 1119–1124.

Pette D & Staron RS. (1997). Mammalian skeletal muscle fiber type transitions. Int Rev Cytol 170, 143–223.

Puigserver P & Spiegelman BM. (2003). Peroxisome proliferator-activated receptor-gamma coactivator 1 alpha (PGC-1 alpha): transcriptional coactivator and metabolic regulator. Endocr Rev 24, 78–90.

Putman CT, Kiricsi M, Pearcey J, MacLean IM, Bamford JA, Murdoch GK, Dixon WT & Pette D. (2003). AMPK activation increases uncoupling protein-3 expression and mitochondrial enzyme activities in rat muscle without fibre type transitions. J Physiol 551, 169–178.

Rommel C, Bodine SC, Clarke BA, Rossman R, Nunez L, Stitt TN, Yancopoulos GD & Glass DJ. (2001). Mediation of IGF-1-induced skeletal myotube hypertrophy by PI(3)K/Akt/mTOR and PI(3)K/Akt/GSK3 pathways. Nat Cell Biol 3, 1009–1013.

Ryder JW, Bassel-Duby R, Olson EN & Zierath JR. (2003). Skeletal muscle reprogramming by activation of calcineurin improves insulin action on metabolic pathways. J Biol Chem 278, 44298–44304.

Sage J, Miller AL, Perez-Mancera PA, Wysocki JM & Jacks T. (2003). Acute mutation of retinoblastoma gene function is sufficient for cell cycle re-entry. Nature 424, 223–228.

Sakamoto K, Arnolds DE, Ekberg I, Thorell A & Goodyear LJ. (2004). Exercise regulates Akt and glycogen synthase kinase-3 activities in human skeletal muscle. Biochem Biophys Res Commun 319, 419–425.

Sakamoto K, Aschenbach WG, Hirshman MF & Goodyear LJ. (2003). Akt signaling in skeletal muscle: regulation by exercise and passive stretch. Am J Physiol Endocrinol Metab 285, E1081–1088.

Sakamoto K, Hirshman MF, Aschenbach WG & Goodyear LJ. (2002). Contraction regulation of Akt in rat skeletal muscle. J Biol Chem 277, 11910–11917.

Salmons S & Vrbova G. (1967). Changes in the speed of mammalian fast muscle following longterm stimulation. J Physiol 192, 39P-40P.

Sandri M, Sandri C, Gilbert A, Skurk C, Calabria E, Picard A, Walsh K, Schiaffino S, Lecker SH & Goldberg AL. (2004). Foxo transcription factors induce the atrophy-related ubiquitin ligase atrogin-1 and cause skeletal muscle atrophy. Cell 117, 399–412.

Sartorelli V & Fulco M. (2004). Molecular and cellular determinants of skeletal muscle atrophy and hypertrophy. Sci STKE 2004, re11.

Schiaffino S, Bormioli SP & Aloisi M. (1976). The fate of newly formed satellite cells during compensatory muscle hypertrophy. Virchows Arch B Cell Pathol 21, 113–118.

Schiaffino S & Reggiani C. (1994). Myosin isoforms in mammalian skeletal muscle. J Appl Physiol 77, 493–501.

Schiaffino S & Reggiani C. (1996). Molecular diversity of myofibrillar proteins: gene regulation and functional significance. Physiol Rev 76, 371–423.

Schiaffino S & Serrano A. (2002). Calcineurin signaling and neural control of skeletal muscle fiber type and size. Trends Pharmacol Sci 23, 569–575.

Serrano AL, Murgia M, Pallafacchina G, Calabria E, Coniglio P, Lomo T & Schiaffino S. (2001). Calcineurin controls nerve activity-dependent specification of slow skeletal muscle fibers but not muscle growth. Proc Natl Acad Sci U S A 98, 13108–13113.

Smerdu V, Karsch-Mizrachi I, Campione M, Leinwand L & Schiaffino S. (1994). Type IIx myosin heavy chain transcripts are expressed in type IIb fibers of human skeletal muscle. Am J Physiol 267, C1723–1728.

Sneddon AA, Delday MI, Steven J & Maltin CA. (2001). Elevated IGF-II mRNA and phosphorylation of 4E-BP1 and p70(S6k) in muscle showing clenbuterol-induced anabolism. Am J Physiol Endocrinol Metab 281, E676–682.

Stitt TN, Drujan D, Clarke BA, Panaro F, Timofeyva Y, Kline WO, Gonzalez M, Yancopoulos GD & Glass DJ. (2004). The IGF-1/PI3K/Akt pathway prevents expression of muscle atrophy-induced ubiquitin ligases by inhibiting FOXO transcription factors. Mol Cell 14, 395–403.

Turinsky J & Damrau-Abney A. (1999). Akt kinases and 2-deoxyglucose uptake in rat skeletal muscles in vivo: study with insulin and exercise. Am J Physiol 276, R277–282.

Vitadello M, Schiaffino MV, Picard A, Scarpa M & Schiaffino S. (1994). Gene transfer in regenerating muscle. Hum Gene Ther 5, 11–18.

Vlahopoulos S, Zimmer WE, Jenster G, Belaguli NS, Balk SP, Brinkmann AO, Lanz RB, Zoumpourlis VC & Schwartz RJ. (2005). Recruitment of the androgen receptor via serum response factor facilitates expression of a myogenic gene. J Biol Chem 280, 7786–7792.

Wang Y, Falasca M, Schlessinger J, Malstrom S, Tsichlis P, Settleman J, Hu W, Lim B & Prywes R. (1998). Activation of the c-fos serum response element by phosphatidyl inositol 3-kinase and rho pathways in HeLa cells. Cell Growth Differ 9, 513–522.

Wang YX, Zhang CL, Yu RT, Cho HK, Nelson MC, Bayuga-Ocampo CR, Ham J, Kang H & Evans RM. (2004). Regulation of muscle fiber type and running endurance by PPARdelta. PLoS Biol 2, e294.

Wu H, Gallardo T, Olson EN, Williams RS & Shohet RV. (2003). Transcriptional analysis of mouse skeletal myofiber diversity and adaptation to endurance exercise. J Muscle Res Cell Motil 24, 587–592.

Wu H, Kanatous SB, Thurmond FA, Gallardo T, Isotani E, Bassel-Duby R & Williams RS. (2002). Regulation of Mitochondrial Biogenesis in Skeletal Muscle by CaMK. Science 296, 349–352.

Wu H, Naya FJ, McKinsey TA, Mercer B, Shelton JM, Chin ER, Simard AR, Michel RN, Bassel-Duby R, Olson EN & Williams RS. (2000). MEF2 responds to multiple calcium-regulated signals in the control of skeletal muscle fiber type. Embo J 19, 1963–1973.

Wu H, Rothermel B, Kanatous S, Rosenberg P, Naya FJ, Shelton JM, Hutcheson KA, DiMaio JM, Olson EN, Bassel-Duby R & Williams RS. (2001). Activation of MEF2 by muscle activity is mediated through a calcineurin- dependent pathway. Embo J 20, 6414–6423.

Yang ZZ, Tschopp O, Baudry A, Dummler B, Hynx D & Hemmings BA. (2004). Physiological functions of protein kinase B/Akt. Biochem Soc Trans 32, 350–354.

Zong H, Ren JM, Young LH, Pypaert M, Mu J, Birnbaum MJ & Shulman GI. (2002). AMP kinase is required for mitochondrial biogenesis in skeletal muscle in response to chronic energy deprivation. Proc Natl Acad Sci U S A 99, 15983–15987.

CHAPTER 5

ACTIVITY DEPENDENT CONTROL
OF THE TRANSCRIPTIONAL REGULATORS NFAT
AND HDAC IN ADULT SKELETAL MUSCLE FIBRES

YEWEI LIU, TIANSHENG SHEN, WILLIAM R. RANDALL
AND MARTIN F. SCHNEIDER
University of Maryland School of Medicine, Baltimore, MD 21201, USA

1.   INTRODUCTION: *IN VITRO* APPROACH TO STUDY MUSCLE
      PLASTICITY

A large number of muscle fiber transformation studies have been carried out on
whole animals *in vivo*, with the advantage that the muscles are under normal
hormonal regulation and physiological environments. However, it is difficult to
control or change ion concentrations, pH, or other cellular factors or to monitor
transcription factor activation in muscle *in situ*. In order to investigate activity
dependent signaling mechanisms underlying fast to slow fiber transformation, our
laboratory developed an *in vitro* system for culturing adult mouse flexor digitorum
brevis (FDB) muscle fibers (predominately fast twitch) with or without electrical
stimulation (Liu and Schneider, 1998). This culture system provides the possibility
of monitoring resting calcium and calcium transients, as well as the quantitative
measurement of the intracellular translocation of expressed fluorescent proteins
following adenoviral transduction. Such measurements would be difficult to obtain
in muscle in situ, as methods for efficient uptake of expression vectors are not
currently available. Using this culture system, we first examined the effects of
5 days of chronic slow fiber type stimulation (5 sec train of 5 Hz stimuli applied
once every min) on resting calcium and calcium transients, as well as on slow fiber
type gene expression. We found that slow fiber type stimulation increased β-myosin
heavy chain mRNA, a slow fiber type marker, while α-skeletal actin mRNA, a
non-fiber type specific protein, remained constant (Liu and Schneider, 1998). Our

121

R. Bottinelli and C. Reggiani (eds.), Skeletal Muscle Plasticity in Health and Disease, 121–135.
© 2006 *Springer.*

early results thus indicated that fast to slow fiber type conversion, at least at the level of mRNA, could be followed in cultured adult FDB fibers.

Activation of slow fiber type genes such as β-myosin heavy chain involves a number of transcription factors, co-activators and co-repressors. Here we focus on more recent studies of the transcriptional regulators NFATc and HDAC, both of which are shown to play important roles in slow fiber gene expression and to respond to electrical stimulation in adult mouse FDB skeletal muscle fibers in culture.

2.   CALCINEURIN ACTIVATION AND NFAT NUCLEAR ENTRY

The importance of calcium for fast to slow fiber type transformation has been underscored by the finding that increased intracellular calcium produced by application of calcium ionophore A23187 increases the level of the slow MyHC I isoform in primary skeletal muscle cell culture (Kubis et al., 1997). This study clearly demonstrated that calcium played critical role in slow fiber gene expression. However, how different types of intracellular calcium transients resulting from different contraction frequencies are decoded by the cell and how the calcium message is converted into the nuclear information necessary to reprogram the genome toward specific responses remained unclear until the discovery that a calcineurin/NFAT-dependent transcriptional pathway was involved in the control of skeletal muscle fiber type (Chin et al., 1998), providing a direct link to calcium pattern dependent signaling. NFAT had been previously studied extensively in lymphocytes (Crabtree, 1999). In unstimulated cells, NFAT is located in the cytoplasm and is phosphorylated. When cells are stimulated, intracellular calcium rises and cytosolic NFAT is dephosphorylated by the calcium dependent phosphatase, calcineurin, which itself is activated by the elevated calcium. Under resting conditions of low calcium, unactivated calcineurin is bound to cytosolic NFAT (Garcia-Cozar et al., 1998). Calcineurin dephosphorylates NFAT at multiple sites in the serine-rich region (SRR) and the serine-proline (SP) repeat regions of the protein (Beals et al., 1997). It is believed that dephosphorylation at these sites unmasks two nuclear localization signals (Beals et al., 1997), allowing its nuclear translocation (Fig. 1).

To determine if NFATc1 nuclear translocation in muscle fibers can be induced by specific patterns of contraction, we used our FDB muscle fiber culture system to study the response of NFATc1 to different patterns of electrical stimulation, characteristic of slow motor neuron (10 Hz continuously or 10 Hz for 5 second every 50 second), fast motor neuron (100 Hz for 100 ms every 50 second), or continuous 1 Hz stimulation, which gives muscle fibers an equal total number of pulses per minute as a 10 Hz train for 5 sec every 50 seconds. FDB cultures were infected with recombinant adenovirus expressing human NFATc1, the primary isoform in adult skeletal muscle (Hoey et al., 1995), tagged with GFP (NFATc1-GFP). In resting conditions without stimulation, NFATc1-GFP is located mainly in cytoplasm and predominantly localized at the sarcomeric z-line, identified by antibody stain of α-actinin. This finding is in agreement with other reports that NFAT docks on

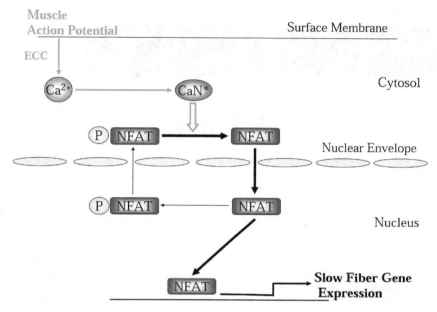

Figure 1. Activation of slow skeletal muscle fiber gene expression by the calcium/calcineurin/NFAT pathway. Elevated cytosolic Ca2+, produced in a muscle fiber in response to an action potential in the external membranes, activates the cytoplasmic Ca2+ dependent phosphatase calcineurin (CaN*), which then dephosphorylates cytoplasmic NFAT (P-NFAT). This exposes a nuclear localization signal and masks the nuclear export signal on NFAT, permitting NFAT nuclear entry via the nuclear pore import system and leading to the activation of slow fiber genes by nuclear NFAT. NFATc1 has up to 11 phosphorylatable serines in the SRR and 10 in the SP repeats, but for simplicity only a single P is indicated for phosphorylated NFAT (P-NFAT). Rectangular boxes denote transcriptional regulatory molecules, eclipses denote kinases or phosphatases, open arrows denote phosphatase or kinase enzymatic activity and circles denote phosphorylated serine groups (yellow) or elevated [Ca2+] (orange). A color figure is freely accessible via the website of the book: http://www.springer.com/1-4020-5176-x

calcineurin under resting conditions and that calcineurin binds selectively to a complex containing calsarcin and FATZ (filamin-, actinin-, and telethonin-binding protein of the z-disc) (Faulkner et al., 2000; Frey et al., 2000; Frey and Olson, 2002) at z-lines in both skeletal and cardiac muscles.

We found in NFATc1-GFP expressing FDB fibers that either 10 Hz continuous stimulation or 5 sec trains of 10 Hz stimuli every 50 sec (slow fiber type paradigm) resulted in similar nuclear translocation of NFATc1-GFP from the cytoplasm (Fig 2). In contrast, 1Hz continuous stimulation (Fig 2) or a 50 Hz train of 100 ms every 50 sec (fast fiber type paradigm) resulted in negligible translocation of NFATc1-GFP to the nucleus. Thus, translocation of NFATc1-GFP in response to electrical activity appears to require a stimulation pattern similar to slow fiber type activity not fast fiber type activity in this cultured adult fiber model system. This

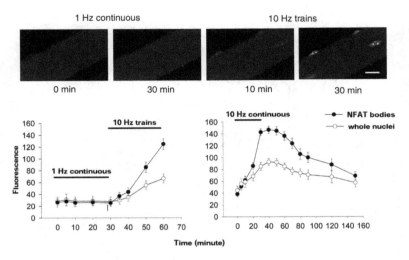

Figure 2. Time course of nuclear translocation of NFATc1-GFP. A fiber was stimulated with 1 Hz for 30 min, and then the same fiber was stimulated with 10 Hz trains for another 30 min. Although 1 Hz continuous stimulation resulted in no detectable change in nuclear fluorescence, subsequent stimulation using 10 Hz trains caused significant increases in whole nuclear fluorescence and resulted in appearance of nuclear NFAT bodies in the same group of fibers. Bar, 10 mm. Reproduced, with permission, from (Liu et al., 2001). © Rockefeller University Press

information provides a link between the activation of NFAT and a physiologic stimulus by the motor neuron in myofibers.

Calcineurin gain-of-function and loss-of-function studies in transgenic mice confirm the importance of calcineurin activation in the maintenance of fiber-type. In line with most *in vitro* experiments (Chin et al., 1998), an expressed constitutively active calcineurin under the control of the muscle creatine kinase enhancer stimulates slow skeletal muscle fiber gene expression in transgenic mice (Naya et al., 2000). In contrast, the loss-of-function of calcineurin by targeted deletion in mice results in muscles exhibiting a reduced oxidative slow muscle fiber-type profile (Parsons et al., 2003).

NFAT-mediated control of fiber-type has also been supported by muscle regeneration studies in adult rats. The induction of the slow gene program in the regenerating rat soleus muscle and the maintenance of the slow gene program in the adult soleus (McCullagh et al., 2004) exhibits higher NFAT activity which is caused by nerve activity. This higher NFAT activity in denervated muscle can be maintained by a tonic low-frequency impulse pattern, typical of slow motor neurons, but not by a phasic high-frequency pattern, typical of fast motor neurons (McCullagh et al., 2004). However, the proposed role for calcineurin and NFAT in slow fiber gene expression is debated. There are reports that calcineurin protein expression in fast muscles (EDL or plantaris) was significantly greater than in slow muscles (soleus) (Swoap et al., 2000; Spangenburg et al., 2001). It was pointed out that NFAT may not be a dominant factor in conferring slow fiber type during development (Calvo et al., 1999). All of these suggests that the role of NFAT in fiber type transformation may not be simple and straightforward. Instead, NFAT may need other co-factors and interactions between NFAT and other transcription factors may be pivotal, as found in cardiac muscle and other tissue (Macian et al., 2001).

3.   NFAT LOCALIZES TO INTRANUCLEAR BODIES

Following translocation in adult FDB fibers, the nuclear distribution of NFATc1, like several other nuclear proteins, is not uniformly partitioned. Expressed recombinant NFATc1 showed a punctuate pattern (Liu et al., 2001) that we termed NFAT bodies (Liu et al., 2005b). The formation of NFAT bodies occurs in both electrically stimulated muscle fibers expressing NFATc1-GFP (Fig 3) and in unstimulated fibers expressing a constuitively nuclear NFATc1-GFP (NFATc1 (S→A)-GFP). The number of NFAT bodies in one muscle fiber nucleus ranges from four to nine (Liu et al., 2001) and their size is similar to bodies of other nuclear proteins (Hendzel et al., 2001; Zimber et al., 2004), varying from about 0.5 μm to 2 μm. Nuclear accumulation of NFATc1 into NFAT bodies can also be demonstrated by inhibiting nuclear rephosphorylation (preventing nuclear export) with certain kinase inhibitors (e.g. GSK-3β inhibitor, LiCl or MAP kinase inhibitor) (Liu et al., 2001). Interestingly, the above mentioned treatments all act without increasing intracellular calcium, indicating that the formation of NFAT bodies is independent of calcium.

Nuclear NFAT bodies do not occur only in skeletal muscle fibers. There are other reports showing a similar punctuate pattern of nuclear distribution in HEK 293T cells expressing NFATc1-GFP (Sheridan et al., 2002) and in mouse cardiac cells with antibody stain to endogenous NFATc1 (Antos et al., 2002).

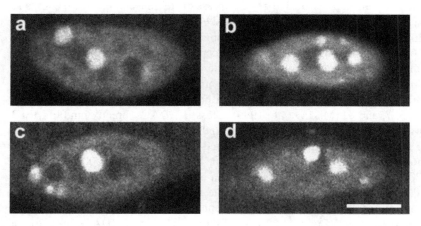

Figure 3. NFAT bodies. a and b present enlarged images of nuclei from individual fibers, expressing NFATc1-GFP and electrically stimulated with 10 Hz trains. c and d are enlarged images of nuclei from fibers expressing NFATc1(S→A)-GFP. Dark nucleoli can be seen in nuclei, suggesting that the nuclear NFATc1 proteins are excluded from nucleolar structures. Bar, 5 μm

It is known that several nuclear proteins are not evenly distributed inside the nucleus. Each kind of nuclear protein can organize into bodies or foci (Misteli T, 2001). One idea is that nuclear foci may facilitate the assembly of multi-molecular regulatory complexes and/or control nucleoplasmic concentrations of assembled complexes (Hendzel et al., 2001). Co-localization of foci with molecules known to function in transcription or in transcript processing has been documented (Hendzel et al., 2001; Zimber et al., 2004). SC35, a splicing factor which has been used to identify active transcription sites, exhibits a non-uniform intranuclear distribution in bodies or "speckles" (Spector et al., 1991). However, antibody staining of SC35 in FDB fibers expressing nuclear targeted NFATc1(S→A)-GFP showed no co-localization between NFAT bodies and SC35 speckles. MEF2, a muscle transcription factor reported to associate with NFAT to activate slow-fiber-specific promoters, did not colocalize with NFATc1(S→A)-GFP in the nucleus of adult FDB fibers (Liu et al., 2001). However, others observed in embryonic myotube cultures that NFATc1 indeed colocalized with MEF2 following 24 hours of electrical stimulation (Kubis et al., 2002) suggesting that there may be different mechanisms of NFATc1 and MEF2 localization at different stages of muscle development.

There remain several questions to be answered regarding the nature of NFAT bodies. Are NFAT bodies general transcription factor storage sites used to regulate nuclear NFAT concentration or are NFAT bodies located in the regions that NFAT binds to chromatin? Are there other nuclear proteins known to associate with NFAT (AP-1 or GATA-4) that co-localize within NFAT bodies?

4. NUCLEAR CAMK ACTIVATION AND HDAC NUCLEAR EFFLUX

The action of MEF2 is critical for the expression of muscle specific genes and contributes to fiber-type transformation in skeletal muscle (Wu et al., 2000; Wu et al., 2001). Analysis of a number of slow fiber type-specific genes showed MEF2 and NFAT consensus binding sequences adjacent or in close proximity to each other within the 5' flanking regions of their promoters (Chin et al., 1998). Unlike NFAT, MEF2 resides solely in the nucleus and control of its transcriptional activity is negatively regulated by the family of class II histone deacetylase proteins (HDAC; HDAC4, 5, 7, and 9) (Miska et al., 1999). Class II HDACs directly bind to MEF2 to repress transcriptional activity and their binding is regulated by phosphorylation of the HDACs. The principal enzyme required for phosphorylation in the nucleus is calmodulin dependent protein kinase (CaMK; Chawla et al., 2003; McKinsey et al., 2000a; McKinsey et al., 2000b; Miska et al., 2001). It is proposed that the elevated nuclear calcium required to activate nuclear CaMK arises from release of calcium during muscle contraction, movement into the nucleus, and subsequent binding to calmodulin (Liu et al., 2005a). Phosphorylation of HDACs leads to the binding of HDAC to 14-3-3 protein, which then masks the nuclear localization signal within HDAC and results in nuclear export of HDAC (McKinsey et al., 2001). The binding of HDAC to 14-3-3 and export from the nucleus physically separates

HDAC from MEF2, relieving its repression and allowing activation of transcription of slow fiber genes (Fig 4).

Using virally expressed HDAC4-GFP and by employing similar techniques to those we used in the NFATc1 studies, we recently demonstrated that HDAC4 translocates from nucleus to cytoplasm in response to both 10 Hz trains and 1 Hz continuously stimulation in adult cultured FDB fibers, but with different rate constants (Fig 5). In our adult skeletal muscle culture model, the CaMK inhibitor KN-62 blocked the HDAC4 nuclear export normally produced from electrical stimulation, demonstrating that CaMK is responsible for phosphorylating HDAC4 in the nucleus in response to muscle activity (Liu et al., 2005a). To determine if endogenous HDACs respond to electrical stimulation of muscle fibers in the same manner (i.e. move out of the nucleus to allow MEF2 binding to MEF2 DNA elements), we infected muscle culture with recombinant adenovirus carrying 6 MEF2 enhancer elements attached to a minimal promotor, driving luciferase. We found that 10 Hz train stimulation enhanced luciferase activity, as well as inducing HDAC4-GFP nuclear export (Liu et al., 2005a), indicating that MEF2

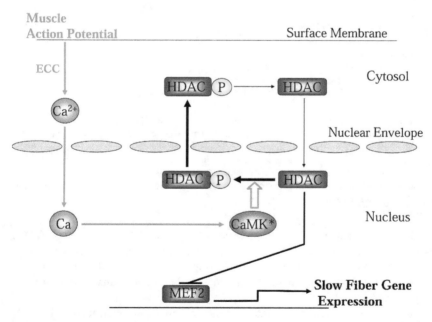

Figure 4. Activation of slow skeletal muscle fiber gene expression by the calcium/CaMK/HDAC pathway. Elevated cytoplasmic Ca2+ in response to muscle activity causes Ca2+ entry into the nucleus via nuclear pores and the activation of intranuclear CaM kinase. Activated nuclear CaMK phosphory-lates HDAC in the nucleus, allowing HDAC to exit from the nucleus via the nuclear export system and thereby removing the HDAC repression of MEF2 activation of slow fiber type gene expression. Symbol shapes are as described in Fig 1. A color figure is freely accessible via the website of the book: http://www.springer.com/1-4020-5176-x

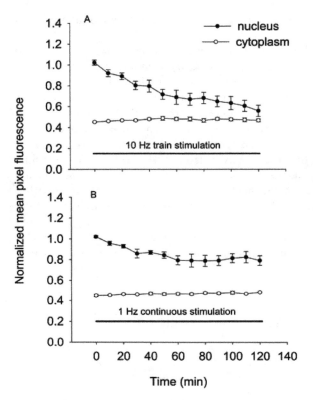

Figure 5. Time course of nuclear to cytoplasmic translocation of HDAC4 during 10 Hz train or 1 Hz continuously stimulation. (A) 5 sec trains of 10 Hz stimuli every 50 sec caused net nuclear to cytoplasmic translocation of HDAC4-GFP. Nuclear fluorescence declined continuously during the 120 min stimulation period. (B) 1 Hz continuous stimulation resulted in nuclear export of HDAC4-GFP, but with a lower export rate compared with 10 Hz train stimulation. The cytoplasmic fluorescence remained constant during the whole stimulation period for both (A) and (B). Reproduced, with permission, from (Liu et al., 2005a). © Rockefeller University Press

was binding the enhancer construct when nuclear HDAC levels dropped. HDACs are also capable of acetylating chromatin and modulating gene expression through these modifications. It is yet not known if the electrical stimulation results in changes in the acetylation status of chromatin associated with transcription of slow fiber-type genes.

In addition to the calcium mediated cytosolic dephosphorylation of NFATc1, the intranuclear phosphorylation of HDAC by CaMK thus provides a second possible calcium-pattern-dependent, phosphorylation-mediated signaling pathway regulating fiber-type gene expression in muscle. Interestingly, HDAC4 moves out of the nucleus in response to both 10 Hz and 1 Hz continuous stimulation, albeit at a faster rate at 10 Hz compared to 1 Hz stimulation, whereas NFATc1 moves into the nucleus only in response to 10 Hz but not to 1 Hz stimulation. This difference in response to stimulation frequency may provide an element of fine-tuning via multiple pathways in the activation of slow fiber type gene expression. The sharper frequency and pattern discrimination of the calcineurin /NFAT pathway compared to the CaMK/HDAC pathway could result from the presence of up to 11 phosphorylatable serine residues per molecule in the SRR domain of NFATc1 (Beals et al., 1997), and 10 in the three SP rich repeat regions (Neil and Clipstone, 2001), implicated in the modulation of NFAT nuclear translocation. In contrast, only 3 serine residues are involved in phosphorylation of HDAC4 for nuclear efflux (Grozinger and Schreiber, 2000). The much larger number of phosphorylation sites in NFATc1 could contribute to a threshold-like activation when a minimum required number is dephosphorylated. These issues, as well as the integration of the calcium signal by NFAT (Tomida et al., 2003), have been considered in theoretical models (Salazar and Hofer, 2003).

CaMK likely mediates the phosphorylation and subsequent translocation of HDACs from the nucleus to cytoplasm in several kinds of cells (Chawla et al., 2003; McKinsey et al., 2000a; McKinsey et al., 2000b; Miska et al., 2001). Thus we examined the intracellular distribution of CaM-YFP in FDB cultures in response to electrical stimulation. Surprisingly, a 10 Hz train or 1 Hz continuously stimulation, both patterns of stimulation resulting in export of HDAC4-GFP from the nucleus, did not alter the subcellular distribution of CaM-YFP (Liu et al., 2005a). Molecular analysis of HDAC4 indicated that amino acids 150-220 were critical for calcium-dependent CaM binding (Youn et al., 2000). This region overlaps with the MEF2 binding domain, so that CaM and MEF2 may compete for binding to HDAC4 and release of MEF2 may be enhanced by activated CaM. Facilitation of MEF2 dissociation from HDAC5 by activated CaM has also been reported (Berger et al., 2003). However, our observation on the subcellular distribution of CaM-YFP does not support this conclusion.

HDACs are also capable of deacetylating histones bound to chromatin and modulating gene expression through these modifications. It is yet not known if the electrical stimulation results in changes in the acetylation status of chromatin associated with transcription of slow fiber-type genes.

Class II HDACs may repress MEF2 transcriptional activity by two mechanisms, simple binding of HDAC to MEF2, or by histone (or other protein) deacetylation via its enzymatic activity (Bertos et al., 2001; Downes et al., 2000; Fischle et al., 2001), or both. The requirement for enzymatic activity of class II HDACs in MEF2 dependent transcriptional repression has not been clearly established. HDAC4-mediated repression of a MEF2 reporter gene requires the integrity of the HDAC4 catalytic domain but is only partially sensitive to the HDAC inhibitor trichostatin A (Miska et al., 1999). The organization of DNA in chromatin is thought to act as a barrier to transcription causing gene repression. The acetylation status of histones is maintained by the dynamic interplay of histone acetyltransferase and deacetylase. The phosphorylation of HDACs and subsequent dissociation from MEF2 does not mean HDACs must translocate out of nucleus. Instead HDACs may simple re-distribute inside nucleus to regions removed from their sites of action (Zhang et al., 2001). The abundance of class II HDACs and their subsequent action also may be modulated at the level of mRNA. MITR (MEF2-interacting transcription repressor), a splice variant of HDAC9, is exclusively nuclear due to the lack of the nuclear export signal (Sparrow et al., 1999; Fischle et al., 2001; Zhang et al., 2001). Denervation lead to a decrease of MITR expression followed by hyperacetylation of histone H3 and up-regulation of the acetylcholine receptor α-subunit in adult rat muscle fibers (Mejat et al., 2005).

In addition to acetylation, chromatin packing also can be regulated by phosphorylation of histone H3, which may act synergistically or in parallel with acetylation (Nowalk and Corces, 2004). Phosphorylation of histone H3 was shown to relieve the chromatin from repressor heterochromatin protein 1 (HP1) (Mateescu et al., 2004). Both histone acetylation and phosphorylation are involved in activation of gene expression and memory formation (Alarcon et al., 2004; Korzus et al., 2004; Levenson et al., 2004). These data pose the question: is there histone phosphorylation in response to different frequencies of electrical stimulation in skeletal muscle? This issue should be addressed to further elucidate these potential pathways in activity dependent gene expression in skeletal muscle.

5.   KINETICS AND SOURCE OF MYOPLASMIC CA^{2+} SIGNALS

In response to an action potential in adult skeletal muscle, calcium is released from sarcoplasmic reticulum (SR) via the opening of ryanodine receptor (RyR) Ca^{2+} release channels, resulting in a brief rise in cytosolic calcium (Carroll et al., 1995; Hollingworth et al., 1996). The released calcium arises solely from SR stores via the RyR, not from extracellular calcium via dihydropyridine receptor calcium channels (Payne et al., 2004). This is in contrast to calcium movement in cardiac myocytes, in which both SR release and extracellular calcum influx contribute to the calcium transients. The influx of extracellular calcium also may occur during contraction in early developing myotubes. How quickly the calcium transient declines is determined by the binding of calcium to myoplasmic sites and its transport out of the cytosol and primarily back into the SR (Klein et al., 1991; Westerblad and Allen, 1994). The major proteins that bind calcium in the fast twitch FDB fibers are troponin C, myosin, and parvalbumin. How fast the calcium transient decays in the cytoplasm and nucleus may respectively affect the frequencies at which the calcineurin/NFAT pathway and the CaMK/HDAC pathway can be activated. In the skeletal muscle primary cultures, myotubes are not completely developed and have much slower decay phase of the calcium transient compared with the adult skeletal muscle fiber cultures (Kubis et al., 2003). Immature myotubes respond to stimulation frequencies different than adult fibers. Repetitive electrical stimulation at 1 Hz is able to cause NFATc1 to translocate into the myotube nuclei and also to cause an increase of the expression of slow fiber myosin heavy chain mRNA (Kubis et al., 2002).

We imaged fiber segments, including nuclei, in order to monitor both cytosolic and nuclear calcium in cultured adult muscle fibers using the non-ratiometric dye fluo-4 (Fig 6). The fibers were stimulated with 5 second duration 10 Hz trains applied every 50 second, the stimulation pattern which triggers both NFATc1 (Liu et al., 2001) and HDAC4 translocation (Liu et al., 2005a). Each stimulus activates the entire fiber cross section uniformly (Fig 6Ab and Af). Within the 10 Hz train of stimuli, cytosolic $[Ca^{2+}]$ rises after each stimulus and then falls between pulses. In contrast, nuclear $[Ca^{2+}]$ seems to rise more slowly and continuously during the train, as anticipated for diffusion of elevated $[Ca^{2+}]$ from cytoplasm into the nucleus. The decay of the cytosolic calcium occurs rapidly after the train of stimuli, whereas nuclear calcium decays more slowly than cytosolic calcium concentration (Fig 6B). Compared to the resting fiber, nuclear $[Ca^{2+}]$ remains elevated a few hundred ms after the end of the 5 s, 10 Hz trains of pulses. This elevated nuclear $[Ca^{2+}]$ then declines during the 45 s interval between successive trains (Fig 6).

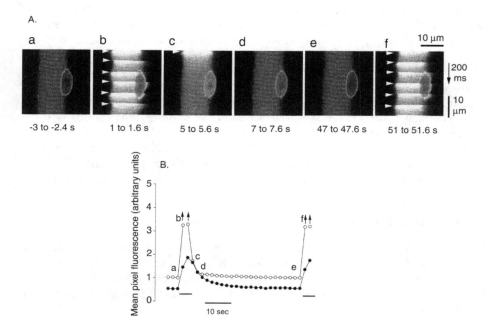

Figure 6. Both nuclear and cytoplasmic calcium is elevated in response to 10 Hz train stimulation. (A) A FDB fiber loaded with fluo-4 was repetitively stimulated with 10 Hz trains every 50 sec. Images were taken in x-y scan mode (a) in the resting condition prior to the start of repetitive stimulation, (b) 1 to 1.6 sec after the start of the first train, (c-e) 0 to 0.6, 2 to 2.6 and 42 to 42.5 sec after the first train, and (f) 1 to 1.6 sec after the start of the second train. The nuclear $[Ca2+]$ was still elevated compared to the resting condition a few hundred ms after the end of the 5 sec train (c). Arrow heads indicate the application of each stimulus within an image. (B) Time course of nuclear (filled circles) or cytosolic (open circles) mean pixel fluorescence in the central 100 μm (recorded during 200 ms) of images in A and others not shown. Arrows indicate underestimation of mean pixel fluorescence due to the inclusion of pixels maxed out due to detector saturation. Horizontal bars below records mark duration of the 5 sec 10 Hz trains of stimuli. Reproduced, with permission, from (Liu et al., 2005a). © Rockefeller University Press

Thus, activation of nuclear CaMK due to elevated nuclear $[Ca^{2+}]$ appears to be a likely mechanism underlying the CaMK dependent HDAC4 efflux during fiber electrical stimulation.

Ca^{2+} release from the nuclear envelope, possibly via IP_3 receptor Ca^{2+} release channels, recently demonstrated in skeletal muscle primary myotube cultures (Powell et al., 2001), might also contribute to intra-nuclear and peri-nuclear $[Ca^{2+}]$. In adult cardiac myocytes, HDAC phosphorylation and export from nucleus was associated with activation of CaMKII at the nuclear envelope, which itself was activated by local calcium release from InsP3 receptor (Wu et al., 2004). In agreement with this report, InsP3 receptors have been found concentrated in the nuclear envelope of cardiac myocytes and associated with CaMKII (Bare et al., 2005). Whether similar mechanism exists in skeletal muscle warrants further investigation. The possible contribution of Ca^{2+} entry from the extracellular medium during chronic repetitive activity, possibly via store operated- (Kurebayashi and Ogawa, 2001; Pan et al., 2002) or TRPC3 channels (Kiselyov et al., 2000; Rosenberg et al., 2004) must also be considered. However, the contribution of calcium entry via these channels to calcium transients during fiber type transformation, which does not involve muscle fatiguing stimulation, must be carefully evaluated. Future studies may clarify the exact time courses, sources, and sinks of cytosolic and nuclear $[Ca^{2+}]$ during the skeletal muscle fiber activity patterns underlying fiber type transformation. Experimental determination of the resulting time courses of Ca^{2+} dependent activation of the cytoplasmic and nuclear Ca^{2+} sensitive phosphatases and kinases, which give rise to the dephosphorylation and observed nuclear entry of NFATc1 and the phosphorylation and observed nuclear efflux of HDAC4, is an important goal for future research in this area of excitation-transcription coupling.

ACKNOWLEDGEMENTS

We thank Ms Carrie Wagner for preparation of figures. This work was supported by National Institutes of Health grant R01-NS33578 from the National Institute of Neurological Disorders and Stroke. T. Shen is supported by NIH Training Grant T32-AR-07592 to the Interdisciplinary Program in Muscle Biology, University of Maryland School of Medicine.

REFERENCES

Alarcon JM, Malleret G, Touzani K, Vronskaya S, Ishii S, Kandel ER, and Barco A (2004). Chromatin acetylation, memory, and LTP are impaired in CBP+/- mice: a model for the cognitive deficit in Rubinstein-Taybi syndrome and its amelioration. *Neuron.* 42, 947–959.
Antos CL, McKinsey TA, Frey N, Kutschke W, McAnally J, Shelton JM, Richardson JA, Hill JA, and Olson EN (2002). Activated glycogen synthase-3 beta suppresses cardiac hypertrophy in vivo. *Proc Natl Acad Sci USA.* 99, 907–912.
Bare DJ, Kettlun CS, Liang M, Bers DM, and Mignery GA (2005). Cardiac type 2 inositol 1,4,5-trisphosphate receptor: interaction and modulation by calcium/calmodulin-dependent protein kinase II. *J Biol Chem.* 280, 15912–15920.

Beals, CR, Clipstone NA, Ho SN, and Crabtree GR (1997). Nuclear localization of NF-ATc by a calcineurin-dependent, cyclosporin-sensitive intramolecular interaction. *Genes Dev.* 11, 824–834.

Berger I, Bieniossek C, Schaffitzel C, Hassler M, Santelli E, and Richmond TJ (2003). Direct interaction of Ca2+/calmodulin inhibits histone deacetylase 5 repressor core binding to myocyte enhancer factor 2. *J Biol Chem.* 278, 17625–17635.

Bertos NR, Wang AH, and Yang X (2001). Class II histone deacetylases: Structure, function, and regulation. *Biochem Cell Biol.* 79, 243–252.

Calvo, S., P. Venepally, J. Cheng, and A. Buonanno. 1999. Fiber-type-specific transcription of the troponin I slow gene is regulated by multiple elements. *Mol. Cell. Biol.* 19, 515–525.

Carroll SL, Klein MG, and Schneider MF (1995). Calcium transients in intact rat skeletal muscle fibers in agarose gel. *Am J Physiol.* 269, C28–C34.

Chawla, S., P. Vanhoutte, F. J. Arnold, C. L. Huang, and H. Bading.2003. Neuronal activity-dependent nucleocytoplasmic shuttling of HDAC4 and HDAC5. *J. Neurochem.* 85, 151–159.

Chin ER, Olson EN, Richardson JA, Yang Q, Humphries C, Shelton JM, Wu H, Zhu W, Bassel-Duby R, and Williams RS (1998). A calcineurin-dependent transcriptional pathway controls skeletal muscle fiber type. *Genes Dev.* 12, 2499–2509.

Crabtree GR (1999). Generic signals and specific outcomes: signaling through Ca^{2+}, calcineurin, and NF-AT. *Cell* 96, 611–614.

Downes M, Ordentlich P, Kao HY, Alvarez JGA, and Evans RM (2000). Identification of a nuclear domain with deactylase activity. *Proc Natl Acad Sci USA.* 97, 10330–10335.

Faulkner G, Pallavicini A, Comelli A, Salamon M, Bortoletto G, Ievolella C, Trevisan S, Kojic' S, Dalla Vecchia F, Laveder P, Valle G, and Lanfranchi G (2000). FATZ, a filamin-, actinin-, and telethonin-binding protein of the Z-disc of skeletal muscle. *J Biol Chem.* 275, 41234–41242.

Fischle, W, Kiermer V, Dequiedt F, and Verdin E (2001). The emerging role of class II histone deacetylases. *Biochem Cell Biol.* 79, 337–348.

Frey N, Richardson JA, and Olson EN (2000). Calsarcins, a novel family of sarcomeric calcineurin-binding proteins. *Proc Natl Acad Sci U S A.* 97, 14632–14637.

Frey N, and Olson EN (2002). Calsarcin-3, a novel skeletal muscle-specific member of the calsarcin family, interacts with multiple Z-disc proteins. *J Biol Chem.* 277, 13998–134004.

Garcia-Cozar, F.J., H. Okamura, J. F. Aramburu, K. T. Shaw, L. Pelletier, R. Showalter, E. Villafranca, A. Rao.1998. Two-site interaction of nuclear factor of activated T cells with activated calcineurin. *J. Biol. Chem.* 273, 23877–23883.

Grozinger CM, and Schreiber SL (2000). Regulation of histone deacetylase 4 and 5 and transcriptional activity by 14-3-3-dependent cellular localization. *Proc Natl Acad Sci USA.* 97, 7835–7840.

Hendzel MJ, Kruhlak MJ, MacLean NA, Boisvert F, Lever MA, and Bazett-Jones DP (2001). Compartmentalization of regulatory proteins in the cell nucleus. *J Steroid Biochem Mol Biol.* 76, 9–21.

Hoey T, Sun YL, Williamson K, and Xu X (1995). Isolation of two new members of the NF-AT gene family and functional characterization of the NF-AT proteins. *Immunity* 2, 461–472.

Hollingworth S, Zhao M, and Baylor SM (1996). The amplitude and time course of the myoplasmic free [Ca2+] transient in fast-twitch fibers of mouse muscle. *J Gen Physiol.* 108, 455–469.

Kiselyov KI, Shin DM, Wang Y, Pessah IN, Allen PD, and Muallem S (2000). Gating of store-operated channels by conformational coupling to ryanodine receptors. *Mol Cell* 6, 421–431.

Klein MG, Kovacs L, Simon BJ, Schneider MF (1991). Decline of myoplasmic Ca2+, recovery of calcium release and sarcoplasmic Ca2+ pump properties in frog skeletal muscle. *J Physiol.* 441, 639–671.

Korzus E, Rosenfeld MG, and Mayford M (2004). CBP histone acetyltransferase activity is a critical component of memory consolidation. *Neuron.* 42, 961–972.

Kubis HP, Haller EA, Wetzel P, and Gros G (1997). Adult fast myosin pattern and Ca2+-induced slow myosin pattern in primary skeletal muscle culture. *Proc Natl Acad Sci USA.* 94, 4205–4210.

Kubis HP, Scheibe RJ, Meissner JD, Hornung G, and Gros G (2002). Fast-to-slow transformation and nuclear import/export kinetics of the transcription factor NFATc1 during electrostimulation of rabbit muscle cells in culture. *J Physiol.* 541, 835–847.

Kubis HP, Hanke N, Scheibe RJ, Meissner JD, Gros G (2003). Ca2+ transients activate calcineurin/NFATc1 and initiate fast-to-slow transformation in a primary skeletal muscle culture. *Am J Physiol Cell Physiol.* 285, C56–C63.

Kurebayashi N, and Ogawa Y (2001). Depletion of Ca2+ in the sarcoplasmic reticulum stimulates Ca2+ entry into mouse skeletal muscle fibres. *J Physiol.* 533, 185–199.

Levenson JM, O'Riordan KJ, Brown KD, Trinh MA, Molfese DL, and Sweatt JD (2004). Regulation of histone acetylation during memory formation in the hippocampus. *J Biol Chem.* 279, 40545–40559.

Liu Y, Cseresnyes Z, Randall WR, and Schneider MF (2001). Activity-dependent nuclear translocation and intranuclear distribution of NFATc in adult skeletal muscle fibers. *J Cell Biol.* 155, 27–39.

Liu Y, Randall WR, and Schneider MF (2005a). Activity-dependent and -independent nuclear fluxes of HDAC4 mediated by different kinases in adult skeletal muscle. *J Cell Biol.* 168, 887–897.

Liu Y, and Schneider MF (1998). Fibre type-specific gene expression activated by chronic electrical stimulation of adult mouse skeletal muscle fibres in culture. *J Physiol.* 512, 337–344.

Liu Y, Shen T, Randall WR, and Schneider MF (2005b). Signaling pathways in activity-dependent fiber type plasticity in adult skeletal muscle. *J Muscle Res Cell Motil.* 26, 13–21.

Macian F, Lopez-Rodriguez C, and Rao A (2001). Partners in transcription: NFAT and AP-1. *Oncogene.* 20, 2476–2489.

Mateescu B, England P, Halgand F, Yaniv M, and Muchardt C (2004). Tethering of HP1 proteins to chromatin is relieved by phosphoacetylation of histone H3. *EMBO Rep.* 5, 490–496.

McCullagh KJ, Calabria E, Pallafacchina G, Ciciliot S, Serrano AL, Argentini C, Kalhovde JM, Lomo T, and Schiaffino S (2004). NFAT is a nerve activity sensor in skeletal muscle and controls activity-dependent myosin switching. *Proc Natl Acad Sci U S A.* 101, 10590–10595.

McKinsey, T.A., C. L. Zhang, J. Lu, and E. N. Olson. (2000a). Signal-dependent nuclear export of a histone deacetylase regulates muscle differentiation. *Nature.* 408, 106–111.

McKinsey, T. A., C. L. Zhang, and E. N. Olson. (2000b). Activation of the myocyte enhancer factor-2 transcription factor by calcium/calmodulin-dependent protein kinase-stimulated binding of 14-3-3 to histone deacatylase 5. *Proc. Natl. Acad. Sci.USA* 97, 14400–14405.

McKinsey TA, Zhang CL, and Olson EN (2001). Control of muscle development by dueling HATs and HDACs. *Curr Opin Genet Dev.* 11, 497–504.

Mejat A, Ramond F, Bassel-Duby R, Khochbin S, Olson EN, and Schaeffer L (2005). Histone deacetylase 9 couples neuronal activity to muscle chromatin acetylation and gene expression. *Nat Neurosci.* 8, 313–321.

Miska EA, Karlsson C, Langley E, Nielsen SJ, Pines J, and Kouzarides T (1999). HDAC4 deacetylase associates with and represses the MEF2 transcription factor. *EMBO J.* 18, 5099–5107.

Miska, E. A., E. Langley, D. Wolf, C. Karlsson, J. Pines, and T. Kouzarides. 2001. Differential localization of HDAC4 orchestrates muscle differentiation. *Nucleic Acids Res.* 29, 3439–3447.

Misteli, T. Protein dynamics: implications for nuclear architecture and gene expression.2001. *Science.* 291, 843–847.

Naya FJ, Mercer B, Shelton J, Richardson JA, Williams RS, and Olson EN (2000). Stimulation of slow skeletal muscle fiber gene expression by calcineurin in vivo. *J Biol Chem.* 275, 4545–4548.

Neal JW, and Clipstone NA (2001). Glycogen synthase kinase-3 inhibits the DNA binding activity of NFATc. *J Biol Chem.* 276, 3666–3673.

Nowak SJ, and Corces VG (2004). Phosphorylation of histone H3: a balancing act between chromosome condensation and transcriptional activation. *Trends Genet.* 20, 214–220.

Pan Z, Yang D, Nagaraj RY, Nosek TA, Nishi M, Takeshima H, Cheng H, and Ma J (2002). Dysfunction of store-operated calcium channel in muscle cells lacking mg29. *Nat Cell Biol.* 4, 379–383.

Parsons SA, Wilkins BJ, Bueno OF, and Molkentin JD (2003). Altered skeletal muscle phenotypes in calcineurin Aalpha and Abeta gene-targeted mice. *Mol Cell Biol.* 23, 4331–4343.

Payne AM, Zheng Z, Gonzalez E, Wang ZM, Messi ML, Delbono O (2004). External Ca(2+).-dependent excitation–contraction coupling in a population of ageing mouse skeletal muscle fibres. *J Physiol.* 560(Pt 1), 137–155.

Powell JA, Carrasco MA, Adams DS, Drouet B, Rios J, Muller M, Estrada M, and Jaimovich E (2001). IP(3) receptor function and localization in myotubes: an unexplored Ca^{2+} signaling pathway in skeletal muscle. *J Cell Sci.* 114, 3673–3683.

Rosenberg P, Hawkins A, Stiber J, Shelton JM, Hutcheson K, Bassel-Duby R, Shin DM, Yan Z, and Williams RS (2004). TRPC3 channels confer cellular memory of recent neuromuscular activity. *Proc Natl Acad Sci U S A.* 101, 9387–9392.

Salazar C and Hofer T (2003). Allosteric regulation of the transcription factor NFAT1 by multiple phosphorylation sites: a mathematical analysis. *J Mol Biol.* 327, 31–45.

Sheridan CM, Heist EK, Beals CR, Crabtree GR, and Gardner P (2002). Protein kinase A negatively modulates the nuclear accumulation of NF-ATc1 by priming for subsequent phosphorylation by glycogen synthase kinase-3. *J Biol Chem.* 277, 48664–48676.

Spangenburg EE, Williams JH, Roy RR, and Talmadge RJ (2001). Skeletal muscle calcineurin: influence of phenotype adaptation and atrophy. *Am J Physiol Regul Integr Comp Physiol.* 280, R1256–R1260.

Sparrow DB, Miska EA, Langley E, Reynaud-Deonauth S, Kotecha S, Towers N, Spohr G, Kouzarides T, Mohun TJ (1999). MEF-2 function is modified by a novel co-repressor, MITR. *EMBO J.* 18, 5085–5098.

Spector, D.L., X. D. Fu, and T. Maniatis. 1991. Associations between distinct pre-mRNA splicing components and the cell nucleus. *EMBO J.* 10, 3467–3481.

Swoap, S.J., R. B. Hunter, E. J. Stevenson, H. M. Felton, N. V Kansagra, J. M. Lang, K. A. Esser, and S. C. Kandarian. 2000. The calcineurin-NFAT pathway and muscle fiber-type gene expression. *Am. J. Physiol.* 279, C915–C924.

Tomida T, Hirose K, Takizawa A, Shibasaki F, Iino M (2003). NFAT functions as a working memory of Ca^{2+} signals in decoding Ca^{2+} oscillation. *EMBO J.* 22, 3825–3932.

Westerblad H, and Allen DG (1994). Relaxation, [Ca2+]i and [Mg2+]i during prolonged tetanic stimulation of intact, single fibres from mouse skeletal muscle. *J Physiol.* 480, 31–43.

Wu H, Naya FJ, McKinsey TA, Mercer B, Shelton JM, Chin ER, Simard AR, Michel RN, Bassel-Duby R, Olson EN, and Williams RS (2000). MEF2 responds to multiple calcium-regulated signals in the control of skeletal muscle fiber type. *EMBO J.* 19, 1963–1973.

Wu H, Rothermel B, Kanatous S, Rosenberg P, Naya F J, Shelton,JM, Hutcheson K A, DiMaio JM, Olson EN, Bassel-Duby R, and Williams RS (2001). Activation of MEF2 by muscle activity is mediated through a calcineurin-dependent pathway. *EMBO J.* 20, 6414–6423.

Wu X, Bossuyt J, Zhang T, McKinsey T, Brown JH, Olson EN, and Bers DM (2004). Excitation-transcription coupling in adult myocytes: local InSP3-dependent perinuclear signaling activates HDAC nuclear export. *Circulation* 110, III–285.

Youn HD, Grozinger CM, and Liu JO (2000). Calcium regulates transcriptional repression of myocyte enhancer factor 2 by histone deacetylase 4. *J Biol Chem.* 275, 22563–22567.

Zhang CL, McKinsey TA, and Olson (2001). The transcriptional corepressor MITR is a signal-responsive inhibitor of myogenesis. *Proc Natl Acad Sci U S A.* 98, 7354–7359.

Zimber A, Nguyen QD, Gespach C (2004). Nuclear bodies and compartments: functional roles and cellular signalling in health and disease. *Cell Signal.* 16, 1085–1104.

CHAPTER 6

THE REGULATION OF SATELLITE CELL FUNCTION IN SKELETAL MUSCLE REGENERATION AND PLASTICITY

MARK A. GILLESPIE, CHET E. HOLTERMAN
AND MICHAEL A. RUDNICKI

Ottawa Health Research Institute, Molecular Medicine Program, and Department of Cellular and Molecular Medicine, University of Ottawa, Ottawa, Ontario, Canada

1.   INTRODUCTION

Postnatal growth and regeneration of skeletal muscle is mediated by satellite cells, which are located adjacent to myofibers, outside the sarcolemma but beneath the basal lamina. Since their identification by Mauro in 1961, satellite cells have been extensively studied in hopes of determining the mechanisms controlling their function. Since skeletal muscle is a dynamic organ, satellite cell regulation is extremely important. Satellite cells can be activated by the everyday use of muscle, with heavy exercise, or even with acute injury. They are also activated during the regeneration observed with chronic muscle injury, which makes them an important target for treatment of muscle wasting diseases. Muscle degeneration results in release of growth factors from damaged muscle and infiltrating immune cells, which are involved in the activation of quiescent satellite cells and maintenance of their proliferative capacity. It is the signalling cascades downstream of these growth factor receptors that regulate such events as cell cycle progression and induction of muscle-specific gene expression and differentiation. These signals are of utmost importance in understanding the regulatory pathways of satellite cells.

The identification of bone marrow- and muscle-derived stem cells that can contribute to the renewal of the satellite cell compartment as well as regeneration of the fibers has also fuelled intensive research. The fact that a cell can be injected into the bloodstream or directly into the muscle and participate in muscle repair has excited many researchers and provides different opportunities for gene therapy of muscle wasting diseases.

137

R. Bottinelli and C. Reggiani (eds.), Skeletal Muscle Plasticity in Health and Disease, 137–172.
© 2006 *Springer.*

This review provides a detailed examination of skeletal muscle regeneration, with emphasis on the signalling mechanisms known to regulate satellite cell function. Initially, we discuss the identification of the satellite cell, its characteristics, its possible embryonic origins, followed by a detailed outline of the regeneration process with focus on the role of the myogenic regulatory factors (MRFs) and other transcription factors, such as Pax7. A comprehensive analysis of the signalling mechanisms known to date to be involved in the activation of quiescent satellite cells and maintenance their proliferation, along with those controlling their self-renewal are provided. We finish the review by highlighting the role of adult stem cells in regenerative myogenesis and the multipotentiality of satellite cells.

2.   SKELETAL MUSCLE STRUCTURE

Skeletal muscle can be divided into two key components, the myofibers that are responsible for the contractile function of the muscle and the connective tissue that provides a supportive framework that keeps the myofiber and surrounding structures intact during contraction (reviewed in Jarvinen et al., 2000). Individual myofibers are primarily composed of actin and myosin filaments that are arranged in a highly organized manner and provide the mechanical force for muscle contraction (Fig. 1). Each multinucleated myofiber is surrounded by an individual plasma membrane or sarcolemma and an extracellular matrix component known as the endomysium (also referred to as the basal lamina or basement membrane) (reviewed in Jarvinen et al., 2000). Located between the basal lamina and sarcolemma are quiescent

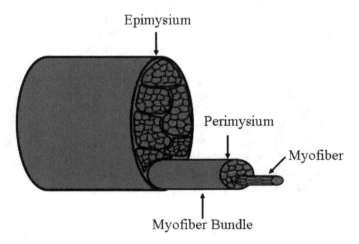

Figure 1. Skeletal muscle morphology. Myofibers are surrounded by a sarcolemma and a basal lamina (endomysium), with quiescent satellite cells located between the two layers. Myofibers are clustered together into a bundle and are surrounded by the perimycium. These bundles of myofibers are held together by the epimysium. A color figure is freely accessible via the website of the book: http://www.springer.com/1-4020-5176-x

mononuclear cells called satellite cells, which are the major effectors of skeletal muscle regeneration (reviewed in Bischoff, 1994; Charge and Rudnicki, 2004). The multinucleated myofibers and associated quiescent satellite cells are clustered into groups of ten to several hundred fibers and are surrounded by a second layer of connective tissue known as the perimysium. This group of myofibers is referred to as a fascicle. Finally, individual muscles are composed of groups of fascicles surrounded by a third layer of connective tissue referred to as the epimysium (reviewed in Huard et al., 2002; Jarvinen et al., 2000).

3.   SATELLITE CELL MORPHOLOGY

Skeletal muscle satellite cells were first described in the skeletal muscles of frogs and have since been identified in mammalian, avian, and reptilian skeletal muscle (Gamble et al., 1978; Hartley et al., 1992; Kahn and Simpson, 1974; Mauro, 1961). At birth, these cells account for 20–30% of sublaminar nuclei associated with mouse skeletal muscle. As mice grow and mature this number rapidly declines to approximately 5% at two months of age (Bischoff, 1994). After sexual maturity, the number of satellite cells continues to decline, albeit at a greatly reduced rate, such that approximately 2% of sublaminar nuclei in senile mice are quiescent satellite cells (Charge and Rudnicki, 2004; Hawke and Garry, 2001). Interestingly, the satellite cell compartment is maintained in adult muscle following multiple rounds of degeneration and regeneration suggesting that a mechanism exists for the self-renewal or maintenance of the quiescent satellite cell population (Morlet et al., 1989; Schultz and Jaryszak, 1985).

Satellite cells are traditionally defined by their unique position adjacent to the skeletal muscle fiber although recent work has identified several molecular markers of these cells such as c-met, Pax7, Syndecan3/4, and CD34 (Beauchamp et al., 2000; Cornelison et al., 2001; Cornelison and Wold, 1997; Seale et al., 2000; Tatsumi et al., 1998). Morphologically, satellite cells are situated in depressions in the muscle fiber, between the plasma membrane and the basal lamina of the muscle such that the basal lamina is continuous over the entire length of the fiber. Satellite cells are further identified by their relatively minute amount of cytoplasm, sparse organelles, and high ratio of heterochromatin to euchromatin, indicative of the inactive state of these cells (Schultz, 1976). These morphological criteria separate quiescent satellite cells from their descendent activated myogenic precursor cells (MPCs), the cells that contribute to muscle growth and repair. When activated in response to trauma, the morphology of the cell changes dramatically. The cytoplasmic volume of the activated cells increases and cytoplasmic extensions become apparent. Furthermore, the amount of heterochromatin decreases and organelles such as the Golgi apparatus, endoplasmic reticulum, ribosomes and mitochondria become apparent (Hawke and Garry, 2001; Schultz and McCormick, 1994).

Quiescent satellite cells are present throughout adult skeletal muscle but their distribution varies greatly with fiber type and muscle groups. In most instances, individual slow twitch fibers have a higher number of associated satellite cells than fast twitch fibers. Therefore, it follows that muscles composed mainly of slow twitch oxidative fibers tend to have more associated satellite cells than fast twitch glycolytic muscles (Charge and Rudnicki, 2004; Hawke and Garry, 2001). This unequal distribution of satellite cells is not only apparent between different fiber types and muscle groups but also manifests itself along individual fibers with higher numbers of satellite cells located at neuromuscular junctions as well as adjacent capillaries (reviewed in Charge and Rudnicki, 2004). One possible explanation for this phenomenon is that these structures may release factors that function in the homing of satellite cells or in the maintenance of satellite cell number. Alternatively, it may be that the surrounding capillaries provide a source of progenitor cells that contribute to the quiescent satellite cell compartment of adult skeletal muscle.

4.   EMBRYONIC ORIGINS OF SATELLITE CELLS

While the temporal appearance of satellite cells has been established in most verte-brates, the developmental origin of these cells has been controversial. Based on results obtained using quail/chick chimeras it was hypothesized that satellite cells are somatic in origin and thus share a common origin with the developing musculature (Armand et al., 1983). Interestingly, the myoblasts that populate the developing limb bud and give rise to skeletal muscle during development do not express MyoD or Myf-5 but do express c-met, an established marker of quiescent satellite cells. These myogenic precursors fail to migrate into the limb buds in the absence of c-met, resulting in a lack of limb musculature (Bladt et al., 1995). Importantly, the migrating myoblasts that give rise to the limb musculature do not proliferate during their migration raising the possibility that satellite cells are derived from migrating myogenic precursors which remain quiescent rather than proliferating and differentiating to contribute to the developing skeletal muscle.

The observation that myogenic cells resembling satellite cell-derived MPCs can be isolated from explants of the embryonic dorsal aorta of mice has challenged the notion that satellite cells are of somatic origin (De Angelis et al., 1999). The myogenic cells isolated from dorsal aorta explants express myogenic markers such as MyoD, Myf5, c-Met, desmin and M-cadherin as well as a variety of vascular endothelial markers. Furthermore, MPCs isolated from adult muscle express a variety of vascular-endothelial markers expressed in myogenic cells derived from the dorsal aorta, including VE-cadherin, VEGF-R2, and PECAM (De Angelis et al., 1999). These results suggest a possible endothelial origin for quiescent adult satellite cells. This hypothesis that satellite cells arise from the dorsal aorta is further supported by the fact that cells with myogenic capacity can be isolated from the developing limbs of Pax3 null and c-Met null mouse embryos (De Angelis et al., 1999). The developing limbs of these animals lack somatically derived migrating myogenic precursors but the vasculature remains undisturbed suggesting a non-somitic origin for MPCs in the limb buds of these mice. Importantly, trans-plantation experiments show that myogenic cells derived from the dorsal aorta are capable of contributing to muscle growth and regeneration in vivo, but whether or not these cells contribute to the quiescent satellite cell compartment remains to be clearly established (De Angelis et al., 1999). While these results draw into question the somatic origin of satellite cells, one must keep in mind that during development the dorsal aorta is colonized by somitic angioblasts (Pardanaud et al., 1996). These cells, derived from the somite, contribute to the endothelial lining of vessels as well as the roof and walls of the aorta raising the possibility that the cells responsible for the appearance of myogenic clones from dorsal aorta are indeed somitic in origin (Pardanaud and Dieterlen-Lievre, 1999; Pardanaud et al., 1996).

Recent observations from two groups have almost unequivocally proven that satellite cells arise from somatic origins. To track the movement of the cells responsible for the formation of skeletal muscle during development, Gros and colleagues used both electroporation to transfect GFP into the somites, along with chick-quail chimeras, and successfully demonstrated that cells originating from the central region of the dermomyotome were responsible for forming the large majority of Pax7+ satellite cells (91–95%) (Gros et al., 2005). At the same time, Relaix and colleagues used Pax7nlacZ and Pax3GFP transgenic mice and identified the same population of cells from the central dermomyotome of the somite, which give rise to Pax7+Pax3+ satellite cells (Relaix et al., 2005). These results are very interesting not only because they define the origin of satellite cells, but because they show that these same progenitor cells are also responsible for formation of the trunk and limb muscles during embryonic and fetal development (Gros et al., 2005; McKinnell and Rudnicki, 2005; Relaix et al., 2005).

Despite these convincing results, it is still possible that a small subset of satellite cells are endothelial in origin. Several lines of evidence point to a heterogeneous satellite cell population. For instance, quiescent satellite cells show heterogeneity in the molecular markers they express. Some satellite cells appear to express CD34 and M-cadherin while others do not (Beauchamp et al., 2000; Cornelison and Wold, 1997). Furthermore, close examination of the kinetics of proliferation reveal that while the majority of satellite cells enter a proliferative phase and eventually fuse to form multinucleated myotubes, a small proportion of satellite cells proliferate at a much slower rate and eventually return to a quiescent state rather than differentiating (Schultz, 1996). The possibility that satellite cells arise from two or more distinct developmental origins is an attractive explanation for the observed heterogeneity within the satellite cell compartment. Clearly though, based on very recent evidence, the majority of satellite cells arise from the central domain of the dermomyotome (Gros et al., 2005; Relaix et al., 2005).

5. CHARACTERISTICS OF SKELETAL MUSCLE REGENERATION

Muscle injury is usually associated with disruption of the actin/myosin filaments resulting in rupture of the myofiber and tearing of the sarcolemma. This results in increased permeability of the sarcolemma and subsequent calcium influx (reviewed in Belcastro et al., 1998). Increased levels of calcium within the myofiber activate calcium responsive proteases such as calpains, which results in tissue degeneration (reviewed in Belcastro et al., 1998). In extreme cases, the endomysium or basal lamina is also damaged. The broken ends that arise pull apart from one another creating a gap within the myofiber, resulting in a loss of contractile strength (Kaariainen et al., 2000). As well, given that individual myofibers are innervated at a single site, one portion of the damaged myofiber becomes denervated (Rantanen et al., 1995). All of these events lead to tissue degeneration and loss of contractile strength.

Following this type of injury, a general pattern of regeneration occurs. Within the first 24 hours following trauma, haemorrhaging occurs at the site of injury resulting in the formation of a haematoma within the gap (Kaariainen et al., 2000). As well, the cytoskeletal material near the ruptured ends of the fiber contracts to form the "contraction band". This band allows for the formation of a new plasma membrane and prevents the spread of necrosis through the entire myofiber (Hurme et al., 1991). The haemorrhaging allows for infiltration of inflammatory cells such as neutrophils and monocytes, which differentiate into macrophages, to the site of damage. Substances released from the necrotic areas also function as chemoattractants for inflammatory cells. The infiltrating neutrophils and monocyte/macrophages phagocytose the necrotic debris and break down the haematoma. At the same time they release cytokines and growth factors that stimulate regeneration in the damaged area mainly through the activation of quiescent satellite cells (Fig. 2A).

5.1   Satellite Cell Activation and Myogenic Regulatory Factor Function

Approximately 24 hours following damage, the majority of quiescent satellite cells adjacent to the damaged area are activated and upregulate the myogenic regulatory factors (MRFs) in a carefully orchestrated manner. Quiescent satellite cells do not express any detectable levels of MRF transcript or protein, and following activation, initially upregulate either MyoD or Myf5 prior to cell cycle entry. This is followed by co-expression of both MRFs, which defines the cell as a myogenic precursor cell (MPC) (Cooper et al., 1999; Cornelison and Wold, 1997; Smith et al., 1994; Yablonka-Reuveni and Rivera, 1994). MPCs continue to proliferate in response to growth factors and cytokines released from inflammatory cells within the damaged area (Fig. 2A,B,D,E).

An important role for MyoD in regulating skeletal muscle regeneration was discovered by analyzing MyoD-deficient muscle. MyoD-/- muscle has elevated numbers of satellite cells, which show upregulated Myf5 expression, and fail to differentiate (Cornelison et al., 2000; Sabourin et al., 1999; Yablonka-Reuveni et al., 1999a). When interbred with the dystrophic mdx mouse, mdx:MyoD-/- muscle is severely defective in regeneration, leading to the hypothesis that MyoD-deficient satellite cells have an increased propensity for self-renewal (Megeney et al., 1996a). Supporting this, MyoD-/- satellite cells express elevated levels of the growth factor IGF-I, which normally signals during the satellite cell proliferation phase

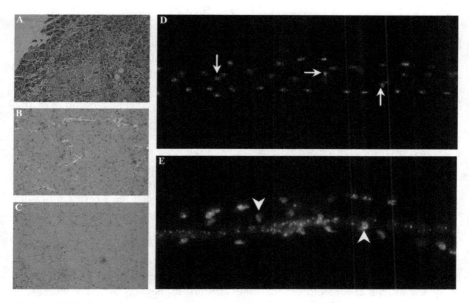

Figure 2. Skeletal muscle regeneration following cardiotoxin injury. (a) Cross-section of mouse tibialis anterior muscle during degeneration/early regeneration showing myofiber necrosis with large numbers of mononuclear cells (activated satellite cells combined with infiltrating immune cells). (b) During the late regeneration stages, muscle precursor cells are proliferating, differentiating and fusing to necrotic fibers to generate new fibers. This process is characterized by the presence of centrally located nuclei. (c) As the muscle finishes repairing itself, most nuclei are again observed along the periphery of the myofibers. (d) Control single myofiber in culture stained with Pax7 (red) to detect satellite cells (arrows). Note the low number of Pax7+ cells compared with the total number of nuclei (DAPI – green). (e) Confocal image of a single myofiber isolated 6 days after injury. Note the increase in Pax7+ cells, marking the activated satellite cells. Some activated satellite cells are in the process of division (arrowheads). [Figures a-c courtesy of A. Scimè and d-e courtesy of S. Kuang] A color figure is freely accessible via the website of the book: http://www.springer.com/1-4020-5176-x

(Sabourin et al., 1999). These studies also highlight the importance of Myf5 in promoting satellite cell renewal (discussed later).

Recently, gene expression profiling during muscle regeneration has allowed Zhao and colleagues to identify the zinc finger transcription factor Slug as a downstream target of MyoD (Zhao et al., 2002). The Slug promoter contains an E-box within a proximal 1kb region, through which MyoD associates and transactivates. Slug expression is increased later during regeneration, coinciding with fiber regeneration, and is also upregulated during C2C12 myoblast differentiation. Slug null muscle is defective in regeneration, with impaired fiber formation and decreased fiber cross-sectional area. Notably, the slug/snail family of transcription factors function both as transcriptional repressors and activators, and it will be interesting to determine which target genes they are regulating and the mechanisms behind this regulation (De Craene et al., 2005).

5.2   Muscle Precursor Cell Differentiation and Fusion

Depending on the severity of the injury and environmental cues, MPCs can undergo several rounds of division prior to differentiation and fusion into multinucleated myotubes. Two to three days following injury MPCs begin to undergo differentiation and fuse to form multinucleated myotubes within the remaining basal lamina of damaged fibers. These myotubes fuse to the surviving portions of the damaged myofiber once again creating an intact myofiber capable of providing contractile strength. As well, MPCs can fuse with each other and form new multinucleated myofibers to repair extreme damage within a muscle. The repaired and newly formed myofibers can be distinguished from undamaged fibers by their centrally located nuclei and lack of striation. By several weeks following injury, the nuclei of regenerated fibers become peripherally located and the fibers acquire the striated appearance rendering them indistinguishable from previously undamaged fibers (Fig. 2C).

One regulator of MPC fusion that appears to have an important role in regulating muscle regeneration is the calcium-dependent transmembrane adhesion protein M-cadherin (Zeschnigk et al., 1995). M-Cadherin is expressed on a small subset of quiescent c-met+ satellite cells, and appears to be concentrated where the satellite cell contacts the muscle fiber, suggesting it may also regulate satellite cell location (Cornelison and Wold, 1997; Irintchev et al., 1994). Expression of M-cadherin is upregulated on satellite cells during muscle regeneration, and is subsequently downregulated following myoblast fusion into nascent or existing fibers (Irintchev et al., 1994; Moore and Walsh, 1993). M-Cadherin expression is also downregulated in MyoD-/- myoblasts, which are differentiation impaired and fail to fuse (Cornelison et al., 2000; Sabourin et al., 1999). These results suggest that M-cadherin is also important for myoblast fusion during muscle regeneration. However, gene targeting of the M-cadherin locus in mouse argues against this function (Hollnagel et al., 2002). M-Cadherin-/- muscle regenerates normally following cardiotoxin injection, while satellite cell-derived myoblasts differentiate and fuse normally in culture. These results would seem to argue against a role for M-cadherin in muscle regeneration, however, there does appear to be a functional compensation for M-cadherin loss by N-cadherin, which is upregulated in M-cadherin-deficient muscle. Studies in mice lacking both these cadherin family members may prove informative in determining if M-cadherin is required for muscle repair.

6.   SATELLITE CELLS AND SKELETAL MUSCLE ADAPTATION

In addition to their role in regeneration following injury, satellite cells are also believed to function in the adaptive response of skeletal muscle to exercise. Resistance training results in the hypertrophy of skeletal muscle, as measured by increases in fiber cross-sectional area. Satellite cell participation in hypertrophy has long been thought to be essential. The myonuclear domain theory states that each myonucleus can only maintain a fixed amount of cytoplasm, implying that any increase in cytoplasmic volume, as seen in hypertrophic fibers, would need to be accompanied by an increase in the number of myonuclei. And the only way to increase the number of myonuclei within a fiber is through satellite cell activation, differentiation, and fusion into existing fibers (reviewed in Allen et al., 1999; Hawke, 2005). Early work demonstrated that proliferating satellite cells eventually fuse to existing fibers during hypertrophy, and that ablation of satellite cells by gamma irradiation actually prevents muscle hypertrophy (Rosenblatt and Parry, 1992; Rosenblatt et al., 1994; Schiaffino et al., 1976).

Recent work has challenged the notion that fusion of satellite cells into existing fibers is necessary for hypertrophy. As expected, resistance training in humans results in significant increases in satellite cell number and fiber cross-sectional area (Kadi et al., 2004). Surprisingly though, these hypertrophic fibers do not have increased numbers of myonuclei despite the expression of p21, a marker for cell cycle withdrawal (Kadi et al., 2004). These results suggest that following activation, satellite cells in this model are returning to a state of quiescence rather than differentiating and fusing into existing fiber. Although this study did not address whether satellite cells are forming nascent fibers, it has previously been documented that hyperplasia is not a mechanism for increased muscle mass under these experimental conditions (reviewed in Allen et al., 1999). Therefore, the existing myonuclei must still possess the ability to increase their respective cytoplasmic volumes enough to account for the level of hypertrophy observed in this exercise model, while still following the rules of the myonuclear domain theory (Kadi et al., 2005; Kadi et al., 2004). It is likely that different models of exercise, which result in significantly more damage and induce greater levels of hypertrophy, would require the fusion of differentiating satellite cells into existing fibers. In fact, Crameri and colleagues have shown that single bouts of high intensity exercise, while sufficient to activate satellite cells, does not induce their differentiation, implying greater damage is needed for these cells to initiate their differentiation program (Crameri et al., 2004).

Chronic low-frequency stimulation (CLFS) results in satellite cell activation, as measured by increased numbers as well as BrdU and PCNA immunochemistry to mark proliferating cells. An interesting observation of this technique is the absence of any fiber degeneration or regeneration (Pette and Dusterhoft, 1992; Putman et al., 1999; Putman et al., 2000). Furthermore, CLFS induces a fast to slow fiber type conversion when administered to fast-twitch muscles, as evidenced by increased expression of the slow myosin heavy chain (MyHC) isoforms MyHC I and MyHC IIa (Putman et al., 1999). Since slow-twitch fibers have higher numbers of satellite cells compared to fast-twitch, it has been hypothesized that satellite cells are involved in fiber-type switching. Putman and colleagues provide circumstantial evidence suggesting that the increase in satellite cell number in fast-twitch fibers resulting from CLFS is actually a prerequisite for MyHC isoform switching (Putman et al., 1999; Putman et al., 2000). By contrast, Rosenblatt and Parry have directly demonstrated that satellite cells are not required for fiber type switching by observing changes in MyHC isoform expression following ablation of the satellite cell population by gamma irradiation (Rosenblatt and Parry, 1992). Further evidence will be required to address the discrepancies between these two models.

7.   PAX7 FUNCTION DURING SKELETAL MUSCLE REGENERATION

Pax7 is a paired box transcription factor that our lab began to study following its identification by representational difference analysis (RDA) as a gene expressed specifically in satellite cell-derived myoblasts (Seale et al., 2004a; Seale et al., 2000). Pax7 mRNA is expressed in proliferating satellite cell-derived and C2C12 myoblasts, albeit weakly in C2C12 myoblasts, and is rapidly downregulated following induction of differentiation (Seale et al., 2000). Further analysis detected Pax7 expression in approximately 5% of wild type muscle nuclei and 22% of MyoD-/- muscle nuclei in vivo, with an upregulation during regeneration in mdx muscle compared to wild type (Seale et al., 2000). These observations are consistent with the notion that Pax7 is expressed in satellite cells.

Mice lacking Pax7 through gene targeting appear normal at birth, but have a 50% reduction in body weight by the time they reach 1 week of age, and manage only to survive up to 2 weeks (Mansouri et al., 1996; Seale et al., 2000). Interestingly, Pax7-deficient mice are completely devoid of satellite cells in the gastrocnemius muscle of 7- to 10-day old mice in vivo, as observed by electron microscopy. This was confirmed by the inability to detect c-met or desmin positive satellite cell-derived myoblasts in culture (Seale et al., 2000). This result, combined with the observation that Pax7-/- muscle-derived stem cells displayed an increased potential towards the hematopoietic lineage, led to the hypothesis that Pax7 was necessary for the specification of satellite cells.

Recently, certain groups have identified the presence of satellite cells in Pax7-deficient muscle, albeit significantly reduced in number. Although their origin is not clear, it is possible they are a subpopulation of satellite cells that express Pax3 alone (Pax3+/Pax7-), or both Pax3 and Pax7 (Pax3+/Pax7+), or simply do not express Pax7 at all (Pax7-/Syndecan4+/Pax3?). Buckingham et al. detect a population of Pax3+ satellite cells in the gracilis muscle using a Pax3-nlacZ transgene, however it is unclear whether they all coexpress Pax7 (Buckingham et al., 2003). The same group has also very recently identified a population of muscle progenitor cells that are Pax7+/Pax3+ and adopt a similar location to satellite cells, juxtaposed between the sarcolemma and the basal lamina, by embryonic day 18.5 (Relaix et al., 2005). Satellite cells which are Pax7-/syndecan-4+ have also been described, however, it is unknown whether these express Pax3 or not (Olguin and Olwin, 2004).

The presence of satellite cells in Pax7-/- muscle has led to the suggestion that Pax7 does not in fact specify this lineage (Oustanina et al., 2004). Cardiotoxin injection into Pax7-/- TA muscle results in fiber necrosis, indicating that regeneration is still impaired in the absence of Pax7. These results suggest that Pax7 may function in the survival and renewal of the satellite cell population, rather than the specification;

however, further experiments are needed to determine the exact role of Pax7 in satellite cell regulation. Examination of the regulatory networks controlling Pax7 function, as well as the identification of Pax7 target genes, should also help to provide further insight into this dilemma.

8.   ROLE OF FOXK1/MNF IN SKELETAL MUSCLE REGENERATION

The forkhead/winged helix transcription factor foxk1 (originally cloned as myocyte nuclear factor/MNF) regulates satellite cell proliferation. Foxk1 expressing nuclei makeup 2–5% of total nuclei and occupy a sublaminar position in adult skeletal muscle, identical to that of quiescent satellite cells (Garry et al., 1997). Following cardiotoxin injury, Foxk1 expression is increased in a temporal fashion that coincides with satellite cell activation and proliferation (Garry et al., 1997). Foxk1-deficient mice display a runted phenotype similar to that of Pax7 mutants, but remain viable and their muscles contain satellite cells (Garry et al., 2000). These mice display muscle regeneration defects, likely due to altered cell cycle regulation. In particular, the cdk inhibitor p21 is upregulated in Foxk1-/- muscle, resulting in G0/G1 arrest (Hawke et al., 2003). Crossing Foxk1-/- with p21-/- mice rescues this proliferation defect and muscle regeneration appears normal by 10 days post-cardiotoxin injection (Hawke et al., 2003). However, no other earlier time points are mentioned, and these are important considering Foxk1 expression is highest early in regeneration. These results do suggest that Foxk1 functions during satellite cell proliferation by regulating cell cycle progression.

9.   GROWTH FACTOR REGULATION OF SATELLITE CELL FUNCTION

Skeletal muscle regeneration requires the activation of quiescent satellite cells, followed by their proliferation, differentiation, and formation into myofibers. The study of growth factors and other signals involved in activating and maintaining satellite cell activation has advanced significantly since the early days where Bischoff described the presence of a mitogen in extracts prepared from crush-injured muscle (Bischoff, 1986; Bischoff, 1990). Several growth factors, including HGF, FGF, IGF, TGF-β, and LIF, along with other signaling molecules such as nitric oxide and heparin sulfate proteoglycans have been identified and shown to function in regulating muscle regeneration. These signals can originate from the damaged fiber, from infiltrating immune cells such as macrophages, or even from satellite cells themselves (Tidball, 2005 and references below). Importantly, the only growth factor which has been proven to activate satellite cells from their quiescent state is HGF (Bischoff, 1986; Tatsumi et al., 1998). This next section discusses the roles of the above-mentioned signaling molecules, including some downstream cascades, during muscle regeneration.

9.1   HGF

Hepatocyte growth factor/scatter factor (HGF/SF) was originally isolated from rat
serum as a mitogen participating in liver regeneration following partial hepatectomy
(Nakamura et al., 1984). Active HGF exists as an α-β heterodimer, produced from
the proteolytic digestion of a biologically inert pro-HGF (reviewed in Zarnegar and
Michalopoulos, 1995). Importantly, HGF is also a responsible for the activation
of quiescent satellite cells during skeletal muscle regeneration. Jennische and
colleagues initially described a role for HGF in adult muscle regeneration by
demonstrating that HGF mRNA was transiently upregulated during the early phases
of regeneration, after which it returned to barely detectable levels (Jennische
et al., 1993).

Bischoff initially observed that a crushed muscle extract (CME) contained a
mitogen that could induce quiescent satellite cells to transiently enter the cell
cycle, however, it was not until 12 years later that this mitogen was identified as
HGF (Bischoff, 1986; Bischoff, 1990; Tatsumi et al., 1998). Addition of anti-HGF
antibody to this CME neutralized its effects and prevented HGF-induced satellite
cell proliferation (Tatsumi et al., 1998). HGF was able to increase the proliferation
of satellite cells grown in vitro by stimulating entry into the cell cycle (Allen
et al., 1995; Gal-Levi et al., 1998; Tatsumi et al., 2001). Moreover, HGF prevented
the induction of genes typically associated with myogenic differentiation (Gal-Levi
et al., 1998). Notably, HGF was able to induce the chemotaxis of satellite cells
in culture, suggesting it also regulates satellite cell migration during regeneration
(Bischoff, 1997).

HGF also activates satellite cells in vivo. Injection of HGF into either normal or
injured tibialis anterior (TA) muscle resulted in increased satellite cell proliferation
(Miller et al., 2000; Tatsumi et al., 1998). Interestingly, the administration of HGF
into injured TA muscle did not enhance regeneration, which can be attributed to
the mitogenic HGF signals blocking myogenic differentiation (Miller et al., 2000).
This suggests that satellite cells must downregulate HGF signaling prior to initiating
differentiation

The proto-oncogene c-met encodes a receptor tyrosine kinase which binds HGF.
As with HGF, c-met is an α-β heterodimer produced by cleavage of a single
precursor polypeptide. It is the β chain of this heterodimer that anchors this
receptor in the membrane and also directly mediates signaling via its cytoplasmic
tyrosine kinase domain (reviewed in Birchmeier and Gherardi, 1998; Zarnegar and
Michalopoulos, 1995). c-met expression is detected in both quiescent and activated
satellite cells, but not in muscle-derived fibroblasts (Cornelison and Wold, 1997;
Tatsumi et al., 1998). Expression of a constitutively active catalytic domain mutant
of c-met prevented expression of MyoD, myogenin, and myosin heavy chain,
further proving that HGF signaling prevents myogenic differentiation (Anastasi
et al., 1997).

HGF is considered both an autocrine and paracrine mitogen for satellite cells.
HGF mRNA is present in activated satellite cells and newly formed myotubes,
while protein is detected in the conditioned media used to grow the satellite cells

(Anastasi et al., 1997; Gal-Levi et al., 1998; Sheehan et al., 2000; Tatsumi et al., 2001). Moreover, activated satellite cells in regenerating mdx muscle coexpress HGF and c-met in vivo (Tatsumi et al., 1998). HGF secreted by activated satellite cells is functional, since regulatory tyrosine residues on the c-met intracellular domain are constitutively phosphorylated (Anastasi et al., 1997). In addition, conditioned media from these activated satellite cells can activate quiescent satellite cells (Tatsumi et al., 2001). HGF is also detected in the extracellular matrix surrounding the muscle fibers, likely sequestered there through interactions with heparin sulfate proteoglycans. Activation of satellite cells by stretch results in release of HGF reserves from the extracellular matrix (Tatsumi et al., 1998; Tatsumi et al., 2001). Taken together, the results of these studies define an essential role for HGF in activating the quiescent satellite cell.

9.1.1   Nitric oxide

Nitric oxide (NO) is a small, freely diffusible molecule that is synthesized by the enzyme neuronal nitric oxide synthase (nNOS) in skeletal muscle. Anderson was the first to provide evidence showing NO was involved in the rapid activation of satellite cells following crush injury (Anderson, 2000). Moreover, pharmacological inhibition of nNOS prevented this immediate activation of satellite cells (Anderson, 2000). Interestingly, Anderson observed satellite cell activation in contralateral muscles, suggesting the existence of a circulating factor that is activated by NO signaling (Anderson, 2000). As it turns out, this factor is likely HGF. Conditioned media from stretched satellite cells was able to activate separate quiescent satellite cells due to the presence of HGF (Tatsumi et al., 2002). However, following inhibition of nNOS activity, conditioned media could no longer activate satellite cells, which was attributed to the lack of HGF in this media (Tatsumi et al., 2002). Together, these results suggest that NO is indirectly involved in the activation of quiescent satellite cells by stimulating the release of HGF.

9.2   FGFs

Fibroblast growth factor (FGF) was first identified in bovine pituitary and brain extracts as a mitogen that could induce the proliferation of fibroblasts in culture (reviewed in Mohammadi et al., 2005). Simplistically speaking, FGF ligands bind their receptors (FGFRs), inducing receptor dimerization, which allows for transphosphorylation of regulatory tyrosine residues by the intracellular kinase domains (reviewed in Mohammadi et al., 2005). These phosphotyrosine residues provide docking sites for signaling molecules involved in regulating the activation of downstream cascades such as the mitogen-activated protein kinase (MAPK) family.

It was originally Allen and colleagues who demonstrated this bovine pituitary FGF could function as a satellite cell mitogen by increasing their rate of proliferation (Allen et al., 1984). Currently, eighteen isoforms of FGF have been identified, with only FGF-1, FGF-2, FGF-4, FGF-6, and FGF-9 were able to increase proliferation of satellite cells in vitro, while FGF-1 and FGF-2 have been shown to prevent their differentiation (Clegg et al., 1987; Kastner et al., 2000; Sheehan and Allen, 1999). Interestingly, only FGF-6 has a skeletal muscle restricted expression pattern, yet its role in regulating muscle regeneration is controversial. Floss et al. noted that FGF-6 gene expression was upregulated during regeneration, and decided to further investigate its role using gene targeting techniques (Floss et al., 1997). Homozygous null FGF-6 mice exhibit impaired regeneration following freeze-crush injury or when interbred with mdx mice. Specifically, loss of FGF-6 results in decreased MyoD and myogenin expression, along with myotube degeneration and increased collagen deposition and fibrosis (Floss et al., 1997). However, when Fiore et al. examined their FGF6-/- mice, they observed no defects in muscle regeneration following notexin injection or crush injury (Fiore et al., 1997; Fiore et al., 2000). There was also no effect when these FGF6-/- mice were interbred with mdx mice. These results suggest that additional redundant factors may be present that can compensate for the lack of FGF-6 signaling in these mutants, as is often observed in mice with genetic disruptions. Fiore et al. suggest that FGF-4 could be this redundant factor, since it is highly similar to FGF-6 and binds to the same receptor. In addition, administration of exogenous FGF-4 stimulates the greatest proliferative response of satellite cells compared to all other FGFs (Sheehan and Allen, 1999). Further experiments crossing different conditional FGF targeted mice should be helpful in solving this controversy.

In addition to FGF-6, research has focused on the role of FGF-2 in promoting muscle regeneration. FGF-2, or basic FGF (bFGF), is released from the cytoplasm of muscle fibers following injury, suggesting it is an important factor is regeneration (Clarke et al., 1993). Furthermore, it is also detected within the extracellular matrix of skeletal muscle, with elevated levels in the highly regenerative mdx muscle (DiMario et al., 1989). In a dose-dependent manner, addition of FGF-2 to satellite cell cultures in vitro increased their proliferation rates (DiMario and Strohman, 1988; Johnson and Allen, 1993; Yablonka-Reuveni et al., 1999b). Injection of varying concentrations of FGF-2 into mdx TA muscle resulted in increased regeneration, which was proportional to FGF-2 concentration and was attributed to increased satellite cell growth (Lefaucheur and Sebille, 1995). By contrast, Mitchell and colleagues observed no enhancement of muscle regeneration with FGF-2 administration following crush injury, denervation, or in muscle from the mdx mouse, suggesting that FGF-2 concentrations are likely not limiting in certain injury models (Mitchell et al., 1996).

There are currently four known FGF receptors, namely FGFR-1 to FGFR-4, with FGFR-1 and FGFR-4 being highly expressed in satellite cells (Kastner et al., 2000; Sheehan and Allen, 1999). Following satellite cell activation in culture, FGFR-1 expression is dramatically upregulated, which is a result of FGF-1 and HGF signaling (Sheehan and Allen, 1999). As expected, treatment of satellite cells with HGF in vitro accelerates this increase in FGFR-1 transcript number (Sheehan and Allen, 1999). As further evidence of a role for FGF signaling in proliferation, myoblasts expressing exogenous FGFR-1 show increased proliferation with a subsequent decrease in the ability to undergo differentiation. This phenotype can be reversed through expression of a truncated mutant of FGFR-1 (Scata et al., 1999).

Unlike HGF-c-met signaling, the pathways functioning downstream of the FGFRs during satellite cell activation have been better defined (Fig. 3). Treatment of satellite cells in culture with FGF-2 results in stimulation of the ERK MAPK signaling cascade, as measured by the activity of MEK and ERK (Campbell et al., 1995). Interestingly, ERK activation was delayed compared with MEK due to the presence of a MAPK phosphatase (Campbell et al., 1995). Pharmacological inhibitors of this pathway prevented activation of both MEK and ERK by FGF-2, which decreased satellite cell proliferation (Jones et al., 2001). Furthermore, attenuation of this pathway using a dominant negative mutant of the upstream kinase Raf-1 also blocked proliferation (Jones et al., 2001). Unexpectedly though, blocking the Raf-MEK-ERK pathway does not promote differentiation, implying that other signaling pathways, perhaps those which are FGF-dependent, function in satellite cells to control proliferation and differentiation (Jones et al., 2001).

Expression of the oncogenic Ras mutant Ha-RasG12V (RasV12) enhances satellite cell proliferation in vitro (Fedorov et al., 2001). RasV12 signaling also increases expression of the atypical protein kinase C (aPKC) isoforms PKCλ and PKCζ, as well as nuclear translocation of PKCλ, a hallmark of its activation (Fedorov et al., 2002). Inhibition of aPKC signaling prevents Ras-mediated repression of differentiation, without affecting MAPK activation (Fedorov et al., 2002). These results suggest that aPKC signaling can cooperate with MAPK signaling to regulate myogenic differentiation.

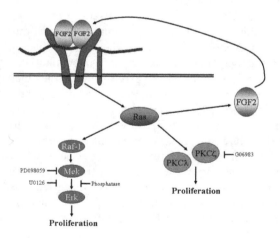

Figure 3. FGF-2 signaling controls satellite cell proliferation. Aided by heparin sulfate (bound to the cell surface by syndecans), FGF-2 associates with and activates the FGFR, resulting in activation of Ras. Ras downstream effectors include the ERK MAPK cascade and the aPKCs, which all promote proliferation. Inhibition of either pathway blocks proliferation, but only aPKC inhibition allows for differentiation to proceed. Ras signaling is also responsible for activating FGF-2, as a positive feedback loop mechanism. A color figure is freely accessible via the website of the book: http://www.springer.com/1-4020-5176-x

Surprisingly, RasV12-induced satellite cell proliferation can be blocked with antibodies directed against FGF-2, suggesting that FGF-2 is in part regulated downstream of Ras (Fedorov et al., 2001). Fedorov et al. propose that Ras is involved in activating extracellularly sequestered FGF-2, however further experiments need to be done to address this hypothesis. Another potential hypothesis could be that Ras downstream effectors activate FGF-2 gene expression, and this is required to sustain FGFR activation in a positive feedback regulatory loop. Taken together, these results demonstrate that both autocrine and paracrine FGF signaling is necessary in controlling satellite cell proliferation.

9.2.1   Heparin sulfate proteoglycans

Both HGF and FGF require heparin sulfate for activation of their respective receptors. The syndecans are a family of heparin sulfate proteoglycans (HSPGs) that share homology between their transmembrane and intracellular domains, and are expressed in skeletal muscle (reviewed in Rapraeger, 2000). It was Cornelison and colleagues who initially noted syndecan-3 and syndecan-4 expression on quiescent and activated satellite cells, and questioned whether they functioned in the activation of satellite cells during muscle regeneration (Cornelison et al., 2001). Furthermore, syndecan-3 expression was upregulated on satellite cells following injury by barium chloride injection (Casar et al., 2004). Treatment of muscle fibers in culture with sodium chlorate, which inhibits protein sulfation and therefore activation of FGFRs and c-met, retards satellite cell proliferation, emphasizing a role for syndecan-3 and syndecan-4 during satellite cell activation (Cornelison et al., 2001).

Mice lacking either syndecan-3 or syndecan-4 display strikingly different phenotypes (Cornelison et al., 2004). Syndecan-3-/- mice have an excess of satellite cells, and undergo chronic regeneration in the absence of damage. Despite this excess of satellite cells, many do not express MyoD, and if they do, it is incorrectly localized, suggesting it is not active. Upon barium chloride injection, these mice appear to regenerate normally, despite altered MyoD expression. On the other hand, syndecan-4-/- mice do not display any chronic regeneration, and are severely impaired in satellite cell activation and proliferation. Treatment of the TA muscle with barium chloride demonstrates these mice significantly lack the ability to regenerate properly because of their failure to activate satellite cells. These results suggest that syndecan-3 may have an inhibitory function with respect to satellite cell activation, whereas syndecan-4 is necessary for activation. There also appears to be no functional redundancy between these two proteins, as no compensation is observed in the null mice. It will also be interesting to determine if either syndecan is specifically required for the function of a particular FGFR or of c-met.

9.3   IGFs

The insulin-like growth factors (IGFs) IGF-I and IGF-II are members of a family of small signaling peptides, similar in structure to proinsulin, which associate with the IGF receptors (IGF-IR and IGF-IIR). IGF-I was initially isolated as a sulfation factor released into the circulation by growth hormone stimulation. Signaling downstream of the IGFRs is complex, with activation of MAPK cascades, as well as regulation of lipid signaling through activation of phosphoinositide 3'-kinase (PI3K) (reviewed in Florini et al., 1996; Florini et al., 1991). The dual role of IGF-I and IGF-II signaling in positively regulating both myoblast proliferation and differentiation has been well documented (Engert et al., 1996; Florini et al., 1996). In fact, introduction of exogenous IGF into muscle results in increased muscle mass, due to the activation of satellite cells and to fiber hypertrophy as a result of increased protein synthesis (Chakravarthy et al., 2000b; Florini et al., 1996; Musaro et al., 2001; Musaro et al., 1999; Semsarian et al., 1999).

Following skeletal muscle injury induced by ischemia or notexin injection, IGF-I expression is upregulated initially, although it does not reach maximal levels until several days after injury (Edwall et al., 1989; Levinovitz et al., 1992). Moreover, IGF-II transcripts are also upregulated, however this occurs following IGF-I upregulation (Levinovitz et al., 1992). In vitro, IGF-I stimulates both the proliferation and differentiation of satellite cells, with a greater enhancement of differentiation (ALLEN 1989). IGF-1 mRNA is also upregulated in MyoD-/- myogenic cells, which are highly proliferative and represent a cell similar to an activated satellite cell (Sabourin et al., 1999). Together, these results favor a role for IGFs during satellite cell activation as well as a potent inducer of differentiation.

IGF-I expression is also able to rescue the "aging muscle phenotype" of senescent muscles. As muscles age, they lose their mass and their regenerative potential, two processes that are regulated by IGF signaling. Transgenic overexpression of IGF-I allows aged muscle to maintain its regenerative capacity following cardiotoxin injection or nerve injury, which is not only a result of fiber hypertrophy, but also of satellite cell proliferation (Musaro et al., 2001; Rabinovsky et al., 2003). IGF-I infusion into aged muscle following lengthy immobilization resulted in increased proliferation of satellite cells, resembling that of a much younger mouse (Chakravarthy et al., 2000b).

These observations led to the hypothesis that IGF signaling may be able to counteract the degeneration of dystrophic muscle. In Duchenne muscular dystrophy, the loss of functional dystrophin protein compromises the structural integrity of the muscle fibers. The muscle can initially regenerate itself, but eventually the satellite cell pool becomes exhausted, resulting in an imbalance between muscle degeneration and muscle repair. Transgenic overexpression of IGF-I in the mdx mouse background results in increased muscle mass and force generation, while showing decreased fiber degeneration (Barton et al., 2002). In addition to IGF signaling promoting satellite cell proliferation and hypertrophy, it is also involved in cell survival. Barton and colleagues detected increased phosphorylation of Akt, an anti-apoptotic effector downstream target of PI3K, in their mdx:mIGFI transgenic mice (Barton et al., 2002). Furthermore, expression of an IGF-II transgene also promotes cell survival in the dystrophic mdx mouse (Smith et al., 2000).

The PI3K-Akt signaling pathway downstream of IGF is also involved in mediating the increase in satellite cell proliferation by controlling specific cell cycle

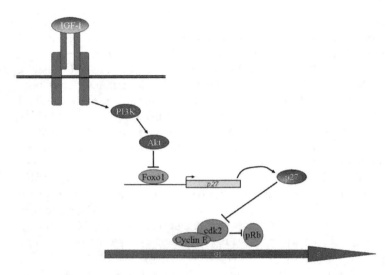

Figure 4. IGF-I signaling promotes G1/S progression. IGF-I binding to the IGF-IR induces activation of PI3K and Akt. Phosphorylation of the transcription factor Foxo1 by Akt prevents it from activating transcription of the cdk inhibitor p27. Without its inhibitor present, cyclin E – cdk2 complexes allow cells to progress into S phase. Cdk2 is also responsible for phosphorylating pRb, which dissociates it from E2Fs, preventing G1 arrest. A color figure is freely accessible via the website of the book: http://www.springer.com/1-4020-5176-x

regulators (Fig. 4). Activation of Akt results in phosphorylation of the forkhead transcription factor Foxo1, inhibiting its function and resulting in a subsequent downregulation in the expression of the cyclin-dependent kinase (cdk) inhibitor p27 (Chakravarthy et al., 2000a; Machida et al., 2003). Inhibition of p27 leads to increased cyclin E-cdk2 activity, along with increased phosphorylation of Rb, thereby promoting progression through the G1/S phase of the cell cycle, and increased cell cycling (Chakravarthy et al., 2000a).

9.4   LIF

Leukemia inhibitory factor (LIF) is a member of the IL-6 family of cytokines, and binds the heterodimeric LIF receptor, which is comprised of the LIF receptor β subunit (LIFR/LIFRβ) and the gp130 subunit (reviewed in Ihle, 1995). LIF signaling in skeletal muscle is pro-mitogenic, and its expression is upregulated in mdx muscle or following injury induced regeneration in wild type muscle (Austin et al., 1992; Barnard et al., 1994; Kami and Senba, 1998; Kurek et al., 1996; Spangenburg and Booth, 2002). Addition of LIF to crush-injured muscle increased the rate of regeneration, demonstrating LIF signaling to be important in regulating satellite cell proliferation (Barnard et al., 1994). Moreover, administration of LIF into mdx muscle results in increased fiber number and size (Austin et al., 2000). LIF-/- mice show impaired regeneration following crush injury, and the addition of LIF back to these muscles reversed this phenotype (Kurek et al., 1997).

Binding of LIF to the LIFR/gp130 receptor heterodimer results in phosphorylation of the receptor bound Janus kinases (JAKs), which in turn phosphorylate tyrosine residues on the receptor, providing docking sites for members of the signal transducer and activator of transcription (STAT) family. Receptor-associated STATs are phosphorylated and activated by JAKs, and can subsequently translocate to the nucleus to regulate transcription (reviewed in Ihle, 1995). Exogenous LIF signaling in satellite cells results in activation of JAK2 and STAT3 (Fig. 5) (Spangenburg and Booth, 2002). Following crush injury, phosphorylated STAT3 is detected in satellite cells early in their activation phase. Active STAT3 continued to be expressed until satellite cells began differentiating into myotubes (Kami and Senba, 2002). In addition to STAT3, it has been shown that LIF signaling activates ERK1 and ERK2 independently of JAK-STAT signaling, which functions to prevent myoblast differentiation (Jo et al., 2005; Megeney et al., 1996b). In contrast to these reports, Spangenburg and Booth failed to detect ERK activation by LIF, suggesting that this result might depend on the dose of exogenous LIF or on the cell type studied (Spangenburg and Booth, 2002).

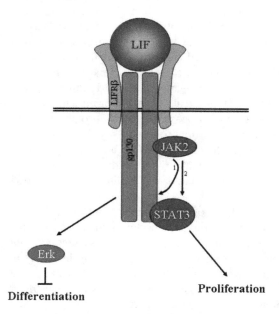

Figure 5. LIF regulates satellite cell proliferation. Association of LIF with the LIFR/gp130 heterodimer results in phosphorylation and activation of JAK2. JAK2 subsequently phosphorylates gp130, which recruits STAT3. Following STAT3 association with gp130, JAK2 phosphorylates and activates STAT3, promoting its dissociation, dimerization, and entry into the nucleus to control proliferation. LIF signaling also activates the ERK MAPK cascade by an unknown mechanism, which prevents cells from differentiating. A color figure is freely accessible via the website of the book: http://www.springer.com/ 1-4020-5176-x

9.5   TGFβ and Myostatin

Members of the transforming growth factor β (TGF-β) growth factor superfamily associate with type I and type II receptors, both of which are actually serine-threonine kinases. TGF-β binding juxtaposes the two receptors, allowing the type II receptor to phosphorylate the type I receptor. Activated type I receptors subsequently phosphorylate the SMAD proteins, which move into the nucleus to regulate target gene transcription (reviewed in Massague, 1998). TGF-β ligands function to negatively regulate the growth and differentiation of skeletal muscle satellite cells in a dose dependent fashion (Allen and Boxhorn, 1987; Allen and Boxhorn, 1989; Massague, 1998). Recently, a new member of the superfamily, myostatin or growth-differentiation factor 8 (GDF-8) has been characterized and also functions in satellite cell regulation.

Myostatin (Mstn) is a secreted factor that is synthesized as a precursor protein, which undergoes multiple modifications, including cleavage events, to finally reach its active form (reviewed in Lee, 2004). Functional myostatin binds the activin type II receptor ActRIIB, a process that can be inhibited by follistatin binding to this receptor (Lee and McPherron, 2001). Disruption of the myostatin gene in mouse results in a considerable increase in muscle mass due to hypertrophy and hyperplasia (McPherron et al., 1997). Interestingly, mutation of the myostatin gene is also responsible for the "double-muscling" phenotype observed in the Belgian Blue and Piedmontese breeds of cattle, and a gross muscle hypertrophy found in a human child (Grobet et al., 1997; McPherron and Lee, 1997; Schuelke et al., 2004).

Following administration of notexin to induce muscle regeneration, myostatin protein is detected within necrotic fibers and connective tissue early during degeneration, prior to satellite cell activation (Kirk et al., 2000). During satellite cell activation and the initiation of the regeneration phase, myostatin protein becomes undetectable. And it is not until later in regeneration, following myotube formation, that myostatin levels begin to return to baseline (Kirk et al., 2000; Mendler et al., 2000). Together, these results indicate that myostatin displays the correct temporal regulation to function as a negative regulator of satellite cell proliferation during skeletal muscle regeneration.

Because myostatin is a negative regulator of muscle growth and satellite cell proliferation, blocking its expression in dystrophic muscle could alleviate some of the pathology. Mdx mice lacking myostatin, either through crossbreeding with the mstn-/- mice or by IP injection of neutralizing antibodies, display increases in muscle mass, size, and force generation and strength (Bogdanovich et al., 2002; Wagner et al., 2002). Their muscles are also less degenerated, resulting in lower serum creatine kinase concentrations (Bogdanovich et al., 2002).

Mechanistically, myostatin controls cell cycle progression with a mechanism almost directly opposite to that of IGF-PI3K-Akt signaling (Fig. 6). Myostatin signaling targets the cdk inhibitor p21, increasing its expression (Thomas et al., 2000). A downregulation of cdk2 expression and activity is also observed, along with an increase in the phosphorylation status of Rb (Thomas et al., 2000). Although it is unknown which of these events are a direct result of myostatin

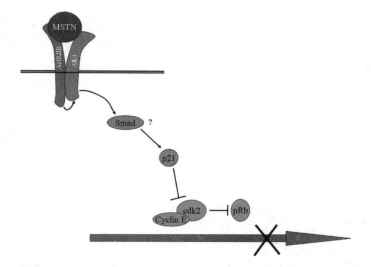

Figure 6. Myostatin signaling negatively regulates satellite cell proliferation. Binding of myostatin to the ActRIIB (type II) receptor results phosphorylation of the type I receptor (Alk4), which then activates the cdk inhibitor p21 by an unknown mechanism. It is likely this activation is mediated through Smads, however this remains to be shown. p21 inhibits cyclin E – cdk2 activity, preventing hyperphosphorylation of Rb, and arresting cells in late G1. A color figure is freely accessible via the website of the book: http://www.springer.com/1-4020-5176-x

signaling and which are a consequence of alterations in the cell cycle, these results do demonstrate that myostatin signaling negatively regulates progression through the G1/S phase of the cell cycle.

An interesting hypothesis was put forward by McCroskery and colleagues stating that myostatin was responsible for maintaining the quiescent state of satellite cells, which could explain why it is downregulated at the same time as satellite cells are activated (McCroskery et al., 2003). These authors show that myostatin is expressed on Pax7-positive satellite cells, and that there is increased numbers of activated satellite cells in mstn-/- mice with altered cell cycle kinetics. However, further experiments are needed to completely address whether myostatin maintains satellite cell quiescence.

9.6   Role of p38α/β MAPK in Satellite Cell Activation

In addition to HGF signaling being responsible for activation of the quiescent satellite cell, p38α/β MAPK has also been hypothesized to function in this activation, although apparently independently of HGF (Fig. 7) (Jones et al., 2005).

p38α/β is mainly activated by two upstream kinases, known as the MAPK kinases MKK6 and MKK3, and has been extensively studied for its role in both early and late skeletal muscle differentiation (reviewed in Gillespie and Rudnicki, 2004; Tapscott, 2005). Jones et al. demonstrate that p38α/β is activated by FGF-2 and

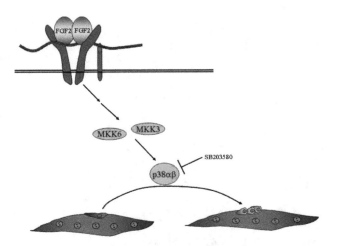

Figure 7. p38 signaling controls satellite cell activation. MKK3/6-p38αβ signaling is activated downstream of FGF-2/FGFR by an unknown mechanism, resulting in the activation and proliferation of quiescent satellite cells. Inhibition of p38αβ signaling with SB203580 prevents satellite cell activation. A color figure is freely accessible via the website of the book: http://www.springer.com/1-4020-5176-x

is also necessary for FGF-2-induced satellite cell proliferation (Jones et al., 2005). Interestingly, p38α/β activity appears to coincide with the induction of MyoD, with the pharmacological p38α/β inhibitor SB203580 blocking the proliferation of syndecan-4+ satellite cells as well as the number of MyoD+ satellite cells (Jones et al., 2005). The target of p38α/β signaling appears to be involved in cell cycle regulation, whereby depending on upstream signals, p38α/β can promote progression through G1 to S phase or arrest cells at the restriction point to allow for differentiation to begin (Jones et al., 2005). These results seem to support an unanticipated role for p38 signaling in activation of quiescent satellite cells; however further experiments are necessary to definitively prove this. Recently, p38 signaling was shown to be a positive target of myostatin signaling, and it will be interesting to decipher how the interaction between these two pathways functions in satellite cell activation (Philip et al., 2005). And it also should be informative to determine if there are any satellite cell activation defects in the p38α null mice.

10.   SATELLITE CELL RENEWAL

In order for skeletal muscle to maintain such a high regenerative capacity, the satellite cell population must be able to renew itself following injury. Schultz observed that proliferating satellite cells in culture could be divided into two populations: those which actively divide, differentiate, and eventually fuse to form myotubes (representing ~80% of cells), and those which grow very slowly (representing ~20% of cells) (Schultz, 1996). These slow growing satellite cells were termed reserve cells and were hypothesized to be responsible for replenishing the satellite cell compartment in vivo (Schultz, 1996). Recent research has begun to elucidate the pathways for satellite cell renewal, and include both symmetric and asymmetric cell division of activated satellite cells (Fig. 8).

The Notch signaling pathway has recently been identified as regulating asymmetric cell division. Notch is a transmembrane receptor expressed on the cell surface and is activated by the receptors Delta and Jagged, which are expressed on different cells and function as Notch ligands. Following activation, Notch can be cleaved such that its intracellular portion can translocate to the nucleus and regulate gene expression (reviewed in Artavanis-Tsakonas et al., 1999). Activation of Notch occurs simultaneously with satellite cell activation in culture or in vivo with injury, making it a candidate for regulating satellite cell proliferation

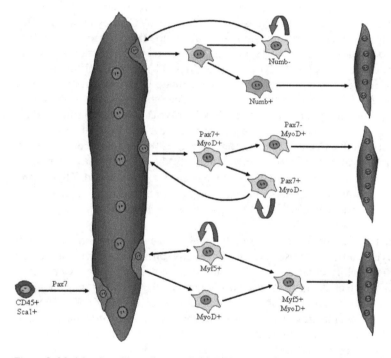

Figure 8. Models of satellite cell renewal. (a) Dividing satellite cells asymmetrically express the Notch inhibitor Numb. Cells lacking Numb expression continue to proliferate as a result of Notch signaling, and will repopulate the satellite cell compartment. Cells with impaired Notch signaling, as a result of Numb expression, differentiate and fuse into fibers. (b) Activated satellite cells express both Pax7 and MyoD. The majority of these cells will downregulate Pax7 expression, differentiate and fuse into fibers. However, a small proportion of cells will downregulate MyoD and continue proliferating. These Pax7+ cells will eventually exit the cell and return to quiescent, thereby renewing the satellite cell population. (c) Activated satellite cells initially express either Myf5 or MyoD. Those which express only Myf5 have an increased propensity to self-renew. (d) Muscle-derived stem cells expressing CD45 and Sca1 can commit to the myogenic lineage by forming satellite cells. A color figure is freely accessible via the website of the book: http://www.springer.com/1-4020-5176-x

(Conboy and Rando, 2002). Interestingly, the Notch inhibitor Numb, which prevents nuclear translocation of activated Notch, is localized asymmetrically in dividing satellite cells, suggesting Notch signaling was involved in asymmetric regulation of division. In fact, Notch activation in satellite cells results in proliferation of Pax3 expressing cells (Pax3+ Myf5- MyoD-), while Notch inhibition results in myogenic commitment and differentiation of Numb+ Pax3- Desmin+ Myf5+ cells (Conboy and Rando, 2002). It is these Notch-activated Pax3+ cells which are postulated to be responsible for satellite cell self-renewal. Further studies by Conboy and colleagues have attributed the age-related decline in muscle regeneration to diminished Notch activation as a result of low expression of Delta combined with high levels of Numb, further proving that Notch activation is necessary for satellite cell proliferation (Conboy et al., 2003). Since satellite cell numbers remain constant from adulthood to old age, these results led to the intriguing hypothesis that Notch could be a candidate for activating quiescent satellite cells (Conboy et al., 2003).

In addition to asymmetric cell division, several groups have noted that satellite cells are able to self-renew following symmetric cell division. Proliferating satellite cells in culture express Pax7 and MyoD (Pax7+ MyoD+). Prior to differentiation, the majority of activated satellite cells will downregulate Pax7 but maintain MyoD expression (Pax7- MyoD+) (Zammit et al., 2004). The other subset of proliferating satellite cells will downregulate only MyoD (Pax7+ MyoD-), while continuing their replication, and will eventually slow down their cell cycle, arrest, and go into a quiescent state (Olguin and Olwin, 2004; Zammit et al., 2004). It has been suggested that Pax7 is responsible for the downregulation of MyoD, however, the role of other transcription factors and signaling pathways cannot be excluded (Olguin and Olwin, 2004). Notably, these quiescent cells can be stimulated to re-enter the cell cycle, proliferate, and even differentiate, suggesting this population is responsible for satellite cell renewal in vivo.

An important role for Myf5 in satellite cell self-renewal originated from work with MyoD-deficient muscle, which is highly proliferative with increased Myf5 expression, while defective in differentiation and regeneration (Cornelison et al., 2000; Sabourin et al., 1999; Yablonka-Reuveni et al., 1999a). It was Megeney and colleagues who initially hypothesized that this inability of MyoD-/- muscle to successfully regenerate resulted from an increased tendency for satellite cell renewal, and Sabourin et al. suggested that MyoD-/- cells represent an intermediate between quiescent satellite cells and committed myogenic precursor cells (Megeney et al., 1996a; Sabourin et al., 1999). As discussed earlier, the activation of quiescent satellite cells results in the initial expression of either MyoD or Myf5 prior to expression of both MRFs (Cornelison and Wold, 1997). Therefore, in MyoD-/- muscle, the only population of satellite cells present following activation would be Myf5+, suggesting this MRF can regulate satellite cell renewal.

The above research demonstrates that satellite cell self-renewal is likely regulated by a combination of Pax7 and/or Myf5 expression, along with asymmetric cell division and Notch signaling. Another interesting possibility is replenishment of the satellite cell compartment by uncommitted adult stem cells. Stem cells isolated from

Pax7-deficient muscle display an increased propensity towards the hematopoietic lineage, suggesting that Pax7 is involved in specifying stem cells to the myogenic lineage (Seale et al., 2000). Furthermore, CD45+ Sca1+ stem cells isolated from skeletal muscle can participate in the regeneration process (Polesskaya et al., 2003). This is mediated through formation of new satellite cells, since CD45+ Sca1+ cells isolated from injured muscle give rise to Pax7+ cells, and Pax7 can induce the muscle specification of CD45+ Sca1+ stem cells (Polesskaya et al., 2003; Seale et al., 2004b).

11.   PARTICIPATION OF ADULT STEM CELLS IN SKELETAL MUSCLE REGENERATION

Both muscle derived and non-muscle derived stem cells have been shown to participate in the muscle regeneration process. Following either bone marrow transplantation or intravenous injection, bone marrow-derived cells are able to migrate to sites of degeneration in skeletal muscle and participate in the regeneration process (Bittner et al., 1999; Ferrari et al., 1998; Gussoni et al., 1999). In two of these studies, the dystrophic mdx mouse was used as a recipient, emphasizing the potential therapeutic value of stem cells in muscle wasting diseases, such as Duchenne muscular dystrophy (DMD). Interestingly, in skeletal muscle from a DMD patient who received a bone marrow transplant for X-linked severe combined immune deficiency (SCID) 13 years earlier, a small proportion of donor nuclei were still detected (under 1%) (Gussoni et al., 2002). Importantly, this observation proves that bone marrow transplanted cells can fuse into skeletal muscle and survive in human patients. There are several possibilities for how stem cells are participating in regeneration, including fusion with existing fibers, however LaBarge and Blau recently proposed an intriguing mechanism (Camargo et al., 2003; Corbel et al., 2003; LaBarge and Blau, 2002). They demonstrated that bone marrow-derived stem cells could form satellite cells following exercise-induced muscle injury. The ability of bone marrow derived cells to replenish the satellite cell compartment is very promising for cell-based therapies.

Skeletal muscle also contains resident populations of stem cells. Side population (SP) cells were originally isolated from bone marrow on the basis of their ability to exclude the dye Hoechst 33342, and these cells were found to be highly enriched for hematopoietic stem cell activity (Goodell et al., 1996). It was later discovered that most tissues, including skeletal muscle, contain this population of cells (Asakura and Rudnicki, 2002; Gussoni et al., 1999). Similar to bone marrow derived SP cells, intravenous injection of muscle SP cells into lethally irradiated mdx mice was not only able to reconstitute the hematopoietic compartment, but partially restored dystrophin expression to muscle fibers as well (Gussoni et al., 1999). Moreover, muscle SP cells were able to give rise to satellite cells under the above conditions and following intramuscular transplantation (Asakura et al., 2002; Gussoni et al., 1999). Interestingly, muscle SP cells from Pax7-deficient muscle were able to form myotubes when co-cultured with primary myoblasts, suggesting a subset of SP cells form muscle independently of satellite cells (Asakura et al., 2002).

The muscle-derived SP cell population was shown to contain stem cells expressing the cell surface markers CD45 and Sca1 (Asakura et al., 2002). As discussed earlier, these cells can participate in muscle regeneration, a process mediated through a satellite cell intermediate (Polesskaya et al., 2003; Seale et al., 2004b). Polesskaya et al. also demonstrated that Wnt signaling was important in specifying CD45+ Sca1+ cells to the myogenic lineage (Polesskaya et al., 2003).

SP and CD45+ Sca1+ cells are not the only population of muscle-derived stem cells that have been isolated. Several groups have isolated apparently different populations of stem cells from skeletal muscle using a preplating technique originally described by Rando and Blau that commonly express both CD34 and Sca1 (Lee et al., 2000; Qu-Petersen et al., 2002; Rando and Blau, 1994; Torrente et al., 2001). CD34+ Sca1+ muscle-derived stem cells (MDSCs) undergo myogenic differentiation in vitro and restore dystrophin expression following injection into mdx mice (Lee et al., 2000; Qu-Petersen et al., 2002; Torrente et al., 2001). MDSCs appear to be unique in that they persist in culture for many passages (up to 200 population doublings) without losing any ability to participate in muscle regeneration (Deasy et al., 2005). Interestingly, these cells initially adhere to capillaries in skeletal muscle prior to being induced to participate in regeneration following injury, suggesting they could have endothelial origins (Torrente et al., 2001).

Although these results show compelling evidence for the participation of both muscle-derived and non-muscle-derived stem cells in skeletal muscle regeneration, it is likely that these events are rare in vivo. Muscle regeneration is primarily mediated through the resident satellite cell population, which has been shown to be fully capable of mediating regeneration (Sherwood et al., 2004; Zammit et al., 2002). Perhaps stem cell involvement only occurs following major damage to skeletal muscle, such as that seen in certain dystrophinopathies. Nonetheless, stem cell-based gene therapy continues to be a promising avenue for future management of muscle wasting diseases.

12.   SATELLITE CELL PLASTICITY

In addition to the formation of skeletal muscle, satellite cells also possess a multipotential mesenchymal stem cell activity. Not only do satellite cell-derived myoblasts express myogenic markers, they also express markers of other mesenchymal lineages, such as the osteogenic determination factor Runx2 or the adipogenic factor PPARγ (Wada et al., 2002). Moreover, treatment of satellite cells with the bone morphogenic proteins (BMPs) BMP-2, BMP-4, or BMP-7, which are known activators of the osteogenic lineage, results in differentiation into osteocytes, as evidenced by expression of alkaline phosphatase and osteocalcin (Asakura et al., 2001; Gersbach et al., 2004; Wada et al., 2002). This result is similar to BMP-2-induced osteogenic differentiation of C2C12 myoblasts, which occurs through repression of MyoD transactivation by BMP signaling (Katagiri et al., 1997; Katagiri et al., 1994).

Interestingly, MyoD-deficient satellite cell-derived myoblasts do not display any increased potential to differentiate into osteocytes or adipocytes compared with wild type muscle (Asakura et al., 2001). These results suggest that MyoD is necessary for the mesenchymal plasticity of satellite cells. In fact, overexpression of MyoD in satellite cells enhanced osteogenic differentiation by cooperating with Runx2 to activate transcription of osteocyte-specific genes following BMP-7 stimulation (Komaki et al., 2004). By contrast, ectopic expression of Runx2 in satellite cell-derived myoblasts downregulates MyoD expression (Gersbach et al., 2004). Clearly, further in vivo work is necessary before any mechanism for BMP/Runx2-induced osteogenic differentiation of satellite cells can be definitively proven.

Treatment of satellite cells with the adipogenic cocktail MDI-I or culturing them in the growth factor rich medium Matrigel results in their differentiation into adipocytes (Asakura et al., 2001; Shefer et al., 2004). Fiber-associated satellite cells cultured in high oxygen concentrations also results in their differentiation into adipocytes (Csete et al., 2001). Moreover, satellite cells appear to display increased adipogenic potential with age, which correlates with increased fibrosis seen in aged muscle (Taylor-Jones et al., 2002). These observations were generally associated with altered levels of the adipogenic regulators PPARγ and C/EBPα. These results support the hypothesis that satellite cells are multipotent, however these events are likely rare in healthy skeletal muscle in vivo.

CONCLUSIONS AND PERSPECTIVES

Satellite cells are the major cell type contributing to postnatal skeletal muscle growth and regeneration. Their regulation is complex, as they must be able to remain in an inactive state, waiting for the correct signals, before proliferating, differentiating, and fusing into muscle fibers. In addition, they must also be able to replenish themselves, either by self-renewal, or by stem cell repopulation of the satellite cell compartment.

As discussed in this review, numerous growth factors have been identified that control satellite cell function during the different phases of muscle regeneration, yet there still remains information to be discovered. One of the major hurdles for identifying further growth factors or other signaling molecules that activate quiescent satellite cells following muscle damage is the lack of a marker specific for quiescent satellite cells. To date, the markers used to try and identify this population are also expressed on activated satellite cells.

Surprisingly little is known about the intracellular signaling pathways regulating satellite cell activation, differentiation, and renewal. Components of these pathways are often dysregulated in diseases affecting skeletal muscle, suggesting they are important therapeutic targets. The molecular regulation of key transcription factors, such as Pax7, is virtually unknown. Furthermore, the molecular details surrounding the role of Pax7 in activating or repressing gene expression are in their infancy, partly due to the lack of known target genes for this transcription factor. Detailed mechanisms for the regulation of muscle differentiation by MyoD are only now being figured out. Perhaps the strategies used to elucidate the regulatory networks for MyoD and its control of differentiation could be applied to studies of Pax7.

ACKNOWLEDGEMENTS

The authors thank Drs. Anthony Scimè and Shihuan Kuang for providing regeneration pictures. M.A.R. holds the Canada Research Chair in Molecular Genetics and is a Howard Hughes Medical Institute International Scholar. This work was supported by grants to M.A.R. from the Muscular Dystrophy Association, the National Institutes of Health, the Canadian Institutes of Health Research, the Howard Hughes Medical Institute, and the Canada Research Chair Program.

REFERENCES

Allen, D. L., Roy, R. R., and Edgerton, V. R. (1999). Myonuclear domains in muscle adaptation and disease. Muscle Nerve 22, 1350–1360.
Allen, R. E., and Boxhorn, L. K. (1987). Inhibition of skeletal muscle satellite cell differentiation by transforming growth factor-beta. J Cell Physiol 133, 567–572.
Allen, R. E., and Boxhorn, L. K. (1989). Regulation of skeletal muscle satellite cell proliferation and differentiation by transforming growth factor-beta, insulin-like growth factor I, and fibroblast growth factor. J Cell Physiol 138, 311–315.
Allen, R. E., Dodson, M. V., and Luiten, L. S. (1984). Regulation of skeletal muscle satellite cell proliferation by bovine pituitary fibroblast growth factor. Exp Cell Res 152, 154–160.
Allen, R. E., Sheehan, S. M., Taylor, R. G., Kendall, T. L., and Rice, G. M. (1995). Hepatocyte growth factor activates quiescent skeletal muscle satellite cells in vitro. J Cell Physiol 165, 307–312.
Anastasi, S., Giordano, S., Sthandier, O., Gambarotta, G., Maione, R., Comoglio, P., and Amati, P. (1997). A natural hepatocyte growth factor/scatter factor autocrine loop in myoblast cells and the effect of the constitutive Met kinase activation on myogenic differentiation. J Cell Biol 137, 1057–1068.
Anderson, J. E. (2000). A role for nitric oxide in muscle repair: nitric oxide-mediated activation of muscle satellite cells. Mol Biol Cell 11, 1859–1874.
Armand, O., Boutineau, A. M., Mauger, A., Pautou, M. P., and Kieny, M. (1983). Origin of satellite cells in avian skeletal muscles. Arch Anat Microsc Morphol Exp 72, 163–181.
Artavanis-Tsakonas, S., Rand, M. D., and Lake, R. J. (1999). Notch signaling: cell fate control and signal integration in development. Science 284, 770–776.
Asakura, A., Komaki, M., and Rudnicki, M. (2001). Muscle satellite cells are multipotential stem cells that exhibit myogenic, osteogenic, and adipogenic differentiation. Differentiation 68, 245–253.
Asakura, A., and Rudnicki, M. A. (2002). Side population cells from diverse adult tissues are capable of in vitro hematopoietic differentiation. Exp Hematol 30, 1339–1345.
Asakura, A., Seale, P., Girgis-Gabardo, A., and Rudnicki, M. A. (2002). Myogenic specification of side population cells in skeletal muscle. J Cell Biol 159, 123–134.
Austin, L., Bower, J., Kurek, J., and Vakakis, N. (1992). Effects of leukaemia inhibitory factor and other cytokines on murine and human myoblast proliferation. J Neurol Sci 112, 185–191.
Austin, L., Bower, J. J., Bennett, T. M., Lynch, G. S., Kapsa, R., White, J. D., Barnard, W., Gregorevic, P., and Byrne, E. (2000). Leukemia inhibitory factor ameliorates muscle fiber degeneration in the mdx mouse. Muscle Nerve 23, 1700–1705.
Barnard, W., Bower, J., Brown, M. A., Murphy, M., and Austin, L. (1994). Leukemia inhibitory factor (LIF) infusion stimulates skeletal muscle regeneration after injury: injured muscle expresses lif mRNA. J Neurol Sci 123, 108–113.

Barton, E. R., Morris, L., Musaro, A., Rosenthal, N., and Sweeney, H. L. (2002). Muscle-specific expression of insulin-like growth factor I counters muscle decline in mdx mice. J Cell Biol *157*, 137–148.

Beauchamp, J. R., Heslop, L., Yu, D. S., Tajbakhsh, S., Kelly, R. G., Wernig, A., Buckingham, M. E., Partridge, T. A., and Zammit, P. S. (2000). Expression of CD34 and Myf5 defines the majority of quiescent adult skeletal muscle satellite cells. J Cell Biol *151*, 1221–1234.

Belcastro, A. N., Shewchuk, L. D., and Raj, D. A. (1998). Exercise-induced muscle injury: a calpain hypothesis. Mol Cell Biochem *179*, 135–145.

Birchmeier, C., and Gherardi, E. (1998). Developmental roles of HGF/SF and its receptor, the c-Met tyrosine kinase. Trends Cell Biol *8*, 404–410.

Bischoff, R. (1986). A satellite cell mitogen from crushed adult muscle. Dev Biol *115*, 140–147.

Bischoff, R. (1990). Cell cycle commitment of rat muscle satellite cells. J Cell Biol *111*, 201–207.

Bischoff, R. (1994). The satellite cell and muscle regeneration. In Myogenesis (New York, McGraw-Hill), pp. 97–118.

Bischoff, R. (1997). Chemotaxis of skeletal muscle satellite cells. Dev Dyn *208*, 505–515.

Bittner, R. E., Schofer, C., Weipoltshammer, K., Ivanova, S., Streubel, B., Hauser, E., Freilinger, M., Hoger, H., Elbe-Burger, A., and Wachtler, F. (1999). Recruitment of bone-marrow-derived cells by skeletal and cardiac muscle in adult dystrophic mdx mice. Anat Embryol (Berl) *199*, 391–396.

Bladt, F., Riethmacher, D., Isenmann, S., Aguzzi, A., and Birchmeier, C. (1995). Essential role for the c-met receptor in the migration of myogenic precursor cells into the limb bud. Nature *376*, 768–771.

Bogdanovich, S., Krag, T. O., Barton, E. R., Morris, L. D., Whittemore, L. A., Ahima, R. S., and Khurana, T. S. (2002). Functional improvement of dystrophic muscle by myostatin blockade. Nature *420*, 418–421.

Buckingham, M., Bajard, L., Chang, T., Daubas, P., Hadchouel, J., Meilhac, S., Montarras, D., Rocancourt, D., and Relaix, F. (2003). The formation of skeletal muscle: from somite to limb. J Anat *202*, 59–68.

Camargo, F. D., Green, R., Capetanaki, Y., Jackson, K. A., and Goodell, M. A. (2003). Single hematopoietic stem cells generate skeletal muscle through myeloid intermediates. Nat Med *9*, 1520–1527.

Campbell, J. S., Wenderoth, M. P., Hauschka, S. D., and Krebs, E. G. (1995). Differential activation of mitogen-activated protein kinase in response to basic fibroblast growth factor in skeletal muscle cells. Proc Natl Acad Sci U S A *92*, 870–874.

Casar, J. C., Cabello-Verrugio, C., Olguin, H., Aldunate, R., Inestrosa, N. C., and Brandan, E. (2004). Heparan sulfate proteoglycans are increased during skeletal muscle regeneration: requirement of syndecan-3 for successful fiber formation. J Cell Sci *117*, 73–84.

Chakravarthy, M. V., Abraha, T. W., Schwartz, R. J., Fiorotto, M. L., and Booth, F. W. (2000a). Insulin-like growth factor-I extends in vitro replicative life span of skeletal muscle satellite cells by enhancing G1/S cell cycle progression via the activation of phosphatidylinositol 3'-kinase/Akt signaling pathway. J Biol Chem *275*, 35942–35952.

Chakravarthy, M. V., Davis, B. S., and Booth, F. W. (2000b). IGF-I restores satellite cell proliferative potential in immobilized old skeletal muscle. J Appl Physiol *89*, 1365–1379.

Charge, S. B., and Rudnicki, M. A. (2004). Cellular and molecular regulation of muscle regeneration. Physiol Rev *84*, 209–238.

Clarke, M. S., Khakee, R., and McNeil, P. L. (1993). Loss of cytoplasmic basic fibroblast growth factor from physiologically wounded myofibers of normal and dystrophic muscle. J Cell Sci *106 (Pt 1)*, 121–133.

Clegg, C. H., Linkhart, T. A., Olwin, B. B., and Hauschka, S. D. (1987). Growth factor control of skeletal muscle differentiation: commitment to terminal differentiation occurs in G1 phase and is repressed by fibroblast growth factor. J Cell Biol *105*, 949–956.

Conboy, I. M., Conboy, M. J., Smythe, G. M., and Rando, T. A. (2003). Notch-mediated restoration of regenerative potential to aged muscle. Science *302*, 1575–1577.

Conboy, I. M., and Rando, T. A. (2002). The regulation of Notch signaling controls satellite cell activation and cell fate determination in postnatal myogenesis. Dev Cell *3*, 397–409.

Cooper, R. N., Tajbakhsh, S., Mouly, V., Cossu, G., Buckingham, M., and Butler-Browne, G. S. (1999). In vivo satellite cell activation via Myf5 and MyoD in regenerating mouse skeletal muscle. J Cell Sci 112 (Pt 17), 2895–2901.

Corbel, S. Y., Lee, A., Yi, L., Duenas, J., Brazelton, T. R., Blau, H. M., and Rossi, F. M. (2003). Contribution of hematopoietic stem cells to skeletal muscle. Nat Med 9, 1528–1532.

Cornelison, D. D., Filla, M. S., Stanley, H. M., Rapraeger, A. C., and Olwin, B. B. (2001). Syndecan-3 and syndecan-4 specifically mark skeletal muscle satellite cells and are implicated in satellite cell maintenance and muscle regeneration. Dev Biol 239, 79–94.

Cornelison, D. D., Olwin, B. B., Rudnicki, M. A., and Wold, B. J. (2000). MyoD(-/-) satellite cells in single-fiber culture are differentiation defective and MRF4 deficient. Dev Biol 224, 122–137.

Cornelison, D. D., Wilcox-Adelman, S. A., Goetinck, P. F., Rauvala, H., Rapraeger, A. C., and Olwin, B. B. (2004). Essential and separable roles for Syndecan-3 and Syndecan-4 in skeletal muscle development and regeneration. Genes Dev 18, 2231–2236.

Cornelison, D. D., and Wold, B. J. (1997). Single-cell analysis of regulatory gene expression in quiescent and activated mouse skeletal muscle satellite cells. Dev Biol 191, 270–283.

Crameri, R. M., Langberg, H., Magnusson, P., Jensen, C. H., Schroder, H. D., Olesen, J. L., Suetta, C., Teisner, B., and Kjaer, M. (2004). Changes in satellite cells in human skeletal muscle after a single bout of high intensity exercise. J Physiol 558, 333–340.

Csete, M., Walikonis, J., Slawny, N., Wei, Y., Korsnes, S., Doyle, J. C., and Wold, B. (2001). Oxygen-mediated regulation of skeletal muscle satellite cell proliferation and adipogenesis in culture. J Cell Physiol 189, 189–196.

De Angelis, L., Berghella, L., Coletta, M., Lattanzi, L., Zanchi, M., Cusella-De Angelis, M. G., Ponzetto, C., and Cossu, G. (1999). Skeletal myogenic progenitors originating from embryonic dorsal aorta coexpress endothelial and myogenic markers and contribute to postnatal muscle growth and regeneration. J Cell Biol 147, 869–878.

De Craene, B., van Roy, F., and Berx, G. (2005). Unraveling signalling cascades for the Snail family of transcription factors. Cell Signal 17, 535–547.

Deasy, B. M., Gharaibeh, B. M., Pollett, J. B., Jones, M. M., Lucas, M. A., Kanda, Y., and Huard, J. (2005). Long-Term Self-Renewal of Postnatal Muscle-derived Stem Cells. Mol Biol Cell.

DiMario, J., Buffinger, N., Yamada, S., and Strohman, R. C. (1989). Fibroblast growth factor in the extracellular matrix of dystrophic (mdx) mouse muscle. Science 244, 688–690.

DiMario, J., and Strohman, R. C. (1988). Satellite cells from dystrophic (mdx) mouse muscle are stimulated by fibroblast growth factor in vitro. Differentiation 39, 42–49.

Edwall, D., Schalling, M., Jennische, E., and Norstedt, G. (1989). Induction of insulin-like growth factor I messenger ribonucleic acid during regeneration of rat skeletal muscle. Endocrinology 124, 820–825.

Engert, J. C., Berglund, E. B., and Rosenthal, N. (1996). Proliferation precedes differentiation in IGF-I-stimulated myogenesis. J Cell Biol 135, 431–440.

Fedorov, Y. V., Jones, N. C., and Olwin, B. B. (2002). Atypical protein kinase Cs are the Ras effectors that mediate repression of myogenic satellite cell differentiation. Mol Cell Biol 22, 1140–1149.

Fedorov, Y. V., Rosenthal, R. S., and Olwin, B. B. (2001). Oncogenic Ras-induced proliferation requires autocrine fibroblast growth factor 2 signaling in skeletal muscle cells. J Cell Biol 152, 1301–1305.

Ferrari, G., Cusella-De Angelis, G., Coletta, M., Paolucci, E., Stornaiuolo, A., Cossu, G., and Mavilio, F. (1998). Muscle regeneration by bone marrow-derived myogenic progenitors. Science 279, 1528–1530.

Fiore, F., Planche, J., Gibier, P., Sebille, A., deLapeyriere, O., and Birnbaum, D. (1997). Apparent normal phenotype of Fgf6-/- mice. Int J Dev Biol 41, 639–642.

Fiore, F., Sebille, A., and Birnbaum, D. (2000). Skeletal muscle regeneration is not impaired in Fgf6 -/- mutant mice. Biochem Biophys Res Commun 272, 138–143.

Florini, J. R., Ewton, D. Z., and Coolican, S. A. (1996). Growth hormone and the insulin-like growth factor system in myogenesis. Endocr Rev 17, 481–517.

Florini, J. R., Ewton, D. Z., and Magri, K. A. (1991). Hormones, growth factors, and myogenic differentiation. Annu Rev Physiol 53, 201–216.

Floss, T., Arnold, H. H., and Braun, T. (1997). A role for FGF-6 in skeletal muscle regeneration. Genes Dev 11, 2040–2051.

Gal-Levi, R., Leshem, Y., Aoki, S., Nakamura, T., and Halevy, O. (1998). Hepatocyte growth factor plays a dual role in regulating skeletal muscle satellite cell proliferation and differentiation. Biochim Biophys Acta *1402*, 39–51.

Gamble, H. J., Fenton, J., and Allsopp, G. (1978). Electron microscope observations on human fetal striated muscle. J Anat *126*, 567–589.

Garry, D. J., Meeson, A., Elterman, J., Zhao, Y., Yang, P., Bassel-Duby, R., and Williams, R. S. (2000). Myogenic stem cell function is impaired in mice lacking the forkhead/winged helix protein MNF. Proc Natl Acad Sci U S A *97*, 5416–5421.

Garry, D. J., Yang, Q., Bassel-Duby, R., and Williams, R. S. (1997). Persistent expression of MNF identifies myogenic stem cells in postnatal muscles. Dev Biol *188*, 280–294.

Gersbach, C. A., Byers, B. A., Pavlath, G. K., and Garcia, A. J. (2004). Runx2/Cbfa1 stimulates transdifferentiation of primary skeletal myoblasts into a mineralizing osteoblastic phenotype. Exp Cell Res *300*, 406–417.

Gillespie, M. A., and Rudnicki, M. A. (2004). Something to SNF about. Nat Genet *36*, 676–677.

Goodell, M. A., Brose, K., Paradis, G., Conner, A. S., and Mulligan, R. C. (1996). Isolation and functional properties of murine hematopoietic stem cells that are replicating in vivo. J Exp Med *183*, 1797–1806.

Grobet, L., Martin, L. J., Poncelet, D., Pirottin, D., Brouwers, B., Riquet, J., Schoeberlein, A., Dunner, S., Menissier, F., Massabanda, J., et al. (1997). A deletion in the bovine myostatin gene causes the double-muscled phenotype in cattle. Nat Genet *17*, 71–74.

Gros, J., Manceau, M., Thome, V., and Marcelle, C. (2005). A common somitic origin for embryonic muscle progenitors and satellite cells. Nature.

Gussoni, E., Bennett, R. R., Muskiewicz, K. R., Meyerrose, T., Nolta, J. A., Gilgoff, I., Stein, J., Chan, Y. M., Lidov, H. G., Bonnemann, C. G., et al. (2002). Long-term persistence of donor nuclei in a Duchenne muscular dystrophy patient receiving bone marrow transplantation. J Clin Invest *110*, 807–814.

Gussoni, E., Soneoka, Y., Strickland, C. D., Buzney, E. A., Khan, M. K., Flint, A. F., Kunkel, L. M., and Mulligan, R. C. (1999). Dystrophin expression in the mdx mouse restored by stem cell transplantation. Nature *401*, 390–394.

Hartley, R. S., Bandman, E., and Yablonka-Reuveni, Z. (1992). Skeletal muscle satellite cells appear during late chicken embryogenesis. Dev Biol *153*, 206–216.

Hawke, T. J. (2005). Muscle stem cells and exercise training. Exerc Sport Sci Rev *33*, 63–68.

Hawke, T. J., and Garry, D. J. (2001). Myogenic satellite cells: physiology to molecular biology. J Appl Physiol *91*, 534–551.

Hawke, T. J., Jiang, N., and Garry, D. J. (2003). Absence of p21CIP rescues myogenic progenitor cell proliferative and regenerative capacity in Foxk1 null mice. J Biol Chem *278*, 4015–4020.

Hollnagel, A., Grund, C., Franke, W. W., and Arnold, H. H. (2002). The cell adhesion molecule M-cadherin is not essential for muscle development and regeneration. Mol Cell Biol *22*, 4760–4770.

Huard, J., Li, Y., and Fu, F. H. (2002). Muscle injuries and repair: current trends in research. J Bone Joint Surg Am *84-A*, 822–832.

Hurme, T., Kalimo, H., Lehto, M., and Jarvinen, M. (1991). Healing of skeletal muscle injury: an ultrastructural and immunohistochemical study. Med Sci Sports Exerc *23*, 801–810.

Ihle, J. N. (1995). Cytokine receptor signalling. Nature *377*, 591–594.

Irintchev, A., Zeschnigk, M., Starzinski-Powitz, A., and Wernig, A. (1994). Expression pattern of M-cadherin in normal, denervated, and regenerating mouse muscles. Dev Dyn *199*, 326–337.

Jarvinen, T. A., Kannus, P., Jarvinen, T. L., Jozsa, L., Kalimo, H., and Jarvinen, M. (2000). Tenascin-C in the pathobiology and healing process of musculoskeletal tissue injury. Scand J Med Sci Sports *10*, 376–382.

Jennische, E., Ekberg, S., and Matejka, G. L. (1993). Expression of hepatocyte growth factor in growing and regenerating rat skeletal muscle. Am J Physiol *265*, C122–128.

Jo, C., Kim, H., Jo, I., Choi, I., Jung, S. C., Kim, J., Kim, S. S., and Jo, S. A. (2005). Leukemia inhibitory factor blocks early differentiation of skeletal muscle cells by activating ERK. Biochim Biophys Acta *1743*, 187–197.

Johnson, S. E., and Allen, R. E. (1993). Proliferating cell nuclear antigen (PCNA) is expressed in activated rat skeletal muscle satellite cells. J Cell Physiol *154*, 39–43.

Jones, N. C., Fedorov, Y. V., Rosenthal, R. S., and Olwin, B. B. (2001). ERK1/2 is required for myoblast proliferation but is dispensable for muscle gene expression and cell fusion. J Cell Physiol *186*, 104–115.

Jones, N. C., Tyner, K. J., Nibarger, L., Stanley, H. M., Cornelison, D. D., Fedorov, Y. V., and Olwin, B. B. (2005). The p38alpha/beta MAPK functions as a molecular switch to activate the quiescent satellite cell. J Cell Biol *169*, 105–116.

Kaariainen, M., Jarvinen, T., Jarvinen, M., Rantanen, J., and Kalimo, H. (2000). Relation between myofibers and connective tissue during muscle injury repair. Scand J Med Sci Sports *10*, 332–337.

Kadi, F., Charifi, N., Denis, C., Lexell, J., Andersen, J. L., Schjerling, P., Olsen, S., and Kjaer, M. (2005). The behaviour of satellite cells in response to exercise: what have we learned from human studies? Pflugers Arch *451*, 319–327.

Kadi, F., Schjerling, P., Andersen, L. L., Charifi, N., Madsen, J. L., Christensen, L. R., and Andersen, J. L. (2004). The effects of heavy resistance training and detraining on satellite cells in human skeletal muscles. J Physiol *558*, 1005–1012.

Kahn, E. B., and Simpson, S. B., Jr. (1974). Satellite cells in mature, uninjured skeletal muscle of the lizard tail. Dev Biol *37*, 219–223.

Kami, K., and Senba, E. (1998). Localization of leukemia inhibitory factor and interleukin-6 messenger ribonucleic acids in regenerating rat skeletal muscle. Muscle Nerve *21*, 819–822.

Kami, K., and Senba, E. (2002). In vivo activation of STAT3 signaling in satellite cells and myofibers in regenerating rat skeletal muscles. J Histochem Cytochem *50*, 1579–1589.

Kastner, S., Elias, M. C., Rivera, A. J., and Yablonka-Reuveni, Z. (2000). Gene expression patterns of the fibroblast growth factors and their receptors during myogenesis of rat satellite cells. J Histochem Cytochem *48*, 1079–1096.

Katagiri, T., Akiyama, S., Namiki, M., Komaki, M., Yamaguchi, A., Rosen, V., Wozney, J. M., Fujisawa-Sehara, A., and Suda, T. (1997). Bone morphogenetic protein-2 inhibits terminal differentiation of myogenic cells by suppressing the transcriptional activity of MyoD and myogenin. Exp Cell Res *230*, 342–351.

Katagiri, T., Yamaguchi, A., Komaki, M., Abe, E., Takahashi, N., Ikeda, T., Rosen, V., Wozney, J. M., Fujisawa-Sehara, A., and Suda, T. (1994). Bone morphogenetic protein-2 converts the differentiation pathway of C2C12 myoblasts into the osteoblast lineage. J Cell Biol *127*, 1755–1766.

Kirk, S., Oldham, J., Kambadur, R., Sharma, M., Dobbie, P., and Bass, J. (2000). Myostatin regulation during skeletal muscle regeneration. J Cell Physiol *184*, 356–363.

Komaki, M., Asakura, A., Rudnicki, M. A., Sodek, J., and Cheifetz, S. (2004). MyoD enhances BMP7-induced osteogenic differentiation of myogenic cell cultures. J Cell Sci *117*, 1457–1468.

Kurek, J. B., Bower, J. J., Romanella, M., Koentgen, F., Murphy, M., and Austin, L. (1997). The role of leukemia inhibitory factor in skeletal muscle regeneration. Muscle Nerve *20*, 815–822.

Kurek, J. B., Nouri, S., Kannourakis, G., Murphy, M., and Austin, L. (1996). Leukemia inhibitory factor and interleukin-6 are produced by diseased and regenerating skeletal muscle. Muscle Nerve *19*, 1291–1301.

LaBarge, M. A., and Blau, H. M. (2002). Biological progression from adult bone marrow to mononucleate muscle stem cell to multinucleate muscle fiber in response to injury. Cell *111*, 589–601.

Lee, J. Y., Qu-Petersen, Z., Cao, B., Kimura, S., Jankowski, R., Cummins, J., Usas, A., Gates, C., Robbins, P., Wernig, A., and Huard, J. (2000). Clonal isolation of muscle-derived cells capable of enhancing muscle regeneration and bone healing. J Cell Biol *150*, 1085–1100.

Lee, S. J. (2004). Regulation of muscle mass by myostatin. Annu Rev Cell Dev Biol *20*, 61–86.

Lee, S. J., and McPherron, A. C. (2001). Regulation of myostatin activity and muscle growth. Proc Natl Acad Sci U S A *98*, 9306–9311.

Lefaucheur, J. P., and Sebille, A. (1995). Basic fibroblast growth factor promotes in vivo muscle regeneration in murine muscular dystrophy. Neurosci Lett *202*, 121–124.

Levinovitz, A., Jennische, E., Oldfors, A., Edwall, D., and Norstedt, G. (1992). Activation of insulin-like growth factor II expression during skeletal muscle regeneration in the rat: correlation with myotube formation. Mol Endocrinol 6, 1227–1234.

Machida, S., Spangenburg, E. E., and Booth, F. W. (2003). Forkhead transcription factor FoxO1 transduces insulin-like growth factor's signal to p27Kip1 in primary skeletal muscle satellite cells. J Cell Physiol 196, 523–531.

Mansouri, A., Stoykova, A., Torres, M., and Gruss, P. (1996). Dysgenesis of cephalic neural crest derivatives in Pax7-/- mutant mice. Development 122, 831–838.

Massague, J. (1998). TGF-beta signal transduction. Annu Rev Biochem 67, 753–791.

Mauro, A. (1961). Satellite cell of skeletal muscle fibers. J Biophys Biochem Cytol 9, 493–495.

McCroskery, S., Thomas, M., Maxwell, L., Sharma, M., and Kambadur, R. (2003). Myostatin negatively regulates satellite cell activation and self-renewal. J Cell Biol 162, 1135–1147.

McKinnell, I. W., and Rudnicki, M. A. (2005). Developmental biology: one source for muscle. Nature 435, 898–899.

McPherron, A. C., Lawler, A. M., and Lee, S. J. (1997). Regulation of skeletal muscle mass in mice by a new TGF-beta superfamily member. Nature 387, 83–90.

McPherron, A. C., and Lee, S. J. (1997). Double muscling in cattle due to mutations in the myostatin gene. Proc Natl Acad Sci U S A 94, 12457–12461.

Megeney, L. A., Kablar, B., Garrett, K., Anderson, J. E., and Rudnicki, M. A. (1996a). MyoD is required for myogenic stem cell function in adult skeletal muscle. Genes Dev 10, 1173–1183.

Megeney, L. A., Perry, R. L., LeCouter, J. E., and Rudnicki, M. A. (1996b). bFGF and LIF signaling activates STAT3 in proliferating myoblasts. Dev Genet 19, 139–145.

Mendler, L., Zador, E., Ver Heyen, M., Dux, L., and Wuytack, F. (2000). Myostatin levels in regenerating rat muscles and in myogenic cell cultures. J Muscle Res Cell Motil 21, 551–563.

Miller, K. J., Thaloor, D., Matteson, S., and Pavlath, G. K. (2000). Hepatocyte growth factor affects satellite cell activation and differentiation in regenerating skeletal muscle. Am J Physiol Cell Physiol 278, C174–181.

Mitchell, C. A., McGeachie, J. K., and Grounds, M. D. (1996). The exogenous administration of basic fibroblast growth factor to regenerating skeletal muscle in mice does not enhance the process of regeneration. Growth Factors 13, 37–55.

Mohammadi, M., Olsen, S. K., and Ibrahimi, O. A. (2005). Structural basis for fibroblast growth factor receptor activation. Cytokine Growth Factor Rev 16, 107–137.

Moore, R., and Walsh, F. S. (1993). The cell adhesion molecule M-cadherin is specifically expressed in developing and regenerating, but not denervated skeletal muscle. Development 117, 1409–1420.

Morlet, K., Grounds, M. D., and McGeachie, J. K. (1989). Muscle precursor replication after repeated regeneration of skeletal muscle in mice. Anat Embryol (Berl) 180, 471–478.

Musaro, A., McCullagh, K., Paul, A., Houghton, L., Dobrowolny, G., Molinaro, M., Barton, E. R., Sweeney, H. L., and Rosenthal, N. (2001). Localized Igf-1 transgene expression sustains hypertrophy and regeneration in senescent skeletal muscle. Nat Genet 27, 195–200.

Musaro, A., McCullagh, K. J., Naya, F. J., Olson, E. N., and Rosenthal, N. (1999). IGF-1 induces skeletal myocyte hypertrophy through calcineurin in association with GATA-2 and NF-ATc1. Nature 400, 581–585.

Nakamura, T., Nawa, K., and Ichihara, A. (1984). Partial purification and characterization of hepatocyte growth factor from serum of hepatectomized rats. Biochem Biophys Res Commun 122, 1450–1459.

Olguin, H. C., and Olwin, B. B. (2004). Pax-7 up-regulation inhibits myogenesis and cell cycle progression in satellite cells: a potential mechanism for self-renewal. Dev Biol 275, 375–388.

Oustanina, S., Hause, G., and Braun, T. (2004). Pax7 directs postnatal renewal and propagation of myogenic satellite cells but not their specification. Embo J 23, 3430–3439.

Pardanaud, L., and Dieterlen-Lievre, F. (1999). Manipulation of the angiopoietic/hemangiopoietic commitment in the avian embryo. Development 126, 617–627.

Pardanaud, L., Luton, D., Prigent, M., Bourcheix, L. M., Catala, M., and Dieterlen-Lievre, F. (1996). Two distinct endothelial lineages in ontogeny, one of them related to hemopoiesis. Development 122, 1363–1371.

Pette, D., and Dusterhoft, S. (1992). Altered gene expression in fast-twitch muscle induced by chronic low-frequency stimulation. Am J Physiol 262, R333–338.

Philip, B., Lu, Z., and Gao, Y. (2005). Regulation of GDF-8 signaling by the p38 MAPK. Cell Signal 17, 365–375.

Polesskaya, A., Seale, P., and Rudnicki, M. A. (2003). Wnt signaling induces the myogenic specification of resident CD45+ adult stem cells during muscle regeneration. Cell 113, 841–852.

Putman, C. T., Dusterhoft, S., and Pette, D. (1999). Changes in satellite cell content and myosin isoforms in low-frequency-stimulated fast muscle of hypothyroid rat. J Appl Physiol 86, 40–51.

Putman, C. T., Dusterhoft, S., and Pette, D. (2000). Satellite cell proliferation in low frequency-stimulated fast muscle of hypothyroid rat. Am J Physiol Cell Physiol 279, C682–690.

Qu-Petersen, Z., Deasy, B., Jankowski, R., Ikezawa, M., Cummins, J., Pruchnic, R., Mytinger, J., Cao, B., Gates, C., Wernig, A., and Huard, J. (2002). Identification of a novel population of muscle stem cells in mice: potential for muscle regeneration. J Cell Biol 157, 851–864.

Rabinovsky, E. D., Gelir, E., Gelir, S., Lui, H., Kattash, M., DeMayo, F. J., Shenaq, S. M., and Schwartz, R. J. (2003). Targeted expression of IGF-1 transgene to skeletal muscle accelerates muscle and motor neuron regeneration. Faseb J 17, 53–55.

Rando, T. A., and Blau, H. M. (1994). Primary mouse myoblast purification, characterization, and transplantation for cell-mediated gene therapy. J Cell Biol 125, 1275–1287.

Rantanen, J., Hurme, T., Lukka, R., Heino, J., and Kalimo, H. (1995). Satellite cell proliferation and the expression of myogenin and desmin in regenerating skeletal muscle: evidence for two different populations of satellite cells. Lab Invest 72, 341–347.

Rapraeger, A. C. (2000). Syndecan-regulated receptor signaling. J Cell Biol 149, 995–998.

Relaix, F., Rocancourt, D., Mansouri, A., and Buckingham, M. (2005). A Pax3/Pax7-dependent population of skeletal muscle progenitor cells. Nature.

Rosenblatt, J. D., and Parry, D. J. (1992). Gamma irradiation prevents compensatory hypertrophy of overloaded mouse extensor digitorum longus muscle. J Appl Physiol 73, 2538–2543.

Rosenblatt, J. D., Yong, D., and Parry, D. J. (1994). Satellite cell activity is required for hypertrophy of overloaded adult rat muscle. Muscle Nerve 17, 608–613.

Sabourin, L. A., Girgis-Gabardo, A., Seale, P., Asakura, A., and Rudnicki, M. A. (1999). Reduced differentiation potential of primary MyoD-/- myogenic cells derived from adult skeletal muscle. J Cell Biol 144, 631–643.

Scata, K. A., Bernard, D. W., Fox, J., and Swain, J. L. (1999). FGF receptor availability regulates skeletal myogenesis. Exp Cell Res 250, 10–21.

Schiaffino, S., Bormioli, S. P., and Aloisi, M. (1976). The fate of newly formed satellite cells during compensatory muscle hypertrophy. Virchows Arch B Cell Pathol 21, 113–118.

Schuelke, M., Wagner, K. R., Stolz, L. E., Hubner, C., Riebel, T., Komen, W., Braun, T., Tobin, J. F., and Lee, S. J. (2004). Myostatin mutation associated with gross muscle hypertrophy in a child. N Engl J Med 350, 2682–2688.

Schultz, E. (1976). Fine structure of satellite cells in growing skeletal muscle. Am J Anat 147, 49–70.

Schultz, E. (1996). Satellite cell proliferative compartments in growing skeletal muscles. Dev Biol 175, 84–94.

Schultz, E., and Jaryszak, D. L. (1985). Effects of skeletal muscle regeneration on the proliferation potential of satellite cells. Mech Ageing Dev 30, 63–72.

Schultz, E., and McCormick, K. M. (1994). Skeletal muscle satellite cells. Rev Physiol Biochem Pharmacol 123, 213–257.

Seale, P., Ishibashi, J., Holterman, C., and Rudnicki, M. A. (2004a). Muscle satellite cell-specific genes identified by genetic profiling of MyoD-deficient myogenic cell. Dev Biol 275, 287–300.

Seale, P., Ishibashi, J., Scime, A., and Rudnicki, M. A. (2004b). Pax7 is necessary and sufficient for the myogenic specification of CD45+:Sca1+ stem cells from injured muscle. PLoS Biol 2, E130.

Seale, P., Sabourin, L. A., Girgis-Gabardo, A., Mansouri, A., Gruss, P., and Rudnicki, M. A. (2000). Pax7 is required for the specification of myogenic satellite cells. Cell 102, 777–786.

Semsarian, C., Wu, M. J., Ju, Y. K., Marciniec, T., Yeoh, T., Allen, D. G., Harvey, R. P., and Graham, R. M. (1999). Skeletal muscle hypertrophy is mediated by a Ca2+-dependent calcineurin signalling pathway. Nature 400, 576–581.

Sheehan, S. M., and Allen, R. E. (1999). Skeletal muscle satellite cell proliferation in response to members of the fibroblast growth factor family and hepatocyte growth factor. J Cell Physiol 181, 499–506.

Sheehan, S. M., Tatsumi, R., Temm-Grove, C. J., and Allen, R. E. (2000). HGF is an autocrine growth factor for skeletal muscle satellite cells in vitro. Muscle Nerve 23, 239–245.

Shefer, G., Wleklinski-Lee, M., and Yablonka-Reuveni, Z. (2004). Skeletal muscle satellite cells can spontaneously enter an alternative mesenchymal pathway. J Cell Sci 117, 5393–5404.

Sherwood, R. I., Christensen, J. L., Conboy, I. M., Conboy, M. J., Rando, T. A., Weissman, I. L., and Wagers, A. J. (2004). Isolation of adult mouse myogenic progenitors: functional heterogeneity of cells within and engrafting skeletal muscle. Cell 119, 543–554.

Smith, C. K., 2nd, Janney, M. J., and Allen, R. E. (1994). Temporal expression of myogenic regulatory genes during activation, proliferation, and differentiation of rat skeletal muscle satellite cells. J Cell Physiol 159, 379–385.

Smith, J., Goldsmith, C., Ward, A., and LeDieu, R. (2000). IGF-II ameliorates the dystrophic phenotype and coordinately down-regulates programmed cell death. Cell Death Differ 7, 1109–1118.

Spangenburg, E. E., and Booth, F. W. (2002). Multiple signaling pathways mediate LIF-induced skeletal muscle satellite cell proliferation. Am J Physiol Cell Physiol 283, C204–211.

Tapscott, S. J. (2005). The circuitry of a master switch: Myod and the regulation of skeletal muscle gene transcription. Development 132, 2685–2695.

Tatsumi, R., Anderson, J. E., Nevoret, C. J., Halevy, O., and Allen, R. E. (1998). HGF/SF is present in normal adult skeletal muscle and is capable of activating satellite cells. Dev Biol 194, 114–128.

Tatsumi, R., Hattori, A., Ikeuchi, Y., Anderson, J. E., and Allen, R. E. (2002). Release of hepatocyte growth factor from mechanically stretched skeletal muscle satellite cells and role of pH and nitric oxide. Mol Biol Cell 13, 2909–2918.

Tatsumi, R., Sheehan, S. M., Iwasaki, H., Hattori, A., and Allen, R. E. (2001). Mechanical stretch induces activation of skeletal muscle satellite cells in vitro. Exp Cell Res 267, 107–114.

Taylor-Jones, J. M., McGehee, R. E., Rando, T. A., Lecka-Czernik, B., Lipschitz, D. A., and Peterson, C. A. (2002). Activation of an adipogenic program in adult myoblasts with age. Mech Ageing Dev 123, 649–661.

Thomas, M., Langley, B., Berry, C., Sharma, M., Kirk, S., Bass, J., and Kambadur, R. (2000). Myostatin, a negative regulator of muscle growth, functions by inhibiting myoblast proliferation. J Biol Chem 275, 40235–40243.

Tidball, J. G. (2005). Inflammatory processes in muscle injury and repair. Am J Physiol Regul Integr Comp Physiol 288, R345–353.

Torrente, Y., Tremblay, J. P., Pisati, F., Belicchi, M., Rossi, B., Sironi, M., Fortunato, F., El Fahime, M., D'Angelo, M. G., Caron, N. J., et al. (2001). Intraarterial injection of muscle-derived CD34(+)Sca-1(+) stem cells restores dystrophin in mdx mice. J Cell Biol 152, 335–348.

Wada, M. R., Inagawa-Ogashiwa, M., Shimizu, S., Yasumoto, S., and Hashimoto, N. (2002). Generation of different fates from multipotent muscle stem cells. Development 129, 2987–2995.

Wagner, K. R., McPherron, A. C., Winik, N., and Lee, S. J. (2002). Loss of myostatin attenuates severity of muscular dystrophy in mdx mice. Ann Neurol 52, 832–836.

Yablonka-Reuveni, Z., and Rivera, A. J. (1994). Temporal expression of regulatory and structural muscle proteins during myogenesis of satellite cells on isolated adult rat fibers. Dev Biol 164, 588–603.

Yablonka-Reuveni, Z., Rudnicki, M. A., Rivera, A. J., Primig, M., Anderson, J. E., and Natanson, P. (1999a). The transition from proliferation to differentiation is delayed in satellite cells from mice lacking MyoD. Dev Biol 210, 440–455.

Yablonka-Reuveni, Z., Seger, R., and Rivera, A. J. (1999b). Fibroblast growth factor promotes recruitment of skeletal muscle satellite cells in young and old rats. J Histochem Cytochem 47, 23–42.

Zammit, P. S., Golding, J. P., Nagata, Y., Hudon, V., Partridge, T. A., and Beauchamp, J. R. (2004). Muscle satellite cells adopt divergent fates: a mechanism for self-renewal? J Cell Biol 166, 347–357.

Zammit, P. S., Heslop, L., Hudon, V., Rosenblatt, J. D., Tajbakhsh, S., Buckingham, M. E., Beauchamp, J. R., and Partridge, T. A. (2002). Kinetics of myoblast proliferation show that resident satellite cells are competent to fully regenerate skeletal muscle fibers. Exp Cell Res *281*, 39–49.

Zarnegar, R., and Michalopoulos, G. K. (1995). The many faces of hepatocyte growth factor: from hepatopoiesis to hematopoiesis. J Cell Biol *129*, 1177–1180.

Zeschnigk, M., Kozian, D., Kuch, C., Schmoll, M., and Starzinski-Powitz, A. (1995). Involvement of M-cadherin in terminal differentiation of skeletal muscle cells. J Cell Sci *108 (Pt 9)*, 2973–2981.

Zhao, P., Iezzi, S., Carver, E., Dressman, D., Gridley, T., Sartorelli, V., and Hoffman, E. P. (2002). Slug is a novel downstream target of MyoD. Temporal profiling in muscle regeneration. J Biol Chem *277*, 30091–30101.

CHAPTER 7

PLASTICITY OF EXCITATION-CONTRACTION COUPLING IN SKELETAL MUSCLE

ANTHONY M. PAYNE AND OSVALDO DELBONO

Wake Forest Unversity School of Medicine, Winston-Salem, NC 27157, USA

Skeletal muscles are plastic organs, undergoing changes during development, through life, and into old age. Among the many changes that occur, the process of excitation-contraction coupling undergoes changes in all stages of life and as a result of several stimuli. The goal of this chapter will be to explain the alterations that occur in excitation-contraction coupling in response to development, changes in activity level, denervation, aging, and trophic stimulation.

1.   EXCITATION-CONTRACTION COUPLING

Skeletal muscle fibers are excitable cells that conduct action potentials which signal for muscle fiber contraction through a series of events – a process collectively called excitation-contraction (EC) coupling. EC coupling actually starts with the conduction of action potentials along the axons of motor neurons causing release of acetylcholine from the nerve terminal at the neuromuscular junction. This is followed by binding of acetylcholine to nicotinic acetylcholine receptors on the end-plate of the muscle fiber membrane, increasing sodium and potassium conductance in the end-plate membrane. End-plate potentials lead to generation of action potentials along the sarcolemmal membrane and into invaginations of the sarcolemma called transverse tubules (T-tubules). Each T-tubule is bordered closely by two sac-like formations of the sarcoplasmic reticulum (SR) called terminal cisternae. This three piece structure is called the triad (Felder et al., 2002), and represents the critical subcellular region where membrane depolarization is translated into intracellular Ca^{2+} elevations.

Two protein complexes at the triad particularly important in EC coupling are the dihydropyridine-sensitive L-type Ca^{2+} channel in the T-tubule membrane – also

173

R. Bottinelli and C. Reggiani (eds.), Skeletal Muscle Plasticity in Health and Disease, 173–211.
© 2006 *Springer.*

called the dihydropyridine receptor (DHPR) (Rios and Brum, 1987) – and the sarcoplasmic reticulum Ca^{2+} release channel in the SR membrane – also called the ryanodine receptor (RyR). The DHPR undergoes a conformational change within the T-tubule membrane upon membrane depolarization to act as a voltage sensor (Schneider and Chandler, 1973). When activated, the DHPR undergoes a proposed mechanical interaction with the skeletal muscle ryanodine receptor (RyR1) to evoke Ca^{2+} release from the SR terminal cisternae (Schneider and Chandler, 1973; Block et al., 1988; Marty et al., 1994). The elevation in cytosolic Ca^{2+} allows Ca^{2+} to bind to troponin C, removing tropomyosin blockade of actin binding sites. Myosin heavy chain heads can now bind to actin binding sites to form cross-bridges and, thus, produce force and active shortening of muscle fibers (Fig. 1) [see (Melzer et al., 1995) for review]. Alterations in one or several of these steps can theoretically result in alterations in EC coupling, and thus, alterations in force production by muscle fibers.

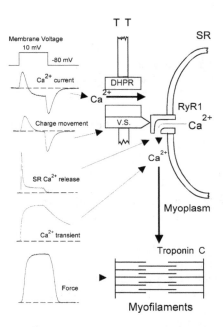

Figure 1. Excitation-Ca^{2+} release-contraction coupling in mammalian skeletal muscle, measured by voltage-clamp and single intact fiber contractility, in mouse flexor digitorum brevis muscle. Inward Ca^{2+} current occurs through the dihydropyridine receptor (DHPR) located in the t-tubule (TT) membrane. Charge movement is a result of movement of the voltage sensor (V.S.) subunit of the DHPR. Ca^{2+} release from the sarcoplasmic reticulum (SR) occurs through the muscle isoform of the ryanodine receptor (RyR1) and triggers a transient increase in myoplasmic $[Ca^{2+}]$ (Ca^{2+} transient). Myoplasmic Ca^{2+} triggers cross-bridge cycling and force generation. Vertical calibration bars: 5nA, 10 μM ms^{-1}, 5 μM, and 200 kPa for Ca^{2+} current and charge movement, SR Ca^{2+} release, Ca^{2+} transient, and force, respectively (Reproduced with permission from Payne and Delbono, 2004. © Lippincott Williams and Wilkins)

1.1   Dihydropyrdine Receptor Structure

DHPRs are L-type voltage-gated Ca^{2+} channels located in skeletal and cardiac muscle sarcolemmal membranes, primarily in the T-tubules. A DHPR is a heteromeric protein complex consisting of $\alpha 1$, $\alpha 2$, β, γ, and δ subunits (Catterall et al., 1988; Catterall, 1991). The $\alpha 1$ subunit of these Ca^{2+} channels – $\alpha 1_S$ in skeletal muscle (Tanabe et al., 1987; Tanabe et al., 1988; Tanabe et al., 1990b) and $\alpha 1_C$ in cardiac muscle (Mikami et al., 1989) – can function alone as a voltage-gated Ca^{2+} channel (Mikami et al., 1989; Perez-Reyes et al., 1989), while the accessory subunits serve modulatory functions [see (Catterall, 1991) for review]. The $\alpha 1$ subunit is a ~ 190 kDa protein that contains four internal domains of sequence homology (I-IV). Each domain contains 6 trans-membrane segments (S1-S6) (Tanabe et al., 1987; Catterall et al., 1991). This structure is very similar to voltage-gated Na^+ channels, and both contain positive charges located (primarily Arginine and Lysine residues) within the S4 membrane-spanning segments of the domains I-IV (Fig. 2). Upon membrane depolarization, the DHPR undergoes a conformational change in the membrane. It is these positively charged Arginine and Lysine residues that allow the DHPR to "sense" membrane depolarization by moving within the confines of the lipid bilayer in response to changes in membrane potential, causing the aforementioned protein conformational change [see (Catterall

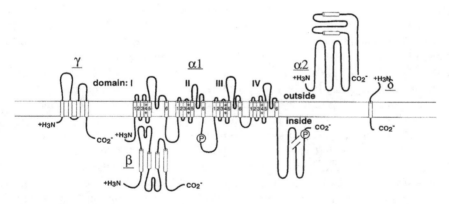

Figure 2. Proposed structure of the skeletal muscle DHPR. This scheme shows the relative position of the five subunits. The $\alpha 1$ subunit is primarily responsible for both the voltage-sensing and Ca^{2+} channel functions. Barrel structures depict α-helix structures. Domains of homology I-IV each contain 6 transmembrane segments labeled 1-6. Transmembrane segment 4 of each homology domain contains positive charges embedded in the membrane which allow for the voltage sensing function. Each homology domain contains a re-entrant loop between segments 5 and 6 that are proposed to form the Ca^{2+} conducting pore. The β subunit is an intracellular subunit associated with the $\alpha 1$ at the intracellular I-II loop at a conserved β-subunit binding domain. The γ subunit contains four transmembrane segments, and is associated with the $\alpha 1$. The $\alpha 2\delta$ subunits are two separate peptides that are connected via disulfide bonds. A single transmembrane segment of the δ subunit anchors this structure, while the $\alpha 2$ subunit exists extracellularly. See text for more information on subunit functions (Reproduced with permission from Catterall, 1991 © AAAS)

et al., 1991) for review]. This movement of charges within the membrane can be measured with electrophysiological methods, and is termed intramembrane charge movement. Relative number of functional DHPRs can be determined by charge movement density by inhibiting the gating function of the remaining voltage-activated ion channels (i.e., Na^+ and K^+ channels) (Adams et al., 1990; Wang et al., 2000). Similarly, inward Ca^{2+} current can be measured upon membrane depolarization, and Ca^{2+} current amplitude also serves as a measure of the pore conducting function of the DHPR (Fig. 1).

The skeletal muscle β subunit, $\beta 1_A$, binds with a 1:1 stoichiometry with the $\alpha 1_S$ subunit (Leung et al., 1987) and with high affinity at the intracellular loop connecting domains of homology I and II (the I-II loop) (Pragnell et al., 1994). Binding occurs at a conserved β-subunit binding domain (De Waard et al., 1994). The β subunit modifies both expression of and Ca^{2+} current through the $\alpha 1_S$. Skeletal myotubes from β-knockout mice exhibit drastically lower charge movement, indicating lower $\alpha 1_S$ expression, and about 7-fold lower Ca^{2+} current density with faster activation kinetics than control myotubes. In addition, the intracellular Ca^{2+} transient and contraction were absent, indicating that the $\beta 1_A$ subunit plays a role in EC coupling (Gregg et al., 1996; Strube et al., 1996; Beurg et al., 1997). However, trans-fection of $\beta 1_A$ cDNA fully restored charge movement, Ca^{2+} current density and kinetics, intracellular Ca^{2+} transient, and contraction to control myotube values (Beurg et al., 1997). Transfection of the $\beta 2_A$ subunit native to heart and brain into β-knockout skeletal myotubes returns I_{Ca} density and activation kinetics to normal, but not charge movement. Ca^{2+} transients were returned with $\beta 2_A$ transfection, but were 3- to 5-fold lower than control (Beurg et al., 1999b). To fully restore EC coupling, an intact $\beta 1_A$ C-terminus was necessary on $\beta 1_A$ and $\beta 2$-$\beta 1$ chimeras (Beurg et al., 1999a), indicating the C-terminus of the $\beta 1_A$ subunit is necessary for skeletal-type EC coupling (see below). Blocking inward Ca^{2+} current in β-knockout myotubes expressing a $\beta 2_A$ or a C-terminal-truncated $\beta 1_A$ eliminates EC coupling (Sheridan et al., 2003a; Sheridan et al., 2003b).

The skeletal muscle $\gamma 1$ subunit is a membrane-spanning protein in close associ-ation with the $\alpha 1_S$ subunit (Catterall, 1991). The $\gamma 1$ subunits of skeletal muscle favor inactivation of inward Ca^{2+} current through the $\alpha 1$ subunit of DHPR. In skeletal myotubes from $\gamma 1$ knockout mice, inactivation of Ca^{2+} current is slower and incom-plete compared to control myotubes, and peak inward Ca^{2+} current is increased due the slower inactivation (Freise et al., 2000; Ahern et al., 2001). However, the absence of the $\gamma 1$ subunit seems to have little or no effect on EC coupling. Muscle from mice lacking the $\gamma 1$ subunit exhibit normal charge movement (Ahern et al., 2001), normal Ca^{2+} release transients (Ahern et al., 2001; Ursu et al., 2001), and normal twitch and tetanic force (Ursu et al., 2001).

The $\alpha 2$ and δ subunits are coded by a single gene, $\alpha 2\delta$-1, and are post-translationally cleaved into two peptide subunits connected by disulfide bonds (De Jongh et al., 1990; Jay et al., 1991). A single transmembrane domain in the δ subunit anchors the $\alpha 2\delta$ subunit to the membrane, while the $\alpha 2$ subunit is extracellular (Gurnett et al., 1996). Co-expression of the $\alpha 2$ subunit with $\alpha 1$ *in vitro* increases Ca^{2+} current (Mikami et al., 1989; Mori et al., 1991). The $\alpha 2$ subunit alone seems to increase the $\alpha 1$ Ca^{2+} current amplitude, while co-expression of the δ with $\alpha 2$ blunts this stimulatory effect on Ca^{2+} current (Gurnett et al., 1996). Co-expression of $\alpha 2$ with $\alpha 1$ also increases activation and inactivation kinetics(Bangalore et al., 1996), and increases the expression and/or membrane targeting of $\alpha 1$ (Brust et al., 1993; Shistik et al., 1995). The majority of this work has been conducted in heart and neuron which contain primarily DHPR $\alpha 1_C$ and $\alpha 1_B$, respectively. However, more recent work using gabapentin as a pharmacological inhibitor of $\alpha 2\delta$, shows that Ca^{2+} current is reduced in both neuronal and skeletal muscle Ca^{2+} channels (Alden and Garcia, 2001). Additionally, gabapentin reduces the rate of rise of intracellular Ca^{2+}, possibly indicating a role for $\alpha 2\delta$ in EC coupling (Alden and Garcia, 2002). The exact role is still unknown at this time.

A common protein that interacts with the DHPR is cAMP-dependent protein kinase, or protein kinase A (PKA), which phosphorylates both the $\alpha 1$ and β subunits of the DHPR (De Jongh et al., 1989; Nunoki et al., 1989; Lai et al., 1990; Rotman et al., 1992; Sculptoreanu et al., 1993; Johnson et al., 1994; De Jongh et al., 1996; Gao et al., 1997). In general, phosphorylation of the DHPR by PKA increases Ca^{2+} channel activity in both the skeletal $\alpha 1_S$ (De Jongh et al., 1989; Nunoki et al., 1989; Johnson et al., 1994) and the cardiac $\alpha 1_C$ (Sculptoreanu et al., 1993; De Jongh et al., 1996; Gao et al., 1997). PKA-mediated phosphorylation of DHPR occurs during repetitive depolarizations (such as during trains of pulses) (Sculptoreanu et al., 1993; Johnson et al., 1994) and occurs very quickly. The rapidity of phosphorylation, and thus, potentiation of channel activity, is due to the fact that PKA is anchored to the membrane in very close proximity to the DHPR in both skeletal (Johnson et al., 1994; Johnson et al., 1997) and cardiac muscle (Gao et al., 1997) via A-kinase anchoring proteins (AKAPs). The role of this phosphorylation is better understood in cardiac muscle than in skeletal muscle. β-adrenergic enhancement of inward Ca^{2+} current, and thus, contraction in cardiac muscle is mediated specifically through anchored PKA phosphorylation of the cardiac DHPR (Gao et al., 1997). The exact role that phosphorylation of the DHPR plays in skeletal muscle EC coupling is not fully understood, but may enhance Ca^{2+} intake into the muscle in order to enhance the available SR Ca^{2+} store during times of increased muscle activity.

1.2 Ryanodine Receptor Structure

RyRs are the intracellular Ca^{2+} release channels located in the membrane of the Ca^{2+} storage organelle, the SR. RyRs are homotetramers, with the four identical subunits forming a structure with rotational symmetry (Serysheva et al., 1995; Serysheva et al., 1999; Paolini et al., 2004). Skeletal and cardiac muscle express distinct types of RyR. RyR1 is the skeletal muscle type (Takeshima et al., 1989) and RyR2 is the cardiac muscle type (Nakai et al., 1990). RyR3 is expressed in smooth muscle, as well as in brain and many other tissues (Giannini et al., 1992; Giannini et al., 1995). Additionally, RyR3 is expressed in developing skeletal muscle (Tarroni et al., 1997; Flucher et al., 1999; Chun et al., 2003) and at very low levels in adult diaphragm (Flucher et al., 1999). RyRs are large proteins (>500 kDa per subunit (Takeshima et al., 1989; Takeshima, 1993; Ottini et al., 1996)) with approximately 90% of each protein subunit existing on the cytoplasmic side of the SR membrane. This large cytoplasmic domain constitutes the "foot" region of the protein, while the C-terminal tenth forms four trans-membrane segments with sequence similarity with the IP_3 receptor Ca^{2+} release channel (Takeshima, 1993). During development of striated muscles, RyRs organize themselves into orthogonal arrays in the SR membrane (Flucher et al., 1993; Flucher et al., 1994; Flucher and Franzini-Armstrong, 1996).However, RyRs can organize into these arrays with or without the presence of DHPRs (Takekura et al., 1995a; Takekura et al., 1995b; Yin and Lai, 2000). Both skeletal muscle RyR1 and cardiac muscle RyR2 are organized into such ordered arrays.

RyR1 activity is regulated by many cellular constituents, including Ca^{2+}, Mg^{2+}, and ATP (Fill and Copello, 2002). Ca^{2+} exerts a biphasic effect on the activation of RyR1. Activation at low $[Ca^{2+}]$ (i.e., micromolar) and inhibition at high $[Ca^{2+}]$ (i.e., millimolar) is due to the presence of two Ca^{2+} binding sites: a high affinity activator site and a low affinity inhibitor site (Meissner et al., 1997). Mg^{2+} at normal physiological cytosolic levels acts to inhibit RyR1 function in skeletal muscles (Lamb and Stephenson, 1991; Laver et al., 1997), competing with Ca^{2+} at the high affinity Ca^{2+} activation site of the RyR1, shifting the Ca^{2+} sensitivity of the RyR1 (Laver et al., 1997; Copello et al., 2002). ATP acts as an activator of the RyR1, but only in its free form ($\sim$300 μM) (Sonnleitner et al., 1997; Copello et al., 2002).

RyR1s interact closely with and are regulated by several proteins, including calmodulin, FK506 binding protein, and calsequestrin. The first protein found to interact with RyRs in lipid bilayers was calmodulin (CaM). One RyR1 has 4 CaM binding sites near to the pore of the RyR1 (Wagenknecht et al., 1997; Stokes and Wagenknecht, 2000) and binds from 4 to 16 molecules of CaM (Tripathy et al., 1995; Rodney et al., 2000). CaM exerts a biphasic effect on RyR1: activation at low $[Ca^{2+}]$ (i.e., nanomolar) and inhibition at high $[Ca^{2+}]$ (i.e., micromolar) (Smith et al., 1989; Tripathy et al., 1995; Fruen et al., 2000). So, while Ca^{2+} alone exerts its biphasic effects on RyR1, CaM seems to shift both Ca^{2+} activation and Ca^{2+} inhibition of channel activity to lower $[Ca^{2+}]$ by enhancing Ca^{2+} binding at the RyR1 Ca^{2+} activation site and shifting its own function to that of a RyR1 inhibitor when bound to Ca^{2+} itself, respectively (Rodney et al., 2000; Rodney et al., 2001). Therefore, CaM serves to modulate Ca^{2+}-dependent effects on RyR1. CaM may even play a role in the physical coupling between DHPR $\alpha 1_S$ and RyR1 (Sencer et al., 2001). Another protein that associates with the RyR1 is the 12 kDa FK506 binding protein, or FKBP12. FKBP12 is a member of the immunophilin family of proteins serving as receptors for immunosuppressant drugs like rapamycin and FK506 (Schreiber, 1991). One FKBP12 binds per RyR1 subunit, 4 per whole RyR1 channel (Jayaraman et al., 1992; Timerman et al., 1993). Much of the available data indicates that the presence of FKBP12 inhibits the RyR1, and that removal of FKBP12 activates the RyR1 (Timerman et al., 1993; Ahern et al., 1994; Brillantes et al., 1994; Chen et al., 1994; Mayrleitner et al., 1994; Barg et al., 1997). Also, there are suggestions that FKBP12 coordinates the four RyR1 subunits openings and closings (Brillantes et al., 1994), because removal of FKBP12 has been shown to induce subconductance states (Ahern et al., 1994; Brillantes et al., 1994; Ahern et al., 1997). FKBP12 has also been suggested to possibly synchronize the opening of adjacent RyR1s (Marx et al., 1998).

Calsequestrin (CSQ) is the main Ca^{2+} buffering protein inside the lumen of the SR. The large number of acidic amino acids allow each CSQ molecule to bind 40-50 Ca^{2+} ions (MacLennan and Wong, 1971; Yano and Zarain-Herzberg, 1994). CSQ forms ribbon-like polymers at the triad junctional membrane of the SR terminal cisternae when $[Ca^{2+}]$ is raised (Wang et al., 1998a) and exists as a polymer at physiological lumen $[Ca^{2+}]$ (Franzini Armstrong, 1970) of ~ 1 mM (Fryer and Stephenson, 1996). CSQ undergoes a conformational change upon T-tubule depolarization but before Ca^{2+} release from the SR (Ikemoto et al., 1991), suggesting that the EC coupling signal is transmitted to CSQ and that it plays a role in Ca^{2+} release (Beard et al., 2004). CSQ is probably not bound directly to the RyR1 but forms a complex with the intrinsic membrane proteins triadin and junctin (Zhang et al., 1997). Conflicting evidence exists on the regulatory effect of CSQ on RyR1. Initial evidence indicated that addition of CSQ to purified RyR1 in lipid bilayers inhibits the channel (Kawasaki and Kasai, 1994; Szegedi et al., 1999). However, more recent evidence indicates that this action occurs when CSQ interacts directly with RyR1 (Beard et al., 2002). When junctional SR membrane is isolated, the effect of luminal CSQ was found to inhibit RyR1 activity, acting through the triadin/junction/RyR1 complex (Beard et al., 2002). The full role of CSQ-RyR1 interaction is still to be elucidated.

1.3   DHPR-RyR Interaction (Coupling, Function)

In both skeletal and cardiac muscle, activation of DHPRs evokes an inward Ca^{2+} current and RyR-mediated Ca^{2+} release from the SR to induce muscular contraction, as mentioned (Schneider and Chandler, 1973; Marty et al., 1994). The structural relationship of DHPRs and RyRs in these two muscle types is different, though. In skeletal muscle, DHPRs are arranged in groups of four, each at the corner of a square – i.e., a tetrad. Each tetrad is arranged such that each of the four DHPRs are lined up with one of the four subunits of the RyR1 cytoplasmic foot (Block et al., 1988; Franzini-Armstrong and Kish, 1995; Protasi et al., 1997; Franzini-Armstrong et al., 1998; Paolini et al., 2004). One DHPR tetrad is found coupled to every other (Block et al., 1988) to every fourth (Renganathan et al., 1997a) RyR1 complex in fast-twitch skeletal muscle (Fig. 3A). The tetrad organization of DHPRs is dictated by the orthogonal array organization of RyR1, as DHPRs in RyR1 null myotubes are not arranged in an orderly fashion (Protasi et al., 1998). However, re-expression of RyR1 in these myotubes restores the ordered tetrad arrangement of DHPRs (Protasi et al., 2000). This structural arrangement and need for both proteins to be expressed in order to exhibit this structural arrangement

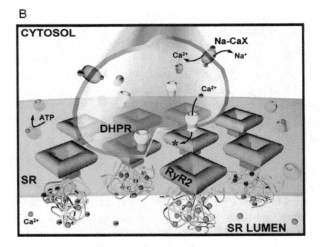

Figure 3. (**A**) Skeletal muscle triad junction. DHPR tetrads (yellow) in the T-tubule membrane (purple) are located at roughly every other RyR1 (red) located in orthogonal arrangement in the sarcoplasmic reticulum (SR) membrane (green). A change in membrane voltage (ΔV_M) activates the DHPR. Ca^{2+} entry through the DHPR is not critical to trigger Ca^{2+} release from the SR via the RyR1. Instead, the ΔV_M signal is thought to be conducted to the RyR1 by a physical protein-protein interaction between the DHPR and RyR1. Calsequestrin (CSQ) located inside the SR lumen near the RyR1 is a low affinity, high capacity Ca^{2+} binding protein that allows the SR to store a large amount of Ca^{2+}. The sarco-endoplasmic reticulum calcium ATPase (SERCA) pumps Ca^{2+} back into the SR lumen to allow relaxation of muscle. (**B**) Cardiac muscle triad junction. DHPR molecules do not form tetrads in cardiac muscle. While the RyR2 does form an orthogonal pattern in cardiac muscle, DHPR complexes are not directly aligned with RyR2. Activation of Ca^{2+} release via the RyR2 is dependent on Ca^{2+} entry through the DHPR. (Reproduced with permission from Fill and Copello, 2002, © American Physiological Society) A color figure is freely accessible via the website of the book: http://www.springer.com/1-4020-5176-x

further supports the mechanical hypothesis of DHPR-RyR1 coupling in skeletal muscle. In cardiac muscle, the RyR2s are arranged into orthogonal arrays similar to skeletal muscle RyR1s (Fill and Copello, 2002), but the DHPRs are not arranged in such an order. Cardiac DHPRs are, indeed, in close proximity to RyR2 (Carl et al., 1995), but do not form tetrads or any other detectable formation (Fig. 3B) (Sun et al., 1995; Protasi et al., 1996). RyR3 in skeletal muscle also does not affect arrangement of DHPRs, as indicated by lack of DHPR organization in response to RyR3 expression in RyR1-knockout 1B5 myotubes (Protasi et al., 2000).

The functional interaction of DHPRs and RyRs is different in the two muscle types, also. In cardiac muscle, Ca^{2+} release from the RyR2 is evoked via Ca^{2+} influx thought the DHPR $\alpha 1_C$, a process termed Ca^{2+}-induced Ca^{2+} release (CICR) (Fabiato, 1983, , 1985; Nabauer et al., 1989). This Ca^{2+} influx has been shown necessary for cardiac EC coupling in experiments in which cardiac DHPR $\alpha 1_C$ subunits were expressed in dysgenic myotubes. Dysgenic myotubes are muscle cells that do not express native DHPRs due to a gene mutation, and EC coupling is absent from these fibers, i.e., these cells do not exhibit charge movement, inward Ca^{2+} current, or contraction (Adams et al., 1990). All these facets of EC coupling – charge movement, inward Ca^{2+} current, and contraction – are returned to dysgenic myotubes when DHPR cDNA is injected into them (Tanabe et al., 1988; Adams et al., 1990; Tanabe et al., 1990b; Garcia et al., 1994). Experiments in which dysgenic myotubes expressed DHPR $\alpha 1_C$, removal of external Ca^{2+} or blockage of inward Ca^{2+} current by Ca^{2+} channel blockers completely eliminated contraction and RyR-mediated Ca^{2+} release transients (Tanabe et al., 1990a; Garcia et al., 1994), indicating that influx of external Ca^{2+} is necessary for DHPR $\alpha 1_C$ to activate RyR-mediated Ca^{2+} release.

In contrast, activation of RyR1 Ca^{2+} release in skeletal muscle is not dependent on the inward Ca^{2+} current through the DHPR $\alpha 1_S$. In skeletal muscle cells, the inward Ca^{2+} current through the DHPR has been calculated to be less than 5% of the intracellular Ca^{2+} transient measured by fluorescent Ca^{2+} indicator dye, suggesting a reduced importance of inward Ca^{2+} current in skeletal muscle compared to cardiac (Brum et al., 1988). Indeed, experiments in which external Ca^{2+} has been removed from (Armstrong et al., 1972; Dulhunty and Gage, 1988) or calcium channel blockers added to (González-Serratos et al., 1982; Dulhunty and Gage, 1988) the bathing medium show EC coupling and contraction persist in skeletal muscle cells. Similarly, experiments in dysgenic myotubes expressing DHPR $\alpha 1_S$ subunits maintained both EC coupling and contraction in the absence of external Ca^{2+} and the presence of Ca^{2+} channel blockers (Tanabe et al., 1990a; Tanabe et al., 1990b). These experiments support the hypothesis of a mechanical protein-protein interaction between the DHPR $\alpha 1_S$ and RyR1 in skeletal muscle (Rios and Brum, 1987; Block et al., 1988).

The exact function of RyR3 in skeletal muscle is not fully understood yet. RyR3 is activated by cytosolic Ca^{2+} in similar fashion to RyR1 and RyR2 (Chen et al., 1997; Jeyakumar et al., 1998). Its role in EC coupling and necessity for efficient contraction of embryonic skeletal muscle has been suggested (Bertocchini et al., 1997), but contractile properties of RyR3-knockout mouse muscle is not different from controls at either embryonic (Barone et al., 1998) or adult stages (Clancy et al., 1999). Furthermore, co-expression of RyR3 with skeletal muscle DHPR in 1B5 myotubes does not return EC coupling (Fessenden et al., 2000) and RyR3-knockout myotubes do not show impaired intracellular Ca^{2+} release in response to voltage depolarization (Dietze et al., 1998).

Alterations in DHPR and/or RyR expression, function, regulation, and/or physical arrangement in relation to each other at the triad in response to a stimulus can theoretically change the effectiveness of EC coupling, and thus force production, in skeletal muscle. The remaining sections of this chapter will primarily focus on alterations that occur at the triad in response to such stimuli as development, changes in activity level, denervation, aging, and trophic stimulation.

2.   NEURAL CONTROL OF EXCITATION-CONTRACTION COUPLING

EC coupling is regulated by several factors, including innervation of muscle fibers. A series of evidence supports the concept that alterations in innervation state alters expression of genes involved in EC coupling, ultimately resulting in EC coupling alterations. During development, muscle fibers are innervated by spinal cord motor neurons. Fiber type is determined by interaction with different subpopulations of these motor neurons that activate contraction at different rates, ranging from 10 Hz (slow-twitch) to 100 Hz (fast-twitch fatigue-resistant) or 150 Hz (fast-twitch fatigue-sensitive) (Buller et al., 1960b, 1960a; Greensmith and Vrbova, 1996). Interaction with motor neurons and induction of activity determines more than just fiber type. Depolarization of myotubes in culture triggers the appearance of dihydropyridine binding sites (Pauwels et al., 1987), suggesting that the induction of muscle activity during innervation induces DHPR expression. During developmental innervation of skeletal muscles, DHPR $\alpha 1_S$ mRNA levels increase while $\alpha 1_C$ levels decrease (Chaudhari and Beam, 1993), shifting toward the adult muscle phenotype.

Experimental denervation in fast-twitch muscle fibers, such as extensor digitorum longus (EDL), results in alterations in EC coupling. Denervated EDL fibers exhibit reduced charge movement (Dulhunty and Gage, 1985; Delbono, 1992) and inward Ca^{2+} current (Delbono, 1992), indicative of reduced number or reduced functional DHPRs in denervated muscle. Additionally, denervation affects DHPR function in other ways, slowing both DHPR activation (Delbono, 1992) and inactivation (Delbono and Stefani, 1993). Denervation also affects RyR1 function and expression. RyR1s from denervated muscles reconstituted into planar lipid bilayers showed increased sensitivity to caffeine (Delbono and Chu, 1995). Morphological evidence from denervated EDL muscles shows a reduction in the number of "indentations" in SR terminal cisternae membranes (Dulhunty et al., 1984). These indentations are the RyR1 feet at triad junctions. Accordingly, following denervation, RyR1 mRNA transcript levels have been shown to decrease (Ray et al., 1995).

DHPR in skeletal muscle is dependent upon the presence of RyR1 to function properly. Dyspedic myotubes, those that do not express RyR1s due to a mutation, display drastically reduced charge movement and inward Ca^{2+} current compared to normal myotubes. When dyspedic myotubes are transfected to express RyR1, charge movement and inward Ca^{2+} current are returned to normal levels (Nakai et al., 1996). This evidence may indicate that denervation-induced reduction of DHPR expression may result from loss of innervation-driven gene expression, RyR1-driven DHPR gene expression, or a combination of both.

However, the full effects of experimental denervation on DHPR and RyR1 expression and function are not completely understood, as some evidence seems to be in contrast with those mentioned above. For example, while reduced DHPR function has been observed in denervated muscle fibers, some investigators have found no reduction in DHPR mRNA expression after denervation (Ray et al., 1995; Pereon et al., 1997b) and no change in RyR1 mRNA expression (Pereon et al., 1997b). Since mRNA levels do not always translate to protein levels, both DHPR and RyR1 protein levels have been shown not to change after denervation (Ray et al., 1995). In further contrast, it has also been found that dihydropyridine binding sites increased in fast-twitch muscle from day 1 to day 10 after sciatic nerve freeze damage, but returned to normal levels by day 14 post-freeze injury (Takekura et al., 2003). Therefore, the type of nerve damage that is induced, the state and extent of ongoing re-innervation, and the time of measurement post-injury may be able to explain some of these differences.

Although denervation removes nerve electrical influence from muscle, evidence indicates that electrical activity of a nerve is not the only controlling factor in muscle gene expression. Chronic blockade of nerve action potentials by infusion of tetrodotoxin, without compromising axonal transport and nerve-muscle contact, apparently increases expression of DHPR $\alpha 1_S$ and RyR1 mRNA expression 4.5- and 2.5-fold, respectively. Accordingly, expression of both proteins was increased by 50% after 10 days of treatment (Ray et al., 1995). In a different experimental paradigm, Albuquerque et al. eliminated axonal transport with a colchicine-containing cuff on the sciatic nerve, without disturbing action potentials, and examined the effects on muscle contraction (Albuquerque et al., 1974). While this report did not examine molecular changes that occurred in the muscle, the investigators found roughly a 25% decrease in nerve-stimulated EDL tetanic force with only a 10% decline in EDL mass over a 7-day period (Albuquerque et al., 1974). Our own calculations based on these data estimate approximately 17% decline in muscle specific force, suggesting that muscle atrophy was not the only reason for force decline. In other words, alterations in EC coupling mechanisms may explain the decline in specific force in response to reduced axonal transport in motor neurons.

While decreases in motor neuron electrical activation of and axonal delivery to skeletal muscle effect changes in gene expression, increases in nerve activity also causes changes in adult muscle gene expression. These changes are generally characterized by a fast-to-slow switch in myosin and Ca^{2+} ATPase isoforms (Ohlendieck et al., 1999). Two weeks of chronic nerve stimulation also causes changes in gene expression in fast-twitch muscle such as transiently decreased expression of DHPR $\alpha 1_S$ mRNA and increased expression of DHPR $\alpha 1_C$ mRNA (Pereon et al., 1997a). However, other investigators have found that prolonged stimulation of fast-twitch skeletal muscle actually decreases the expression of DHPR, RyR1, triadin, and sarcalumenin proteins in skeletal muscle Ohlendieck et al., 1991; Hicks et al., 1997; Ohlendieck et al., 1999). In apparent contrast to these findings, a 12-week endurance training program in rats, an increase in activity which can also lead to fast-to-slow transformations in muscle, was shown to increase DHPR expression as much as 60% in EDL muscles (Saborido et al., 1995). In another mode of increased muscle activity, surgical ablation of synergist muscles to induce hypertrophy of the plantaris muscle in rats causes no change in DHPR or RyR1 expression in this predominately fast-twitch skeletal muscle (Kandarian et al., 1996). The possibility exists that the "protocol" of increased activity plays an important role in regulating gene expression: Endurance exercise training allows for long quiescent recovery periods between bouts of increased activity, while chronic low-frequency stimulation does not; and mechanical loading of muscle that induces hypertrophy apparently differs from these two protocols altogether.

Decreases in muscle activity, without compromise of nerve-muscle interaction, can be induced by mechanical unloading of muscle through the use of hindlimb unweighting (Kandarian et al., 1992; Schulte et al., 1993; Kandarian et al., 1996). Hindlimb unloading is generally characterized by a slow-to-fast transition in fiber type, as evidenced by the increase in fast Ca^{2+} ATPase (Schulte et al., 1993) and myosin isoforms (Thomason et al., 1987; Diffee et al., 1991) as well as speed of unloaded muscle fiber shortening (Gardetto et al., 1989). These decreases in activity lead to marked atrophy of slow-twitch soleus muscles, but increases in DHPR mRNA (Kandarian et al., 1992) and DHPR and RyR1 protein expression (Kandarian et al., 1996; Bastide et al., 2000) in the same soleus muscles. Additionally, RyR3 expression declines in unloaded soleus muscles (Bastide et al., 2000). These results, taken together, indicate that electrical activity of the nerve, pattern of nerve activation, axonal delivery of trophic factors to the muscle, muscle fiber type, fiber type transition direction (i.e., fast-to-slow or slow-to-fast), and type of muscle activity (i.e., endurance vs. strength training) are all variables that exist in tenuous balance to regulate muscle gene expression.

3.   EXCITATION-CONTRACTION COUPLING ALTERATIONS IN AGING SKELETAL MUSCLE

3.1   EC Uncoupling

Aging leads to decreased muscle strength in elderly humans (Frontera et al., 1991; Lindle et al., 1997) that can lead to impairment of daily living activities (Bassey et al., 1992; Brown et al., 1995). Strength decline with age cannot be fully accounted for by age-related muscle atrophy alone, as strength decreases are greater than loss of muscle mass in elderly humans (Jubrias et al., 1997; Lynch et al., 1999). Accordingly, *in vitro* study of skeletal muscle contractility has also shown that when lower maximum isometric force in aged mice is normalized to the smaller muscle fiber cross-sectional area from aged mice, a significant deficit in force remains unexplained by atrophy (Brooks and Faulkner, 1988; González et al., 2000). In other words, the specific force of the muscle is reduced with age, indicating some impairment in the muscle itself is acquired with age. In addition to alterations in contractile protein function with age (Lowe et al., 2002), we have proposed that age-related alterations in DHPR and RyR1 expression and function contribute to age-related decline in muscle specific force – a process we refer to as EC uncoupling (Delbono and Chu, 1995; Renganathan et al., 1997a; Wang et al., 2000).

In addition to study of whole muscles (Brooks and Faulkner, 1988; Barton-Davis et al., 1998; Payne et al., 2003), single intact fiber experiments have also revealed age-related decline in specific force (González et al., 2000; González et al., 2003). Electrophysiological studies have revealed a reduced charge movement and inward Ca^{2+} current from both aged human (Delbono et al., 1995) and aged mouse (Wang et al., 2000) fast-twitch muscle fibers, compared with young (Fig. 4). These studies suggest a reduction in the number of DHPRs in aged skeletal muscle fibers. Consistent with these findings, radioligand binding studies show a reduction in the number of dihydropyridine binding sites in aged muscle compared to young (Fig. 5) (Renganathan et al., 1997a, , 1998). Similarly, immunoblot analysis revealed a decrease in the amount of DHPR $\alpha 1_S$ in aged muscle (Ryan et al., 2000). In addition to reduced DHPR in aged muscle, binding studies also reveal reduction in the number of ryanodine binding sites (Fig. 5) and a reduction in the dihydropyridine-to-ryanodine binding site ratio (Renganathan et al., 1997a, , 1998) in skeletal muscle from aged compared to young. Age-related reduction in the number of DHPRs and RyR1s and a reduction in the DHPR/RyR1 ratio in skeletal muscles indicates fewer RyR1s available for SR Ca^{2+} release and a greater percentage of uncoupled RyR1s, resulting in reductions in voltage-activated Ca^{2+} release from the SR. Indeed, use of fluorescent Ca^{2+} indicator dyes has shown that, in addition to reduced charge movement and inward Ca^{2+} current, the voltage-activated intracellular SR Ca^{2+} release transient ($[Ca^{2+}]_i$) is also reduced in aged muscle fibers compared with young (Fig. 6) (Delbono et al., 1995; Wang et al., 2000). To determine if this reduction in $[Ca^{2+}]_i$ contributes to age-related decline in specific force, González et al. (González et al., 2003) measured the $[Ca^{2+}]_i$ and maximum isometric specific force simultaneously in single intact fibers, finding a direct relationship between the age-related decreases in these two variables (Fig. 7 and 8). Interestingly, addition of caffeine to the bathing medium (a RyR1 activator) was able to increase both $[Ca^{2+}]_i$ and isometric specific force in fibers from old mice to levels equal with fibers from young mice before caffeine treatment. This suggests that availability of Ca^{2+} in the SR is not reduced with aging, but that reduction in DHPR and RyR1 levels in aged skeletal muscle result in impaired voltage-activated Ca^{2+} release from the SR – i.e., an impaired EC coupling process. However, caffeine increased isometric specific

A

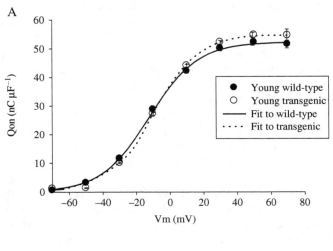

B

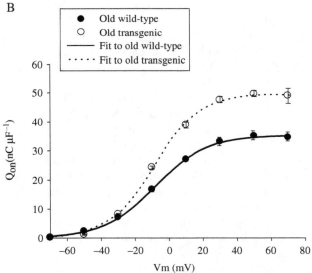

Figure 4. Charge movement (Q_{on}) – membrane voltage (Vm) relationship for (A) young wild-type and transgenic mice and for old wild-type and transgenic mice (B). Transgenic mice overexpressed hIGF-1 exclusively in striated muscle. Muscle fibers from the flexor digitorum brevis were voltage-clamped in the whole-cell configuration of the patch-clamp technique. Data were fit to a Boltzmann equation. Muscle fibers from old mice exhibited lower Qon than fibers from young mice. Q_{on} was not higher in young transgenic mice, but overexpression of hIGF-1 prevented decline in Q_{on} in fibers from aged mice. (Reproduced with permission from Wang et al., 2002, © Biphysical Society)

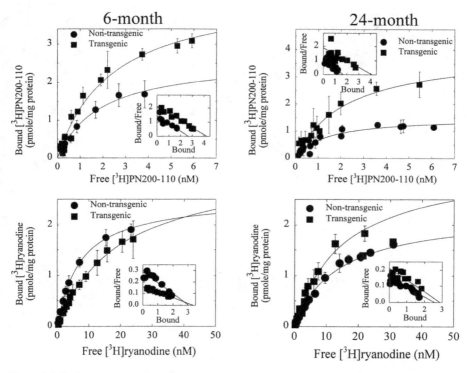

Figure 5. Dihydropyridine binding (top panels) and ryanodine binding (bottom panels) in young adult (6 month old, left panels) and old mice (24 month old, right panels). Aging leads to a decline in dihydropyridine and ryanodine binding sites. Targeted overexpression of hIGF-1 in skeletal muscle increases dihydropyridine binding in both young and aged mice, and prevents age-related decline in ryanodine binding. Inset boxes display Scatchard analyses from which B_{max} and K_d values were obtained. (Reproduced with permission from Renganathan et al., 1998, © Amercian Society for Biochemistry and Molecular Biology)

force in young fibers, also; and the caffeine-enhanced specific force of fibers from young mice remained higher than fibers from old mice (Fig. 9). Therefore, we must consider other potential factors for reduced specific force in aged skeletal muscle fibers. Other investigators have shown reduced specific force in skinned single fibers from humans and rodents (Li and Larsson, 1996; Larsson et al., 1997; Thompson and Brown, 1999; Frontera et al., 2000; Lowe et al., 2002), indicating that the fully activated contractile apparatus is impaired with age in addition to alterations in EC coupling. Interestingly, though, the caffeine-mediated increase in specific force reported by González et al. (González et al., 2003) was greater in fibers from old mice (∼50%) compared to fibers from young mice (∼20%), providing further support that voltage-activated Ca^{2+} release is impaired in aged muscle fibers under normal conditions. These findings further support the EC uncoupling hypothesis in aged skeletal muscle.

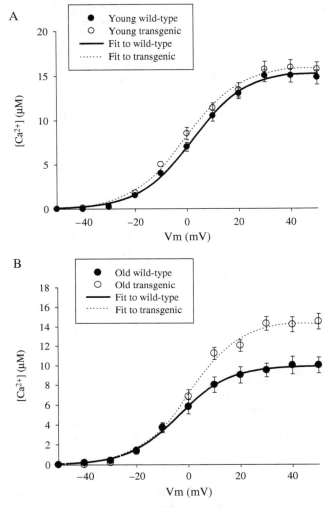

Figure 6. Peak intracellular Ca^{2+} concentration ($[Ca^{2+}]$) – membrane voltage (V_m) relationship for young wild-type and transgenic (**A**) and old wild-type and transgenic mice (**B**). Muscle fibers from the flexor digitorum brevis were voltage-clamped in the whole-cell configuration of the patch-clamp technique. Data were fit to a Boltzmann equation. Muscle fibers from old mice exhibited lower $[Ca^{2+}]$ than fibers from young mice. $[Ca^{2+}]$ was not higher in young transgenic mice, but overexpression of hIGF-1 prevented decline in $[Ca^{2+}]$ in fibers from aged mice. (Reproduced with permission from Wang et al., 2002, © Biophysical Society)

3.2 Ca^{2+}-Dependent EC Coupling

In addition to reduced numbers of DHPRs and RyR1s and reduced specific force, some aged muscle fibers exhibit Ca^{2+}-dependent EC coupling. As mentioned above, cardiac EC coupling is fully dependent on the influx of external Ca^{2+} ions

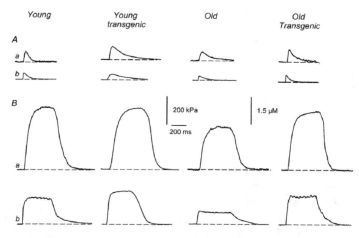

Figure 7. Representative records of twitch (**A**) and tetanus (**B**) contractions and intracellular Ca^{2+}. Traces in (a) and (b) correspond to contraction and intracellular Ca^{2+} transients, respectively, recorded in single intact flexor digitorum brevis fibers from young and old wild-type and transgenic mice. (Reproduced with permission from González et al., 2003, © The Physiological Society, Blackwell Publishing)

through the DHPR $\alpha 1_C$ to activate SR Ca^{2+} release from RyR2 for contraction (Fabiato, 1983, , 1985; Nabauer et al., 1989). However, skeletal muscle EC coupling (in muscles from young adult mammals) is dependent on a physical protein-protein coupling between DHPR $\alpha 1_S$ and RyR1 (Rios and Brum, 1987; Block et al., 1988), and is therefore not dependent on the influx of external Ca^{2+} ions to activate RyR1 (Tanabe et al., 1990a; Tanabe et al., 1990b). However, a recent finding from our lab has shown that nearly half of the fibers from a fast-twitch muscle, the flexor digitorum brevis (FDB), from aged mice show decreases in force output when the bathing medium is changed to a Ca^{2+}-free solution (Fig. 10) (Payne et al., 2004), in a similar fashion to cardiac muscle fibers or dysgenic myotubes expressing DHPR $\alpha 1_C$ (Tanabe et al., 1990a; Tanabe et al., 1990b; Garcia et al., 1994). Charge movement was lower in fibers from old mice, but was unaffected by Ca^{2+}-free solution in either young or old, so force did not decline due to reduced DHPR activation. $[Ca^{2+}]_i$ was reduced in roughly half the fibers from old mice during voltage-clamp experiments, and was reduced in parallel with force during contraction experiments, indicating the removal of external Ca^{2+} ions reduces force in this population of fibers by way of reducing SR Ca^{2+} release. The initial hypothesis was that some aging fibers express DHPR $\alpha 1_C$, removing the mechanical coupling with the adjacent RyR1, and causing EC coupling to become dependent on influx of Ca^{2+} through the DHPR. However, although some differences were found in activation time constant, Ca^{2+} dependent inactivation was not present, and DHPR $\alpha 1_C$ mRNA and protein were not found in aging muscle. These data indicate that DHPR $\alpha 1_C$ is not expressed in aging muscle. RyR3 protein, which is expressed in developing skeletal muscle (Flucher et al., 1999; Chun et al., 2003) and does not mechanically couple to

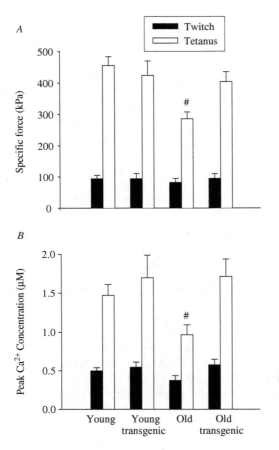

Figure 8. Maximum twitch and tetanic specific force (**A**) and peak intracellular Ca^{2+} (**B**) in young and old wild-type and transgenic mice. (**A**) Tetanic specific force declined in old wild-type compared to young wild-type. hIGF-1 overexpression in muscle prevented the age-related decline in tetanic specific force. (**B**) Peak intracellular Ca^{2+} recorded during tetanic contractions declined in old wild-type muscle fibers. hIGF-1 overexpression in muscle prevented the age-related decline in peak intracellular Ca^{2+}. (Reproduced with permission from González et al., 2003. © The Physiological Society, Blackwell Publishing)

DHPRs (Fessenden et al., 2000; Protasi et al., 2000), was also not found in aged mouse skeletal muscle. The primary culprit for the age-related "shift" toward Ca^{2+}-dependent EC coupling in skeletal muscle is still unknown. Current investigations are examining the possibilities of DHPR $\alpha 1_S$ splice variants, alterations in DHPR subunits other than the $\alpha 1$, RyR1 splice variants, and/or alterations to membrane structures, such as the T-tubule or vacuolization of the SR terminal cisternae, as contributors to the age-related Ca^{2+}-dependent EC coupling phenomenon (Payne et al., 2004).

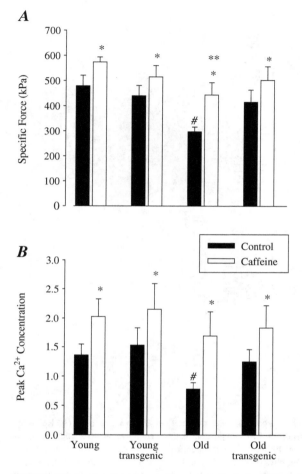

Figure 9. Effects of caffeine on maximum specific force (**A**) and peak intracellular Ca^{2+} (**B**) in single intact fibers from young and old wild-type and transgenic mice. (**A**) Incubation with caffeine (5 mM) increased maximum specific force in all groups. Specific force after caffeine treatment increased in muscle fibers from old wild-type mice to levels not different from young wild-type and transgenic and old transgenic before caffeine treatment. Although caffeine-enhanced specific force in old wild-type was not different from young and old transgenic caffeine enhanced force, a deficit remained compared to young wild-type. (**B**) Caffeine treatment enhanced Ca^{2+} release in all groups studied and elevated peak intracellular Ca^{2+} in old wild-type to levels equal to the other three groups. (Reproduced with permission from González et al., 2003. © The Physiological Society, Blackwell Publishing)

3.3   Age-Related Denervation of Skeletal Muscle

As mentioned above, innervation is one of the primary regulators of EC coupling. We have previously proposed that a slow age-related denervation of skeletal muscle contributes to age-related alterations in EC coupling (Delbono, 2003; Payne and

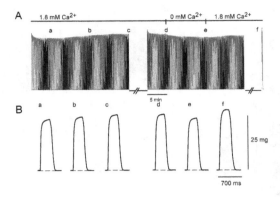

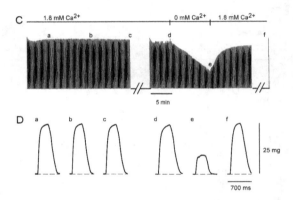

Figure 10. Ca^{2+} dependent contraction in a population of muscle fibers from aging mice. **(A)** Reference (left traces) and test (right traces) trials of 150 tetanic contractions each, set at 10-second intervals, for an 'unaffected' single intact flexor digitorum brevis fiber from an old mouse. The reference trial consisted of 150 contractions in normal physiological solution. During the test trial, the bathing solution was switched to a Ca^{2+}-free solution from contraction 30 to 90. No significant force decline was seen in this fiber during Ca^{2+}-free solution perfusion. A contraction (far right trace) was elicited 5 minutes after the test trial to ensure full recovery. **(C)** Reference and test trials for an 'affected' single intact flexor digitorum brevis fiber from an old mouse. The reference and test trials were the same as in (A). However, during the Ca^{2+}-free solution perfusion of the test trial, force declined steadily in this fiber. Upon return to normal physiological solution, most of the force recovered quickly. The recovery contraction 5 minutes after the test trial indicates full recovery of force. **(B and D)** Selected traces (a-f) corresponding to the same letters in (A and B): Contraction 30 of 150 (a and d, immediately preceding change to Ca^{2+}-free solution in the test trial), contraction 90 of 150 (b and e, immediately preceding return to normal physiological solution in test trial), contraction 150 of 150 (c, end of reference trial), and the recovery contraction (f, 5 minutes after the end of test trial). (Reproduced with permission from Payne et al., 2004. © The Physiological Society, Blackwell Publishing)

Delbono, 2004). Several groups have reported a skeletal muscle denervation and re-innervation process and motor unit remodeling or loss in aging rodents and humans (Hashizume et al., 1988; Kanda and Hashizume, 1989; Einsiedel and Luff, 1992b, 1992a; Kanda and Hashizume 1992; Doherty et al., 1993; Hashizume and Kanda, 1995; Zhang et al., 1995; Zhang et al., 1996). NCAM protein, a marker for denervated muscle fibers, has been shown to increase in aged rodent muscles compared with young (Andersson et al., 1993). A recent report combines force measurements with immunostaining for NCAM-positive fibers, and indicates that denervated fibers account partially for the age-related decline in nerve-stimulated whole muscle specific force (Urbanchek et al., 2001). Age-related motor unit remodeling seems to involve denervation of primarily fast-twitch muscle fibers and re-innervation by axonal sprouting from motor neurons innervating nearby slow-twitch muscle fibers (Larsson, 1995; Lexell, 1995; Kadhiresan et al., 1996; Frey et al., 2000). This has been shown by analysis of synapse type after re-innervation (Frey et al., 2000); by decreased fast motor unit contractile force and increased slow motor unit contractile force (representing increased relative number of fibers innervated by single "slow" motor neurons) (Kadhiresan et al., 1996); and by "fiber type grouping" of slow-twitch muscle fibers in particular (Lexell, 1995) in aged muscle compared to young.

During this age-related denervation and re-innervation process, if denervation outpaces re-innervation a population of fibers becomes functionally excluded and, if not re-innervated, will be eliminated altogether. Indeed, aging muscle shows decreased numbers of muscle fibers compared to young (Lexell, 1995; Dutta, 1997). Several methods have been used to study the re-innervation process following denervation in aged rodent muscle. Following crush injury in peripheral nerve, re-innervation and muscle regeneration has been shown to occur at a slower rate in aged compared to young (Kawabuchi et al., 2001). Alterations in the spatiotemporal relationships between Schwann cells, axons, and post-synaptic acetylcholine receptor regions were implicated in the age-related differences. Using transplantation of the EDL muscle from young to old rats or old to young, using young-to-young and old-to-old transplants as control, Carlson et al. (Carlson and Faulkner, 1989; Carlson et al., 2001) has found that the muscles from old transplanted into young hosts regenerated just as well and young-to-young autotransplants, and young-to-old transplants regenerated as poorly as old-to-old autotransplants. Therefore, the ability of the muscle itself to regenerate seems to be unchanged with age, whereas the ability of the peripheral motor neuron axons to re-innervate the target muscle is blunted with age (Carlson and Faulkner, 1989). Indeed, damage directly to muscle without damaging the motor axon supply through the use of a local anesthetic, bupivicaine, illustrates that muscle regeneration occurs at a similar rate in both young and old rats as long as nerve supply is available (Carlson et al., 2001).

The age-related slowing of re-innervation may be due to several factors. Age-related changes in peripheral nerves such as accumulation of collagen in the perineurium, lipid droplets in the perineural sheath, and increased macrophages and mast cells have been found in tibial nerves of aged mice (Ceballos et al., 1999). The myelin sheath of peripheral nerves deteriorates due to a demyelination and remyelination process that causes wide incisures and infolded or outfolded myelin loops resulting in morpho-logical irregularities with age (Ceballos et al., 1999). This process may impede re-innervation of denervated muscle fibers due to a reduction in the endoneurial conduits for axonal regeneration (Cederna et al., 2001). In fact, nerve-intact EDL autotrans-plants, in which the neural sheath is still intact, allow re-innervation, and thus, regen-eration of muscle fibers to occur in both young and old rats (Cederna et al., 2001). In addition to the need for an axonal conduit, re-innervation requires anterograde axonal delivery of cytoskeletal proteins to the growing axon, many of which are trans-ported in the "slow component b" (SCb) of anterograde axonal transport (Jacob and McQuarrie, 1991, 1993). Evidence has shown that aging leads to retardation of the slow component of axonal transport (McQuarrie et al., 1989). In addition, proteins which exert trophic regulation on the neuromuscular junction, such as calcitonin gene-related peptide, are transported slower in aging motor neurons than in young (Fernandez and Hodges-Savola, 1994). Indeed, both axonal atrophy (Verdu et al., 2000) and impairment of nerve regeneration (Kerezoudi and Thomas, 1999) seen with age have been correlated with diminished axonal transport.

In summary, because of the aforementioned effects of denervation on EC coupling combined with the evidence of age-related alterations in skeletal muscle innervation, we believe that age-related alterations in EC coupling (i.e., EC uncoupling and Ca^{2+}-dependent EC coupling) are due to the slow age-related denervation and re-innervation process.

4.   INFLUENCE OF TROPHIC FACTORS
ON EXCITATION-CONTRACTION COUPLING

4.1   Effects of Insulin-Like Growth Factor-1 on EC Coupling
in Skeletal Muscle

The role of insulin-like growth factor-1 (IGF-1) and its related binding proteins in muscle EC coupling and fiber type composition is currently under investigation. Systemic overexpression of human IGF-1 cDNA in transgenic mice resulted in IGF-1 overexpression in a broad range of visceral organs and increased serum concentration of IGF-1, resulting in organomegaly and increased body weight, but only modest increase in muscle mass (Mathews et al., 1988). Because of the possible confounding effects of systemic IGF-1 overexpression, human liver IGF-1 (hIGF-1) overexpression was targeted specifically to striated muscle using a myogenic expression vector containing regulatory elements from the avian skeletal α-actin gene (Coleman et al., 1995). Transgenic mice carrying one copy of the hybrid skeletal α-actin/hIGF-1 transgene had hIGF-1 mRNA levels that were approximately half those of the endogenous murine skeletal α-actin gene on a per allele basis, conferring substantial tissue-specific overexpression of hIGF-1 without elevated circulating serum levels. This specific local overexpression of hIGF-1 causes cultured muscle cells to undergo precocious alignment and fusion into myoblasts, and causes significant hypertrophy of myofibers *in vivo* (Coleman et al., 1995). These data suggest that hIGF-1 has a more potent hypertrophic effect on muscle when derived from a sustained autocrine/paracrine release compared to elevated serum IGF-1. Intramuscular injections of IGF-1 in adult rats confirm this hypothesis, showing similar level of muscle hypertrophy as local IGF-1 overexpression (Adams and McCue, 1998).

The effects of IGF-1 overexpression on skeletal muscle are also being studied in aging animals. Apparently, the capacity of hIGF-1 to induce muscle hypertrophy declines in senescent mice (Chakravarthy et al., 2001; Fiorotto et al., 2003). However, the overexpression of another form of IGF-1, the muscle-specific splice variant mIGF-1 (which corresponds to the human IGF-1Ea gene) driven off the myosin light-chain 1/3 promoter/enhancer, has been found to sustain hypertrophy and satellite cell replicative capacity more consistently throughout the lifespan of mice (Musaro et al., 2001). Viral-mediated muscle overexpression of mIGF-1 has been shown to maintain muscle mass and force and even specific force in old animals to the level of young counterparts and this animal model along with hIGF-1 targeted overexpression have both been shown to prevent age-related fast-twitch muscle fiber loss (Barton-Davis et al., 1998; Messi and Delbono, 2003). As mentioned above, fast-twitch muscle fibers are lost with age due to a slow age-related denervation process. Whether fast fiber maintenance was due to a direct effect of mIGF-1 on the muscle fibers to maintain them as targets for motor neurons or by paracrine secretion of IGF-1 from muscle cells to the motor neurons to maintain innervation is unknown at this time (see below) (Barton-Davis et al., 1998; Messi and Delbono, 2003).

Targeted overexpression of IGF-1 in muscle, whether genetically (Renganathan et al., 1998; González et al., 2003) or virally (Barton-Davis et al., 1998) prevents the age-related decrease in skeletal muscle specific force. The mechanisms of IGF-1 overexpression-mediated maintenance of specific force in muscle of aging animals are being explored. Even though muscle-specific overexpression of hIGF-1 does not sustain hypertrophy into senescence like mIGF-1 overexpression, they both maintain specific force of fast-twitch muscles in aged mice to the level of young (Barton-Davis et al., 1998; Renganathan et al., 1998; González et al., 2003), suggesting that the pathways evoked by IGF-1 to control fiber hypertrophy and force-generating capacity diverge. In our lab, we have found that muscle overexpression of hIGF-1 increases DHPR binding site number and the DHPR-to-RyR1 binding ratio in adult mice (Renganathan et al., 1997b), and increases functional measures of the DHPR – charge movement and inward Ca^{2+} current – as well as mRNA expression of skeletal DHPR in cultured muscle cells from adult mice (Wang et al., 1999). Importantly, the enhanced expression of DHPR is continued into senescence in these animals, indicated by increased DHPR and RyR1 binding sites (Fig. 5), increased DHPR-to-RyR1 binding ratio, and increased specific force in muscles from aged transgenic compared to aged wild-type mice (Renganathan et al., 1998). Subsequently, we found that charge movement (Fig. 4) and $[Ca^{2+}]_i$ (Fig. 6) are increased in voltage-clamped muscle fibers from old transgenic compared to old wild-type mice (Wang et al., 2002). To examine if these effects on DHPR expression and $[Ca^{2+}]_i$ were functionally significant in old transgenic mice, simultaneous measurements of specific force and $[Ca^{2+}]_i$ were made in single intact muscle fibers from young and old wild-type and transgenic mice (González et al., 2003). While hIGF-1 overexpression did not increase muscle fiber tetanic specific force or $[Ca^{2+}]_i$ in young animals, muscle fibers from old transgenic mice exhibited both increased tetanic specific force and $[Ca^{2+}]_i$ compared to muscle fibers from old wild-type mice. In fact, specific force and $[Ca^{2+}]_i$ in fibers from old transgenic were not different from the same measures in young wild-type or transgenic (Fig. 7 and 8) (González et al., 2003). Taken together, these data indicated that IGF-1 plays a crucial role in regulating the expression of the DHPR and in maintaining EC coupling, and that IGF-1 overexpression in muscle can prevent the previously described age-related process of EC uncoupling.

To explore the role of IGF-1 on DHPR $\alpha 1_S$ expression, our laboratory isolated and sequenced the 1.2-kb 5'-flanking region immediately upstream of the mouse DHPR $\alpha 1_S$ gene (Zheng et al., 2002). Luciferase reporter constructs driven by different regions of the DHPR $\alpha 1_S$ gene were used for transfection in muscle C2C12 cells. These preparations showed that three regions corresponding to CREB, GATA-2, and SOX-5 consensus sequences within the 5'-flanking region of the DHPR $\alpha 1_S$ gene are important for transcription. Antisense oligonucleotides against these three consensus sequences significantly reduced charge movement in C2C12 cells (Zheng et al., 2002). Subsequently, the effects of IGF-1 were studied on various promoter-luciferase constructs. These were transfected into C2C12 cells and effects of IGF-1 were measured by luciferase activity. IGF-1 enhanced DHPR $\alpha 1_S$ transcription in the constructs carrying the CREB binding site but not in CREB binding site mutants. Gel mobility shift assay using a double stranded oligonucleotide for the CREB binding site in the promoter region, and competition experiments with excess unlabelled or mutated promoter oligonucleotide, and unlabelled consensus CREB oligonucleotide indicate that IGF-1 induces CREB binding to the DHPR $\alpha 1_S$ promoter. IGF-1 enhanced charge movement C2C12 cells, and this effect was prevented by incubating cells with antisense oligonucleotides against CREB. These results indicate that IGF-1 regulates DHPR $\alpha 1_S$ transcription in muscle cells by acting on the CREB element of the promoter (Zheng et al., 2002). These results also provide mechanistic support to the findings that targeted overexpression of hIGF-1 in mouse skeletal muscle prevents age-related decline in DHPR binding sites (Renganathan et al., 1998), charge movement, inward Ca^{2+} current (Wang et al., 2000), and intracellular Ca^{2+} transients (Wang et al., 2000; González et al., 2003).

4.2   Effects of Insulin-Like Growth Factors on Skeletal Muscle Innervation

The role of IGF-1 in motor neuron survival has been examined during embryonic life (Neff et al., 1993). During the embryonic stage of vast motor neuron removal and cell death, treatment with IGF-1 maintained a larger number of living motor neurons. Also, treatment of embryos with IGF-1 following axotomy rescued approximately 50% of the motor neurons that die in untreated axotomized embryos (Neff et al., 1993). Similarly, in young rodents, IGF-1 is upregulated in Schwann cells and astrocytes and IGF-binding protein 6 is upregulated in motor neurons following spinal cord and peripheral nerve injury; these agents possibly serving to promote nerve regeneration (Hammarberg et al., 1998).

Neurotrophic theory (Davies, 1996) describes a well-established role for target-derived neurotrophic factors in the regulation of survival of developing neurons in the peripheral and central nervous systems. Therefore, neurotrophic factors produced in skeletal muscle, such as IGF-1, may play an important role in maintaining motor neuron innervation in adult and aged animals as well as re-innervation of denervated muscle fibers. Injections of IGF-1 or IGF-2 into adult rat and mouse skeletal muscle produce motor neuron nerve sprouting in innervated skeletal muscle, and increase expression of nerve-specific growth-associated protein-43 (GAP-43) (Caroni and Grandes, 1990). GAP-43 is a critical protein in axonal growth cones of motor neurons (Skene et al., 1986), and is downregulated upon synapse formation with muscle fibers (Caroni and Becker, 1992). This downregulation of GAP-43 was found to be controlled by IGF-1 expression in the muscle. Upon synapse formation and, subsequently, nerve-dependent muscle activity, IGF-1 expression declines in developing muscle, followed by a decline in motor neuron expression of GAP-43. Counteracting the activity-dependent decline in IGF-1 with local IGF-1 injections prevented the downregulation of GAP-43 (Caroni and Becker, 1992).

Following nerve injury IGF-2 is elevated for two weeks or more in skeletal muscle (Czerwinski et al., 1993; Krishan and Dhoot, 1996; Pu et al., 1999), while IGF-1 has been found to increase transiently (Caroni and Schneider, 1994; Krishan and Dhoot, 1996). However, conflicting evidence has been found regarding IGF-1, which has also been shown not to increase following nerve damage (Czerwinski et al., 1993; Pu et al., 1999). These increases in IGFs in denervated or partially denervated skeletal muscle are reactions presumably aimed at restoring a functional neuromuscular system (Caroni and Schneider, 1994). To restore function, nerve-muscle interaction must be restored, requiring axonal growth of motor neurons and formation of new synapses, i.e., requiring upregulation of GAP-43 and other growth-associated proteins in the motor neuron. Indeed, GAPs are re-upregulated upon regenerative nerve growth following nerve damage (Schreyer and Skene, 1993). The function of increased IGF-1 and -2 expression following nerve inactivation was studied with the local injection of high-affinity and specific IGF-binding proteins IGFBP-4 and IGFBP-5 to effectively remove the actions of upregulated IGF-1 and IGF-2, respectively (Caroni et al., 1994). Injection of IGFBP-4 in botulinum toxin A-paralyzed muscle prevented nerve sprouting in the muscle. These data indicate that upregulation of muscle IGF-1 and -2 following nerve damage is essential to turn on GAP expression in motor neuron axons in order to promote axonal growth and re-innervation of muscle fibers.

Evidence indicates that by elevating IGF-1 in skeletal muscle (via injections or transgenic overexpression) the process of re-innervation is enhanced following motor neuron damage. Systemic injections of IGF-1 in young rats reduces motor neuron death and enhances re-innervation of muscle following sciatic nerve axotomy, thus improving muscle morphometry and force generation in treated compared to untreated animals (Contreras et al., 1995; Vergani et al., 1998). Similarly, targeted local overexpression of hIGF-1 (animal model described above) in skeletal muscle has been shown to increase the rate and extent of muscle re-innervation following sciatic nerve crush (Rabinovsky et al., 2003), as evidenced by increased neurofilament 150-positive axons and nerve conduction velocity in nerve-injured transgenic mice compared to control. The same gene overexpression induced by a muscle-specific non-viral vector has been shown to significantly enhance muscle fiber re-innervation in a model of rat laryngeal paralysis (Shiotani et al., 1998).

Since the level of resting and exercise-induced IGF-1 in skeletal muscle is known to decline with age (Hameed et al., 2002; Hameed et al., 2003) and the rate of re-innervation is slower in muscle from aged animals (Kawabuchi et al., 2001), we can speculate that this important target-derived trophic factor is in short supply for proper re-innervation of aged skeletal muscle. Therefore, more studies need to be conducted to examine the effects of targeted overexpression of IGF-1 in muscle on inhibition of the aforementioned age-related skeletal muscle denervation. However, some indirect evidence does exist to suggest that targeted overexpression of IGF-1 in muscle does prevent or attenuate age-related denervation. Since experimental denervation and aging lead to many similar changes in skeletal muscle (i.e., alterations in EC coupling and fast-twitch muscle fiber loss), the finding that IGF-1 overexpression in muscle prevents such changes in aged animals (such as EC uncoupling and fast-twitch fiber loss) suggests that locally elevated IGF-1 may be preventing age-related denervation of these fibers, enhancing re-innervation by axonal sprouting and growth, or a combination of the two.

These results prompted our laboratory to directly study the innervation of skeletal muscle in young and aged mice with targeted overexpression of hIGF-1 in skeletal muscle (Messi and Delbono, 2003). Combined silver-cholinesterase staining of neuromuscular junctions showed that IGF-1 overexpression increased the number of nerve branches and nerve thickness in cholinesterase stained zones, as well as increased the size of the cholinesterase stained zone, in old transgenic compared to old wild-type mice. Immunocytochemical examination of the post-terminal with fluorescent-labeled α-bungarotoxin revealed that IGF-1 overexpression completely prevents the age-related decline in post-terminal surface area. Additionally, electron microscopy of the post-terminal also reveals that IGF-1 overexpression completely prevents the age-related decline in the number of post-synaptic membrane folds, length of the folds, and density of the folds (Messi and Delbono, 2003). These results, along with the finding of preservation of fast-twitch muscle fibers, suggest that elevated skeletal muscle IGF-1, serving as a target-derived trophic factor, can prevent or attenuate age-related denervation of muscle fibers (Messi and Delbono, 2003). With the previously mentioned importance of muscle fiber innervation in EC coupling, these results provide an additional potential mechanism for IGF-1 overexpression maintenance of EC coupling in aged skeletal muscle.

4.3   Involvement of Other Neurotrophic Factors in Skeletal Muscle Innervation

Although most available evidence describes functions of neurotrophic factors in developing neurons, some evidence points to a continued role for target-derived neurotrophic factors other than IGF-1 in the maintenance and plasticity of adult and aged neurons (Cowen and Gavazzi, 1998; Bergman et al., 2000). In the neuromuscular system, in particular, several neurotrophins have been implicated in neuronal and synaptic plasticity at the neuromuscular junction. For example, brain-derived neurotrophic factor (BDNF) and neurotrophin-3 (NT-3) both potentiate spontaneous and impulse-evoked synaptic activity of developing neuromuscular synapses in cell culture (Lohof et al., 1993). A recent formulation of target-derived neurotrophic theory proposes that neurotrophins participate in muscle activity-induced modification of synaptic transmission (Schinder and Poo, 2000). For example, BDNF potentiation of neuromuscular synapse efficacy is facilitated by presynaptic depolarization at developing neuromuscular synapses (Boulanger and Poo, 1999). Using a system of nerve-muscle co-culture in which NT-4 is overexpressed in a population of myocytes, presynaptic potentiation was restricted to those synapses formed on myocytes overexpressing NT-4. Nearby synapses formed by the same neurons on control myocytes were not potentiated (Wang et al., 1998b). Intramuscular administration of NT-4 in adult rat skeletal muscle has been shown to induce motor neuron axon sprouting. Additionally, NT-4 has been found to be an activity-dependent regulator of motor neuron growth. NT-4 mRNA increases upon stimulation of skeletal muscle in a dose-responsive manner, and decreases upon neuromuscular blockade by α-bungarotoxin. Therefore, NT-4 may serve to mediate the effects of exercise and electrical stimulation on neuromuscular performance (Funakoshi et al., 1995). The findings that several muscle-derived neurotrophic factors can enhance innervation and/or re-innervation of muscle is important in that, as mentioned earlier, innervation is a crucial regulator of skeletal muscle EC coupling. However, since IGF-1 directly enhances DHPR expression in skeletal muscle in addition to maintaining muscle fiber innervation, targeted overexpression of other neurotrophic factors, such as NT-4, may provide more insight as to the importance of age-related denervation in EC uncoupling. Also, as IGF-1 overexpression in muscle has been shown to enhance muscle re-innervation after nerve injury and attenuate age-related denervation, overexpression of other neurotrophic factors in muscle may produce similar results. One of these strategies may, therefore, provide direction to future clinical trials on aging or treatment of nerve damage.

CONCLUSIONS

This review has provided evidence to support the notion of plasticity in a critical skeletal muscle process, EC coupling. Primarily, alterations in expression of DHPR and RyR1 account for changes in the efficiency of EC coupling. Evidence indicates that muscle innervation and trophic signaling are the primary regulators of EC coupling in skeletal muscle. Accordingly, changes in innervation status, whether during development, physical denervation, or age-related slow denervation, can alter the efficiency of the EC coupling signal. According to neurotrophic theory, a decline in muscle expression of IGF-1 and other neurotrophins during aging may contribute to the age-related slow denervation process, ultimately leading to EC uncoupling with age. In addition, a decline in muscle IGF-1 with age may directly contribute to EC uncoupling via reduced IGF-1—CREB signaling, thereby reducing DHPR $\alpha 1_S$ gene expression. The initial trigger for this age-related denervation process is unknown at this time, but reduction in motor neuron function may precede and, in fact, lead to the reduction in muscle IGF-1 production, thus causing a vicious cycle of reduced target-derived trophism, and, ultimately, neurodegeneration and EC uncoupling. This evidence is further supported by the findings that targeted overexpression of IGF-1 in skeletal muscle prevents fiber type alterations, degeneration of the neuromuscular junction, and EC uncoupling in aged mice, and enhances muscle regeneration and re-innervation following injury. These findings may, therefore, provide strategies for combating muscle weakness and frailty associated with aging and peripheral nerve damage.

REFERENCE

Adams BA & Beam KG. (1990). Muscular dysgenesis in mice: a model system for studying excitation-contraction coupling. *Faseb J* 4, 2809–2816.

Adams BA, Tanabe T, Mikami A, Numa S & Beam KG. (1990). Intramembrane charge movement restored in dysgenic skeletal muscle by injection of dihydropyridine receptor cDNAs. *Nature* 346, 569–572.

Adams GR & McCue SA. (1998). Localized infusion of IGF-I results in skeletal muscle hypertrophy in rats. *J Appl Physiol* 84, 1716–1722.

Ahern CA, Powers PA, Biddlecome GH, Roethe L, Vallejo P, Mortenson L, Strube C, Campbell KP, Coronado R & Gregg RG. (2001). Modulation of L-type Ca2+ current but not activation of Ca2+ release by the gamma1 subunit of the dihydropyridine receptor of skeletal muscle. *BMC Physiol* 1, 8.

Ahern GP, Junankar PR & Dulhunty AF. (1994). Single channel activity of the ryanodine receptor calcium release channel is modulated by FK-506. *FEBS Lett* 352, 369–374.

Ahern GP, Junankar PR & Dulhunty AF. (1997). Subconductance states in single-channel activity of skeletal muscle ryanodine receptors after removal of FKBP12. *Biophys J* 72, 146–162.

Albuquerque EX, Warnick JE, Sansone FM & Onur R. (1974). Trophic functions of the neuron. 3. Mechanisms of neurotrophic interactions. The effects of vinblastine and colchicine on neural regulation of muscle. *Ann N Y Acad Sci* 228, 224–243.

Alden KJ & Garcia J. (2001). Differential effect of gabapentin on neuronal and muscle calcium currents. *J Pharmacol Exp Ther* 297, 727–735.

Alden KJ & Garcia J. (2002). Dissociation of charge movement from calcium release and calcium current in skeletal myotubes by gabapentin. *Am J Physiol Cell Physiol* 283, C941–949.

Andersson AM, Olsen M, Zhernosekov D, Gaardsvoll H, Krog L, Linnemann D & Bock E. (1993). Age-related changes in expression of the neural cell adhesion molecule in skeletal muscle: a comparative study of newborn, adult and aged rats. *Biochem J* 290 (Pt 3), 641–648.

Armstrong C, Benzanilla F & Horowicz P. (1972). Twitches in the presence of ethylene glycol bis(-aminoethyl ether)-N, N-9-tetraacetic acid. *Biochim Biophys Acta* 267, 605–608.

Bangalore R, Mehrke G, Gingrich K, Hofmann F & Kass RS. (1996). Influence of L-type Ca channel alpha 2/delta-subunit on ionic and gating current in transiently transfected HEK 293 cells. *Am J Physiol* 270, H1521–1528.

Barg S, Copello JA & Fleischer S. (1997). Different interactions of cardiac and skeletal muscle ryanodine receptors with FK-506 binding protein isoforms. *Am J Physiol* 272, C1726–1733.

Barone V, Bertocchini F, Bottinelli R, Protasi F, Allen PD, Franzini Armstrong C, Reggiani C & Sorrentino V. (1998). Contractile impairment and structural alterations of skeletal muscles from knockout mice lacking type 1 and type 3 ryanodine receptors. *FEBS Lett* 422, 160–164.

Barton-Davis ER, Shoturma DI, Musaro A, Rosenthal N & Sweeney HL. (1998). Viral mediated expression of insulin-like growth factor I blocks the aging-related loss of skeletal muscle function. *Proc Natl Acad Sci U S A* 95, 15603–15607.

Bassey E, Fiatarone M, O'Neill E, Kelly M, Evans W & Lipsitz L. (1992). Leg extensor power and functional performance in very old men and women. *Clin Sci* 82, 321–327.

Bastide B, Conti A, Sorrentino V & Mounier Y. (2000). Properties of ryanodine receptor in rat muscles submitted to unloaded conditions. *Biochem Biophys Res Commun* 270, 442–447.

Beard NA, Laver DR & Dulhunty AF. (2004). Calsequestrin and the calcium release channel of skeletal and cardiac muscle. *Prog Biophys Mol Biol* 85, 33–69.

Beard NA, Sakowska MM, Dulhunty AF & Laver DR. (2002). Calsequestrin is an inhibitor of skeletal muscle ryanodine receptor calcium release channels. *Biophys J* 82, 310–320.

Bergman E, Ulfhake B & Fundin BT. (2000). Regulation of NGF-family ligands and receptors in adulthood and senescence: correlation to degenerative and regenerative changes in cutaneous innervation. *Eur J Neurosci* 12, 2694–2706.

Bertocchini F, Ovitt CE, Conti A, Barone V, Scholer HR, Bottinelli R, Reggiani C & Sorrentino V. (1997). Requirement for the ryanodine receptor type 3 for efficient contraction in neonatal skeletal muscles. *Embo J* 16, 6956–6963.

Beurg M, Ahern CA, Vallejo P, Conklin MW, Powers PA, Gregg RG & Coronado R. (1999a). Involvement of the carboxy-terminus region of the dihydropyridine receptor beta1a subunit in excitation-contraction coupling of skeletal muscle. *Biophys J* 77, 2953–2967.

Beurg M, Sukhareva M, Ahern CA, Conklin MW, Perez-Reyes E, Powers PA, Gregg RG & Coronado R. (1999b). Differential regulation of skeletal muscle L-type Ca2+ current and excitation-contraction coupling by the dihydropyridine receptor beta subunit. *Biophys J* 76, 1744–1756.

Beurg M, Sukhareva M, Strube C, Powers PA, Gregg RG & Coronado R. (1997). Recovery of Ca2+ current, charge movements, and Ca2+ transients in myotubes deficient in dihydropyridine receptor beta 1 subunit transfected with beta 1 cDNA. *Biophys J* 73, 807–818.

Block BA, Imagawa T, Campbell KP & Franzini-Armstrong C. (1988). Structural evidence for direct interaction between the molecular components of the transverse tubule/sarcoplasmic reticulum junction in skeletal muscle. *J Cell Biol* 107, 2587–2600.

Boulanger L & Poo MM. (1999). Presynaptic depolarization facilitates neurotrophin-induced synaptic potentiation. *Nat Neurosci* 2, 346–351.

Brillantes AB, Ondrias K, Scott A, Kobrinsky E, Ondriasova E, Moschella MC, Jayaraman T, Landers M, Ehrlich BE & Marks AR. (1994). Stabilization of calcium release channel (ryanodine receptor) function by FK506-binding protein. *Cell* 77, 513–523.

Brooks SV & Faulkner JA. (1988). Contractile properties of skeletal muscles from young, adult and aged mice. *J Physiol* 404, 71–82.

Brown M, Sinacore DR & Host HH. (1995). The relationship of strength to function in the older adult. *J Gerontol A Biol Sci Med Sci* 50 Spec No, 55–59.

Brum G, Rios E & Stefani E. (1988). Effects of extracellular calcium on calcium movements of excitation-contraction coupling in frog skeletal muscle fibres. *J Physiol* 398, 441–473.

Brust PF, Simerson S, McCue AF, Deal CR, Schoonmaker S, Williams ME, Velicelebi G, Johnson EC, Harpold MM & Ellis SB. (1993). Human neuronal voltage-dependent calcium channels: studies on subunit structure and role in channel assembly. *Neuropharmacology* 32, 1089–1102.

Buller AJ, Eccles JC & Eccles RM. (1960a). Differentiation of fast and slow muscles in the cat hind limb. *J Physiol* 150, 399–416.

Buller AJ, Eccles JC & Eccles RM. (1960b). Interactions between motoneurones and muscles in respect of the characteristic speeds of their responses. *J Physiol* 150, 417–439.

Carl SL, Felix K, Caswell AH, Brandt NR, Ball WJ, Jr., Vaghy PL, Meissner G & Ferguson DG. (1995). Immunolocalization of sarcolemmal dihydropyridine receptor and sarcoplasmic reticular triadin and ryanodine receptor in rabbit ventricle and atrium. *J Cell Biol* 129, 672–682.

Carlson BM, Dedkov EI, Borisov AB & Faulkner JA. (2001). Skeletal muscle regeneration in very old rats. *J Gerontol A Biol Sci Med Sci* 56, B224–233.

Carlson BM & Faulkner JA. (1989). Muscle transplantation between young and old rats: age of host determines recovery. *Am J Physiol* 256, C1262–1266.

Caroni P & Becker M. (1992). The downregulation of growth-associated proteins in motoneurons at the onset of synapse elimination is controlled by muscle activity and IGF1. *J Neurosci* 12, 3849–3861.

Caroni P & Grandes P. (1990). Nerve sprouting in innervated adult skeletal muscle induced by exposure to elevated levels of insulin-like growth factors. *J Cell Biol* 110, 1307–1317.

Caroni P & Schneider C. (1994). Signaling by insulin-like growth factors in paralyzed skeletal muscle: rapid induction of IGF1 expression in muscle fibers and prevention of interstitial cell proliferation by IGF-BP5 and IGF-BP4. *J Neurosci* 14, 3378–3388.

Caroni P, Schneider C, Kiefer MC & Zapf J. (1994). Role of muscle insulin-like growth factors in nerve sprouting: suppression of terminal sprouting in paralyzed muscle by IGF-binding protein 4. *J Cell Biol* 125, 893–902.

Catterall WA. (1991). Functional subunit structure of voltage-gated calcium channels. *Science* 253, 1499–1500.

Catterall WA, Scheuer T, Thomsen W & Rossie S. (1991). Structure and modulation of voltage-gated ion channels. *Ann N Y Acad Sci* 625, 174–180.

Catterall WA, Seagar MJ & Takahashi M. (1988). Molecular properties of dihydropyridine-sensitive calcium channels in skeletal muscle. *J Biol Chem* 263, 3535–3538.

Ceballos D, Cuadras J, Verdu E & Navarro X. (1999). Morphometric and ultrastructural changes with ageing in mouse peripheral nerve. *J Anat* 195 (Pt 4), 563–576.

Cederna PS, Asato H, Gu X, van der Meulen J, Kuzon WM, Jr., Carlson BM & Faulkner JA. (2001). Motor unit properties of nerve-intact extensor digitorum longus muscle grafts in young and old rats. *J Gerontol A Biol Sci Med Sci* 56, B254–258.

Chakravarthy MV, Fiorotto ML, Schwartz RJ & Booth FW. (2001). Long-term insulin-like growth factor-I expression in skeletal muscles attenuates the enhanced in vitro proliferation ability of the resident satellite cells in transgenic mice. *Mech Ageing Dev* 122, 1303–1320.

Chaudhari N & Beam KG. (1993). mRNA for cardiac calcium channel is expressed during development of skeletal muscle. *Dev Biol* 155, 507–515.

Chen SR, Li X, Ebisawa K & Zhang L. (1997). Functional characterization of the recombinant type 3 Ca2+ release channel (ryanodine receptor) expressed in HEK293 cells. *J Biol Chem* 272, 24234–24246.

Chen SR, Zhang L & MacLennan DH. (1994). Asymmetrical blockade of the Ca2+ release channel (ryanodine receptor) by 12-kDa FK506 binding protein. *Proc Natl Acad Sci U S A* 91, 11953–11957.

Chun LG, Ward CW & Schneider MF. (2003). Ca2+ sparks are initiated by Ca2+ entry in embryonic mouse skeletal muscle and decrease in frequency postnatally. *Am J Physiol Cell Physiol* 285, C686–697.

Clancy JS, Takeshima H, Hamilton SL & Reid MB. (1999). Contractile function is unaltered in diaphragm from mice lacking calcium release channel isoform 3. *Am J Physiol* 277, R1205–1209.

Coleman ME, DeMayo F, Yin KC, Lee HM, Geske R, Montgomery C & Schwartz RJ. (1995). Myogenic vector expression of insulin-like growth factor I stimulates muscle cell differentiation and myofiber hypertrophy in transgenic mice. *J Biol Chem* 270, 12109–12116.

Contreras PC, Steffler C, Yu E, Callison K, Stong D & Vaught JL. (1995). Systemic administration of rhIGF-I enhanced regeneration after sciatic nerve crush in mice. *J Pharmacol Exp Ther* 274, 1443–1449.

Copello JA, Barg S, Sonnleitner A, Porta M, Diaz-Sylvester P, Fill M, Schindler H & Fleischer S. (2002). Differential activation by Ca2+, ATP and caffeine of cardiac and skeletal muscle ryanodine receptors after block by Mg2+. *J Membr Biol* 187, 51–64.

Cowen T & Gavazzi I. (1998). Plasticity in adult and ageing sympathetic neurons. *Prog Neurobiol* 54, 249–288.

Czerwinski SM, Novakofski J & Bechtel PJ. (1993). Modulation of IGF mRNA abundance during muscle denervation atrophy. *Med Sci Sports Exerc* 25, 1005–1008.

Davies AM. (1996). The neurotrophic hypothesis: where does it stand? *Philos Trans R Soc Lond B Biol Sci* 351, 389–394.

De Jongh KS, Merrick DK & Catterall WA. (1989). Subunits of purified calcium channels: a 212-kDa form of alpha 1 and partial amino acid sequence of a phosphorylation site of an independent beta subunit. *Proc Natl Acad Sci U S A* 86, 8585–8589.

De Jongh KS, Murphy BJ, Colvin AA, Hell JW, Takahashi M & Catterall WA. (1996). Specific phosphorylation of a site in the full-length form of the alpha 1 subunit of the cardiac L-type calcium channel by adenosine 3',5'-cyclic monophosphate-dependent protein kinase. *Biochemistry* 35, 10392–10402.

De Jongh KS, Warner C & Catterall WA. (1990). Subunits of purified calcium channels. Alpha 2 and delta are encoded by the same gene. *J Biol Chem* 265, 14738–14741.

De Waard M, Pragnell M & Campbell KP. (1994). Ca2+ channel regulation by a conserved beta subunit domain. *Neuron* 13, 495–503.

Delbono O. (1992). Calcium current activation and charge movement in denervated mammalian skeletal muscle fibres. *J Physiol* 451, 187–203.

Delbono O. (2003). Neural control of aging skeletal muscle. *Aging Cell* 2, 21–29.

Delbono O & Chu A. (1995). Ca2+ release channels in rat denervated skeletal muscles. *Exp Physiol* 80, 561–574.

Delbono O, O'Rourke KS & Ettinger WH. (1995). Excitation-calcium release uncoupling in aged single human skeletal muscle fibers. *J Membr Biol* 148, 211–222.

Delbono O & Stefani E. (1993). Calcium current inactivation in denervated rat skeletal muscle fibres. *J Physiol* 460, 173–183.

Dietze B, Bertocchini F, Barone V, Struk A, Sorrentino V & Melzer W. (1998). Voltage-controlled Ca2+ release in normal and ryanodine receptor type 3 (RyR3)-deficient mouse myotubes. *J Physiol* 513 (Pt 1), 3–9.

Diffee GM, Caiozzo VJ, Herrick RE & Baldwin KM. (1991). Contractile and biochemical properties of rat soleus and plantaris after hindlimb suspension. *Am J Physiol* 260, C528–534.

Doherty TJ, Vandervoort AA, Taylor AW & Brown WF. (1993). Effects of motor unit losses on strength in older men and women. *J Appl Physiol* 74, 868–874.

Dulhunty A & Gage P. (1988). Effects of extracellular calcium concentration and dihydropyridines on contraction mammalian skeletal muscle. *J Physiol* 399, 63–80.

Dulhunty AF & Gage PW. (1985). Excitation-contraction coupling and charge movement in denervated rat extensor digitorum longus and soleus muscles. *J Physiol* 358, 75–89.

Dulhunty AF, Gage PW & Valois AA. (1984). Indentations in the terminal cisternae of denervated rat EDL and soleus muscle fibers. *J Ultrastruct Res* 88, 30–43.

Dutta C. (1997). Significance of sarcopenia in the elderly. *J Nutr* 127, 992S–993S.

Einsiedel LJ & Luff AR. (1992a). Alterations in the contractile properties of motor units within the ageing rat medial gastrocnemius. *J Neurol Sci* 112, 170–177.

Einsiedel LJ & Luff AR. (1992b). Effect of partial denervation on motor units in the ageing rat medial gastrocnemius. *J Neurol Sci* 112, 178–184.

Fabiato A. (1983). Calcium-induced release of calcium from the cardiac sarcoplasmic reticulum. *Am J Physiol* 245, C1–14.

Fabiato A. (1985). Simulated calcium current can both cause calcium loading in and trigger calcium release from the sarcoplasmic reticulum of a skinned canine cardiac Purkinje cell. *J Gen Physiol* 85, 291–320.

Felder E, Protasi F, Hirsch R, Franzini-Armstrong C & Allen PD. (2002). Morphology and molecular composition of sarcoplasmic reticulum surface junctions in the absence of DHPR and RyR in mouse skeletal muscle. *Biophys J* 82, 3144–3149.

Fernandez HL & Hodges-Savola CA. (1994). Axoplasmic transport of calcitonin gene-related peptide in rat peripheral nerve as a function of age. *Neurochem Res* 19, 1369–1377.

Fessenden JD, Wang Y, Moore RA, Chen SR, Allen PD & Pessah IN. (2000). Divergent functional properties of ryanodine receptor types 1 and 3 expressed in a myogenic cell line. *Biophys J* 79, 2509–2525.

Fill M & Copello JA. (2002). Ryanodine receptor calcium release channels. *Physiol Rev* 82, 893–922.

Fiorotto ML, Schwartz RJ & Delaughter MC. (2003). Persistent IGF-I overexpression in skeletal muscle transiently enhances DNA accretion and growth. *Faseb J* 17, 59–60.

Flucher BE, Andrews SB & Daniels MP. (1994). Molecular organization of transverse tubule/sarcoplasmic reticulum junctions during development of excitation-contraction coupling in skeletal muscle. *Mol Biol Cell* 5, 1105–1118.

Flucher BE, Conti A, Takeshima H & Sorrentino V. (1999). Type 3 and type 1 ryanodine receptors are localized in triads of the same mammalian skeletal muscle fibers. *J Cell Biol* 146, 621–630.

Flucher BE & Franzini-Armstrong C. (1996). Formation of junctions involved in excitation-contraction coupling in skeletal and cardiac muscle. *Proc Natl Acad Sci U S A* 93, 8101–8106.

Flucher BE, Takekura H & Franzini-Armstrong C. (1993). Development of the excitation-contraction coupling apparatus in skeletal muscle: association of sarcoplasmic reticulum and transverse tubules with myofibrils. *Dev Biol* 160, 135–147.

Franzini-Armstrong C & Kish JW. (1995). Alternate disposition of tetrads in peripheral couplings of skeletal muscle. *J Muscle Res Cell Motil* 16, 319–324.

Franzini-Armstrong C, Protasi F & Ramesh V. (1998). Comparative ultrastructure of Ca2+ release units in skeletal and cardiac muscle. *Ann N Y Acad Sci* 853, 20–30.

Franzini Armstrong C. (1970). Studies of the triad. I. Structure of the junction in frog twitch fibers. *J Cell Biol* 47, 488–498.

Freise D, Held B, Wissenbach U, Pfeifer A, Trost C, Himmerkus N, Schweig U, Freichel M, Biel M, Hofmann F, Hoth M & Flockerzi V. (2000). Absence of the gamma subunit of the skeletal muscle dihydropyridine receptor increases L-type Ca2+ currents and alters channel inactivation properties. *J Biol Chem* 275, 14476–14481.

Frey D, Schneider C, Xu L, Borg J, Spooren W & Caroni P. (2000). Early and selective loss of neuromuscular synapse subtypes with low sprouting competence in motoneuron diseases. *J Neurosci* 20, 2534–2542.

Frontera WR, Hughes VA, Lutz KJ & Evans WJ. (1991). A cross-sectional study of muscle strength and mass in 45- to 78-yr-old men and women. *J Appl Physiol* 71, 644–650.

Frontera WR, Suh D, Krivickas LS, Hughes VA, Goldstein R & Roubenoff R. (2000). Skeletal muscle fiber quality in older men and women. *Am J Physiol Cell Physiol* 279, C611–618.

Fruen BR, Bardy JM, Byrem TM, Strasburg GM & Louis CF. (2000). Differential Ca(2+) sensitivity of skeletal and cardiac muscle ryanodine receptors in the presence of calmodulin. *Am J Physiol Cell Physiol* 279, C724–733.

Fryer MW & Stephenson DG. (1996). Total and sarcoplasmic reticulum calcium contents of skinned fibres from rat skeletal muscle. *J Physiol* 493 (Pt 2), 357–370.

Funakoshi H, Belluardo N, Arenas E, Yamamoto Y, Casabona A, Persson H & Ibanez CF. (1995). Muscle-derived neurotrophin-4 as an activity-dependent trophic signal for adult motor neurons. *Science* 268, 1495–1499.

Gao T, Yatani A, Dell'Acqua ML, Sako H, Green SA, Dascal N, Scott JD & Hosey MM. (1997). cAMP-dependent regulation of cardiac L-type Ca2+ channels requires membrane targeting of PKA and phosphorylation of channel subunits. *Neuron* 19, 185–196.

Garcia J, Tanabe T & Beam KG. (1994). Relationship of calcium transients to calcium currents and charge movements in myotubes expressing skeletal and cardiac dihydropyridine receptors. *J Gen Physiol* 103, 125–147.

Gardetto PR, Schluter JM & Fitts RH. (1989). Contractile function of single muscle fibers after hindlimb suspension. *J Appl Physiol* 66, 2739–2749.

Giannini G, Clementi E, Ceci R, Marziali G & Sorrentino V. (1992). Expression of a ryanodine receptor-Ca2+ channel that is regulated by TGF-beta. *Science* 257, 91–94.

Giannini G, Conti A, Mammarella S, Scrobogna M & Sorrentino V. (1995). The ryanodine receptor/calcium channel genes are widely and differentially expressed in murine brain and peripheral tissues. *J Cell Biol* 128, 893–904.

González-Serratos H, Valle-Aguilera R, Lathrop DA & Garcia MC. (1982). Slow inward calcium currents have no obvious role in muscle excitation-contraction coupling. *Nature* 298, 292–294.

González E, Messi ML & Delbono O. (2000). The specific force of single intact extensor digitorum longus and soleus mouse muscle fibers declines with aging. *J Membr Biol* 178, 175–183.

González E, Messi ML, Zheng Z & Delbono O. (2003). Insulin-like growth factor-1 prevents age-related decrease in specific force and intracellular Ca2+ in single intact muscle fibres from transgenic mice. *J Physiol* 552, 833–844.

Greensmith L & Vrbova G. (1996). Motoneurone survival: a functional approach. *Trends Neurosci* 19, 450–455.

Gregg RG, Messing A, Strube C, Beurg M, Moss R, Behan M, Sukhareva M, Haynes S, Powell JA, Coronado R & Powers PA. (1996). Absence of the beta subunit (cchb1) of the skeletal muscle dihydropyridine receptor alters expression of the alpha 1 subunit and eliminates excitation-contraction coupling. *Proc Natl Acad Sci U S A* 93, 13961–13966.

Gurnett CA, De Waard M & Campbell KP. (1996). Dual function of the voltage-dependent Ca2+ channel alpha 2 delta subunit in current stimulation and subunit interaction. *Neuron* 16, 431–440.

Hameed M, Harridge SD & Goldspink G. (2002). Sarcopenia and hypertrophy: a role for insulin-like growth factor-1 in aged muscle? *Exerc Sport Sci Rev* 30, 15–19.

Hameed M, Orrell RW, Cobbold M, Goldspink G & Harridge SD. (2003). Expression of IGF-I splice variants in young and old human skeletal muscle after high resistance exercise. *J Physiol* 547, 247–254.

Hammarberg H, Risling M, Hokfelt T, Cullheim S & Piehl F. (1998). Expression of insulin-like growth factors and corresponding binding proteins (IGFBP 1–6) in rat spinal cord and peripheral nerve after axonal injuries. *J Comp Neurol* 400, 57–72.

Hashizume K & Kanda K. (1995). Differential effects of aging on motoneurons and peripheral nerves innervating the hindlimb and forelimb muscles of rats. *Neurosci Res* 22, 189–196.

Hashizume K, Kanda K & Burke RE. (1988). Medial gastrocnemius motor nucleus in the rat: age-related changes in the number and size of motoneurons. *J Comp Neurol* 269, 425–430.

Hicks A, Ohlendieck K, Gopel SO & Pette D. (1997). Early functional and biochemical adaptations to low-frequency stimulation of rabbit fast-twitch muscle. *Am J Physiol* 273, C297–305.

Ikemoto N, Antoniu B, Kang JJ, Meszaros LG & Ronjat M. (1991). Intravesicular calcium transient during calcium release from sarcoplasmic reticulum. *Biochemistry* 30, 5230–5237.

Jacob JM & McQuarrie IG. (1991). Axotomy accelerates slow component b of axonal transport. *J Neurobiol* 22, 570–582.

Jacob JM & McQuarrie IG. (1993). Acceleration of axonal outgrowth in rat sciatic nerve at one week after axotomy. *J Neurobiol* 24, 356–367.

Jay SD, Sharp AH, Kahl SD, Vedvick TS, Harpold MM & Campbell KP. (1991). Structural characterization of the dihydropyridine-sensitive calcium channel alpha 2-subunit and the associated delta peptides. *J Biol Chem* 266, 3287–3293.

Jayaraman T, Brillantes AM, Timerman AP, Fleischer S, Erdjument-Bromage H, Tempst P & Marks AR. (1992). FK506 binding protein associated with the calcium release channel (ryanodine receptor). *J Biol Chem* 267, 9474–9477.

Jeyakumar LH, Copello JA, O'Malley AM, Wu GM, Grassucci R, Wagenknecht T & Fleischer S. (1998). Purification and characterization of ryanodine receptor 3 from mammalian tissue. *J Biol Chem* 273, 16011–16020.

Johnson BD, Brousal JP, Peterson BZ, Gallombardo PA, Hockerman GH, Lai Y, Scheuer T & Catterall WA. (1997). Modulation of the cloned skeletal muscle L-type Ca2+ channel by anchored cAMP-dependent protein kinase. *J Neurosci* 17, 1243–1255.

Johnson BD, Scheuer T & Catterall WA. (1994). Voltage-dependent potentiation of L-type Ca2+ channels in skeletal muscle cells requires anchored cAMP-dependent protein kinase. *Proc Natl Acad Sci U S A* 91, 11492–11496.

Jubrias SA, Odderson IR, Esselman PC & Conley KE. (1997). Decline in isokinetic force with age: muscle cross-sectional area and specific force. *Pflugers Arch* 434, 246–253.

Kadhiresan VA, Hassett CA & Faulkner JA. (1996). Properties of single motor units in medial gastrocnemius muscles of adult and old rats. *J Physiol* 493 (Pt 2), 543–552.

Kanda K & Hashizume K. (1989). Changes in properties of the medial gastrocnemius motor units in aging rats. *J Neurophysiol* 61, 737–746.

Kanda K & Hashizume K. (1992). Factors causing difference in force output among motor units in the rat medial gastrocnemius muscle. *J Physiol* 448, 677–695.

Kandarian S, O'Brien S, Thomas K, Schulte L & Navarro J. (1992). Regulation of skeletal muscle dihydropyridine receptor gene expression by biomechanical unloading. *J Appl Physiol* 72, 2510–2514.

Kandarian SC, Peters DG, Favero TG, Ward CW & Williams JH. (1996). Adaptation of the skeletal muscle calcium-release mechanism to weight-bearing condition. *Am J Physiol* 270, C1588–1594.

Kawabuchi M, Zhou CJ, Wang S, Nakamura K, Liu WT & Hirata K. (2001). The spatiotemporal relationship among Schwann cells, axons and postsynaptic acetylcholine receptor regions during muscle reinnervation in aged rats. *Anat Rec* 264, 183–202.

Kawasaki T & Kasai M. (1994). Regulation of calcium channel in sarcoplasmic reticulum by calsequestrin. *Biochem Biophys Res Commun* 199, 1120–1127.

Kerezoudi E & Thomas PK. (1999). Influence of age on regeneration in the peripheral nervous system. *Gerontology* 45, 301–306.

Krishan K & Dhoot GK. (1996). Changes in some troponin and insulin-like growth factor messenger ribonucleic acids in regenerating and denervated skeletal muscles. *J Muscle Res Cell Motil* 17, 513–521.

Lai Y, Seagar MJ, Takahashi M & Catterall WA. (1990). Cyclic AMP-dependent phosphorylation of two size forms of alpha 1 subunits of L-type calcium channels in rat skeletal muscle cells. *J Biol Chem* 265, 20839–20848.

Lamb GD & Stephenson DG. (1991). Effect of Mg2+ on the control of Ca2+ release in skeletal muscle fibres of the toad. *J Physiol* 434, 507–528.

Larsson L. (1995). Motor units: remodeling in aged animals. *J Gerontol A Biol Sci Med Sci* 50 Spec No, 91–95.

Larsson L, Li X & Frontera WR. (1997). Effects of aging on shortening velocity and myosin isoform composition in single human skeletal muscle cells. *Am J Physiol* 272, C638–649.

Laver DR, Baynes TM & Dulhunty AF. (1997). Magnesium inhibition of ryanodine-receptor calcium channels: evidence for two independent mechanisms. *J Membr Biol* 156, 213–229.

Leung AT, Imagawa T & Campbell KP. (1987). Structural characterization of the 1,4-dihydropyridine receptor of the voltage-dependent Ca2+ channel from rabbit skeletal muscle. Evidence for two distinct high molecular weight subunits. *J Biol Chem* 262, 7943–7946.

Lexell J. (1995). Human aging, muscle mass, and fiber type composition. *J Gerontol A Biol Sci Med Sci* 50 Spec No, 11–16.

Li X & Larsson L. (1996). Maximum shortening velocity and myosin isoforms in single muscle fibers from young and old rats. *Am J Physiol* 270, C352–360.

Lindle RS, Metter EJ, Lynch NA, Fleg JL, Fozard JL, Tobin J, Roy TA & Hurley BF. (1997). Age and gender comparisons of muscle strength in 654 women and men aged 20–93 yr. *J Appl Physiol* 83, 1581–1587.

Lohof AM, Ip NY & Poo MM. (1993). Potentiation of developing neuromuscular synapses by the neurotrophins NT-3 and BDNF. *Nature* 363, 350–353.

Lowe DA, Thomas DD & Thompson LV. (2002). Force generation, but not myosin ATPase activity, declines with age in rat muscle fibers. *Am J Physiol Cell Physiol* 283, C187–192.

Lynch NA, Metter EJ, Lindle RS, Fozard JL, Tobin JD, Roy TA, Fleg JL & Hurley BF. (1999). Muscle quality. I. Age-associated differences between arm and leg muscle groups. *J Appl Physiol* 86, 188–194.

MacLennan DH & Wong PT. (1971). Isolation of a calcium-sequestering protein from sarcoplasmic reticulum. *Proc Natl Acad Sci U S A* 68, 1231–1235.

Marty I, Robert M, Villaz M, De Jongh K, Lai Y, Catterall WA & Ronjat M. (1994). Biochemical evidence for a complex involving dihydropyridine receptor and ryanodine receptor in triad junctions of skeletal muscle. *Proc Natl Acad Sci U S A* 91, 2270–2274.

Marx SO, Ondrias K & Marks AR. (1998). Coupled gating between individual skeletal muscle Ca2+ release channels (ryanodine receptors). *Science* 281, 818–821.

Mathews LS, Hammer RE, Behringer RR, D'Ercole AJ, Bell GI, Brinster RL & Palmiter RD. (1988). Growth enhancement of transgenic mice expressing human insulin-like growth factor I. *Endocrinology* 123, 2827–2833.

Mayrleitner M, Timerman AP, Wiederrecht G & Fleischer S. (1994). The calcium release channel of sarcoplasmic reticulum is modulated by FK-506 binding protein: effect of FKBP-12 on single channel activity of the skeletal muscle ryanodine receptor. *Cell Calcium* 15, 99–108.

McQuarrie IG, Brady ST & Lasek RJ. (1989). Retardation in the slow axonal transport of cytoskeletal elements during maturation and aging. *Neurobiol Aging* 10, 359–365.

Meissner G, Rios E, Tripathy A & Pasek DA. (1997). Regulation of skeletal muscle Ca2+ release channel (ryanodine receptor) by Ca2+ and monovalent cations and anions. *J Biol Chem* 272, 1628–1638.

Melzer W, Herrmann-Frank A & Luttgau HC. (1995). The role of Ca2+ ions in excitation-contraction coupling of skeletal muscle fibres. *Biochim Biophys Acta* 1241, 59–116.

Messi ML & Delbono O. (2003). Target-derived trophic effect on skeletal muscle innervation in senescent mice. *J Neurosci* 23, 1351–1359.

Mikami A, Imoto K, Tanabe T, Niidome T, Mori Y, Takeshima H, Narumiya S & Numa S. (1989). Primary structure and functional expression of the cardiac dihydropyridine-sensitive calcium channel. *Nature* 340, 230–233.

Mori Y, Friedrich T, Kim MS, Mikami A, Nakai J, Ruth P, Bosse E, Hofmann F, Flockerzi V, Furuichi T, Tikoshiba K, Imoto K, Tanabe T & Numa S. (1991). Primary structure and functional expression from complementary DNA of a brain calcium channel. *Nature* 350, 398–402.

Musaro A, McCullagh K, Paul A, Houghton L, Dobrowolny G, Molinaro M, Barton ER, Sweeney HL & Rosenthal N. (2001). Localized Igf1 transgene expression sustains hypertrophy and regeneration in senescent skeletal muscle. *Nat Genet* 27, 195–200.

Nabauer M, Callewaert G, Cleemann L & Morad M. (1989). Regulation of calcium release is gated by calcium current, not gating charge, in cardiac myocytes. *Science* 244, 800–803.

Nakai J, Dirkesen R, Nguyen H, Pessah I, Beam K & Allen P. (1996). Enhanced dihydropyridine receptor channel activity in the presence of ryanodine receptor. *Nature* 380, 72–75.

Nakai J, Imagawa T, Hakamata Y, Shigekawa M, Takeshima H & Numa S. (1990). Primary structure and functional expression from cDNA of the cardiac ryanodine receptor/calcium release channel. *FEBS Lett* 271, 169–177.

Neff NT, Prevette D, Houenou LJ, Lewis ME, Glicksman MA, Yin QW & Oppenheim RW. (1993). Insulin-like growth factors: putative muscle-derived trophic agents that promote motoneuron survival. *J Neurobiol* 24, 1578–1588.

Nunoki K, Florio V & Catterall WA. (1989). Activation of purified calcium channels by stoichiometric protein phosphorylation. *Proc Natl Acad Sci U S A* 86, 6816–6820.

Ohlendieck K, Briggs FN, Lee KF, Wechsler AW & Campbell KP. (1991). Analysis of excitation-contraction-coupling components in chronically stimulated canine skeletal muscle. *Eur J Biochem* 202, 739–747.

Ohlendieck K, Fromming GR, Murray BE, Maguire PB, Leisner E, Traub I & Pette D. (1999). Effects of chronic low-frequency stimulation on Ca2+-regulatory membrane proteins in rabbit fast muscle. *Pflugers Arch* 438, 700–708.

Ottini L, Marziali G, Conti A, Charlesworth A & Sorrentino V. (1996). Alpha and beta isoforms of ryanodine receptor from chicken skeletal muscle are the homologues of mammalian RyR1 and RyR3. *Biochem J* 315 (Pt 1), 207–216.

Paolini C, Protasi F & Franzini-Armstrong C. (2004). The relative position of RyR feet and DHPR tetrads in skeletal muscle. *J Mol Biol* 342, 145–153.

Pauwels PJ, Van Assouw HP & Leysen JE. (1987). Depolarization of chick myotubes triggers the appearance of (+)-[3H]PN 200–110-binding sites. *Mol Pharmacol* 32, 785–791.

Payne AM & Delbono O. (2004). Neurogenesis of excitation-contraction uncoupling in aging skeletal muscle. *Exerc Sport Sci Rev* 32, 36–40.

Payne AM, Dodd SL & Leeuwenburgh C. (2003). Life-long calorie restriction in Fischer 344 rats attenuates age-related loss in skeletal muscle-specific force and reduces extracellular space. *J Appl Physiol* 95, 2554–2562.

Payne AM, Zheng Z, Gonzalez E, Wang ZM, Messi ML & Delbono O. (2004). External Ca2+ -dependent excitation-contraction coupling in a population of ageing mouse skeletal muscle fibres. *J Physiol* 560, 137–155.

Pereon Y, Navarro J, Hamilton M, Booth FW & Palade P. (1997a). Chronic stimulation differentially modulates expression of mRNA for dihydropyridine receptor isoforms in rat fast twitch skeletal muscle. *Biochem Biophys Res Commun* 235, 217–222.

Pereon Y, Sorrentino V, Dettbarn C, Noireaud J & Palade P. (1997b). Dihydropyridine receptor and ryanodine receptor gene expression in long-term denervated rat muscles. *Biochem Biophys Res Commun* 240, 612–617.

Perez-Reyes E, Kim HS, Lacerda AE, Horne W, Wei XY, Rampe D, Campbell KP, Brown AM & Birnbaumer L. (1989). Induction of calcium currents by the expression of the alpha 1-subunit of the dihydropyridine receptor from skeletal muscle. *Nature* 340, 233–236.

Pragnell M, De Waard M, Mori Y, Tanabe T, Snutch TP & Campbell KP. (1994). Calcium channel beta-subunit binds to a conserved motif in the I-II cytoplasmic linker of the alpha 1-subunit. *Nature* 368, 67–70.

Protasi F, Franzini-Armstrong C & Allen PD. (1998). Role of ryanodine receptors in the assembly of calcium release units in skeletal muscle. *J Cell Biol* 140, 831–842.

Protasi F, Franzini-Armstrong C & Flucher BE. (1997). Coordinated incorporation of skeletal muscle dihydropyridine receptors and ryanodine receptors in peripheral couplings of BC3H1 cells. *J Cell Biol* 137, 859–870.

Protasi F, Sun XH & Franzini-Armstrong C. (1996). Formation and maturation of the calcium release apparatus in developing and adult avian myocardium. *Dev Biol* 173, 265–278.

Protasi F, Takekura H, Wang Y, Chen SR, Meissner G, Allen PD & Franzini-Armstrong C. (2000). RYR1 and RYR3 have different roles in the assembly of calcium release units of skeletal muscle. *Biophys J* 79, 2494–2508.

Pu SF, Zhuang HX, Marsh DJ & Ishii DN. (1999). Insulin-like growth factor-II increases and IGF is required for postnatal rat spinal motoneuron survival following sciatic nerve axotomy. *J Neurosci Res* 55, 9–16.

Rabinovsky ED, Gelir E, Gelir S, Lui H, Kattash M, DeMayo FJ, Shenaq SM & Schwartz RJ. (2003). Targeted expression of IGF-1 transgene to skeletal muscle accelerates muscle and motor neuron regeneration. *Faseb J* 17, 53–55.

Ray A, Kyselovic J, Leddy JJ, Wigle JT, Jasmin BJ & Tuana BS. (1995). Regulation of dihydropyridine and ryanodine receptor gene expression in skeletal muscle. Role of nerve, protein kinase C, and cAMP pathways. *J Biol Chem* 270, 25837–25844.

Renganathan M, Messi ML & Delbono O. (1997a). Dihydropyridine receptor-ryanodine receptor uncoupling in aged skeletal muscle. *J Membr Biol* 157, 247–253.

Renganathan M, Messi ML & Delbono O. (1998). Overexpression of IGF-1 exclusively in skeletal muscle prevents age-related decline in the number of dihydropyridine receptors. *J Biol Chem* 273, 28845–28851.

Renganathan M, Messi ML, Schwartz R & Delbono O. (1997b). Overexpression of hIGF-1 exclusively in skeletal muscle increases the number of dihydropyridine receptors in adult transgenic mice. *FEBS Lett* 417, 13–16.

Rios E & Brum G. (1987). Involvement of dihydropyridine receptors in excitation-contraction coupling in skeletal muscle. *Nature* 325, 717–720.

Rodney GG, Krol J, Williams B, Beckingham K & Hamilton SL. (2001). The carboxy-terminal calcium binding sites of calmodulin control calmodulin's switch from an activator to an inhibitor of RYR1. *Biochemistry* 40, 12430–12435.

Rodney GG, Williams BY, Strasburg GM, Beckingham K & Hamilton SL. (2000). Regulation of RYR1 activity by Ca(2+) and calmodulin. *Biochemistry* 39, 7807–7812.

Rotman EI, De Jongh KS, Florio V, Lai Y & Catterall WA. (1992). Specific phosphorylation of a COOH-terminal site on the full-length form of the alpha 1 subunit of the skeletal muscle calcium channel by cAMP-dependent protein kinase. *J Biol Chem* 267, 16100–16105.

Ryan M, Carlson BM & Ohlendieck K. (2000). Oligomeric status of the dihydropyridine receptor in aged skeletal muscle. *Mol Cell Biol Res Commun* 4, 224–229.

Saborido A, Molano F, Moro G & Megias A. (1995). Regulation of dihydropyridine receptor levels in skeletal and cardiac muscle by exercise training. *Pflugers Arch* 429, 364–369.

Schinder AF & Poo M. (2000). The neurotrophin hypothesis for synaptic plasticity. *Trends Neurosci* 23, 639–645.

Schneider MF & Chandler WK. (1973). Voltage dependent charge movement of skeletal muscle: a possible step in excitation-contraction coupling. *Nature* 242, 244–246.

Schreiber SL. (1991). Chemistry and biology of the immunophilins and their immunosuppressive ligands. *Science* 251, 283–287.

Schreyer DJ & Skene JH. (1993). Injury-associated induction of GAP-43 expression displays axon branch specificity in rat dorsal root ganglion neurons. *J Neurobiol* 24, 959–970.

Schulte LM, Navarro J & Kandarian SC. (1993). Regulation of sarcoplasmic reticulum calcium pump gene expression by hindlimb unweighting. *Am J Physiol* 264, C1308–1315.

Sculptoreanu A, Rotman E, Takahashi M, Scheuer T & Catterall WA. (1993). Voltage-dependent potentiation of the activity of cardiac L-type calcium channel alpha 1 subunits due to phosphorylation by cAMP-dependent protein kinase. *Proc Natl Acad Sci U S A* 90, 10135–10139.

Sencer S, Papineni RV, Halling DB, Pate P, Krol J, Zhang JZ & Hamilton SL. (2001). Coupling of RYR1 and L-type calcium channels via calmodulin binding domains. *J Biol Chem* 276, 38237–38241.

Serysheva, II, Orlova EV, Chiu W, Sherman MB, Hamilton SL & van Heel M. (1995). Electron cryomicroscopy and angular reconstitution used to visualize the skeletal muscle calcium release channel. *Nat Struct Biol* 2, 18–24.

Serysheva, II, Schatz M, van Heel M, Chiu W & Hamilton SL. (1999). Structure of the skeletal muscle calcium release channel activated with Ca2+ and AMP-PCP. *Biophys J* 77, 1936–1944.

Sheridan DC, Carbonneau L, Ahern CA, Nataraj P & Coronado R. (2003a). Ca2+-dependent excitation-contraction coupling triggered by the heterologous cardiac/brain DHPR beta2a-subunit in skeletal myotubes. *Biophys J* 85, 3739–3757.

Sheridan DC, Cheng W, Ahern CA, Mortenson L, Alsammarae D, Vallejo P & Coronado R. (2003b). Truncation of the carboxyl terminus of the dihydropyridine receptor beta1a subunit promotes Ca2+ dependent excitation-contraction coupling in skeletal myotubes. *Biophys J* 84, 220–237.

Shiotani A, O'Malley BW, Jr., Coleman ME, Alila HW & Flint PW. (1998). Reinnervation of motor endplates and increased muscle fiber size after human insulin-like growth factor I gene transfer into the paralyzed larynx. *Hum Gene Ther* 9, 2039–2047.

Shistik E, Ivanina T, Puri T, Hosey M & Dascal N. (1995). Ca2+ current enhancement by alpha 2/delta and beta subunits in Xenopus oocytes: contribution of changes in channel gating and alpha 1 protein level. *J Physiol* 489 (Pt 1), 55–62.

Skene JH, Jacobson RD, Snipes GJ, McGuire CB, Norden JJ & Freeman JA. (1986). A protein induced during nerve growth (GAP-43) is a major component of growth-cone membranes. *Science* 233, 783–786.

Smith JS, Rousseau E & Meissner G. (1989). Calmodulin modulation of single sarcoplasmic reticulum Ca2+-release channels from cardiac and skeletal muscle. *Circ Res* 64, 352–359.

Sonnleitner A, Fleischer S & Schindler H. (1997). Gating of the skeletal calcium channel by ATP is inhibited by protein phosphatase 1 but not by Mg2+. *Cell Calcium* 21, 283–290.

Stokes DL & Wagenknecht T. (2000). Calcium transport across the sarcoplasmic reticulum: structure and function of Ca2+-ATPase and the ryanodine receptor. *Eur J Biochem* 267, 5274–5279.

Strube C, Beurg M, Powers PA, Gregg RG & Coronado R. (1996). Reduced Ca2+ current, charge movement, and absence of Ca2+ transients in skeletal muscle deficient in dihydropyridine receptor beta 1 subunit. *Biophys J* 71, 2531–2543.

Sun XH, Protasi F, Takahashi M, Takeshima H, Ferguson DG & Franzini-Armstrong C. (1995). Molecular architecture of membranes involved in excitation-contraction coupling of cardiac muscle. *J Cell Biol* 129, 659–671.

Szegedi C, Sarkozi S, Herzog A, Jona I & Varsanyi M. (1999). Calsequestrin: more than 'only' a luminal Ca2+ buffer inside the sarcoplasmic reticulum. *Biochem J* 337 (Pt 1), 19–22.

Takekura H, Nishi M, Noda T, Takeshima H & Franzini-Armstrong C. (1995a). Abnormal junctions between surface membrane and sarcoplasmic reticulum in skeletal muscle with a mutation targeted to the ryanodine receptor. *Proc Natl Acad Sci U S A* 92, 3381–3385.

Takekura H, Takeshima H, Nishimura S, Takahashi M, Tanabe T, Flockerzi V, Hofmann F & Franzini-Armstrong C. (1995b). Co-expression in CHO cells of two muscle proteins involved in excitation-contraction coupling. *J Muscle Res Cell Motil* 16, 465–480.

Takekura H, Tamaki H, Nishizawa T & Kasuga N. (2003). Plasticity of the transverse tubules following denervation and subsequent reinnervation in rat slow and fast muscle fibres. *J Muscle Res Cell Motil* 24, 439–451.

Takeshima H. (1993). Primary structure and expression from cDNAs of the ryanodine receptor. *Ann N Y Acad Sci* 707, 165–177.

Takeshima H, Nishimura S, Matsumoto T, Ishida H, Kangawa K, Minamino N, Matsuo H, Ueda M, Hanaoka M, Hirose T & et al. (1989). Primary structure and expression from complementary DNA of skeletal muscle ryanodine receptor. *Nature* 339, 439–445.

Tanabe T, Beam KG, Adams BA, Niidome T & Numa S. (1990a). Regions of the skeletal muscle dihydropyridine receptor critical for excitation-contraction coupling. *Nature* 346, 567–569.

Tanabe T, Beam KG, Powell JA & Numa S. (1988). Restoration of excitation-contraction coupling and slow calcium current in dysgenic muscle by dihydropyridine receptor complementary DNA. *Nature* 336, 134–139.

Tanabe T, Mikami A, Numa S & Beam KG. (1990b). Cardiac-type excitation-contraction coupling in dysgenic skeletal muscle injected with cardiac dihydropyridine receptor cDNA. *Nature* 344, 451–453.

Tanabe T, Takeshima H, Mikami A, Flockerzi V, Takahashi H, Kangawa K, Kojima M, Matsuo H, Hirose T & Numa S. (1987). Primary structure of the receptor for calcium channel blockers from skeletal muscle. *Nature* 328, 313–318.

Tarroni P, Rossi D, Conti A & Sorrentino V. (1997). Expression of the ryanodine receptor type 3 calcium release channel during development and differentiation of mammalian skeletal muscle cells. *J Biol Chem* 272, 19808–19813.

Thomason DB, Herrick RE, Surdyka D & Baldwin KM. (1987). Time course of soleus muscle myosin expression during hindlimb suspension and recovery. *J Appl Physiol* 63, 130–137.

Thompson LV & Brown M. (1999). Age-related changes in contractile properties of single skeletal fibers from the soleus muscle. *J Appl Physiol* 86, 881–886.

Timerman AP, Ogunbumni E, Freund E, Wiederrecht G, Marks AR & Fleischer S. (1993). The calcium release channel of sarcoplasmic reticulum is modulated by FK-506-binding protein. Dissociation and reconstitution of FKBP-12 to the calcium release channel of skeletal muscle sarcoplasmic reticulum. *J Biol Chem* 268, 22992–22999.

Tripathy A, Xu L, Mann G & Meissner G. (1995). Calmodulin activation and inhibition of skeletal muscle Ca2+ release channel (ryanodine receptor). *Biophys J* 69, 106–119.

Urbanchek MG, Picken EB, Kalliainen LK & Kuzon WM, Jr. (2001). Specific force deficit in skeletal muscles of old rats is partially explained by the existence of denervated muscle fibers. *J Gerontol A Biol Sci Med Sci* 56, B191–197.

Ursu D, Sebille S, Dietze B, Freise D, Flockerzi V & Melzer W. (2001). Excitation-contraction coupling in skeletal muscle of a mouse lacking the dihydropyridine receptor subunit gamma1. *J Physiol* 533, 367–377.

Verdu E, Ceballos D, Vilches JJ & Navarro X. (2000). Influence of aging on peripheral nerve function and regeneration. *J Peripher Nerv Syst* 5, 191–208.

Vergani L, Di Giulio AM, Losa M, Rossoni G, Muller EE & Gorio A. (1998). Systemic administration of insulin-like growth factor decreases motor neuron cell death and promotes muscle reinnervation. *J Neurosci Res* 54, 840–847.

Wagenknecht T, Radermacher M, Grassucci R, Berkowitz J, Xin HB & Fleischer S. (1997). Locations of calmodulin and FK506-binding protein on the three-dimensional architecture of the skeletal muscle ryanodine receptor. *J Biol Chem* 272, 32463–32471.

Wang S, Trumble WR, Liao H, Wesson CR, Dunker AK & Kang CH. (1998a). Crystal structure of calsequestrin from rabbit skeletal muscle sarcoplasmic reticulum. *Nat Struct Biol* 5, 476–483.

Wang X, Berninger B & Poo M. (1998b). Localized synaptic actions of neurotrophin-4. *J Neurosci* 18, 4985–4992.

Wang ZM, Messi ML & Delbono O. (2000). L-Type Ca(2+) channel charge movement and intracellular Ca(2+) in skeletal muscle fibers from aging mice. *Biophys J* 78, 1947–1954.

Wang ZM, Messi ML & Delbono O. (2002). Sustained overexpression of IGF-1 prevents age-dependent decrease in charge movement and intracellular Ca(2+) in mouse skeletal muscle. *Biophys J* 82, 1338–1344.

Wang ZM, Messi ML, Renganathan M & Delbono O. (1999). Insulin-like growth factor-1 enhances rat skeletal muscle charge movement and L-type Ca2+ channel gene expression. *J Physiol* 516 (Pt 2), 331–341.

Yano K & Zarain-Herzberg A. (1994). Sarcoplasmic reticulum calsequestrins: structural and functional properties. *Mol Cell Biochem* 135, 61–70.

Yin CC & Lai FA. (2000). Intrinsic lattice formation by the ryanodine receptor calcium-release channel. *Nat Cell Biol* 2, 669–671.

Zhang C, Goto N, Suzuki M & Ke M. (1996). Age-related reductions in number and size of anterior horn cells at C6 level of the human spinal cord. *Okajimas Folia Anat Jpn* 73, 171–177.

Zhang C, Goto N & Zhou M. (1995). Morphometric analyses and aging process of nerve fibers in the human spinal posterior funiculus. *Okajimas Folia Anat Jpn* 72, 259–264.

Zhang L, Kelley J, Schmeisser G, Kobayashi YM & Jones LR. (1997). Complex formation between junctin, triadin, calsequestrin, and the ryanodine receptor. Proteins of the cardiac junctional sarcoplasmic reticulum membrane. *J Biol Chem* 272, 23389–23397.

Zheng Z, Wang ZM & Delbono O. (2002). Insulin-like growth factor-1 increases skeletal muscle dihydropyridine receptor alpha 1S transcriptional activity by acting on the cAMP-response element-binding protein element of the promoter region. *J Biol Chem* 277, 50535–50542.

CHAPTER 8

MUSCLE PLASTICITY AND VARIATIONS IN MYOFIBRILLAR PROTEIN COMPOSITION OF MAMMALIAN MUSCLE FIBERS

LAURENCE STEVENS, BRUNO BASTIDE AND YVONNE MOUNIER
Universite' des Sciences et Technologies de Lille I, Villeneuve d'Ascq Cedex, France

1.   MYOSIN AND THICK FILAMENT PROTEINS

Myosin is the most abundant protein expressed in striated muscle cells, comprising ~25% of the total protein pool. Each myosin molecule (~500 kDa) is a hexamere composed of two heavy chains (MHC) and four light chains (MLC), which are assembled to form two globular heads with a long tail. Each globular head is composed of the S1 sub fragment of a myosin heavy chain and two non-identical myosin light chains (essential MLC 25 kDa and regulatory MLC 20 kDa) (fig. 1). The S1 sub fragment contains a catalytic site for binding and hydrolyzing ATP, as well as a site for binding actin in the thin filament.

Thick filaments in mammalian striated muscles are composed of ~600 molecules of myosin, which are arranged in such a way that the tails point towards the center of the sarcomere, and the heads decorate the outer ends of each emi-filament. This arrangement of the myosin molecules implies that the myosin heads protrude from the thick filament every 60°, maximizing the opportunities for binding with actin. Myosin is a mechanochemical enzyme, or motor protein, that converts, via ATPase activity, chemical into mechanical energy in order to move along actin filaments. Many cellular movements, such as organelle/vesicle transport and muscle contraction, depend on the interactions between myosin and actin filaments. In skeletal muscle fibers, in cardiomyocytes and in smooth muscle cells the actin-myosin interaction represents the molecular basis for force generation and contractile functions.

Striated muscle myosin has not only the extraordinary capability to fulfill different functional requirements, corresponding to various functions of skeletal muscle,

R. Bottinelli and C. Reggiani (eds.), Skeletal Muscle Plasticity in Health and Disease, 213–264.
© 2006 *Springer.*

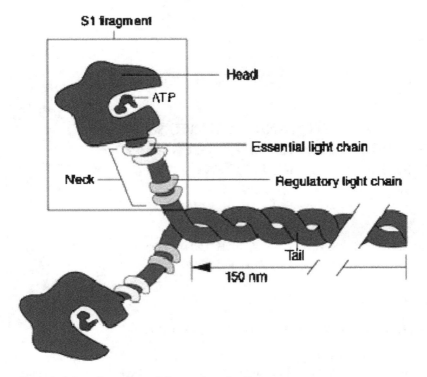

Figure 1. Schematic structure of the myosin molecule

but also to respond to a variety of stimuli coming from changes of functional demands. Both of these properties can be referred to as "muscle plasticity" (Buller et al., 1960) and are based, in the first place, on the existence of multiple MHC and MLC isoforms. The myosin isoforms have distinct ATPase activity and shortening velocity, which also provide the molecular basis of muscle fiber functional diversity. They can be exchanged (isoform transitions) under the influence of various internal or external factors such as neuromuscular activity, loading conditions and hormonal status.

1.1 Myosin Heavy Chain Polymorphism

The MHC molecule is a key marker of muscle plasticity, in view of its major role in determining muscle contractility, its high abundance sequence and its heterogeneity. MHCs are encoded by a multigene family and all members of this family exhibit specific temporal and spatial expression patterns.

In mammalian striated muscles, there are at least nine distinct known isoforms of MHC expressed in striated muscles and encoded by specific genes (see for a comprehensive review Berg et al., 2001): MHC-1 (or β/slow), expressed in slow skeletal muscle fibers and ventricular myocardium; MHC-α (or 1α), a slow isoform expressed in atrial myocardium and in some skeletal muscle fibers of specialized muscles such as extraocular and masticatory muscles; MHC-2A, 2D/X and 2B, expressed in fast skeletal muscle fibers; MHC-eom, a fast isoform only found in extraocular and laryngeal muscle fibers; the fast MHC-2M expressed in masticatory muscle fibers; two developmental isoforms, MHC-emb and MHC-peri, expressed predominantly in muscles during embryonic and perinatal development, respectively (Vikstrom et al., 1997), but also in adult muscles in regenerating fibers and in specialized fibers in masticatory and extraocular muscles (Lompre et al., 1994).

1.2   Muscle Plasticity and Variations of MHC Expression

The plasticity at the MHC expression level has been demonstrated using a variety of animal models and experimental treatments. Indeed, it can be modulated by many factors, including but not limited to: fetal/embryonic developmental programs, neural stimulation and associated neuronal firing patterns, hormonal factors, muscular activity, loading conditions (see for a reviews Schiaffino and Reggiani, 1996; Pette and Staron, 1997). In the following paragraphs we will focus on few different key models that induce muscle contractile protein plasticity by decreased or increased neuromuscular activity (i.e., decreased- increased-neural activation and/or loading) (Pette and Staron, 1997; Talmadge, 2000; Fitts et al., 2000; Ohira, 2000; Huey et al., 2001). Although similarities exist in the phenotypic shift in MHC isoforms in different models inducing transformations, the molecular mechanisms leading to these transformations are specific to the imposed perturbation and the type of MHC isoform regulation.

Among the "inactivity" models, the following models will be examined: space-flight (SF) and parabolic flights (both inducing microgravity conditions); hindlimb unloading (HU) and bed rest, which are ground-based models of microgravity; ankle immobilization in a shortened position; denervation, i.e., de-efferentation and de-afferentation; spinal transection (ST) and spinal isolation (SI).

Spaceflight and hindlimb unloading, in which the lower limb muscles remain active, but are unloaded (i.e., they are required to produce less force output), induce a reduction in the electrical activity, reflected by a significant decrease in the EMG signals until 7 days of HU (Alford et al., 1987; Leterme and Falempin, 1998)), or even until 28 days (Blewett and Elder, 1993). Reduced activity elicits muscle atrophy, linked to a significant loss of mass and myofibrillar proteins in hindlimb muscles, due to decreased protein synthesis and increased protein degradation (Thomason and Booth, 1990). A significant decrease in the fiber content of myofibrillar proteins has been observed in the soleus (Steffen and Musacchia, 1986) and Haddad et al. (1993) have suggested that the decline in myofibrillar proteins can primarily be attributed to a loss of slow myosins. Muscle phenotype was thus affected, resulting in a slow-to-fast fiber type transitions (Desplanches et al. 1987; Thomason and Booth, 1990). In rodents, however, a decrease of the slow type 1 with an increase of the fast type 2D/X fibers was observed, with postural muscles such as soleus, vastus intermedius and gastrocnemius being affected to a greater extent than fast phasic muscles as extensor digitorum longus (Ilyina-Kakueva et al., 1976; Riley et al., 1990; Ohira et al., 1992; Stevens et al., 1993; Kischel et al., 2001b).

The changes in MHC isoform distribution in unloaded skeletal muscles has been extensively studied (for a reviews, see Thomason and Booth, 1990; Morey-Holton et al., 2005). In small mammals, slow muscles express MHC1 isoform with minor amounts of MHC2A, whereas fast muscles express various combinations of MHC-2A, 2D/X and 2B (with or without MHC1 depending on the muscle). After SF or HU, the slow-to-fast shift in MHC isoform distribution is characterized by the new expression of the fast MHC-2D/X and MHC-2B isoforms, the fast predominant isoform being the MHC2D/X. For example, after 6 days of SF, the soleus has been shown to contain significantly less MHC-2A and a newly expressed MHC-2D/X (Caiozzo et al., 1994). After 10-14 days of SF, the soleus MHC-2D/X content is increased to 4-11% respectively, whereas MHC-2A is either decreased or unchanged (Caiozzo et al., 1994; Staron et al., 1998) and MHC-1 is significantly reduced.

Stevens et al. (1999) showed that in slow soleus muscle MHC1 content was significantly decreased after 7 of HU, and MHC2A increased, reaching a transient maximum at 15 days (Fig. 2). MHC-2D/X and 2B were progressively induced from 7 until 28 days, where they reached 18 and 8%, respectively. An additional interesting observation in that study was that MHC1a, a second slow MHC isoform previously described and characterized by a mobility higher than MHC1β (Fauteck and Kandarian, 1995; Galler et al., 1997), was present at significant levels in the 28-day unloaded muscles (Stevens et al., 1999). Moreover, the MHC-1α isoform, normally present in intrafusal fibers, was detected at small, but significant, amounts in unloaded soleus muscle fibers, both as mRNA and as protein (Stevens et al., 1999). It was suggested that upregulation of MHC-1α could reflect a transient intermediate step during the transition from MHC-1β to MHC-2A during unloading. Longer periods of inactivity by unloading have been investigated thanks to HU. These studies revealed changes in MHC expression quantitatively greater than with shorter periods. For instance, 56-day- HU reduced the amount of slow isomyosin to 33% (Thomason et al., 1987).

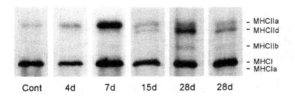

Figure 2. Silver-stained electrophoretical separation of MHC isoforms in soleus muscles from control rats (Cont) and rats exposed to hindlimb unloading for various periods. MHCIa, the fastest moving band, is clearly separated in one of the two samples from 28-day unloaded soleus muscles (Reprinted with permission from Stevens et al., 1999, © Elsevier B.V.)

Data on human muscles exposed to unloading have been collected with the model of bed-rest (Berg et al., 1991; Trappe et al., 2001; Tesch et al., 2004), which has provided an opportunity to analyze some characteristics of MHC plasticity in humans. The changes in MHC isoform distribution were very dependent on the bed-rest duration, but greatly differed from one study to the other. For instance, no change in MHC isoform pattern was seen after 14, 30 and 37 days of bed-rest in vastus lateralis as well as in soleus muscles by Hikida et al. (1989). In another study, however, 17 days of bed-rest induced a significant decrease in type I fibers expressing MHC-1 in soleus muscle (Widrick et al., 1999). Recently, Trappe et al. (2004) showed that, in vastus lateralis muscle, after 84 days bed-rest the proportions of fibers expressing MHC-1 and MHC-2A were decreased (from 64 to 35%, and from 23 to 17%, respectively), while hybrid fibers expressing [MHC-1 + MHC-2A], [MHC-2A + MHC-2D/X] and [MHC-1 + MHC-2A + MHC-2D/X] were significantly increased.

Limb immobilization in a shortened position also leads to a decrease in muscle load and atrophy, but it is accompanied by a greater reduction in electrical activity (EMG). A reduced protein synthesis occurs very early, i.e., 6 hours after starting the immobilization. A rapid slow-to-fast shift in MHC isoform profile takes place in both slow and fast (plantaris and gastrocnemius) muscles. For instance, after only 2 days of immobilization, the levels of fast isoforms are already increased (even MHC-2B) with no change in MHC-1 expression (Loughna et al., 1990). Although less data are available, the changes in MHC isoform expression pattern for longer periods of immobilization are no more pronounced than after a short period (Jakubiec-Puka et al., 1992). It seems that the age at which the immobilization is induced plays an important role in the adaptation of muscle phenotype. When immobilization was performed during polyneural innervation (between 6 and 12 days), the postnatal maturation of the soleus was delayed (i.e., a high level of MHCperi was maintained), whereas in the presence of adult monosynaptic innervation, the immobilized soleus evolves towards a fast phenotype (i.e., fast MHC isoforms are expressed) (fig. 3) (Picquet et al., 1998).

Denervation results in complete loss of neuromuscular activity, but does not necessarily eliminate the mechanical load. Moreover, it removes the possible role of neurotrophic factors at the neuromuscular junction. The effects of denervation are generally age, time and species dependent (Gutmann et al., 1972). Fourteen-day denervation causes a decrease to about 50% of the weight of fast and slow muscles like soleus, extensor digitorum longus and gastrocnemius (Jakubiec-Puka et al., 1999). In slow muscles a slow-to-fast MHC isoform transition is induced by denervation (Michel et al., 1996; Huey and Bodine, 1998; Jakubiec-Puka et al., 1999). Huey and Bodine (1998) followed the time course of the changes induced by soleus denervation and found, after only 30 days, a measurable expression of MHC2D/X (9%). Huey et al. (2003) showed that, after 4 weeks of denervation, MHC2A and MHC2D/X decreased or increased depending on the muscle type, whereas MHC1 and MHC2B content was reduced in all muscles.

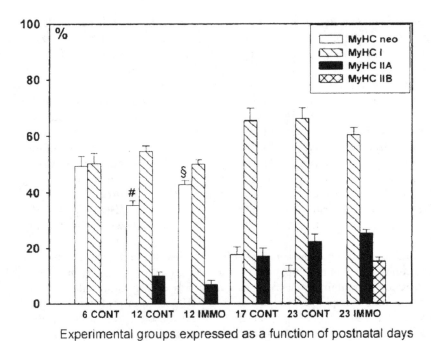

Experimental groups expressed as a function of postnatal days

Figure 3. MHC isoform expressions in muscle bundles of control and immobilized rat soleus. The relative proportions of MHC isoform were obtained from densitometry analysis of SDS-PAGE gel electrophoresis (Reprinted with permission from Picquet et al., 1998, © Springer-Verlag). The results are shown as means ±SE. # indicates a statistical difference (p<0.05) with day 6 CONT group, $ with day 12 CONT group

More precisely, MHC2A isoform content decreased in soleus and increased in fast muscles, whereas the proportion of MHC2D/X changed inversely.

SI (spinal cord isolation) and ST (spinal cord transection) have the advantage over denervation that motoneuron-muscle connection is maintained, as confirmed by EMG recording (Talmadge, 2000). In SI conditions, the cord is severed in two places and afferent (sensory) input is eliminated between the two lesions. It is thus an animal model that removes neuromuscular activation, sensory input and loading while the motoneurons remain connected with muscles and can deliver neurotrophic factors (Roy et al., 1992). In SI, rodent muscles undergo rapid atrophy during the first 2 weeks, with a progressive downregulation of MHC1 and an upregulation of MHC2D/X isoforms, which can last for a period of 60 (Grossman et al., 1998) or 90 days (Huey et al., 2000). Recently, Huey et al. (2001) quantified more precisely the time courses of changes in MHC isoforms after SI in soleus and adductor longus (Fig. 4). In soleus, they found that from 4 to 30 days of SI, the MHC1 content remained at 90% of total MHC, the remainder being MHC2A. After 30 days, there was a decrease in MHC1, which was offset by an increase in MHC2D/X isoform content. MHC2B was not detected until 90 days of SI. In adductor longus,

SI resulted in an increase in MHC2D/X at 30 days and a decrease in MHC1 by 60 days. Ninety days of SI transformed both the soleus and the adductor longus into muscles composed of predominantly MHC2D/X. The MHC shifts induced by SI are quantitatively similar to those observed in other models of inactivity like HU or denervation, they differ, however, in the increased level of MHC2D/X isoform.

The ST (spinal cord transection) model results in the reduction of electrical activity and loading of both slow and fast muscles, although motoneurons can still be activated by sensory input. In rats, the ST induces a slow-to-fast shift of MHC isoforms similar to that observed in SI (Talmadge et al., 1995). However, the extent of the transformation is greater in SI conditions, as the decrease in MHC1 was greater (+15%) and the increase in MHC2D/X lower (−12%) with ST than with SI. Surprisingly, in humans suffering for spinal cord injuries (a condition comparable the ST), the slow-to-fast transformation of MHC profiles is delayed when compared with that observed in animal models (Andersen et al., 1996; Castro et al., 1999; Talmadge, 2000; Baldwin and Haddad, 2001).

The importance of the neural activation in the determination and mainte-nance of muscle phenotype could also be studied thanks to two models of specific suppression of the motor (de-efferentation) or sensory (de-afferentation) muscle innervation. After 2 weeks of bilateral spinal cord ventral root section

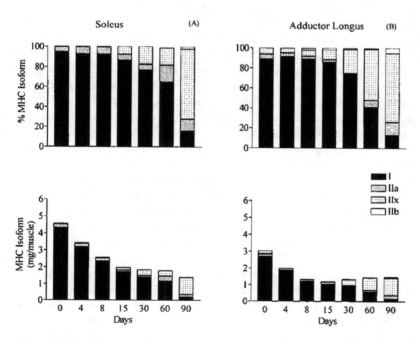

Figure 4. Time course of changes in the relative distribution of MHC protein (top row) and in the absolute amounts of each isoform (bottom row) in the soleus (A) and adductor longus (B) of rats at 0 (control), 4, 8, 15, 30, 60, or 90 days after spinal cord isolation surgery (Reprinted with permission from Huey et al., 2001, © John Willey & Sons, Inc.)

(from L3 to L6), rat soleus muscles exhibited a dramatic increase of MHC2D/X (15%) and MHC2B (26%) expression, while MHC2A did not change (Stevens et al., 2000). The finding that the changes in MHC isoform distribution were greater after de-efferentation than after denervation points to the motor inner-vation as a key factor for triggering transitions in MHC isoform pattern. After 2 weeks of de-afferentation (section of the dorsal roots L3 to L5 after laminectomy), Picquet et al. (2003) showed that the changes at the MHC level were very small, when compared to HU conditions. They consisted only in a slight decrease in MHC1 isoform and increased levels of MHC2D/X. It was concluded that some of the modifications induced by HU (such as slow-to-fast MHC isoform shift) were primarily of motor origin, and that other parameters like muscle atrophy were nevertheless more dependent on the afferent input integrity.

The tetrodotoxine (TTX) application, which blocks the motoneuron action potential, also provides interesting insights into the control of muscle plasticity. Indeed, it has been often speculated that the slow-to-fast type transformations observed after inactivity could be, specifically, a consequence of the changes in the neural activity (Pette and Vrbova, 1985; Buonanno et al., 1996). Falempin and Fodili (1995) have shown that chronic perfusion of TTX during 2 weeks of HU can prevent the decrease in type I fibers and in MHC1 isoforms observed in the unloaded soleus. The effects of TTX alone were studied by Michel et al. (1996), who reported that 2 weeks of TTX-induced paralysis caused an increase in rat soleus fibers expressing MHC1 and MHC2D/X, while the proportions of MHC2A decreased. MHCemb expression was found to be induced by TTX application (Schiaffino et al., 1988), suggesting activation of satellite cells for fusion with existing fibers, or formation of new fibers. An interesting finding is that MHC2B expression is never induced by TTX. This would suggest that the expression of this isoform is not regulated by the same factors as the other ones.

When the mechanical load and/or electric activity on the muscles are increased, the muscle adaptation may include changes in fiber type and changes in fiber size. This response is regulated by processes resulting in contractile protein expression in fast muscles reflecting slower phenotypes, thereby enabling the muscle to better support load-bearing activity, for instance. A number of models have been used to increase neuromuscular activity including, but not limited to, chronic electrical stimulation, functional overload, endurance and resistance training, and hypergravity.

One of the most frequently used models for the study of increased neuromuscular activity is chronic low-frequency stimulation (CLFS). CLFS, by the application of a tonic stimulus pattern, allows one to mimic the impulse pattern of a slow motoneuron, thus increasing the level of electrical stimulation and inducing a wide range of muscle adaptations (see Pette, 2001, for a review). As a general rule, CLFS results in MHC isoform transitions in the fast-to-slow direction. For instance, Pette and coworkers (Leeuw and Pette, 1993; Hämäläinen and Pette, 1997; Peuker et al., 1998) described that in rabbit EDL and TA muscles, where MHC2D/X isoform is predominant, isoform transitions in a sequential order from MHC2D/X to MHC2A, MHC1α and, ultimately, MHC1β were induced by CLFS. Interestingly, MHC1α isoform was not observed during the MHC2A→MHC1β transition in low frequency stimulated rat EDL (Pette and Vrbova, 1999), but was observed in the opposite transition from MHC1β →MHC2A in rat soleus in HU conditions (Stevens et al., 1999).

The functional overload model, whereby a given muscle is forced to adapt to compensate for the surgical removal of its synergists, results in increasing levels of both electrical activation and load bearing (Gardiner et al., 1986). A shift towards an increased expression of slow MHC has been generally demonstrated in rat plantaris overloaded muscle (Morgan and Loughna, 1989; Fauteck and Kandarian, 1995), as an increase in MHC1, 2A, 2D/X and a decrease in MHC2B has been reported (Diffee et al., 1993; Rosenblatt and Parry, 1993). Dunn and Michel (1999) have analyzed the model of overload to extend the knowledge about muscle plasticity and have reported that, in plantaris muscle, overload shifted MHC expression in the order MHC2B→MHC2D/X→ MHC2A→MHC1β + 1α, with a transient re-expression of embryonic and neonatal MHC isoforms. Changes were less pronounced in muscles with a predominance of slow fibers, such as soleus, where functional overload causes fast-to-slow MHC transitions only in the population of fast fibers present in that muscle (for a review, see Pette and Staron, 1997).

Endurance training increases the level of muscle electrical activity with or without increasing the level of load bearing. Previous experiments clearly demonstrated the capacity of skeletal muscle to adapt to endurance training by transitions in MHC isoforms, in the order MHC2B→MHC2D/X→MHC2A→MHC1 (for a review see Pette, 2001). However, although endurance training results in qualitatively similar transition processes as chronic stimulation, transitions are generally limited to the fast subtypes and thus consist of a decrease in the faster MHC2B isoform with a corresponding increase in the slower MHC2A isoform (Demirel et al., 1999; Wada et al., 2003). Only if running is extended for extreme durations (more than 90 min/day), is it possible to induce increased expression of the slow MHC-1 (Green et al., 1987). Resistance (or strength) training, which increases the level of load bearing with minor increases in the level of electrical stimulation, appears to resemble the changes observed in endurance training, but the regulation occurs at different MHC levels. Strength training in the rat has been shown to increase in MHC2D/X and decrease in MHC2B isoforms (Caiozzo et al., 1996, 1997). By contrast, resistance training in humans causes a down-regulation of MHC2D/X and the up-regulation of MHC2A (Adams et al., 1993; Carroll et al., 1998).

Hypergravity (HG) condition has also been used to enhance muscle activity, generating a condition of overloading in all limb muscles; for example, during centrifugation at 2G, body weight doubles. There are controversial views on whether electrical activation was increased in hypergravity. Indeed, some studies have demonstrated that EMG activity of the rat soleus muscle was immediately strongly increased during the two hypergravity phases of the parabolic flight

(Leterme and Falempin, 1998), whereas other studies have described that locomotor activity was reduced in hypergravity (D'Amelio et al., 1998). In most studies rats were born in normal gravity and placed in hypergravity when adult (Amtmann and Oyama, 1976; Chi et al., 1998; Roy et al., 1996; Stevens et al., 2003). Adult rats exposed to 14 days (Roy et al., 1996) or more recently to 19 days in HG (Stevens et al., 2003) showed a partial slow-to-fast transformation of the slow soleus, whereas fast muscles were not (plantaris) or little (gastrocnemius) modified. The lack of changes in MHC isoform does not imply lack of adaptation. Stevens et al. (2003) showed that in HG soleus, the composition in MHC isoforms was not modified, but the regulatory proteins were changed.

The effects of hypergravity have also been studied in young rats reared in HG. Martin (1980) has shown an increase in slow oxidative fibers in soleus and plantaris muscles of 30-day-old rats centrifuged at 2G for 2 weeks. In the rectus capitis muscle (responsible for head posture) of rats conceived, born and reared until the 7th postnatal day at a 1.8G, Martrette et al. (1998) also showed an increase in slow MHC1, accompanied by a an increase in perinatal MHC. In more recent studies, Picquet et al. (2002) and Bozzo et al. (2004) have shown that the slow soleus muscle underwent up-regulation of MHC1 and complete disappearance of MHC2A in rats reared in HG, while the fast plantaris was less affected.

1.3   MHC Isoform Heterogeneity in Single Muscle Fibers

Even if it was formerly held that each mature skeletal muscle fiber expressed only a given MHC isoform (pure fibers), in the last 10-15 years a number of studies have shown that multiple MHC isoforms can be expressed in single muscle fibers (hybrid fibers, see Staron and Pette, 1987; Termin et al., 1989; Wu et al., 2000; Rosser and Bandmann, 2002). Actually, skeletal muscle fibers are elongated multinucleated cells, containing hundreds to thousands of myonuclei, each controlling protein synthesis within its surrounding cytoplasm. A fiber can be considered, therefore, as a series of myonuclear domains, each responding to distinct localized signals that may result in differential gene expression along the fiber.

It is now well established that hybrid fibers co-expressing at least two MHC isoforms exist in normal as well as in transforming muscles (for a review, see Stephenson, 2001). The number of hybrid fibers in normal skeletal muscles is relatively low, but there are differences among species and among muscles. Based on functional properties like calcium and strontium activations, Cordonnier et al., (1995) found that the soleus of normal monkeys contained 61% slow fibers and 39% fast fibers. When fibers were classified using MHC isoform electrophoretical separation, the slow fibers (61%) could be grouped into 41% pure slow fibers (expressing only MHC1) and 20% hybrid fibers (co-expressing MHC1 and minor amounts of MHC2A, MHC1>MHC2A). The fast fibers (39%) were composed of 17% pure fast fibers (expressing MHC2A) and 22% hybrid fibers (MHC2A>MHC1). In normal adult rat soleus, according to Caiozzo et al. (2000), 70% of the fibers expressed only MHC1, while the remaining population was made of hybrid fibers co-expressing MHC1 and MHC2A in various proportions. As shown by Stevens et al. (1999), up to 4% of the hybrid fibers in rat soleus expresses small amounts of MHC2D/X with MHC1 and 2A.

The number of these hybrid fibers increase in muscles undergoing molecular and functional transitions due to activity/inactivity conditions (Talmadge et al., 1995, 1999; Caiozzo et al., 2000; Stevens et al., 2004). It is not really known if the hybrid fibers represent transitional fiber types (Stephenson, 2001; Lutz and Lieber, 2000) or if altered functional demand triggers a "dynamic" process leading to the existence of hybrid fibers, each containing a mix of MHC isoforms (Pette and Staron, 2000; Baldwin and Haddad, 2001). For instance, unloading the soleus for 7 days results in pronounced changes in the pattern of MHC isoforms in single fibers, with decrease in pure (from 68 to 40%) and increase in hybrid fibers when compared with control (Stevens et al., 1999). After HU, the hybrid fibers contained MHC1 and MHC2A in combination with MHC2D/X (23%) and MHC2B (13%). In control and HU fibers, the MHC isoforms at the protein level generally coexisted in combinatorial patterns according to the next neighbor rule, previously established by Pette and Staron (1997). However, at mRNA level, atypical MHC isoform combination (i.e., fibers in which an isoform was missing) were detected after HU in pure fibers containing only MHC1 as well as in hybrid fibers coexpressing slow and fast MHC proteins [table 1 (A) and (B)].

1.4   Myosin Light Chain Polymorphism

Skeletal muscle myosin is a hexamer consisting of two MHC and four MLC. Each MHC is associated with one essential light chain (ELC, also named alkali light chain) and one regulatory light chain (RLC, also called phosphorylable light chain). Thus, in each myosin molecule, myosin light chains consist of a pair of RLC (called MLC2 in skeletal muscle) and a pair of ELC (MLC1 and/or MLC3) (Matsuda, 1983). Both RLC and ELC are non-covalently associated with the MHC of myosin at the base of the head part where the C-terminal rod portion begins. RLC and ELC belong to the CTER family, which also comprises the subfamilies of TnC, calmodulin and probably parvalbumin. All vertebrates express a specific subfamily of ELC and RLC isoforms in their skeletal muscles. MLC isoforms are generally express in combination with MHC isoforms, as described in the following paragraph.

In adult skeletal muscles, type I fibers express MHC-1 together with a slow isoform of ELC, MLC1s, and a slow isoform of RLC, MLC2s, while in fast fibers, fast MHC isoforms are associated with two distinct isoforms of ELC, MLC1f or MLC3f, and one isoform of RLC, MLC2f. The MLC genes are dispersed in different chromosomes and each MLC is coded by a distinct gene. There is an exception for MLC1f and MLC3f isoforms, which originate from a single gene by alternative promoters and splicing of the first exons (Barton and Buckingham, 1985). Developing muscles contain embryonic and neonatal MHCs associated with MLC1emb

Table 1. MHC isoform composition of single muscle fibers from rat soleus in control conditions and after 4 and 7 days of HU (from Stevens et al., 1999, with permission)

(A) MHC mRNA complements of pure MHCI fibers from control, 4-day and 7-day hindlimb unloaded (HU) soleus muscles. Atypical combinations of MHC mRNA isoforms are underlined. (n) is the number of the fibers.

MHC mRNA complement	Control (n)	HU-4d (n)	HU-7d (n)
MHCIβ	7	–	–
MHCIβ+MHCIIa	10	1	–
MHCIβ+MHCIIa + MHCIId(x)	6	–	2
MHCIβ+MHCIIa+MHCIId(x) + MHCII	6	24	7
MHCIβ+MHCIId	–	2	4
MHCIβ+MHCIId(x) + MHCIIb	–	–	1
MHCIß + MHCIIb	–	–	1
MHCIIa + MHCIId(x) + MHCIIb	1	–	–
MHCIIa + MHCIIb	–	1	–
MHCIId(x) + MHCIIb	–	–	1

(B) The same as in (A) but MHC mRNA complements are from hybrid fibers of control, 4-day and 7-day HU soleus muscles characterized by coexistence of MHCI and MHCII protein isoforms.

MHC mRNA complement	Control (n)	HU-4d (n)	HU-7d (n)
MHCIβ+MHCIIa	9	–	–
MHCIβ+MHCIIa + MHCIId(x)	2	2	1
MHCIβ+MHCIIa+MHCIId(x)+MHCII	4	7	8
MHCIβ+MHCIIa + MHCIIb	–	–	4
MHCIβ+MHCIId(x)	–	1	3
MHCIβ+MHCIId(x) + MHCIIb	–	–	1
MHCIIa	–	–	2
MHCIIa + MHCIId(x)	1	–	–
MHCIIa + MHCIId(x) + MHCIIb	1	1	1
MHCIIa + MHCIIb	–	–	2

and/or MLC1f and MLC2f. MLC1emb also persists in adult extraocular and masseters muscles, in intrafusal nuclear chain fibers and in regenerating and denervated muscles.

1.5   Variations of MHC and/or MLC Expressions

In a given species, variable associations of multiple MHC and MLC isoforms result in populations of muscle fibers showing a large range of specific functional properties. There is less data on the expression, during muscle adaptations, of the different isoforms of MLC, or of MHC and MLC considered in parallel (Bortolotto et al., 1999; Lutz et al., 2001). The studies on MHC and MLC isoforms in transformed fibers were based essentially on increased neuromuscular activity experiments, for instance chronic low frequency stimulation (Seedorf et al., 1983) or endurance training (Wahrmann et al., 2001). Important observations on the parallel MHC-MLC transitions come also from HU experiments. For example, a complex panel of hybrid fibers was found in soleus undergoing transitions in myosin isoform pattern after hindlimb unloading in rats (Stevens et al., 2004). This study showed that the slow-to-fast transitions induced by HU are characterized at fiber level by the presence of various hybrids of both MHC and MLC, even if the pure slow MHC1-MLC1s, MLC2s population remained predominant (93% for control, 78% for HU). "Matched" MHC-MLC hybrid fibers formed only 7% of the control fibers, while they reached 13% in HU fibers. A second type of hybrid fibers consisted of fibers where the profile of MHC isoforms did not match that of MLC isoforms. These fibers, called «mismatched» MHC-MLC fibers, expressed either one MHC isoform and a hybrid MLC profile, or a hybrid MHC profile and a pure MLC set. As previously reported (Stephenson, 2001), the mismatched fibers were less numerous than the matched hybrid fibers, but as described above, they increased in HU ($\sim$8%) muscles, i.e., in adapting muscles, when compared to control ($\sim$1%).

The MLC isoform profiles evolve in parallel to the MHC isoform profiles, with varying proportions and combinations (Stevens et al., 2004). In pure slow MHC1 fibers, the appearance of the hybrid profile was first limited to the fast MLC2 isoform, present at lower levels than MLC2s, and to the appearance (only in "transformed" fibers) of both MLC1f and MLC2f, with MLC2f becoming expressed more than MLC2s. The MHC1>MHC2 hybrid slow fiber group (see above) was characterized by the expression of MHC2A accompanied by the expression of MLC1f isoform (only in "transformed" fibers) and by an increased proportion of MLC2f isoform. Finally, the MHC2>MHC1 hybrid fast fibers population was characterized by the fact that, when MHC2 was predominant, only hybrid combinations of MLC were present, as the MLC2f>2s combination was present not only in the "transformed" but also in the control fibers and significant and constant amounts of MLC3f appeared. Thus, the predominance of fast MHC and the higher degree of muscle fiber transformation may somehow induce a higher degree of MLC isoform co-expressions. It did not seem too unreasonable to suppose that a majority of fibers had the transcriptional and/or translational signals necessary for MLC isoform expression at the protein level (Bär et al., 1989; Caiozzo et al., 1998).

A recent and relevant information is that the MHC2/MHC1 ratio varied in parallel to fast/slow ratio of regulatory and essential MLC isoforms, but the correlation with MLC2f/2s was significantly higher than with MLC1f+3/MLC1s (see fig. 5, from Stevens et al., 2004). MLC1 and MLC2 turnover rates are considered to be similar (Zak et al., 1977) and thus cannot be responsible for the greater expression of MLC2f. This would be in agreement with the fact that regulatory and essential MLC are encoded by different genes, MYCL2 for RLC and MYCL3 for ELC, and differentially regulated (Poetter et al., 1996; Schiaffino and Reggiani, 1996). Transcriptional and translational processes, as well as protein degradation are likely all important steps in controlling the levels of the different MLC isoforms.

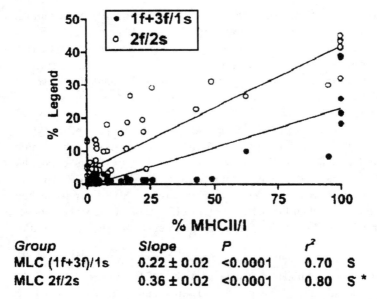

Group	Slope	P	r^2	
MLC (1f+3f)/1s	0.22 ± 0.02	<0.0001	0.70	S
MLC 2f/2s	0.36 ± 0.02	<0.0001	0.80	S *

Figure 5. Relationships between the percentages of fast or slow MLC levels of RLC and ELC and the corresponding relative amounts of MHCII/I in soleus single fibers. % Legend (ordinate) refers to ELC (MLC1f + MLC3f/MLC1s) or RLC (MLC2f/MLC2s) percentages. ELC and RLC amounts were measured by quantitative densitometry of gels after SDS-PAGE in single fibers of slow, hybrid and fast MHC types pooled from control and HU groups. The slopes, significances (P) and correlation coefficients (r2) are given for each combination.* r2 for 2f/2s significantly different from r2 for 1f + 3f/1s. (Reprinted with permission from Stevens et al., 2004, © Springer-Verlag)

1.6 Functional Impact of MHC and MLC Isoforms

It is generally accepted that both MHC and MLC isoforms determine skeletal muscle contractile function. While correlative studies at whole muscle level are very complex because of heterogeneous composition, single fiber studies have permitted the correlation of biochemical and physiological properties with myosin isoform composition.

It has been extensively shown that MHC isoforms can be considered as the major determinants of the diversity in shortening velocity between slow and fast muscle fibers (see Fig. 6 from Bottinelli et al., 1991; for a review see Schiaffino and Reggiani, 1996).

For instance, fibers containing MHC2A were found to display a lower maximal shortening velocity (Vmax) than fibers containing MHC2B (Bottinelli et al., 1991; Cordonnier et al., 1995). However, since the study of Larsson and Moss (1993) on human skeletal muscles, it has been established that the MLC isoforms, in association with MHC, significantly influence maximal shortening velocity. Indeed, MHC2A fibers that contain MLC2f have a higher Vmax than hybrid fibers containing both MLC2f and MLC2s. Moreover, Bottinelli et al. (1994)

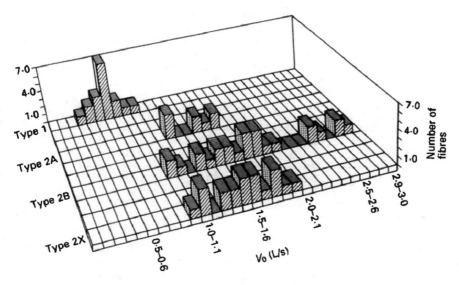

Figure 6. Distribution of maximum shortening velocity (V0) values in the four types of rat single skinned muscle fibers identified by monoclonal antibodies on the basis of their MHC isoform composition. The height of each vertical bar represents the number of fibers. V0 is measured in fiber length per second. (Reprinted with permission from Bottinelli et al., 1991; © The Physiological Society, Blackwell Publishing)

studied the correlation between Vmax and the MLC isoform composition of single fibers from rat fast muscle, and showed, in different fast fiber types, direct positive correlation between Vmax and the MLC3f content expressed as MLC3f/MLC2f. Recently, distinct values of sliding filament velocity in in-vitro motility assay has been found to be associated with different MHC isoforms (Pellegrino et al., 2003). Force transients induced by small stepwise stretches in rat and rabbit skinned fibers containing different MHC and MLC isoforms have been studied (Andruchov et al., 2005). The results showed that, while the force generation kinetics is mainly determined by the MHC isoforms, it is affected by the slow and fast MLC isoforms, but not by the relative content of MLC1f and MLC3f.

Many studies have analyzed the relative contribution of MHC and MLC isoforms to the myosin ATPase activity and its kinetic parameters. MHC isoforms are the main determinant of myosin ATPase activity in muscle fibers as in purified myosin, while the ELC isoforms play a significant regulatory role only when ATPase activity is determined at low ionic strength (for a review, see Schiaffino and Reggiani, 1996).

Although the sensitivity of myofibrillar apparatus to calcium is generally determined by the diversity of the regulatory protein isoforms, myosin can also influence the tension / pCa relationship. Indeed, slow fibers from different muscles show similar calcium sensitivities when containing only the slow set of regulatory

Table 2. Kinetics parameters of maximally activated muscle fibers from hind limb muscles of control rats (CONT) and rats with Hindlimb Suspension (HS). Values are means ± SE. Abbreviations for muscle fibers: CONT SOL, control soleus fibers; HSS SOL, HS slow soleus fibers; HSF SOL, HS fast soleus fibers, CONT EDL, control EDL fibers. V_0, shortening velocity expressed in fiber lengths per second; τ, time constant of tension recovery during the first 10 ms after a quick release (0.97% of the fiber length L_0) of a maximally activated fiber. Compliance was deduced from the x-axis intercept of the linear regression through data of T1 curves. * Significant difference between HS groups, † significantly different from CONT SOL, ‡ significantly different from CONT EDL. (from Toursel et al., 1999)

	CONT SOL (n = 18)	HSS SOL (n = 19)	HSF SOL (n = 17)	CONT EDL (n = 15)
V_0(FL s^{-1})	0.71 ± 0.03‡	0.72 ± 0.06‡*	1.02 ± 0.03 † ‡*	1.21 ± 0.06†
τ (ms)	3.92 ± 0.11‡	3.74 ± 0.12‡	3.43 ± 0.14†	3.31 ± 0.10†
Compliance ($\Delta L/L_0 \times 10^{-2}$)	1.73 ± 0.06‡	1.88 ± 0.07 ‡ *	2.21 ± 0.05 † *	2.23 ± 0.08†

proteins, MHC and MLC (Danieli-Betto et al., 1990). It was proposed that when both fast and slow isoforms of MHC and MLC are present in a muscle fiber, calcium sensitivity is dictated mainly by the fast isoforms (Danieli-Betto et al., 1990). Alternatively, Cordonnier et al., (1995) showed that in monkey soleus fibers, where both slow and fast isoforms of MHC and regulatory proteins are expressed, the position of the tension / pCa curve is mainly governed by the predominant isoform.

Since muscle adaptations generally involve changes in myosin isoform composition and both MHC and MLC isoforms control contractile muscle function, it is expected that muscle adaptations will also change contractile function. A specific role in this respect might be played by hybrid fibers, which become more abundant both in muscle disuse as well as in muscle increased activity. In "transforming muscles", hybrid fibers, generally considered as "transitional fibers", could contribute to the functional plasticity of muscles by rapidly tuning functional parameters as Vmax, tension / pCa relationship and maximal force. For instance, training-induced MHC transitions in the order MHC2B→MHC2D/X→MHC2A result in a 7-10% decrease in Vmax of whole muscle and this decrease is accompanied by a significant decrease in the relative concentration of MLC3f (Wada et al., 2003). Vmax changes after HU in single fibers co-expressing MHC1 and MHC-2A: Vmax is increased and τ, the time constant of tension transients, is decreased and the compliance is increased (Table 2 from Toursel et al., 1999). The tension / pCa curve of the same fibers is shifted to the right after HU, indicating a decrease in calcium affinity.

Changes in maximal specific force may also reflect alterations in MHC isoform composition induced by a variety of conditions, including development, aging, unloading exercise and various diseases (Sieck and Regnier, 2001). Indeed, the effect of denervation on maximal force is more pronounced in fibers expressing fast MHC isoforms, as maximum force of fibers expressing MHC2A and MHC1 decreased by only 20%, while that of fibers expressing MHC2D/X decreased by 40% (Geiger et al., 2001).

2.   MYOFIBRILLAR PROTEINS OF THE THIN FILAMENT

2.1   Structure and Function of the Thin Filament

Calcium regulates actomyosin interactions in striated skeletal muscles via tropomyosin (Tm) and troponin (Tn) complex (Farah and Reinach, 1995; Gordon et al., 2000). Contraction is initiated by the binding of Ca^{2+} to TnC and this triggers a number of conformational changes in the thin filament, which ultimately allow cross bridge formation and tension generation. The stoichiometry of thin filament proteins is one Tm and one Tn complex for every seven actin monomers (A7 Tn Tm, fig. 7). The geometry of the sarcomere is such that up to four myosin heads potentially interact with each A7 Tn Tm structural unit (Gordon et al., 2000).

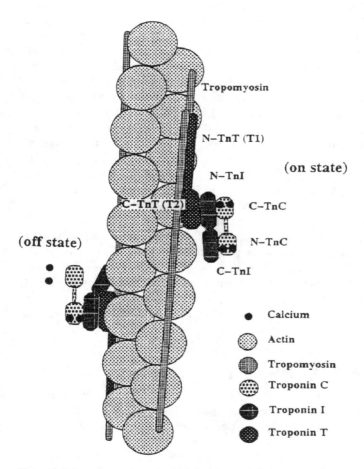

Figure 7. Schematic representation of the molecular architecture of the thin filament (Reproduced with permission from Reinach et al., 1997, © Japan Science and Technology Agency)

Moreover, cross-bridge formation influences the position of Tm on the thin filament (Vibert et al., 1997; Xu et al., 1999).

In this context, it is clear that the structural integrity of the thin filament and its protein composition are critical factors for force development and filament sliding. For instance, a loss in thin filaments has been described as a result of spaceflight (Fitts et al., 2000) and related to the increase in the maximal velocity of shortening in the type I and IIa fiber types.

2.2 Actin Structure and Polymorphism

Actin is a globular protein (G actin, molecular weight $\sim$43 kDa) composed of 375 amino acids. Actin comprises $\sim$20% of myofibrillar proteins and 12% of total protein in skeletal muscle (Ebashi and Nonomura, 1973). It polymerizes in saline solutions and forms actin F filaments, which represent the back-bone of the thin filament with a two-stranded long-pitch helical structure. Each actin monomer comprises four subdomains; the smaller ones designed 1 and 2 are located at the periphery of the filament, on the outside of the helix and are available for interaction with myosin. In particular, subdomain 1 contains the NH_2 and COOH termini of actin and plays a prominent role in the interactions with myosin. Subdomains 3 and 4 close to helix axis interact with subdomains 3 and 4 of other actin monomers.

There are two actin isoforms in sarcomeric muscles, α-skeletal actin and α-cardiac actin (Whalen et al., 1976), which are different from the smooth and non-muscle actins, each variant being coded by a distinct gene. The amino acid sequences of the two-sarcomeric actins differ only by 4 substitutions (Vandekerckhove and Weber, 1979), is highly conserved during vertebrate evolution and has a differential NH_2-terminal sequence so that Asp-Asp-Glu-Glu is present in α-cardiac actin and Asp-Glu-Asp-Glu in α-skeletal actin (Vandekerckhove et al., 1984). This difference may be of importance for the interaction since myosin heads bind mainly to subdomain 1 of actin (Sutoh, 1982), very close to the Tm binding site. The expression of the two actin isoforms varies according to species and stage of development. In mouse and rat, the two-isoforms are coexpressed during development in both skeletal and cardiac muscles, whereas adult skeletal muscles contain only α-skeletal actin and adult heart only α-cardiac actin (Vandekerckhove et al., 1986).

2.3 Variations in Actin Expression

The most common variations described for actin deal with its level of expression and not with changes in isoforms. Thus, loss of weight-bearing function of skeletal muscle has been described as a major factor affecting the levels of α-actin during muscle atrophy.

After spaceflight (14 days, cosmos 2044), vastus intermedius and lateral gastrocnemius rat muscles exhibited significant decreases in skeletal α-actin mRNA

(25 and 36%, respectively). At the protein level, spaceflight causes skeletal muscle atrophy with a higher loss of actin thin filaments than myosin thick filaments. (Fitts et al., 2000). Thus, the thick filament density remains unchanged whereas the thin filament density also decreases after a 17-day spaceflight (Riley et al., 2002) as after a 17-day bed rest (Riley et al., 1998). The presence of fewer thin filaments increases thick to thin filament spacing. This causes earlier cross-bridge detachment and faster cycling and may explain the increase in the maximal velocity of shortening in type I and IIa fibers reported after a period of weight-lessness. Thus, fiber shortening velocity appears inversely related to thin filament density.

In hindlimb unloading conditions, α-actin mRNA was reduced in soleus (Howard et al., 1989) after 1 (37%), 3 (28%) and 7 (59%) days of HU and in gastrocnemius after 3 (44%) and 7 (41%) days. No change was observed in EDL. Similar decreases, although less important, were noted after 7 days of HU in the lateral gastrocnemius (-33%) and the soleus (-29%) (Babij and Booth, 1988). At protein level, a 17% decrease of actin expression was observed in the soleus only after 6 weeks of HU and not in the EDL (Chopard et al., 2001).

During immobilization, an early decline in actin synthesis was described in the gastrocnemius muscle. It was attributed, however, to an alteration in the translation of α-actin specific mRNA, since the decrease in mRNA begins only after 7 days and was not observed after 6 or 72 hours (Watson et al., 1984). In soleus muscle, denervation was found to induce not only a decrease in α-actin mRNA (Babij and Booth, 1988), but also changes in F-actin conformation; a decrease of the flexibility of the C-terminus of the actin polypeptide might explain differences in the actin-myosin interaction between the normal and the denervated muscle. The changes are induced by the lack of connection with motoneurons rather than by the muscle inactivity (Szczepanowska et al., 1998).

Finally, in the opposite condition, i.e., with increased loading of human vastus lateralis muscle due to heavy leg resistance exercise sufficient to induce myofiber hypertrophy, α-actin expression remains unchanged (Bamman et al., 2004).

2.4   Structure and Function of Troponin C Isoforms

Contraction of the skeletal muscle in vivo is elicited by a transient rise in the cytosolic calcium level, which is detected by the calcium sensor protein Troponin C (TnC). TnC is an acidic protein (18 kDa, see Parmacek and Leiden, 1991) belonging to the Calcium Binding Protein family. It is a dump bell shaped protein, composed of two globular heads, each including two EF hand Ca^{2+} binding sites, linked by a central α - Helix (see for a review Tobacman, 1996). The two carboxy-terminal sites III and IV keep the TnC bound to the thin filament, bind Ca^{2+} with a high affinity ($Ka \approx 10^7 \ M^{-1}$) and low specificity such that they bind Mg^{2+} with a binding constant of $\approx 10^3 \ M^{-1}$ and are occupied by Mg^{2+} under physiological conditions in relaxed muscle (Zot and Potter, 1982).

The two amino-terminal sites I and II bind Ca^{2+} specifically with a relatively low affinity ($Ka \approx 3.10^5 M^{-1}$) and regulate the muscular switch (Potter and Gergely, 1975). Stability of the TnC molecule as its calcium affinity are modulated by the degree of hydration of the protein (Suarez et al., 2003). There is no strict independence between regulatory (I and II) and structural (III and IV) sites.

There are two TnC isoforms with tissue dependent expression (Van Eerd and Takahashi, 1975). A slow isoform is expressed in slow skeletal muscles (TnCs) and in the myocardium (Wilkinson, 1980) and has an inactive site I (Burtnick and Kay, 1977; Leavis and Kraft, 1978) due to an insertion (position 28) and a double substitution of critical amino acids (Leu 29 – Asp 30; Ala 31 – Asp 32). In contrast, the two regulatory sites are functional in the fast isoform (TnCf), which is expressed in fast skeletal muscles. Slow and fast isoforms have about 70% homology, the main differences being in the forty-first amino acids.

Since the introduction of the Herzberg-Moult-James model, based on studies of fast TnC, it has been established that calcium fixation on regulatory sites induces conformational changes at the origin of muscular contraction (Herzberg et al., 1986). From Nuclear Magnetic Resonance data, it has been shown that Ca^{2+} fixation to site II which has a higher affinity primarily occurs. This binding induces minor but essential conformational changes (Spyracopoulos et al., 1997), which trigger Ca^{2+} binding to site I. This induces a large structural opening of the N-lobe (Sia et al., 1997). Therefore, in slow fibers, which express only TnCs isoform, i.e., with non-functional site I, Ca^{2+} fixation on site II leaves the amino-terminal lobe partially closed. Mutants with active site I and inactive site II do not function in slow fibers (Gulati et al., 1991; Sweeney et al., 1990) and site I seems to be only a modulator of site II activity (Putkey et al., 1989).

Upon Ca^{2+} activation, the movement of the α-helices of the amino-terminal EF hand domains leads to an open configuration of the TnC molecule with an exposition of a hydrophobic sequence previously not accessible due to its closed conformation. The new conformation of the Ca^{2+} saturated amino-terminal domain presents a high TnI affinity. Thus, the peptide 115-131 of the skeletal TnI, which corresponds to its second binding site to TnC (Tripet et al., 1997), induces an opening of the amino-terminal domain of TnCs similar to that produced by Ca^{2+} binding to TnCf isoform. The residue 29 of the TnC molecule, close to the Ca^{2+} binding site I, should strongly modulate the Ca^{2+} affinity of site I (Chandra et al., 1994; Pearlstone et al., 1992; Valencia et al., 2003). This result could be related to the role of this site in the hydrophobic sequence exposition when TnC shifts from its closed state to an opened state due to Ca^{2+} binding (Valencia et al., 2003).

The functional diversity between slow and fast TnC acquires special interest when pharmacological tools are used. Indeed, TnC represents a major target for compounds called "Ca^{2+} sensitizers" which enhance the Ca^{2+} responsiveness of the contractile system. Although those compounds are more often used as cardioactive drugs (Solaro et al., 1986), bepridil (BPD) was tested in skeletal muscle to check out a differential action on the two structurally and functionally distinct TnC isoforms. BPD is a polycyclic molecule that can attach hydrophobic parts of TnC buried in close conformation ("apo" state = no binding of Ca^{2+}) but accessible to solvent upon Ca^{2+} fixation. As a result, BPD is believed to stabilize the conformational changes, causing an increase in apparent affinity by decreasing the Ca^{2+} off-rate (Mac Lachlan et al., 1990). Slow skeletal fibers were more responsive than the fast ones and showed a BPD response dependent on the Ca^{2+} concentration, the higher tension enhancement being obtained at low activation levels. On the contrary, the sensitizing effect of the drug in fast fibers was quite independent on the Ca^{2+} concentration (fig. 8) (Kischel et al., 1999). This differential effect of BPD in slow and fast fibers was specifically correlated to the TnC isoforms, as demonstrated by extraction-replacement experiments of TnC or use of mutant (Kischel et al., 2000). Indeed, fast tibialis fibers reconstituted with TnCs exhibited a higher BPD reactivity and tension / pCa curves became similar to those obtained with slow fibers; in the same way, slow soleus fibers reincorporated with TnCf displayed a typically fast response to PBD and tension / pCa curves more similar to those of fast fibers. These results confirmed i) the involvement of TnC in the tension / pCa relationships and ii) the role of TnC isoforms in the differential effect of BPD in slow and fast fibers. Study with the VG2 mutant, lacking the regulatory site II, substituted for endogenous fast TnC of tibialis fibers, permitted one to define the role of site I. Ca^{2+} activation of these reconstituted fibers without site II was more difficult and BPD induces a Ca^{2+} affinity increase without altering the intrinsic cooperativity. This suggests that site I appears essential for keeping the cooperativity intact and, when only site II is active, the cooperativity is greatly affected.

Another mean to functionally differentiate slow and fast fibers is based on their differential activation with another divalent cation: strontium (Sr^{2+}). Although the strontium ion does not play a direct role in vertebrate skeletal muscle contraction in vivo, it has been used frequently in experiments as an activator of contractile

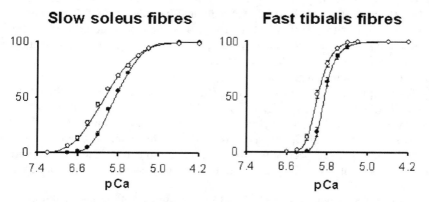

Figure 8. Differential effects of the Ca2+ sensitizer, bepridil (BPD) on Ca2+ activation of skinned slow (from soleus) and fast (from tibialis anterior) fibers, as expressed by the Tension /pCa relationship. BPD was used at a concentration of 100 μ M (Reprinted with permission from Kischel et al., 1999 © British Pharmacological Society)

processes in reconstituted actomyosin, myofibrils and skinned muscle fibers. Sr^{2+} activation properties constitute valuable indicators for identifying slow and fast-twitch fibers. Thus, a much greater affinity for Sr^{2+} has been demonstrated in slow fibers compared to fast fibers (Kerrick et al., 1980), the TnC isoform being the main determinant of such differential Sr^{2+} sensitivity (Babu et al., 1987; Hoar et al., 1988; Kischel et al., 2000; Moss et al., 1986).

2.5   Regulation of Muscle Tension by TnC Isoforms

The initial step in tension generation is Ca^{2+} binding to the low-affinity Ca^{2+} specific sites (I and II) of the TnC subunit (Farah and Reinach, 1995; Grabarek et al., 1992). The kinetics of Ca^{2+} binding to the high-affinity C-terminal sites is limited by Ca^{2+} / Mg^{2+} exchange, which is too slow to account for the kinetics of Ca^{2+} dependent tension development. As mentioned above, the role of sites I and II in muscle contraction have been demonstrated especially by extraction / replacement experiments, which clearly showed that TnC isoforms contribute to differences in Ca^{2+} activation properties. In chemically skinned fibers, where the contractile proteins are directly exposed to the activating Ca^{2+}, the tension / pCa relationship of fast fibers reconstituted with a TnC slow (or cardiac) become similar to that found for slow fibers (Moss et al., 1986). The reverse has also been established for replacement of TnC fast in slow fibers, which thus acquire Ca^{2+} activation properties similar to fast fibers (Babu et al., 1987; Gulati et al., 1998).

When compared to slow fibers, fast fibers exhibit a higher Ca^{2+} activation threshold (lower values of pCa_{thr}), a decrease in Ca^{2+} affinity (lower value of pCa_{50}) and a higher cooperativity between the proteins of the thin filament (larger values of the Hill coefficient n_H). Thus, differences in Ca^{2+} sensitivity of these parameters may be related to different types of fibers and among different proteins to properties of the TnC (Danieli-Betto et al., 1990; Bottinelli and Reggiani, 2000). More specifically, the position of the tension / pCa curve is assumed to be controlled by the Ca^{2+} binding properties and the values of pCa_{50} are supposed to be determined by TnC properties (Gulati et al., 1998; Kischel et al., 1999).

In rat soleus muscle, three types of fibers can be identified on the basis of TnC isoform expression (fig. 9): slow fibers (S) which express TnCs alone, hybrid fibers which co-express TnCs and TnCf divided into hybrid slow or HS (predominant expression of TnCs) and hybrid fast or HF (predominant expression of TnCf). The proportions of S, HS and HF fibers in the control soleus muscle were 74, 10 and 16% respectively. No fibers expressing only TnCf are found in control soleus but they are present in fast muscles such as tibialis anterior (Kischel et al., 2001a).

The S and HS fibers have similar Ca^{2+} and Sr^{2+} affinities and similar sensitivity to BPD and their tension / pCa curves (Fig. 10, upper and middle curves) are identical. This indicates that in HS fibers, the presence of a minor amount of TnCf does not influence the Ca^{2+}-activation characteristics, which are identical to those found in

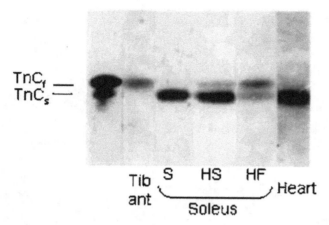

TnC_f —
TnC_s —

Tib
ant
S   HS   HF
Soleus
Heart

Figure 9. Expression profiles of TnC in single fibers. Immuno-detection of TnC isoforms (only the region of TnC is shown) in fast tibialis anterior fiber, soleus slow (S) hybrid slow (HS) and hybrid fast (HF) fibers and myocardium from rats. The migration of TnCf and TnCs isoforms is confirmed using TnC extracts (first left lane). (Reprinted from Kischel et al., 2001a © American Physiological Society)

pure slow fibers (with TnCs alone). Therefore, functional properties related to TnC are dependent on the TnC isoform predominantly expressed (Kischel et al., 2001a). Accordingly, the HF soleus fibers (TnCf > TnCs) exhibit Ca^{2+} activation properties (fig. 10, lower curve) similar to those of pure fast fibers, which express only TnCf as found in tibialis anterior muscle (see fig. 8).

TnC isoforms also control the way by which changes in Ca^{2+} concentration affect the rate of tension redevelopment (ktr) after a fast shortening-re-lengthening maneuver (Regnier et al., 1996, 1999). Increase (or decrease) in Ca^{2+} affinity of TnC have been described as being dependent on increased (or decreased) rate of Ca^{2+} dissociation from TnC. Moreover, a strong coupling between the Ca^{2+} regulatory sites of TnC and cross-bridge interactions with the thin filament has been demonstrated (Güth and Potter, 1987). The affinity of TnC for Ca^{2+} increases the cross-bridge attachment (Bremel and Weber, 1972) and TnC is sensitive to the structural changes of myosin during the force generation (Li and Fajer, 1998).

Symmetrically, strong force-generating cross bridges induce changes in the amino-terminal configuration of TnC (Zot and Potter, 1989) and regulate the off rate of Ca^{2+} from TnC (Wang and Kerrick, 2002). Rigor cross-bridges also increase Ca^{2+} affinity of TnC (Bremel and Weber, 1972; Güth and Potter, 1987; Hofmann and Fuchs, 1987). Changes in Ca^{2+} sensitivity also result from intrinsic modifications in the cross-bridge kinetic rates (Robinson et al., 2002). For instance, by a decrease in the dissociation rate (parameter "g" in Huxley's model), force generating cross-bridges might retain Ca^{2+} on the thin filament and thus contribute to an apparent increase in Ca^{2+} sensitivity.

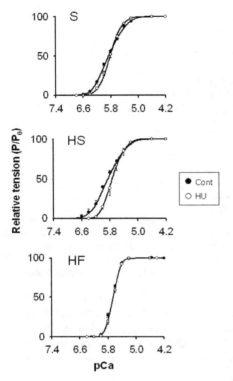

Figure 10. Tension / pCa relationships of slow (S), hybrid slow (HS) and hybrid fast (HF) soleus in control (filled symbols) and HU conditions (open symbols). Data points are fitted with the Hill equation. (Redrawn and simplified with permission from Kischel et al., 2001a, © American Physiological Society)

2.6   Variations in TnC Isoform Expression

There are many reports that changes in neuron-muscular activity can trigger variations in TnC isoform expression. Denervation of fast twitch rabbit muscles induces a fast-to-slow transition which can be assessed by the changes in MHC isoform expression and is also accompanied by similar fast-to-slow changes of the isoforms of the three troponin subunits (TnC, TnT, TnI, Leeuw et al., 1994). The hindlimb suspension model (designated as HU for hindlimb unloading) (Morey et al., 1979; Musacchia et al., 1980) has three main effects on limb muscles. The first effect is a pronounced atrophy of antigravitational muscles (Holy and Mounier, 1990; Stevens et al., 1993). The loss in mass is associated to a decrease in force and both the changes become greater with the duration of unloading. However, the decrease in force goes beyond the values expected on the basis of the loss in muscle mass and proteins (Gardetto et al., 1989; Kischel et al., 2001a; Stevens et al., 1990, 1993, 1996). The second effect is a slow-to-fast shift in the kinetics of the contractile system related to changes in MHC expression (see above).

The slow-to-fast shift was also found in the expression of the regulatory proteins, among them TnC in the soleus muscle. TnCs isoform which represented about 92% of total TnC in control soleus muscle was partially replaced by TnCf during HU. The first changes were detectable only after 15 days and the increase in TnCf remained moderate, so that TnCs was still the predominant isoform ($\approx$ 80%), even after 28 days of HU (fig. 11A) (Stevens et al., 2002). Therefore, relatively small changes of TnC isoforms at the protein level were found in the slow-to-fast transition due to unloading. More pronounced changes were described at the mRNA level after a 9-day spaceflight (Esser and Hardemann, 1995).

As stated above, three groups of fibers can be identified in control rat soleus, S, HS and HF, representing 74, 10 and 16% respectively, and change their distribution after HU becoming 45, 10, 45% respectively. Fibers expressing TnCf alone cannot be found after HU as in control soleus, confirming the limits of the transitions in the TnC expression in soleus. Thus, the about twofold higher TnCf expression after HU in the whole muscle might be attributed to the increase in the number of hybrid, especially HF fibers. It is possible that the transitions of TnC isoforms appear limited because of their kinetics slower than those of other myofibrillar proteins. Actually, there are no data relative to changes in TnC expression for HU longer than 28 days.

The functional significance of the changes in TnC could be analyzed in terms of Ca^{2+} activation. The changes in Ca^{2+} activation constitute the third main effect of muscle unloading, reported after HU in rats as well as after spaceflight in rats or monkeys or after bed rest in humans. In these conditions, the tension / pCa relationship of slow fibers which remained slow was significantly shifted to the right: the pCa_{thr} and pCa_{50} parameters were lowered, whereas the cooperativity (n_H) increased. These modifications were described after 2 weeks of HU in rats (Kischel et al., 2001b), after a 14-day spaceflights in monkeys (Kischel et al., 2001a) and in one astronaut who presented a large fiber atrophy (quoted by Fitts et al., 2000). Kischel et al. (2001b) demonstrated that the TnC-dependent properties (Sr^{2+} affinity and BPD response) of those slow HU fibers were identical to those of control fibers in spite of the changes in the position and the shape of the tension / pCa curve. Therefore, HU induced changes in Ca^{2+} activation of slow soleus fibers that could not be attributed to the TnC isoforms.

For the hybrid slow (HS fibers), the tension / pCa curve obtained after HU was intermediate between the curve of slow fibers and that of HF unloaded fibers. However, the TnC-dependent properties (Sr^{2+} and BPD) were close to those of the slow fibers. Hybrid fast (HF) fibers had Ca^{2+} activation identical to that of control fast fibers: their tension / pCa curves, Sr^{2+} affinity and BPD responsiveness did not differ. Therefore, the effects of HU consisted in a transition from a slow type (TnCs only) to a hybrid fast type (TnCf coexpressed with TnCs in a predominant proportion), the decrease in Ca^{2+} sensitivity being related to the increased expression of fast TnC isoform. The changes in Ca^{2+} activated properties in slow and hybrid slow fibers after HU, with TnCs expressed alone or predominantly, might therefore be attributed to changes in other regulatory proteins. Moreover, unloading induced

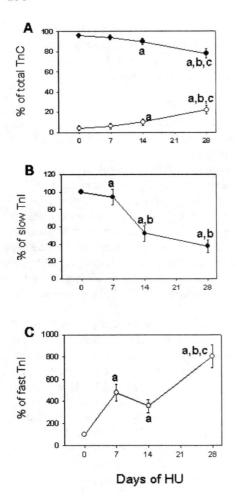

Figure 11. Changes in the proportion of slow (•) and fast (o) isoforms of TnC (A) and TnI (B and C) subunits in rat soleus muscle after 7, 15 and 28 days of hindlimb unloading (HU).TnCs and TnCf were recognized by the same antibody. Both isoforms were studied on the same gels and their proportions were determined as percentages of total TnC. On the contrary, TnIs and TnIf were identified by different antibodies and their respective relative amounts were related to actin signal. a: Significantly different from control soleus (0-day HU). b: Significantly different from 7-day HU soleus. c: Significantly different from 15-day HU soleus. (Reproduced with permission from Stevens et al., 2002, © American Physiological Society)

in all transformed fibers an increased cooperativity (higher n_H value) between the proteins of the thin filament, which can, at least in HF fibers, partly be attributed to the expression of TnCf. This increase has been described as a discriminating factor in monkey triceps muscle between restraint and 0G conditions during a spaceflight

(Kischel et al., 2001b), although Fitts and coworkers, (2000) have shown that no change in cooperativity is determined by muscle unloading.

Changes in TnC isoform expression associated with increases in neuromuscular activity or increased load were observed after chronic low frequency stimulation (CLFS) and in hypergravity conditions, respectively.

CLFS applied for 80 days to the rabbit EDL muscle induced a remarkable increase of TnCs expression from 2% in the control EDL to 80% in the electro-stimulated EDL (Härtner and Pette, 1990). In rabbit tibialis anterior, the progressive replacement of the fast with the slow isoform took place at a nearly constant rate but the transition was more moderate than on EDL and the fast isoform still represented 55% after 60 days of CLFS (Leeuw and Pette, 1993).

Centrifugation used to produce an artificially induced gravity (hypergravity 2G, HG) results in an increased load on muscles and causes a hypertrophic response. Increases in muscle mass (absolute and relative to body mass) were described in the soleus, whereas no change or slight decreases were found in fast muscles of rats exposed to HG (Chi et al., 1998; Roy et al., 1996; Stevens et al., 2003; Vasques et al., 1998). In the soleus, TnC showed a slow-to-fast transition, the level of TnCf being increased more after HG than after microgravity. The functional consequence of this increased expression of TnCf could be found in the rightward shift in the tension / pCa relationship (i.e., decrease in Ca^{2+} sensitivity) and to the large increase in the steepness of the curve. The latter changes, however, could be also attributed to other proteins involved in the cooperative activation along the thin filament and among them tropomyosin, since the expression of the TnT subunit was unchanged after 19 days of 2G centrifugation. Thus, it is clear that there is a continuum for muscle adaptation from 0G to 2G with respect to muscle mass but not with respect to regulatory and contractile proteins, since similar changes can be observed at 0G or 2G. Interestingly, in rats conceived, born and reared at 2G, the higher gravity has an opposite effect, as it apparently interferes with muscle development causing an increase of Ca^{2+} affinity and a slow-to-slower TnC isoform transition (Bozzo et al., 2004; Picquet et al., 2002).

2.7   Structure and Inhibitory Function of Troponin I

Troponin I (TnI) is a basic globular protein composed of 180 amino acids (21 kDa, see Perry, 1999), which has permanent Ca^{2+} independent interactions between its N-terminal region and TnC (Sheng et al., 1992) and TnT (Chong and Hodges, 1982) (fig.12). These interactions are responsible for the structural integrity of the troponin complex. In addition, TnI molecule has Ca^{2+}-dependent interactions between its C-terminal region and TnC, actin and tropomyosin.

TnI binds with actin and thus inhibits the ATPase activity of the acto-myosin complex Greaser and Gergely, (1971). This inhibitory activity is potentiated by Tropomyosin (Tm) (Perry et al., 1972). The peptide 96-115 contains the minimal sequence 104-115, called Ip (Inhibitory peptide) necessary for the inhibition of the contractile process (Syska et al., 1976; Talbot and Hodges, 1981). Upon binding

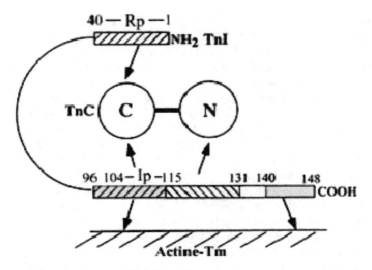

Figure 12. Schematic representation of the interactions of TnI residues (1 to 40 and 96 to 148) with TnC and Actin-Tm. The interactions between TnI regions and other proteins are indicated by arrows. Regions of TnI proposed to interact with the N and C domains of TnC are hatched. Regions of TnI that bind the Actin-Tm filament and are responsible for the full inhibitory activity are shaded. (Reproduced with permission from Tripet et al., 1997, © Elsevier B.V.)

to actin, Ip shows an inhibitory activity which ranges between 45 and 75% of that of the whole protein. Residues 135-181 of TnI are also involved in the interaction with actin and Tm (Takeda et al., 1997) and residues 96-148 potentiates the effects of Ip, thus producing an inhibitory activity similar to that of the intact TnI molecule (Tripet et al., 1997). The residues 140-148 correspond to the second binding site to [actin-Tm] complex. Thus, in absence of Ca^{2+}, the inhibitory function of TnI is due to binding of its sequences 104-115 and 140-148 to the actin filament.

The amino-terminal part of TnI is not necessary for TnI inhibitory function but has an important function, as residues 1-40 (Rp peptide) bind to the C-terminal region of TnC, in a Ca^{2+}-independent manner (Farah and Reinach, 1995) and are responsible for the fixation of TnC in the ternary structure (Sheng et al., 1992; Farah et al., 1994; Potter et al., 1995).

A specific role of TnI in the regulation of Ca^{2+} activated properties has been described in crustacean muscles in a recent paper, which showed that differences in TnI expression in two fast muscles coincide with differences in Ca^{2+}-sensitivity (Koenders, 2004). It is, however, well known that the interactions between TnC and TnI are of central importance to the Ca^{2+} regulation of muscle contraction. In the presence of Ca^{2+}, the N-terminal domain of TnC is opened and its hydrophobic residues are exposed. Peptide 114-125 in the C-terminal region of TnI binds the N-terminal region of TnC (Farah et al., 1994; Ngai et al., 2001). The formation of the complex TnC-TnI modifies the interactions between TnI and the thin filament, such that TnI becomes dissociated from actin (Tao et al., 1990). Moreover, in the presence of Ca^{2+}, the peptide 48-81 of TnI interacts with TnT, and peptide 90-113 with the central-helix of TnC (Li et al., 2001; Luo et al., 2000). Together with TnT the N-terminal region of TnI regulates acto-myosin ATPase activity (Van Eyk et al., 1997). Finally, TnI has been proposed to interact with Tm in relation with a periodic distribution of TnI along Tm-actin filaments (Ohtsuki and Shiraishi, 2002). On the whole, the interactions of TnI with actin, TnC and TnT undergo a re-organization when the free calcium concentration in the muscle fiber rises, resulting in the loss of the inhibitory activity.

2.8   Polymorphism of TnI and Variations in TnI Isoform Expression

Three TnI isoforms are expressed in striated muscles: a slow isoform (TnIs) and a fast isoform (TnIf) are present in slow and fast skeletal muscles, respectively, whereas the expression of the third or cardiac isoform (TnIc) is restricted to myocardium (Härtner and Pette, 1990). The three isoforms have a high homology in their C-terminal domain, whereas the N-terminal part is more variable. A highly conserved region corresponds to the Ip segment (residues 105-114) and the few amino acids that are different in the three isoforms of TnI are considered responsible for functional diversity (Ball et al., 1994). TnIc is characterized by a long amino-terminal extension, which can be phosphorylated on two serines after β-adrenergic stimulation by cAMP-dependent protein - kinase A.

Loading conditions have been shown to induce changes in TnI expression both in case of unloading during real or simulated microgravity and in case of increased load in hypergravity conditions.

During HU (Stevens et al., 2002), TnIs isoform expression in the soleus muscle was reduced, moderately after 7 days and much more after 15 days (by ∼50%) and 28 days (by 60%). Conversely, TnIf was ≈ 4.5-fold elevated as soon as after 7 days of HU and continued to increase with prolonged HU up to 8-fold elevated after 28 days. This increase in TnIf is in agreement with previous qualitative observations (Campione et al., 1993). The moderate decay of TnIs during the initial 7-day period of HU coincides with a steep rise in TnIf (see fig. 5B and C). The discrepancy suggests that downregulation of TnIs and upregulation of TnIf are independently regulated. It has been proposed that the steep rise of TnIf is neurally induced and coincides with a change in EMG activity which has been reported to turn from a tonic in a phasic pattern after 7 days of HU (Alford et al., 1987; Canu and Falempin, 1996). At the transcript level, Criswell et al. (1998) reports a TnIs mRNA downregulation after 7 days of HU, whereas TnIf mRNA has been found to increase after 9 days of spaceflight (Esser and Hardemann, 1995). Criswell et al. (1998) underlined the fact that unloading involves a large downregulation of the TnIs gene, while expression of the fast gene is simply preserved at control values. Thus, the TnIs gene seems to be very sensitive to the loading condition of the muscle. In contrast, the TnIf gene may not only be independent of loading, but may also be maintained near normal expression during muscle atrophy.

In HG conditions (2G centrifugation), the TnIs expression was reduced whereas TnIf was increased, in the soleus muscle of adult rats, i.e., the isoform transitions were in the same slow-to-fast direction as after unloading (Stevens et al., 2003). However, TnIf increased less after hypergravity ($\times 1.5$) than after unloading ($\times 4$). As described above for TnC, the slow-to-fast transition was not observed in rats submitted to hypergravity from conception to adult stage (Bozzo et al., 2004). In rats born and maintained in HG, the proportion of TnIs in the soleus was even higher than in control rats, i.e., a slow-to-slow TnI isoform transition occurred, opposite to that found at 0G. This effect was attributed to a change in motor innervation (Picquet et al., 2002), since HG induces an increase in EMG activity (Leterme and Falempin, 1998). During development in hypergravity, an increase in the recruitment of new motor neurons, with a shift toward higher-amplitude units, was observed and this likely influenced protein expression. The gene coding for TnIs is sensitive to changes in load (see above), shows a developmental regulation (Zhu et al., 1995) and includes a nerve-responsive promoter (Lewitt et al., 1995). Indeed, a selective nerve-dependent activation and repression of TnI genes expression has been described (Eldridge et al., 1984) as denervation causes a down-regulation of the either TnIs or TnIf, in slow or fast muscles, respectively, whereas specific patterns of electrical stimulation can differentially activate the expression of either TnIs or TnIf (Buonanno and Rosenthal, 1996; Calvo et al., 1996). On the whole, the transition of TnI isoform expression seems to be more pronounced at the mRNA level than at the protein level, as observed by Härtner and Pette, 1990. This suggests that transcriptional regulation is the main mechanism that control the TnI isoform expression in response to various factors, in the first place motor innervation. At the protein level, changes in the distribution of fast and slow TnI isoforms become manifest only after prolonged low-frequency stimulation as, for example, in rabbit tibialis anterior TnIs did not exceed 10% of total TnI after 35 days and reached 40% in 60-day stimulated muscles (Leeuw and Pette, 1993). Fast-to-slow transitions for the TnI isoforms remain less complete than those for TnT and TnC.

Skeletal muscle TnI isoforms have been used as an initial, specific marker of exercise-induced muscle damage (Sorichter et al., 1997) when compared to different other proteins including MHC. Cardiac TnI, which has earlier been described as a highly cardiac-specific marker never expressed in skeletal muscles, has been found in the skeletal muscles of patients with Duchenne muscular dystrophy. In these patients, the TnIc mRNA is expressed at various levels and translated into protein, which at later stages of the disease can be detected in the serum. The mRNA of TnIc has been also found in the skeletal muscles of patients with other myopathies, suggesting that TnIc might be considered as a sensitive and reliable marker for the detection of muscle diseases (Messner et al., 2000).

2.9   Structure and Polymorphism of Troponin T

Troponin T (TnT) is a polypeptide with a molecular weight varying between 31–36 kDa, composed of 250 to 300 amino acids. Different genes and differential RNA splicing generate multiple isoforms, the diversity being primarily localized in the N-terminal region. Three genes code for slow skeletal or TnTs (Gahlmann et al., 1987), fast skeletal or TnTf (Breitbart and Nadal-Ginard, 1986) and cardiac or TnTc (Cooper and Ordahl, 1984) isoforms.

For a long time, only two isoforms were described in adult rat and rabbit muscles (Härtner et al., 1989; Sabry et al., 1991a,b). Later, two isoforms of slow TnT (TnTs) were discovered in murine muscles, resulting from an alternative splicing of the exons 5 and 6 in the N-terminal region of the slow TnT gene (Jin et al., 1998). More recently, Bastide et al. (2002) have demonstrated the existence of a third TnTs isoform in adult rat soleus muscle fibers. Yonemura et al. (2000) have also shown the presence of 6 transcripts of TnTs in chicken muscles, suggesting a larger heterogeneity in TnTs isoforms than previously described.

The TnT isoform diversity is best illustrated by fast skeletal TnT (TnTf) for which more than seven (adult mammals) exons encoding the N-terminal region can be alternatively spliced to generate a large number of isoforms (Ogut and Jin, 2000). Thirteen isoforms of TnTf have been identified in mice muscles (Wang and Jin, 1998) and 11 in chicken muscles (Ogut and Jin, 1998). In rabbit muscle, five fast isoforms of TnT have been separated and called TnT1f to TnT5f (Härtner et al., 1989; Sabry and Dhoot, 1991a,b). Actually, the TnT-fast gene can potentially produce up to 64 variants due to the alternative splicing of five exons at the 5' end and one exon at the 3' end. As shown by Briggs and Schachat (1996), only a limited number of splicing patterns occurs in mammalian muscles: this gives origin to a limited number of isoforms: five or six in fast rabbit muscles (EDL, see Härtner et al., 1989; Sabry and Dhoot, 1991a) and four or five in rat (Briggs and Schachat, 1996). These isoforms are generally identified with numbers increasing with decreasing molecular weight: TnT1f, TnT2f, TnT3f, TnT4f. Preferential associations between TnT1f and MHC-2X, TnT3f and MHC-IIA, TnT4f and MHC-IIB have been observed in rat single fibers (Galler et al., 1997). Electrophoretical and immunochemical studies have revealed the presence of four major variants of TnT fast in human limb skeletal muscles (Anderson et al., 1991; Sabry and Dhoot, 1991a).

Isoforms derived from TnT cardiac are expressed in atrial and ventricular myocardium. However, there are indications that two isoforms derived from TnT-cardiac are present in fetal skeletal muscles of rats (Saggin et al., 1990) and humans (Anderson et al., 1991). Re-expression of cardiac TnT isoforms in skeletal muscles of patients affected by renal disease has been also demonstrated (McLaurin et al., 1997).

The expression of TnT isoforms varies with muscle development (Yonemura et al., 2002). The N-terminal domains of embryonic and neonatal TnT isoforms have more acidic isoelectric points when compared to adult TnT more basic isoelectric points. Moreover, cardiac and slow skeletal muscle TnT have more acidic N-terminal domains than fast skeletal TnT ones. This reflects the role of TnT isoforms in the functional adaptation of developing striated muscles and the role of the charge of the N-terminal region as an important parameter influencing the regulatory properties of TnT (Jin et al., 2000).

Troponin T is part of the troponin complex and binds to TnC, TnI, also Tm. Binding to Tm occurs in two different regions. One of the two binding sites for

Tm (TnT2) is close to the binding sites for TnC and TnI and is calcium sensitive, whereas the other (TnT1) is calcium insensitive (see for reviews Perry, 1998 and Solaro and Rarick, 1998).

Attempts to investigate the role of TnT isoforms have been made by comparing pCa-tension curves of fibers that contain the same TnC and TnI and various TnT isoforms (Greig et al., 1994, Schachat et al., 1987; Tobacman and Lee, 1987). However, because TnT isoforms also vary in a coordinated fashion with Tm isoforms, the relative role of TnT and Tm isoforms remains to be established. In fast fibers of the rabbit, the highest values of pCa_{50} and n might be associated with the presence of TnT-2f and TM-a-fast. The variations in n have been related with the diversity between fast isoforms of TnT in the region corresponding to exons 4-6-7 close to the N-terminus (Briggs et al., 1987), namely, the portion of the TnT molecule that binds to C terminal region of Tm molecules. In a reconstituted system, the pCa-ATPase curve has been shifted by 0.15 by exchanging different cardiac TnT isoforms (Tobacman and Lee, 1987).

2.10   Structure and Polymorphism of Tropomyosin

Tropomyosin (Tm) is an elongated dimeric protein, 40 nm long, with molecular weight of about 35kDa and consisting of two-stranded α-helical coiled-coils and arranged in parallel orientation. Tropomyosin molecules are assembled in a head-to-tail fashion (Flicker et al., 1982; Smillie, 1979), to form a filament that binds to the actin double helix and the troponin complex. The coiled-coil structure is based on a repeated pattern of seven amino acids with hydrophobic residues at the first and fourth positions and is highly conserved in all Tm isoforms found in eukaryotic organisms from yeast to man. Each Tm molecule (284 amino acids) spans seven actin monomers (Cummins and Perry, 1973; Mac Lachlan and Stewart, 1976; Phillips et al., 1986). Together, these periodic features promote the winding of a Tm filament around the actin helix. The C-terminal region of the molecule, including the head-to-tail overlap, has been widely investigated. A fragment of 22 residues (263-284) does not form the two-stranded α-helical coiled-coil as does the rest of the molecule, but here the two α-helices splay apart and are stabilized by association with another Tm molecule (Li et al., 2002). Each C-terminal domain of Tm has a calcium-independent binding site of the Tn complex (Hammell and Hitchcock-De Gregori, 1996) which constitutes the permanent interaction of the Tn complex to the thin filament (Hinkle et al., 1999). In striated muscles, this region has a specific recognition site for TnT which contributes to maintaining the binding between two Tm molecules (Maytum et al., 1999; Li et al., 2002). Formation of a ternary complex including TnT fragment that binds to the tropomyosin overlap region takes place only when the overlap complex sequences are those found in striated muscle tropomyosins. A second site of interaction between TnT and Tm exists and is located in the proximity of Cys-190 (region 175-190 of Tm). This second interaction between Tm and TnT is calcium-dependent.

Tropomyosins are encoded by a four-member multigene family: TPM1, TPM2, TPM3, TPM4 (Lees-Miller and Helfman, 1991). The four genes generate a multitude of tissue and developmental specific isoforms through alternative exon splicing, the use of different promoters and differential 3' end processing. Three of the four genes are expressed in striated muscles: TPM1, which codes Tm-α-fast, TPM3, which codes for Tm-α-slow and TPM2, which codes for Tm-β.

The two α isoforms of Tm are smaller than Tm-β isoforms ($\sim$34 and 36 kDa, respectively) and the diversity is mainly localized in the C-terminal domain, i.e., the region of TnT interaction. In spite of the diversity, they can assemble as $\alpha\alpha$- and $\beta\beta$-homodimers or $\alpha\beta$-heterodimers as shown by electrophoretic analyses. The expression of Tm isoforms is coordinated with those MHC isoforms and Tn isoforms and, as a first approximation, slow fibers express predominantly Tm-α-slow and Tm-β, whereas fast fibers generally express Tm-α-fast along with Tm-β (Salviati et al., 1982; Salviati et al., 1983; Schachat et al., 1985; Danieli-Betto et al., 1990; Bottinelli et al., 1998).

Although the striated muscle Tm isoforms are highly homologous (Tm-α-fast is 93% identical to Tm-α-slow and 86% toTm-β are 86% at the amino acid level), their unique expression patterns suggest that each isoform may have functional differences. For example, in striated skeletal muscles, Tm-α-fast is predominantly expressed in fast-twitch skeletal and cardiac muscles, whereas slow-twitch skeletal muscle has a higher amount of Tm-β (Muthuchamy et al., 1995, 1997). Further diversity in striated muscle Tm is created by phosphorylation of serine 283 by a tropomyosin-kinase and this post-translational modification is supposed to affect head-to-tail polymerization to Tm (Heeley et al., 1989). Diversity among Tm isoforms has been suggested to confer different contractile parameters to muscle fibers (Muthuchamy et al., 1995).

2.11   Role of Tropomyosin in the Regulation of Muscle Contraction

In the resting state, the Tm-troponin complex inhibits the actin-myosin interaction; following an increase in Ca^{2+} concentration in the myofilament space and the binding of Ca^{2+} to TnC, Tm moves away from its blocking position, and goes deeper towards the groove of the actin double helix (see for a review Gordon et al., 2001). Such movement of Tm results in a non-blocking position, which removes the inhibition and allows actin to bind myosin heads leading to muscle contraction. Thus, Tm can be positioned on actin in two states designed "on" and "off", which determine the activation state of the thin filament (Lehman et al., 1994). The relation between ATPase activity of the [Actin-S1 head-Tm] complex and the number of S1 heads is sigmoid indicating that Tm is necessary to produce the cooperativity process within the thin filament (Lehrer and Morris, 1982). The cooperativity is a consequence of the interaction of the Tm dimer with seven actin monomers and of the overlap of the Tm dimers (Lehrer, 1994) and can be measured by the Hill coefficient "n" (i.e., the slope of the tension / pCa relationship). The Hill coefficient is dependent on the expression of Tm isoforms in different muscles

(Schachat et al., 1987). Calcium sensitivity is lower and Hill's slope is significantly higher when Tm-α shows high expressions in fast muscles and when Tm-α is predominant compared to Tm-β.

Some post-translational modifications in the Tm molecule influences the regulation of the contractile mechanism. Thus, N-terminal modifications of Tm affects the affinity for TnT, the highest affinity being achieved when Tm is N-acetylated (Palm et al., 2003). Deletion of the Tm-α amino-terminal regions results in a severe loss of regulatory function through either an inability to bind actin and / or impaired troponin binding. Deletion of the last 11 amino acids in the C-terminal region results in loss of actin affinity and impaired the cooperative activity (Pan et al., 1989). Again at the C terminal end of Tm, two other important charge diversity of Tm-β compared to Tm-α are the substitutions [Ser229Glu] and [His276Asn], which give a more negative charge. This difference in charge appears responsible for the weaker binding of Tm-β to TnT compared to Tm-α (Thomas and Smillie, 1994). Such differences in interaction might contribute to making myofilaments more sensitive to Ca^{2+}. Finally, the different Tm isoforms affect the sarcomeric performance by a modulation of its Ca^{2+} sensitivity as a function of its length (Pieples et al., 2002), the specific carboxyl terminus being a critical determinant for structural and functional properties (Jagatheesan et al., 2003).

2.12   Variations of Tropomyosin isoform Expression

Expression of Tm isoforms has been shown to be affected by neuron-muscular activity. In cross-innervation experiments between rabbit soleus and EDL, the fraction of the β subunit remain constant and the relative proportions of α and two other forms (designed as γ and δ in that study) change as expected from the overall muscle transformation (Heeley et al., 1983).

In denervated rat diaphragm muscle where a loss of slow myofibrillar components occurs, the expression of Tm-β isoform also decreases (Carraro et al., 1981). Transitions between Tm isoforms were also found on denervated neonatal or mature muscles in chicken breast (Obinata et al., 1984).

During the SLS1 flight (9 days), atrophy was associated with a general increase in mRNA of fast isoforms of myofibrillar proteins, including Tm (Esser and Hardemann, 1995). In hindlimb unloading conditions (14 days), we measured a decrease in the rat soleus in the predominant isoform Tm-β by about 10% at the benefit of Tm-α fast, the TPM3 isoform being maintained at its normal level ($\approx 35\%$) (unpublished data). On the contrary, compensatory hypertrophy due to overload of a muscle after elimination of its synergist, leads to a decrease of the α / β subunits ratio both in soleus and plantaris muscles, i.e., with transition of Tm towards a slower type (Kaasik et al., 1998).

Transitions in the expressions of Tm were also described in some pathologies. Myotonia, resulting from a defective functioning of voltage-gated chloride channel is characterized by fiber type transitions (such as reductions in the fiber IIb and MHC2b mRNA levels) and muscle properties in the slow direction. A reduced level of the mRNA encoding Tm-α was described (Jockusch, 1990). Mutations in the α-fast Tm, with other mutations in different contractile proteins, were also described in Familial Hypertrophic Cardiomyopathy (FHC) (Thierfelder et al., 1994). These patients have an Asp175 Asn α-fast Tm mutation that can be studied from biopsies of vastus lateris muscles, since α-fast Tm is expressed in heart and fast skeletal muscles. Thus, in humans it was demonstrated that muscle fibers containing the Tm mutation exhibited an increased Ca^{2+} sensitivity, indicating that FHC might be not associated with a decrease in contractile performance (Bottinelli et al., 1998). Importantly, those data confirmed that myofilament activation by Ca^{2+} and strong cross-bridge formation are influenced by Tm isoforms. Indeed, in transgenic mouse where Tm-α isoform is replaced by Tm-β, the ability of strong cross-bridge binding to activate the thin filament was increased, the transgenic Tm-β myofilaments being more sensitive to Ca^{2+} (Muthuchamy et al., 1999).

3.   MUSCLE PROTEINS AND POST-TRANSLATIONAL MODIFICATIONS

There are many ways a cell can increase the complexity of its proteome along the pathways from genes to proteins. One of these ways, post-translational modifications such as phosphorylation, glycosylation or acetylation, provide additional levels of functional complexity to the cell's proteome. Post-translational modifications can be involved in the modulation of the activity of many proteins and could be involved in the muscle contractile proteins plasticity and adaptation to physiological environment. Thus, changes in the neural pattern of stimulation can result in trans-formations of muscle phenotype and contractile properties via post-translational modifications.

3.1   Phosphorylation (MLC2, tropomyosin and troponins)

Regulatory myosin light chain (RLC or MLC2) is a component of the myosin molecule that contains sites suitable for phosphorylation and is able to modulate myosin-actin interaction in relation to its phosphorylation (Perrie et al., 1973). While MLC2 phosphorylation is a key step to induce the contractile response in smooth muscle, it can only modulate the interaction between contractile proteins in striated muscles (Manning and Stull, 1982). MLC2 phosphorylation increases the force of development at submaximal calcium concentration, conferring to the fibers a higher calcium sensitivity (Persechini et al., 1985; Stephenson and Stephenson, 1993; Sweeney and Stull, 1990; Szczesna et al., 2002). An increase of the rate constant of the cross bridge transition from non force-generating to force-generating states is a likely explanation of the effect of MLC2 phosphory-lation on force development (Sweeney and Stull, 1990). MLC2 exists in several isoforms and encodes by distinct genes. Two isoforms are expressed in mammalian skeletal muscles and correspond to slow MLC2 and fast MLC2, which are predominantly expressed in slow and fast muscles, respectively (Schiaffino and Reggiani, 1996). In both isoforms phosphorylation occurs only at one site at the amino terminal end, the serine residue 14 (or 15, depending on the isoform), while two distinct sites have been described in smooth muscle MLC2. The existence of a second site of phosphorylation remains controversial for the sarcomeric MLC2. The phosphorylation is catalyzed by the MLC kinase, which is activated by Ca^{2+}/Calmodulin. Thus, phosphorylation increases when intracellular Ca^{2+} rises, for example during a tetanic or repetitive stimulation enhancing force development, a phenomenon called "post-tetanic potentiation". While long-term changes in myosin phosphorylation have been described in cardiac and smooth muscles, only recently have reports described variations of MLC2 phosphorylation in skeletal muscle associated with changes in physiological conditions. A recent study has analyzed the phosphorylation states of MLC2 before and after 14 days of HU as well as after clenbuterol administration (Bozzo et al., 2003). Both models induce slow-to-fast transformations of the slow soleus muscle with over-expression of MHC and MLC fast isoforms. Two variants (MLC2s and MLC2s1) of the slow MLC2 isoform and one variant (MLC2f) of fast MLC2, differing by their phosphorylation states, have been identified in control rat soleus with 2-dimensional gel electrophoresis. After HU and clenbuterol administration, a decrease of the less acidic unphosphorylated spots of MLC2s occurred, associated with the appearance of a third phosphorylated spot (MLC2s2). In addition a phosphorylated variant (MLC2f1) of MLC2fast appeared (see fig.13). Thus, a clear correlation can be observed between increase of MLC2 phosphorylation and slow-to-fast phenotype transition.

In contrast to MLC2, little is know about the phosphorylation states and regulation of tropomyosin and troponin. While the phosphorylation of α and β tropomyosin was demonstrated in adult human and rat skeletal muscles (Edwards and Romero-Herrera, 1983; Heeley et al., 1982), the role of this post-translational modification remains under discussion. A structural role has been proposed. Since tropomyosin (TM) polymerizes in a head-to-tail manner and binds cooperatively to actin, phosphorylation of residue serine 283 would modify the stability of $\alpha\alpha$ and $\alpha\beta$ dimers of tropomyosin (Mak et al., 1978) and strengthen the head-to-tail interaction in the tropomysosin filament (Sano et al., 2000). However, there is also evidence in favor of a functional role of tropomyosin phosphorylation (Heeley, 1994), as the activity of myosin-S1 into a complete reconstituted regulatory system was doubled in presence of phosphorylated $\alpha\alpha$-TM, while dephosphorylation of the phosphorylated $\alpha\alpha$-TM reduced the rates to control values. Thus, developmental changes in the steady-state phosphorylation of $\alpha\alpha$-TM observed in rat muscles (Heeley et al., 1985) could be a mechanism to improve the thin filament activation especially by modifying the functional properties of the T1 section of TnT (Heeley, 1994).

Less is known about the post-translational modifications of troponin. While phosphorylated TnI is known to play a major role in the regulation of cardiac muscle contraction, only TnT has been demonstrated to be a phosphoprotein in the skeletal

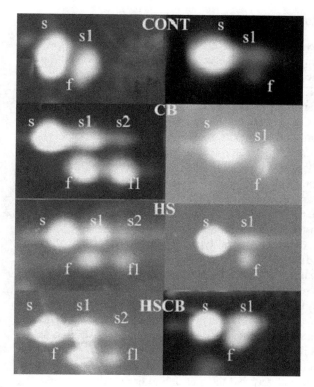

Figure 13. MLC2 variant distribution in soleus muscle in control rats (CONT) and after chronic clenbuterol administration (CB), hindlimb suspension (HS) and combined treatment (HS-CB). Variants are separated with two-dimensional gel electrophoresis: acidic pH is on the right-hand side. Left-hand panels: MLC2 variants in control and treated (CB, HS, HS-CB) soleus. Right-hand panels: MLC2 variants in each group after incubation with alkaline phosphatase. The variants of slow MLC2 are indicated as s, s1, and s2, and the variants of fast MLC2 are indicated as f and f1. Note that all groups (Cont, CB, HS, HS-CB) exhibit similar patterns after incubation with phosphatase, except for the relative proportion of fast and slow MLC2 isoforms (Reproduced with permission from Bozzo et al., 2003, © The Company of Biologists Ltd)

muscle troponin complex. Thus, fast and slow TnT isoforms have been found to exist in different charge variants which were phosphorylated (Härtner et al., 1989) because alkaline phosphatase treatment reduced the 19 fast and 7 slow variants to 12 and 3, respectively. These authors demonstrated that a transition from a fast to a slow TnT phenotype followed chronic stimulation of rabbit tibialis anterior as well as extensor digitorum longus muscles but did not analyze the changes in charge variant expression after chronic stimulation. However, changes in the phosphorylated states associated to the transition of expressions might give skeletal muscles a subtle contractile adaptation to various physiological conditions.

3.2   Glycosylation of Muscle Proteins

Carbohydrates present a real structural diversity and are attached to proteins through two main types of linkage, respectively asparagine (N-) or serine/threonine (O-) residues. O-linked N-acetylglucosaminyl moieties (O-GlcNAc) constitute an abundant glycosylation, which concerns cytosolic and nuclear proteins. Glycosylation is a dynamic and reversible process resulting from the enzyme activities of O-GlcNAc transferase and O-GlcNAcase, the latter for deglycosylation. All the identified O-GlcNAc proteins are also phosphoproteins. Indeed phosphorylation and O-GlcNAc modification are often reciprocal, occurring at the same or adjacent hydroxyl moieties, and O-GlcNAcylation appears as a regulatory modification that has a complex, dynamic interplay with phosphorylation (Lefebvre et al., 2003; Cheng et al., 2000; Griffith and Schmitz, 1999). This interplay between O-GlcNAc and phosphorylation called the "Yin-Yang" process, has been demonstrated on isolated proteins but also on total cellular proteins.

While glycosylation was studied in several types of pathology, such as diabetes, cancer, neurodegenerative diseases (Comer and Hart, 2000), the question of a role of such post-translational modification in the regulation of muscle physiology was not considered. However, muscle adaptive changes result in transitions in myofibrillar and other protein isoform expression in slow or fast directions as well as in metabolic changes in aerobic-oxidative or glycolytic directions (Pette and Staron, 1990). Moreover, the muscle metabolism and the O-GlcNAc binding are strongly dependent on the cellular glucose level (Yki-Järvinen et al., 1998). Indeed, Cieniewski-Bernard et al., 2004 clearly showed that several proteins of the energetic metabolism and the glycolytic pathway and the muscle motor protein, myosin, were found as O-GlcNAc proteins (Table 3). Since some of these proteins are involved in the contractile and metabolic adaptations that occur during muscle atrophy, it has been proposed that O-GlcNAc could play a key role in the development of atrophy as well as in contractile impairment (Cieniewski-Bernard et al., 2004). The functional implication of O-GlcNAc level remains to be analyzed. O-GlcNAc glycosylation could protect against the proteasomal degradation by modifying the target proteins as has been demonstrated for murine estrogen receptor β (Cheng et al., 2000), Sp1 (Han and Kudlow, 1997) and plakoglobin (Hatsell et al., 2003). In fact, O-GlcNAc sites on both ER-α and ER-β are within regions of high PEST scores which is considered a signal responsible for rapid protein degradation. O-GlcNAc binding at these sites might block degradation while phosphorylation will activate it (Cheng et al., 2000). It has recently been demonstrated that in cell cultures, levels of O-GlcNAc increase rapidly in response to different forms of stress, such as heat shock, ethanol, UV, hypoxia, reductive, oxidative and osmotic stress (Zacchara et al., 2004). Suppression of O-GlcNAc binding renders cells more sensitive to stress, causing a decrease in cell survival, while an increase of O-GlcNAc levels protects cells against stress. Disuse in hindlimb suspension increases oxidative stress in relation to an imbalance in the antioxidant enzyme systems (Lawler et al., 2003). On this ground, a role of O-GlcNAc in the development of atrophy can be envisaged.

Table 3. O-GlcNAc proteins identified in skeletal muscle, after purification on a WGA column and separation on 2-D electrophoresis and MALDI-TOF mass spectrometry analysis (from Cieniewski-Bernard et al., 2004)

Name	M_r (Da)	pI	Score (NCBI)	Score (Swiss-Prot)	Number of matched peptides	Percentage of coverage sequence
Proteins implicated in signal transduction, nuclear transport, and structural proteins						
αB-crystallin	19958	6.8	1.66e+5	3.23e+5	14	66
PI-3-kinase reg. subunit, p85	85532	5.9	1.24e+4	9.16e+3	8	13
Protein phosphatase 2A	45555	5.4	2.01e+4	2.70e+4	16	43
MAP kinase kinase kinase 8	52808	5.7	1.09e+5	1.05e+5	9	16
Homolog of yeast nuclear protein local. 4	68 057	6.0	1.18e+5	–	10	15
Serine protease inhibitor III	45555	5.3	2.54e+6	8.83e+6	16	43
Proteins of glycolytic pathway and energetic metabolism						
Muscle specific ß-enolase	46 961	7.6	9.61e+3	5.09e+3	15	24
			3.00e+8	1.41e+8	24	48
Fructose bisphosphatase aldolase, muscle specific	39 352	8.3	2.55e+10	3.01e+9	19	65
			2.74e+10	4.07e+9	13	50
Creatine kinase, M form	43 019	6.6	3.05e+9	1.258e+9	26	60
			1.42e+9	5.55e+8	30	70
			9.98*e*+5	7.22e+5	18	53
Triose phosphate isomerase	26 921	6.4	6.35e+3	2.23e+4	12	42
Glyceraldehyde-3-phosphate dehydrogenase	35 836	8.4	2.78e+12	5.39e+11	21	47
Malate dehydrogenase mitochondrial	35 656	8.9	6.23e+5	1.962e+5	11	22
Carbonic anhydrase III	29 401	6.9	1.14e+6	2.99e+6	13	53
Contractile protein						
Myosin heavy chain	103 583	5.3	2.72e+4	–	14	14

In conclusion, increasing evidence points to post-translational modifications as a way, which in addition to transcriptional regulation, can enhance the capacity of skeletal muscles to adapt to new functional requirements.

REFERENCES

Adams GR, Hather BM, Baldwin KM, Dudley GA. (1993) Skeletal muscle myosin heavy chain composition and resistance training. J Appl Physiol 74, 911–915,

Andruchov O, Andruchova O, Wang Y, Galler S. (2005) Cross-bridge kinetics in rabbit and rat skeletal muscle fibers depending on myosin light chain isoforms. J Physiol 2006 in press.

Alford EK, Roy RR, Hodgson JA, Edgerton VR. (1987) Electromyography of rat soleus, medial gastrocnemius, and tibialis anterior during hind limb suspension. Exp Neurol 96: 635–649.

Amtmann E, Oyama J. (1976) Effect of chronic centrifugation on the structural development of the musculoskeletal system of the rat. Anat Embryol 149: 47–70.

Andersen JL, Mohr T, Biering-Sorensen F, Galbo H, Kjaer M. (1996)Myosin heavy chain isoform transformation in single fibers from m. vastus lateralis in spinal cord injured individuals: effects of long-term functional electrical stimulation (FES). Pflügers Arch 431: 513–518.

Anderson PA, Malouf NN, Oakeley AE, Pagani ED, Allen PD. (1991) Troponin T isoform expression in humans. A comparison among normal and failing adult heart, fetal heart, and adult and fetal skeletal muscle. Circ Res. 69: 1226–33.

Babij P, Booth FW (1988) Clenbuterol prevents or inhibits loss of specific mRNAs in atrophying rat skeletal muscle. Am J Physiol 254: C657-C660.

Babu A, Scordilis SP, Sonnenblick EH, Gulati J. (1987) The control of myocardial contraction with skeletal fast muscle troponin C. J Biol Chem 262: 5815–5822.

Baldwin KM, Haddad F. (2001) Effects of different activity and inactivity paradigms on myosin heavy chain gene expression in striated muscle. J Appl Physiol 90: 345–57.

Ball KL, Johnson MD and Solaro RJ.(1994) Isoform specific interactions of troponin I and troponin C determine sensitivity of myofibrillar Ca^{2+} activation. Biochemistry 33: 8464–8471.

Bamman MM, Ragan RC, Kim JS, Cross JM, Hill VJ, Tuggle SC, Allman RM. (2004) Myogenic protein expression before and after resistance loading in 20-35-64-yr-old men and women. J Appl Physiol 97:1329–1337.

Bär A, Simoneau JA, Pette D.(1989) Altered expression of myosin light-chain isoforms in chronically stimulated fast-twitch muscle of the rat. Eur J Biochem 178: 591–594.

Barton PJ, Buckingham ME. (1985)The myosin alkali light chain proteins and their genes. Biochem J 231: 249–61.

Bastide B, Kischel P, Puterflam J, Stevens L, Pette D, Jin J, Mounier Y. (2002) Expression and functional implications of troponin T isoforms in soleus muscle fibers of rat after unloading. Pflugers Arch 444: 345–35.

Berg, J. S., B. C. Powell, and R. E. Cheney. (2001). A millennial myosin census. Mol. Biol. Cell 12: 780–794.

Berg HE, Dudley GA, Haggmark T, Ohlsen H, Tesch PA. (1991) Effects of lower limb unloading on skeletal muscle mass and function in humans. J Appl Physiol 70: 1882–5.

Blewett C, Elder GC. (1993) Quantitative EMG analysis in soleus and plantaris during hindlimb suspension and recovery. J Appl Physiol 74: 2057–66.

Bortolotto SK, Stephenson DG, Stephenson GMM. (1999) Fiber type populations and Ca-activation properties of single fibers in soleus muscles from SHR and WKY rats. Am J Physiol 276: C628–C637.

Bottinelli R, Schiaffino S, Reggiani C. (1991) Force-velocity relations and myosin heavy chain isoform compositions of skinned fibers from rat skeletal muscle. J Physiol (Lond) 437: 655–672.

Bottinelli R, Betto R, Schiaffino S, Reggiani C. (1994) Unloaded shortening velocity and myosin heavy chain and alkali light chain composition in rat skeletal muscle fibers. J Physiol (Lond) 478: 341–349.

Bottinelli R, Coviello DA, Redwood CS, Pellegrino MA, Maron BJ, Spirito P, Watkins H, Reggiani C. (1998) A mutant tropomyosin that causes hypertrophis cardiomyopathy is expressed in vivo and associated with an increased calcium sensitivity.Circ Res 82: 106–115.

Bottinelli R, Reggiani C. (2000) Human skeletal muscle fibers: molecular and functional diversity. Prog Biophys Mol Biol 73: 195–262.

Bozzo C, Stevens L, Toniolo L, Mounier Y, Reggiani C. (2003) Increased phosphorylation of myosin light chain associated with slow to fast transition in rat soleus. Am J Physiol 285: C575–C583.

Bozzo C, Stevens L, Bouet V, Montel V, Picquet F, Falempin F, Lacour M, Mounier Y. (2004) Hypergravity from conception to adult stage: effects on contractile properties and skeletal muscle phenotype. J Exp Biol 207: 2793–2802.

Breitbart RE, Nadal-Ginard B. (1986) Complete nucleotide sequence of the fast skeletal troponin T gene. Alternatively spliced exons exhibit unusual interspecies divergence. J Mol Biol 188: 313–24.

Bremel RD, Weber A. (1972) Cooperation within actin filament in vertebrate skeletal muscle. Nat New Biol 238: 97–101.

Briggs MM, Li JJ, Schachat F (1987) The extent of amino-terminal heterogeneity in rabbit fast skeletal muscle troponin T. J Muscle Res Cell Motil. 8, 1–12.

Briggs MM, Schachat F. (1996) Physiologically regulated alternative splicing patterns of fast troponin T RNA are conserved in mammals. Am J Physiol. 270, C298–305

Buller AJ, Eccles JC, Eccles RM. (1960) Interactions between motoneurones and muscles in respect of the characteristic speed of their responses. J Physiol (Lond) 150: 417–439.

Buonanno A, Rosenthal N. (1996) Transcriptional control of muscle plasticity: differential regulation of troponin i genes by electrical activity. Dev Genet 19: 95–107.

Burtnick LD, Kay CM. (1977) The calcium-binding properties of bovine cardiac troponin C. FEBS Lett 75: 105–110.

Caiozzo VJ, Baker M, Herrick RE, Baldwin KM. (1994) Effect of spaceflight on skeletal muscle: mechanical properties and myosin isoform content of a slow antigravity muscle. J Appl Physiol 76: 1764–1773.

Caiozzo VJ, Haddad F, Baker MJ, Herrick RE, Prietto N, Baldwin KM. (1996) Microgravity-induced transformations of myosin isoforms and contractile properties of skeletal muscle. J Appl Physiol 81: 123–132.

Caiozzo VJ, Baker MJ, McCue SA, Baldwin KM. (1997) Single-fiber and whole muscle analyses of MHC isoform plasticity: interaction between T_3 and unloading. Am J Physiol Cell Physiol 273: C944–C952.

Caiozzo VJ, Baker MJ, Baldwin KM. (1998) Novel transitions in MHC isoforms: separate and combined effects of thyroid hormone and mechanical unloading. J Appl Physiol 85: 2237–2248.

Caiozzo VJ, Haddad F, Baker M, McCue S, Baldwin KM. (2000) MHC polymorphism in rodent plantaris muscle: effects of mechanical overload and hypothyroidism. Am J Physiol Cell Physiol 278: C709–C717.

Calvo S, Stauffer J, Nakayama M, Buonanno A. (1996) Transcriptional control of muscle plasticity: differential regulation of troponin I genes by electrical activity. Dev Genet 19: 169–181.

Campione M, Ausoni S, Guezennec CY, Schiaffino S. (1993) Myosin and troponin changes in rat soleus muscle after hindlimb suspension. J Appl Physiol 74: 1156–1160.

Canu MH, M Falempin. Effect of hindlimb unloading on locomotor strategy during treadmill locomotion in the rat. Eur J Appl Physiol Occup Physiol 74: 297–304, 1996.

Carraro U, Catani C, Dalla Libera L, Vascon M, Zanella G. (1981) Differential distribution of tropomyosin subunits in fast and slow rat muscles and its changes in long-term denervated hemidiaphragm. FEBS Lett 128: 233–236.

Carroll TJ, Abernethy PJ, Logan PA, Barber M, McEniery MT. Resistance training frequency: strength and myosin heavy chain responses to two and three bouts per week. Eur J Appl Physiol 78: 270–275, 1998.

Castro MJ, Apple DF, Jr, Staron RS, Campos GE, Dudley GA. (1999) Influence of complete spinal cord injury on skeletal muscle within 6 mo of injury. J Appl Physiol 86: 350–358.

Chandra M, da Silva EF, Sorenson MM, Ferro JA, Pearlstone JR, Nash BE, Borgford T, Kay CM, Smillie LB. (1994) The effects of N helix deletion and mutant F29W on the Ca^{2+} binding and functional properties of chicken skeletal muscle troponin. J Biol Chem 269: 14988–14994.

Cheng X, Cole RN, Zaia J, Hart GW. (2000) Alternative O-glycosylation/O-phosphorylation of the murine estrogen receptor beta. Biochemistry 39: 11609–20.

Chi MM, Manchester JK, and Lowry OH. (1998) Effect of centrifugation at 2G for 14 days on metabolic enzymes of the tibialis anterior and soleus muscles. Aviat Space Environ Med 69(6 Suppl): A9–11.

Chong PC, Hodges R.S. (1982) Photochemical cross-linking between rabbit skeletal troponin subunits. Troponin I-troponin T interactions. J Biol Chem 257: 11667–11672.

Chopard A, Pons F, Marini J. (2001) Cytoskeletal protein contents before and after hindlimb suspension in a fast and slow rat skeletal muscle. Am J Physiol Regulatory Integrative Comp Physiol 280: 323–330.

Cieniewski-Bernard C, Bastide B, Lefebvre T, Lemoine J, Mounier Y, Michalski JC. (2004) Identification of O-linked N-acetylglucosamine proteins in rat skeletal muscle using two-dimensional gel electrophoresis and mass spectrometry. Mol Cell Proteomics 3: 77–85.

Cooper TA, Ordahl CP. (1984) A single troponin T gene regulated by different programs in cardiac and skeletal muscle development. Science 226: 979–82.

Comer FI, Hart GW. (2000) O-Glycosylation of nuclear and cytosolic proteins. Dynamic interplay between O-GlcNAc and O-phosphate. J Biol Chem 275: 29179–29182.

Cordonnier C, Stevens L, Picquet F, Mounier Y. (1995) Structure function relationship of soleus muscle fibers from the rhesus monkey. Pflügers Arch 430: 19–25.

Criswell DS, Hodgson VRM, Hardeman EC, Booth FW. (1998) Nerve-responsive troponin I slow promoter does not respond to unloading. J Appl Physiol 84: 1083–1087.

Cummins P, Perry SV. (1973) The subunits and biological activity of polymorphic forms of tropomyosin. Biochem J 133: 765–777.

D'Amelio F, Wu LC, Fox RA, Daunton NG, Corcoran ML, Polyakov I. (1998) Hypergravity exposure decreases gamma-aminobutyric acid immunoreactivity in axon terminals contacting pyramidal cells in the rat somatosensory cortex: a quantitative immunocytochemical image analysis. J Neurosci Res 53: 135–42.

Danieli-Betto D, Betto R, Midrio M. (1990) Calcium sensitivity and myofibrillar protein isoforms of rat skinned skeletal muscle fibers. Pflugers Arch 417: 303–8.

Demirel HA, Powers SK, Naito H, Hughes M, and Coombes JS. (1999) Exercise-induced alterations in skeletal muscle myosin heavy chain phenotype: dose-response relationship. J Appl Physiol 86: 1002–1008.

Desplanches D, Mayet MH, Sempore B, Flandrois R. (1987) Structural and functional responses to prolonged hindlimb suspension in rat muscle. J Appl Physiol 63: 558–63.

Diffee GM, McCue S, Larosa A, Herrick RE, Baldwin KM. (1993) Interaction of various mechanical activity models in regulation of myosin heavy chain isoform expression. J Appl Physiol 74: 2517–2522.

Dunn SE, Michel RN. (1999) Differential sensitivity of myosin-heavy-chain-typed fibers to distinct aggregates of nerve-mediated activation. Pflugers Arch 437: 432–40,.

Ebashi S, Nonomura Y. (1973) Proteins of the myofibril. The structure and function of muscle 3: 288.

Edwards BF, Romero-Herrera AE. (1983) Tropomyosin from adult human skeletal muscle is partially phosphorylated. Comp Biochem Physiol 76: 373–5.

Eldridge L, Dhoot GK, Mommaerts WF. (1984) Neural influences on the distribution of troponin I isotypes in the cat. Exp Neurol 83: 328–46.

Esser KA, Hardemann EC. (1995) Changes in contractile protein mRNA accumulation in response to spaceflight.Am J Physiol 37: 466–471.

Falempin M, Fodili S. (1995) Effect of the elimination of neural influences in the rat soleus muscle during unweighting. BAM 5: 155–161.

Farah CS, Miyamoto CA, Ramos CH, Da Silva AC, Quaggio RB, Fujimori K, Smillie LB, Reinach FC. (1994) Structural and regulatory functions of the NH2- and COOH-terminal regions of skeletal muscle troponin I. J Biol Chem 269: 5230–5240.

Farah CS, Reinach FC. (1995) The troponin complex and regulation of muscle contraction. FASEB J 9: 755–767.

Fauteck SP, Kandarian SC. (1995) Sensitive detection of myosin heavy chain composition in skeletal muscle under different loading conditions. Am J Physiol 268: C419-C424.

Fitts RH, Riley DR, Widrick JJ. (2000) Physiology of a microgravity environment invited review: microgravity and skeletal muscle. J Appl Physiol 89: 823–839.

Flicker PF, Phillips GN, Cohen C. (1982) Troponin and its interactions with tropomyosin. An electron microscope study. J. Mol Biol 162: 495–501.

Gahlmann R, Troutt AB, Wade RP, Gunning P, Kedes L. (1987) Alternative splicing generates variants in important functional domains of human slow skeletal troponin T. J Biol Chem 262: 16122–6.

Galler S, Hilber K, Gohlsch B, Pette D. (1997) Two functionally distinct myosin heavy chain isoforms in slow skeletal muscle fibers. FEBS Lett 410: 150–152.

Gardetto PR, Schluter JM, Fitts RH. (1989) Contractile function of single muscle fibers after hindlimb suspension. J Appl Physiol 66: 2739–2749.

Gardiner P, Michel R, Browman C, Noble E. (1986) Increased EMG of rat plantaris during locomotion following surgical removal of its synergists. Brain Res 380: 114–21.

Geiger PC, Cody MJ, Macken RL, Bayrd ME, Sieck GC. (2001) Effect of unilateral denervation on maximum specific force in rat diaphragm muscle fibers. J Appl Physiol 90: 1196–204.

Gordon AM., Homsher E, Regnier M. (2000) Regulation of contraction in striated muscle. Physiological Reviews 80: 853–924.

Gordon AM, Regnier M, Homsher E. (2001) Skeletal and cardiac muscle contractile activation:tropomyosin "rocks and rolls". News Physiol Sci 16:49–55.

Grabarek Z, Tao T, Gergely J. (1992) Molecular mechanism of troponin C function. J Muscle Res 13: 383–393.

Greaser ML, Gergely J. (1971) Reconstitution of troponin activity from three protein components. J Biol Chem 246: 4226–4233.

Greig A, Hirschberg Y, Anderson PA, Hainsworth C, Malouf NN, Oakeley AE, Kay BK (1994) Molecular basis of cardiac troponin T isoform heterogeneity in rabbit heart. Circ Res 74: 41–7.

Green HJ, Thomson JA, Houston ME. (1987) Supramaximal exercise after training-induced hypervolemia. II. Blood/muscle substrates and metabolites. J Appl Physiol 62: 1954–61.

Griffith LS, Schmitz B. (1999) O-linked N-acetylglucosamine levels in cerebellar neurons respond reciprocally to pertubations of phosphorylation. Eur J Biochem., 262: 824–831.

Grossman, EJ, Roy RR, Talmadge RJ, Zhong H, and Edgerton VR. (1998) Effects of inactivity on myosin heavy chain composition and size of rat soleus fibers. Muscle Nerve 21: 375–389.

Gulati J, Scordilis S, Babu A. (1998) Effect of troponin C on the cooperativity in Ca^{2+} activation of cardiac muscle. FEBS Lett 236: 441–444.

Gulati J, Sonnenblick E, Babu A. (1991) The role of troponin C in the length dependence of Ca^{2+}-sensitive force of mammalian skeletal muscle and cardiac muscles. J Physiol 441: 305–324.

Guth K, Potter JD. (1987) Effect of rigor and cycling cross-bridges on the structure of troponin C and on the Ca^{2+} affinity of the Ca^{2+}-specific regulatory sites in skinned rabbit psoas fibers. J Biol Chem 262: 13627–13635.

Gutmann E, Melichna J, Syrovy (1972) Contraction properties and ATPase activity in fast and slow muscle of the rat during denervation. Exp Neurol 36: 488–97.

Haddad F, Bodell PW, McCue SA, Herrick RE, and Baldwin KM. (1993) Food restriction-induced transformations in cardiac functional and biochemical properties in rats. J Appl Physiol 74: 606–612.

Hamalainen N, Pette D. (1997) Expression of an alpha-cardiac like myosin heavy chain in diaphragm, chronically stimulated, and denervated fast-twitch muscles of rabbit. J Muscle Res Cell Motil 18: 401–11.

Hammell RL, Hitchcock-DeGregori SE. (1996) Mapping the functional domains within the carboxyl terminus of alpha-tropomyosin encoded by the alternatively spliced ninth exon. J Biol Chem 271: 4236–4242.

Han, I. and Kudlow, J.E. (1997) Reduced O-glycosylation of Sp1 is associated with increased proteasome susceptibility. Mol Cell Biol 17: 2550–6.

Härtner KT, Pette D. (1990) Fast and slow isoforms of troponin I and troponin C. Distribution in normal rabbit muscles and effects of chronic stimulation. Eur J Biochem 188: 261–267.

Hartner KT, Kirschbaum BJ, Pette D. (1989) The multiplicity of troponin T isoforms. Distribution in normal rabbit muscles and effects of chronic stimulation. Eur J Biochem 179: 31–38.

Hatsell, S., Medina, L., Merola, J., Haltiwanger, R. and Cowin, P. (2003) Plakoglobin is O-glycosylated close to the N-terminal destruction box. J Biol Chem 278: 37745–52.

Heeley A, Moir AJ, Perry SV. (1982) Phosphorylation of tropomyosin during development in mammalian striated muscle. FEBS Lett 146: 115–8.

Heeley DH, Dhoot GK, Frearson N, Perry SV, Vrbova G. (1983) The effect of cross-innervation on the tropomyosin composition of rabbit skeletal muscle. FEBS Lett 152: 282–286.

Heeley DH, Dhoot GK, Perry SV. (1985) Factors determining the subunit composition of tropomyosin in mammalian skeletal muscle. Biochem J 226: 461–8.

Heeley DH, Watson MH, Mak AS, Dubord P, Smillie LB. (1989) Effect of phosphorylation on the interaction and functional properties of rabbit striated muscle alpha alpha-tropomyosin. J Biol Chem 264: 2424–30.

Heeley DH. (1994) Investigation of the effects of phosphorylation of rabbit striated muscle alpha alpha-tropomyosin and rabbit skeletal muscle troponin-T. Eur J Biochem 221: 129–137.

Herzberg O, Moult J, James MN. (1986) A model for the Ca^{2+}-induced conformational transition of troponin C. A trigger for muscle contraction. J Biol Chem 261: 2638–2644.

Hikida RS, Gollnick PD, Dudley GA, Convertino VA, Buchanan P. (1989) Structural and metabolic characteristics of human skeletal muscle following 30 days of simulated microgravity. Aviat Space Environ Med 60: 664–70.

Hinkle A, Goranson A, Butters CA, Tobacman LS.(1999) Roles for the troponin tail domain in thin filament assembly and regulation. A deletional study of cardiac troponin T. J Biol Chem 274: 7157–7164.

Hitchcock-DeGregori SE, Song Y, Greenfield NJ. (2002) Functions of tropomyosin's periodic repeats. Biochemistry 41: 15036–15044.

Hoar PE, Potter JD, Kerrick WG. (1988) Skinned ventricular fibers: troponin C extraction is species-dependent ansd its replacement with skeletal troponin C changes Sr^{2+} activation properties. J Muscles Res Cell Motil 9: 165–173.

Hofmann PA, Fuchs F. (1987) Evidence for a force-dependent component of calcium binding to cardiac troponin C. Am J Physiol 253: C541–C546.

Holmes KC, Popp D, Gebhard W, Kabsch W. (1990) Atomic model of the actin filament. Nature 347: 44–49.

Howard G, Steffer JM, Geoghegan TE. (1989) Transcriptional regulation of decreased protein synthesis during skeletal muscle unloading. J Appl Physiol. 66: 1093–8.

Holy X, Stevens L, Mounier Y. (1990) Compared effects of a 13 day spaceflight on the contractile protiens of soleus and plantaris rat muscles. Physiologist. 33: S80–81.

Holy X, Mounier Y. (1991) Effects of short spaceflights on mechanical characteristics of rat muscles. Muscle Nerve 14: 70–78.

Huey KA, Bodine SC. (1998) Changes in myosin mRNA and protein expression in denervated rat soleus and tibialis anterior. Eur J Biochem 256: 45–50.

Huey KA, Roy RR, Edgerton VR, Baldwin KM.(2000) In vivo regulation of type I MHC gene expression in the soleus of spinal cord isolated rats. Physiologist 43: 346.

Huey KA, Roy RR, Baldwin KM, Edgerton VR. (2001) Temporal effects of inactivty on myosin heavy chain gene expression in rat slow muscle. Muscle Nerve 24: 517–26.

Huey KA, Haddad F, Qin AX, Baldwin KM. (2003) Transcriptional regulation of the type I myosin heavy chain gene in denervated rat soleus. Am J Physiol Cell Physiol 284: C738–48.

Ilyina-Kakueva EI, Portugalov VV, Krivenkova NP. (1976) Space flight effects on the skeletal muscles of rats. Aviat Space Environ Med 47: 700–3.

Jagatheesan G, Rajan S, Petrashevskaya N, Schwartz A, Boivin G, Vahebi S, Detombe P, Labitzke E, Hilliard G, Wieczorek DF. (2003) Functional importance of the carboxyl-terminal region of striated muscle tropomyosin. J Biol Chem 278: 23204–23211.

Jakubiec-Puka A, Catani C, Carraro U. (1992) Myosin heavy-chain composition in striated muscle after tenotomy. Biochem J 282: 237–42.

Jakubiec-Puka A, Ciechomska I, Morga J, Matusiak A. (1999) Contents of myosin heavy chains in denervated slow and fast rat leg muscles. Comp Biochem Physiol 122: 355–362.

Jin JP, Chen A, Huang QQ. (1998) Three alternatively spliced mouse slow skeletal muscle troponin T isoforms: conserved primary structure and regulated expression during postnatal development. Gene 214: 121–129.

Jin JP, Chen A, Ogut O, Huang QQ.(2000) Conformational modulation of slow skeletal muscle troponin T by an NH(2)-terminal metal-binding extension. Am J Physiol Cell Physiol 279: C1067–77.

Jockusch H. (1990) Muscle fiber transformations in myotonic mouse mutants.the Dynamic State of Muscle Fibers:429–443.

Kaasik P, Alev K, Pehme A, Seene T. (1998) Composition of tropomyosin subunits in different types of skeletal muscle and effect of compensatory hypertrophy. J Muscle Res Cell Motil19: 296.

Kerrick WG, Malencik DA, Hoar PE, Potter JD, Coby RL, Pocinwong S, Fischer EH. (1980) Ca^{2+} and Sr^{2+} activation: comparison of cardiac and skeletal muscle contraction models. Pflugers Arch 386: 207–213.

Kischel P, Stevens L, Mounier Y. (1999) Differential effects of bepridil on functional properties of troponin C in slow and fast skeletal muscles. Brit J Pharmacol 128: 767–773.

Kischel P, Bastide B, Potter JD, Mounier Y. (2000) The role of Ca^{2+} regulatory sites of skeletal troponin C in modulating muscle fiber reactivity to the Ca^{2+} sensitizer bepridil. British Journal of Pharmacology 131: 1496–1502.

Kischel P, Stevens L, Montel V, Picquet F, Mounier Y. (2001a) Plasticity of monkey triceps muscle fibers in microgravity conditions. J Appl Physiol 90: 1825–1832.

Kischel P, Bastide B, Stevens L, Mounier Y. (2001b) Expression and functional behavior of troponin C in soleus muscle fibers of rat after hindlimb unloading. J Appl Physiol 90: 1095–1101.

Koenders A, Lamey TM, Medler S, West JM, Mykles DL. (2004) Two fast-type fibers in claw closer and abdominal deep muscles of the Australian Freshwater crustacaen, Cherax destructor, differ in Ca^{2+} sensitivity and troponin-I isoforms. J Exp Zoolog A Comp Exp Biol 301: 588–598.

Larsson L, Moss RL. (1993) Maximum velocity of shortening in relation to myosin isoform composition in single fibers from human skeletal muscles. J Physiol (Lond) 472: 595–614.

Lawler, J.M., Song, W. and Demaree, S.R. (2003) Hindlimb unloading increases oxidative stress and disrupts antioxidant capacity in skeletal muscle. Free Radic Biol Med, 35: 9–16.

Leavis PC, Kraft EL. (1978) Calcium binding to cardiac troponin C. Arch.Biochem.Biophys. 186: 411–415.

Lees-Miller JP, Helfman DM. (1991) The molecular basis for tropomyosin isoform diversity. Bioessays 13: 429–437.

Lefebvre T, Ferreira S, Dupont-Wallois L, Bussiere T, Dupire M.J, Delacourte A, Michalski JC, Caillet-Boudin ML. (2003) Evidence of a balance between phosphorylation and O-GlcNAc glycosylation of Tau proteins–a role in nuclear localization. Biochim Biophys Acta 1619: 167–176.

Lehman W, Craig R, Vibert P. (1994) Ca(2+)-induced tropomyosin movement in Limulus thin filaments revealed by three-dimensional reconstruction. Nature 368: 65–67.

Lehrer SS. (1994) The regulatory switch of the muscle thin filament: Ca^{2+} or myosin heads? J Muscle Res Cell Motil 15: 232–236.

Lehrer SS, Morris EP. (1982) Dual effects of tropomyosin and troponin-tropomyosin on actomyosin subfragment 1 ATPase. J Biol Chem 257: 8073–8080.

Leeuw T, Pette D. (1993) Coordinate changes in the expression of troponin subunit and myosin heavy-chain isoforms during fast-to-slow transition of low-frequency-stimulated rabbit muscle. Eur J Biochem 213: 1039–1046.

Leeuw T, Kapp M, Pette D. (1994) Role of innervation for development and maintenance of troponin subunit isoform patterns in fast- and slow-twitch muscles of the rabbit. Differentiation 55: 193–201.

Leterme D, Falempin M. (1998) EMG activity of three rat hindlimb muscles during microgravity and hypergravity phase of parabolic flight. Aviat Space Environ. Med 69: 1065–1070.

Lewitt LK, O'Mahoney JV, Brennan KJ, Joya J, Zhu L, Wade R, Hardeman EC. (1995) The human troponin I solw promoter directs slow fiber-specific expression in transgenic mice. DNA Cell Biol 14: 599–607.

Li HC, Fajer PG. (1998) Structural coupling of troponin C and actomyosin in muscle fibers. Biochemistry 37: 6628–6635.

Li Y, Mui S, Brown JH, Strand J, Reshetnikova L, Tobacman LS, Cohen C. (2002) The crystal structure of the C-terminal fragment of striated-muscle alpha-tropomyosin reveals a key troponin T recognition site. Proc Natl Acad Sci USA 99: 7378–7383.

Li Z, Gergely J, Tao T. (2001) Proximity relationships between residue 117 of rabbit skeletal troponin-I and residues in troponin-C and actin. Biophys J 81: 321–333.

Lompre AM, Anger M, Levitsky D. (1994) Sarco(endo)plasmic reticulum calcium pumps in the cardiovascular system: function and gene expression. J Mol Cell Cardiol 26: 1109–21.

Loughna PT, Izumo S, Goldspink G, Nadal-Ginard B. (1990) Disuse and passive stretch cause rapid alterations in expression of developmental and adult contractile protein genes in skeletal muscle. Development 109: 217–23.

Luo Y, Leszyk J, Li B, Gergely J, Tao T. (2000) Proximity relationships between residue 6 of troponin I and residues in troponin C: further evidence for extended conformation of troponin C in the troponin complex. Biochemistry 39: 15306–15315.

Lutz GJ, Lieber RL. (2000) Myosin isoforms in anuran skeletal muscle: their influence on contractile properties and in vivo muscle function. Microsc Res Tech 50: 443–57.

Lutz GJ, Bremner SN, Bade MJ, Lieber RL. (2001) Identification of myosin light chains in Rana pipiens skeletal muscle and their expression patterns along single fibers. J Exp Biol 204: 4237–4248.

Mac Lachlan D, Stewart M. (1976) The 14-fold periodicity in α-tropomyosin and the interaction with actin. J Mol Biol 103: 271–298.

Mac Lachlan LK, Reid DG, Mitchell RC, Salter CJ, Smith SJ. (1990) Binding of calcium sensitizer, bepridil, to cardiac troponin C. A fluorescence stopped-flow kinetic, circular dichroism and proton nuclear magnetic resonnance study. J Biol Chem 265: 9764–9770.

McLaurin MD, Apple FS, Voss EM, Herzog CA, Sharkey SW. (1997) Cardiac troponin I, cardiac troponin T, and creatine kinase MB in dialysis patients without ischemic heart disease: evidence of cardiac troponin T expression in skeletal muscle. Clin Chem. 43: 976–982

Mak A, Smillie LB, Barany M. (1978) Specific phosphorylation at serine-283 of alpha tropomyosin from frog skeletal and rabbit skeletal and cardiac muscle. Proc Natl Acad Sci USA 75: 3588–92.

Manning DR, Stull JT. (1982) Myosin light chain phosphorylation-dephosphorylation in mammalian skeletal muscle. Am J Physiol, 242: C234–C241.

Martin WD. (1980) Effects of chronic centrifugation on skeletal muscle fibers in young developing rats. Aviat Space Environ Med, 51: 473–479.

Martrette JM, Hartmann N, Vonau S, and Westphal A. (1998) Effects of pre- and perinatal exposure to hypergravity on muscular structure development in rat. J Muscle Res Cell Motil, 19: 689–694.

Matsuda G. (1983) The light chains of muscle myosin: its structure, function, and evolution.Adv Biophys 16: 185–218.

Maytum R, Lehrer SS, Geeves MA. (1999) Cooperativity and switching within the three-state model of muscle regulation. Biochemistry 38: 1102–1110.

Messner B, Baum H, Fischer P, Quasthoff S, Neumier D. (2000) Expression of messenger RNA of the cardiac isoforms of troponin T and I in myophatic skeletal muscle. Am J Clin Pathol 114: 544–549.

Michel RN, Parry DJ, Dunn SE. (1996) Regulation of myosin heavy chain expression in adult rat hindlimb muscles during short-term paralysis: comparison of denervation and tetrodotoxin-induced neural inactivation. FEBS Lett 391: 39–44.

Morgan MJ, Loughna PT. (1989) Work overload induced changes in fast and slow skeletal muscle myosin heavy chain gene expression. FEBS Lett 255: 427–30.

Morey ER, Sabelman EE, Turner RT, Baylink DJ. (1979) A new rat model simulating some aspects of space flight. Physiologist 22: 23–24.

Morey-Holton E, Globus RK, Kaplansky A, Durnova G. (2005) The hindlimb unloading rat model: literature overview, technique update and comparison with space flight data. Adv Space Biol Med 10: 7–40.

Moss RL, Lauer MR, Giulian GG, Greaser ML. (1986) Altered Ca^{2+} dependence of tension development in skinned skeletal muscle fibers following modification of troponin by partial substitution with cardiac troponin C. J Biol Chem 261: 6096–6099.

Musacchia XJ, Deavers DR, Meininger GA, Davis TP. (1980) A model for hypokinesia: effects on muscle atrophy in the rat. J.Appl.Physiol 48: 479–486.

Muthuchamy M, Grupp IL, Grupp G, O'Toole BA, Kier AB, Boivin GP, Neumann J, Wieczorek DF. (1995) Molecular and physiological effects of overexpressing striated muscle beta-tropomyosin in the adult murine heart. J Biol Chem 270: 30593–30603.

Muthuchamy M, Pieples K, Rethinasamy P, Hoit B, Grupp IL, Boivin GP, Wolska B, Evans C, Solaro RJ, Wieczorek DF.(1999) Mouse model of a familial hypertrophic cardiomyopathy mutation in α-tropomyosin manifests cardiac dysfunction. Circ Res 85: 47–56.

Ngai SM, Pearlstone JR, Farah CS, Reinach FC, Smillie LB, Hodges RS. (2001) Structural and functional studies on Troponin I and Troponin C interactions. J Cell Biochem. 83: 33–46.

Obinata T, Saitoh O, Takano-Ohmura H. (1984) Effect of denervation on the isoform transitions of tropomyosin, troponin T, and myosin isozyme in chiken breast muscle. J Biochem 95: 585–588.

Ogut O, Jin JP. (1998) Developmentally regulated, alternative RNA splicing-generated pectoral muscle-specific troponin T isoforms and role of the NH2-terminal hypervariable region in the tolerance to acidosis. J Biol Chem 273: 27858–27866.

Ogut O, Jin JP. (2000) Cooperative interaction between developmentally regulated troponin T and tropomyosin isoforms in the absence of F-actin. J Biol Chem 275: 26089–26095.

Ohira Y, Jiang B, Roy RR, Oganov V, Ilyina-Kakueva E, Marini JF, Edgerton VR. (1992) Rat soleus muscle fiber responses to 14 days of spaceflight and hindlimb suspension. J Appl Physiol 73: 51S–57S.

Ohira Y. (2000) Neuromuscular adaptation to microgravity environment. Jpn J Physiol 50: 303–14.

Ohtsuki I, Shiraishi F. (2002) Periodic binding of troponin C.I and troponin I to tropomyosin-actin filaments. J Biochem (Tokyo) 131: 739–43.

Palm T, Greefield NJ, Hitchcock-Degregori SE. (2003) Tropomyosin ends determine the stability and functionality of overlab and troponin T complexes. Biophys J 84: 3181–3189.

Pan B, Gordon A, Luo Z. (1989) Removal of tropomyosin overlap modifies cooperative binding of myosin S-1 to reconstituted thin filaments of rabbit striated muscle. J Biol Chem 264: 8495–8498.

Parmacek MS, Leiden JM. (1991) Structure, function, and regulation of troponin C. Circulation 84: 991–1003.

Pearlstone JR, Borgford T, Chandra M, Oikawa K, Kay CM, Herzberg O, Moult J, Herklotz A, Reinach FC, Smillie LB. (1992) Construction and characterization of a spectral probe mutant of troponin C: application to analyses of mutants with increased Ca^{2+} affinity. Biochemistry 31: 6545–6553.

Pellegrino M.A., Canepari M., Rossi R., D'Antona G., Reggiani C., Bottinelli R. (2003) Orthologous myosin isoforms and scaling of shortening velocity with body size in mouse, rat, rabbit and human muscles. J. Physiol 546: 676–689.

Perry SV, Cole HA, Head JF, Wilson FJ. (1972) Localization and mode of action of the inhibitory component of the troponin complex. Cold Spring Harbor Symp Quant Biol 37: 251–262.

Perry SV. (1998) Troponin T: genetics, properties and function. J Muscle Res Cell Motil. 19: 575–602

Perry SV. (1999) Troponin I: inhibitor or facilitator. Mol Cell Biochem 190: 9–32.

Perrie WT, Smillie LB, Perry SB. (1973) A phosphorylated light-chain component of myosin from skeletal muscle. Biochem J 135: 151–164.

Persechini A, Stull JT, Cooke R. (1985) The effect of myosin phosphorylation on the contractile properties of skinned rabbit skeletal muscle fibers. J BiolChem 260: 7951–7954.

Pette D. (2001) Historical perspectives: plasticity of mammalian skeletal muscle. J Appl Physiol 90: 1119–1124.

Pette D, Staron RS. (1990) Cellular and molecular diversities of mammalian skeletal muscle fibers. Rev Physiol Biochem Pharmacol 116: 1–76.

Pette D, Staron RS. (1997) Mammalian skeletal muscle fiber type transitions. Int Rev Cytol 170: 143–223.

Pette D, Staron RS. (2000) Myosin isoforms, muscle fiber types and transitions. Microsc Res Tech 50: 500–509.

Pette D, Vrbova G. (1985) Invited review: neural control of phenotypic expression in mammalian muscle fibers. Muscle Nerve 8: 676–689.

Pette D, Vrbova G. (1999) What does chronic electrical stimulation teach us about muscle plasticity? Muscle Nerve 22: 666–677.

Peuker H, Conjard A, Pette D. (1998) Alpha-cardiac-like myosin heavy chain as an intermediate between MHCIIa and MHCI beta in transforming rabbit muscle. Am J Physiol 274: C595–602.

Phillips, G. N., Jr., J. P. Fillers, and C. Cohen. (1986) Tropomyosin crystal structure and muscle regulation. J.Mol.Biol. 192: 111–131.

Picquet F, Stevens L, Butler-Browne GS, Mounier Y. (1998) Differential effects of a six-day immobilization on newborn rat soleus muscles at two developmental stages. J Muscle Res Cell Motil 19: 743–55.

Picquet F, Bouet V, Canu MH, Stevens L, Mounier Y, Lacour M, Falempin M. (2002) Contractile properties and myosin expression in rats born and reared in hypergravity. Am J Physiol 282: R1687–R1695.

Picquet F, Falempin M. (2003) Compared effects of hindlimb unloading versus terrestrial deafferentation on muscular properties of the rat soleus. Exp Neurol 182: 186–94.

Pieples K, Arteaga G, Solaro RJ, Grupp I, Lorenz JN, Boivin GP, Jagatheesan G, Labitzke E, DeTombe PP, Konhilas JP, Irving TC, Wieczorek DF. (2002) Tropomyosin 3 expression leads to hypercontractility and attenuates myofilament length-dependent Ca^{2+} activation. Am J Physiol Heart Circ Physiol 283: H1344–H1353.

Potter JD, Gergely J. (1975) The calcium and magnesium binding sites on troponin and their role in the regulation of myofibrillar adenosine triphosphatase. J.Biol.Chem. 250: 4628–4633.

Potter JD, Sheng Z, Pan BS, Zhao J. (1995) A direct regulatory role for troponin T and a dual role for troponin C in the Ca2+ regulation of muscle contraction. J Biol Chem 270: 2557–2562.

Poetter K, Jiang H, Hassanzadeh S, Master SR, Chang A, Dalakas MC, Rayment I, Sellers JR, Fananapazir L, Epstein ND. (1996) Mutations in either the essential or regulatory light chains of myosin are associated with a rare myopathy in human heart and skeletal muscle. Nat Genet 13: 63–69.

Putkey JA, Sweeney HL, Campbell ST. (1989) Site-directed mutation of the trigger calcium-binding sites in cardiac troponin C. J Biol Chem 264: 12370–12378.

Regnier M, Martyn DA, Chase PB. (1996) Calmidazolium alters Ca^{2+} regulation of tension redevelopment rate in skinned skeletal muscle. Biophys J 71: 2786–2794.

Regnier M, Rivera AJ, Chase PB, Smillie LB, Sorenson MM. (1999) Regulation of skeletal muscle tension redevelopment by troponin C constructs with different Ca^{2+} affinities. Biophys J 76: 2664–2672.

Regnier M, Rivera AJ, Wang CK, Bates MA, Chase PB, Gordon AM. (2002) Thin filament near-neighbour regulatory unit interactions affect rabbit skeletal muscle steady-state force-Ca^{2+} relations. J Physiol 540: 485–497.

Reinach FC, Farah CS, Monteiro PB, Malnic B. (1997) Structural interactions responsible for the assembly of the troponin complex on the muscle thin filament. Cell Structure and Function 22: 219–223.

Riley DA, Slocum GR, Bain JL, Sedlak FR, Sowa TE, Mellender JW. (1990) Rat hindlimb unloading: soleus histochemistry, ultrastructure, and electromyography. J Appl Physiol 69: 58–66.

Riley DA., Bain JLW, Thompson JL, Fitts RH, Widrick JJ, Trappe SW, Trappe TA, Costill DL. (1998) Disproportionate loss of thin filaments in human soleus muscle after 17-day bed rest. Muscle Nerve 21: 1280–1289.

Riley DA., Bain JLW, Thompson JL, Fitts RH, Widrick JJ, Trappe SW, Trappe TA, Costill DL. (2002) Thin filament diversity and physiological properties of fast and slow fiber types in astronaut leg muscle. J Appl Physiol 92: 817–825.

Robinson JM, Wang Y, Kerrick WGL, Kawai R, Cheung HC. (2002) Activation of striated muscle:nearest-neighbour regulatory-unit and cross-bridge influence on myofilament kinetics. J Mol Biol 322: 1065–1088.

Rosenblatt JD, Parry DJ. (1993) Adaptation of rat extensor digitorum longus muscle to gamma irradiation and overload. Pflugers Arch 423: 255–64.

Rosser BW, Dean MS, Bandman E. Myonuclear domain size varies along the lengths of maturing skeletal muscle fibers. Int J Dev Biol 46: 747–54, 2002.

Roy RR, Hodgson JA, Lauretz SD, Pierotti DJ, Gayek RJ, Edgerton VR. (1992) Chronic spinal cord-injured cats: surgical procedures and management. Lab Anim Sci 42: 335–343.

Roy RR, Roy ME, Talmadge RJ, Mendoza R, Grindeland RE, and Vasques M. (1996) Size and myosin heavy chain profiles of rat hindlimb extensor muscle fibers after 2 weeks at 2G. Aviat Space Environ Med. 67: 854–858.

Sabry MA, Dhoot GK. (1991a) Identification and pattern of transitions of some developmental and adult isoforms of fast troponin T in some human and rat skeletal muscles. J Muscle Res Cell Motil 12: 447–54.

Sabry MA, Dhoot GK. (1991b) Identification of and pattern of transitions of cardiac, adult slow and slow skeletal muscle-like embryonic isoforms of troponin T in developing rat and human skeletal muscles. J Muscle Res Cell Motil 12: 262–70.

Saggin L, Gorza L, Ausoni S, Schiaffino S (1990) Cardiac troponin T in developing, regenerating and denervated rat skeletal muscle.Development. 110: 547–54.

Salviati G, Betto R, Danieli-Betto D. (1982) Polymorphism of myofibrillar proteins of rabbit skeletal muscle muscle fibers. An electrophoretic study of single fibers. Biochem J 207: 261–272.

Salviati G, Betto R, Danieli-Betto D, Zeviani M. (1983) Myofibrillar-protein isoforms and sarcoplasmic-reticulum Ca^{2+}-transport activity of single human muscle fibers. Biochem J 224: 215–225.

Sano K, Maeda K, Oda T, Maeda Y. (2000) The effect of single residue substitutions of serine-283 on the strength of head-to-tail interaction and actin binding properties of rabbit skeletal muscle alpha-tropomyosin. J Biochem (Tokyo)., 127: 1095–102.

Schachat FH, Bronson DD, McDonald OB. (1985) Heterogeneity of contractile proteins. A continuum of troponin-tropomyosin expression in mammalian skeletal muscle. J Biol Chem 260: 1108–13.

Schachat FH, Diamond MS, Brandt PW. (1987) Effect of different troponin T-tropomyosin combinations on thin filament activation. J Mol Biol 198: 551–554.

Schiaffino S, Reggiani C. (1996) Molecular diversity of myofibrillar proteins: gene regulation and functional significance. Physiol Rev 76: 371–423.

Schiaffino S, Ausoni S, Gorza L, Saggin L, Gundersen K, Lomo T. (1988) Myosin heavy chain isoforms and velocity of shortening of type 2 skeletal muscle fibers. Acta Physiol Scand 134: 575–6.

Seedorf K, Seedorf U, Pette D. (1983) Coordinate expression of alkali and DTNB myosin light chains during transformation of rabbit fast muscle by chronic stimulation. FEBS Lett 158: 321–324.

Sheng, Z., B. S. Pan, T. E. Miller, and J. D. Potter. (1992) Isolation, expression, and mutation of a rabbit skeletal muscle cDNA clone for troponin I. The role of the NH2 terminus of fast skeletal muscle troponin I in its biological activity. J.Biol.Chem. 267: 25407–25413.

Sia SK, Li MX, Spyracopoulos L, Gagne SM, Liu W, Putkey JA, Sykes BD. (1997) Structure of cardiac muscle troponin C unexpectedly reveals a closed regulatory domain. J.Biol.Chem. 272: 18216–18221.

Sieck GC, Regnier M. (2001) Invited Review: plasticity and energetic demands of contraction in skeletal and cardiac muscle. J Appl Physiol 90: 1158–64.

Smillie LB. (1979) Structure and functions of tropomyosins from muscle and non-muscle sources. Trends Biochem Sci 4: 151–155.

Solaro RJ., Bousquet P, Johnson D. (1986) Stimulation of cardiac myofilament force, ATPase avtivity and troponin C Ca^{2+} binding by bepridil. J Pharmacol Exp Ther 238: 502–507.

Solaro RJ, Rarick HM (1998) Troponin and tropomyosin: proteins that switch on and tune in the activity of cardiac myofilaments.Circ Res. 783: 471–480.

Sorichter S, Mair J, Koller A, Gebert W, Rama D, Calzolari C, Artner-Dworzak E, Puschendorf B. (1997) Skeletal troponin I as a marker of exercise-induced muscle dammage. J Appl Physiol 83: 1076–1082.

Spyracopoulos L, Li MX, Sia SK, Gagne SM, Chandra M, Solaro RJ, Sykes BD. (1997) Calcium-induced structural transition in the regulatory domain of human cardiac troponin C. Biochemistry 36: 12138–12146.

Staron RS, Pette D. (1987) The multiplicity of myosin light and heavy chain combinations in histo-chemically typed single fibers. Rabbit soleus muscle. Biochem J 243: 687–693.

Staron RS, Kraemer WJ, Hikida RS, Reed DW, Murray JD, Campos GE, Gordon SE. (1998) Comparison of soleus muscles from rats exposed to microgravity for 10 versus 14 days. Histochem Cell Biol 110: 73–80.

Steffen JM, Musacchia XJ. (1986) Spaceflight effects on adult rat muscle protein, nucleic acids, and amino acids. Am J Physiol 251: R1059–63.

Stephenson GM, Stephenson DG. (1993) Endogenous MLC2 phosphorylation and Ca^{2+}-activated force in mechanically skinned skeletal fibers of the rat. Pflugers Arch 424: 30–38.

Stephenson GM. (2001) Hybrid skeletal muscle fibers: a rare or a common phenomenon? Clin Exp Pharmacol Physiol 28: 692–702.

Stevens L, Mounier Y, Holy X, Falempin M (1990) Contractile properties of rat soleus muscle after fifteen days of hindlimb suspension. J Appl Physiol 68: 334–340.

Stevens L, Mounier Y. (1990) Evidences for slow to fast changes in the contractile proteins of rat soleus muscle after hindlimb suspension: studies on skinned fibers. Physiologist 33: S90–1.

Stevens L, Mounier Y, Holy X. (1993) Functional adaptation of different rat skeletal muscles to weightlessness. Am J Physiol 264: R770–6.

Stevens L, Picquet F, Catinot MP, Mounier Y. (1996) Differential adaptation to weightlessness of functional and structural characteristics of rat hindlimb muscles. J Gravit Physiol 3: 54–57.

Stevens L, Gohlsch B, Mounier Y, Pette D. (1999) Changes in myosin heavy chain mRNA and protein isoforms in single fibers of unloaded rat soleus muscle. FEBS Lett 463: 15–18.

Stevens L, Sultan KR, Peuker H, Gohlsch B, Mounier Y, Pette D. (2000) Time-dependent changes in myosin heavy chain mRNA and protein isoforms in unloaded soleus muscle of rat. Am J Physiol Cell Physiol 277: C1044–C1049.

Stevens L, Bastide B, Kischel P, Pette D, Mounier Y. (2002) Time-dependent changes in expression of troponin subunit isoforms in unloaded rat soleus muscle. Am J Physiol Cell Physiol 282: 1025–1030.

Stevens L, Bozzo C, Nemirovskaya T, Montel V, Falempin M, Mounier Y. (2003) Alterations in contractile properties and expression pattern of myofibrillar proteins in rat muscles after hypergravity. J Appl Physiol 94: 2398–405.

Stevens L, Bastide B, Bozzo C, Mounier Y. (2004) Hybrid fibers under slow-to-fast transformations: expression is of myosin heavy and light chains in rat soleus muscle. Pflugers Arch. 448: 507–14.

Suarez MC, Machado CJ, Lima LM, Smillie LB, Pearlstone JR, Silva JL, Sorenson MM, Foguel D. (2003) Role of hydration in the closed-to-open transition involved in Ca2+ binding by troponin C. Biochemistry 42: 5522–5530.

Sutoh K. (1982) Identification of myosin-binding sites on the actin sequence. Biochemistry 21: 3654–3661.

Sweeney HL, Brito RM, Rosevear PR, Putkey JA. (1990) The low-affinity Ca^{2+}-binding sites in cardiac/slow skeletal muscle troponin C perform distinct functions: site I alone cannot trigger contraction. Proc Natl Acad Sci USA 87: 9538–9542.

Sweeney HL, Stull JT. (1990) Alteration of cross-bridge kinetics by myosin light chain phosphorylation in rabbit skeletal muscle: implications for regulation of actin-myosin interaction. Proc Natl Acad Sci USA 87: 414–418.

Syska H, Wilkinson JM, Grand RJ, Perry SV. (1976) The relationship between biological activity and primary structure of troponin I from white skeletal muscle of the rabbit. Biochem J 153: 375–387.

Szczepanowska J, Borovikov YS, Jakubiec-Puka A. (1998) Effect of denervation and muscle inactivity on the organization of f-actin. Muscle Nerve 21: 309–317.

Szczesna D, Zhao J, Jones, M, Zhi G, Stull J, Potter JD. (2002) Phosphorylation of the regulatory light chains of myosin affects Ca^{2+} sensitivity of skeletal muscle contraction. J Appl Physiol 92: 1661–70.

Takeda S, Kobayashi T, Taniguchi H, Hayashi H, Maeda Y. (1997) Structural and functional domains of the troponin complex revealed by limited digestion. Eur J Biochem. 246: 611–617.

Talbot, J. A. and R. S. Hodges. (1981) Synthetic studies on the inhibitory region of rabbit skeletal troponin I. Relationship of amino acid sequence to biological activity. J Biol Chem 256: 2798–2802.

Talmadge RJ, Roy RR, Edgerton VR. (1995) Prominence of myosin heavy chain hybrid fibers in soleus muscle of spinal cord-transected rats. J Appl Physiol 78: 1256–1265.

Talmadge, RJ, Roy RR, and Edgerton VR. (1999) Persistence of hybrid fibers in rat soleus after spinal cord transection. Anat Rec 255: 188–201.

Talmadge, RJ. (2000) Myosin heavy chain isoform expression following reduced neuromuscular activity: potential regulatory mechanisms. Muscle Nerve 23: 661–679.

Tao T, Gong BJ, Leavis PC. (1990) Calcium-induced movement of troponin-I relative to actin in skeletal muscle thin filaments. Science 247: 1339–1341.

Termin, A, Staron RS, and Pette D. (1989) Changes in myosin heavy chain isoforms during chronic low frequency stimulation of rat fast hindlimb muscles. A single-fiber study. Eur J Biochem 186: 749–754.

Tesch PA, Trieschmann JT, Ekberg A. (2004) Hypertrophy of chronically unloaded muscle subjected to resistance exercise. J Appl Physiol 96: 1451–8.

Thierfelder L, Watkins H , MacRae C. (1994) α-tropomyosin and cardiac tropnin T mutations cause familial hypertrophic cardiomyopathy:a disease of the sarcomere. Cell 77: 701–712.

Thomas L, Smillie L. (1994) Comparison of the interaction and functional properties of dephosphorylated hetero- and homo-dimers of rabbit striated muscle tropomyosins. Biophys J 66: 310–339.

Thomason DB, Herrick RE, Surdyka D, Baldwin KM (1987). Time course of soleus muscle myosin expression during hindlimb suspension and recovery. J Appl Physiol. 63: 130–137.

Thomason DB, Morrison PR, Oganov V, Ilyina-Kakueva E, Booth FW, Baldwin KM. (1992) Altered actin and myosin expression in muscle during exposure to microgravity. J Appl Physiol 73: 90S-93S.

Thomason DB, Booth FW. (1990) Atrophy of the soleus muscle by hindlimb unweighting. J Appl Physiol 68: 1–12.

Tobacman LS. (1996) Thin filament-mediated regulation of cardiac contraction. Annual Rev Physiol 58: 447–481.

Tobacman LS, Lee R (1987) Isolation and functional comparison of bovine cardiac troponin T isoforms. J Biol Chem. 262: 4059–64.

Toursel T, Stevens L, Mounier Y. (1999) Differential response to unloading conditions of contractile and elastic properties of rat soleus. Exp Physiol 84: 93–107.

Tripet B, Van Eyk JE, Hodges RS. (1997) Mapping of a second actin-tropomyosin and a second troponin C binding site within the C terminus of troponin I, and their importance in the Ca^{2+}-dependent regulation of muscle contraction. J Mol Biol 271: 728–750.

Trappe S, Godard M, Gallagher P, Carroll C, Rowden G, Porter D. (2001) Resistance training improves single muscle fiber contractile function in older women. Am J Physiol Cell Physiol. 281: C398–406.

Trappe S, Trappe T, Gallagher P, Harber M, Alkner B, Tesch PT. (2004) Human single muscle fiber function with 84 day bed-rest and resistance exercise. J Physiol. 557: 501–13.

Valencia FF, Paulucci AA, Quaggio RB, Da Silva AC, Farah CS, and Reinach FF. (2003) Parallel measurement of Ca^{2+}-binding and fluorescence emission upon Ca^{2+} titration of recombinant skeletal muscle troponin C: Measurement of sequential calcium binding to the regulatory sites. J Biol Chem. 278: 11007–14.

Vandekerckhove, J., G. Bugaisky, and M. Buckingham. (1986) Simultaneous expression of skeletal muscle and heart actin proteins in various striated muscle tissues and cells. A quantitative determination of the two actin isoforms. J.Biol.Chem. 261: 1838–1843.

Vandekerckhove J, Weber K. (1979) The complete amino acid sequence of actins from bovine aorta, bovine heart, bovine fast skeletal muscle and rabbit skeletal muscle. Differentation 14: 123–133.

Vandekerckhove J, Weber K. (1984) Chordate muscle actins differ distinctly from invertebrate muscle actins. J Mol Biol 179: 391–413.

Van Eerd JP, Takahashi K. (1975) The amino acid sequence of bovine cardiac tamponin-C. Comparison with rabbit skeletal troponin-C. Biochem Biophys Res Commun 64: 122–127.

Van Eyk JE, Thomas LT, Tripet B, Wiesner RJ, Pearlstone JR, Farah CS, Reinach FC, Hodges RS. (1997) Distinct region of troponin I regulate Ca^{2+}-dependent activation and Ca^{2+} sensitivity of the acto-S1-TM ATPase activity of the thin filament. J Biol Chem 272: 10529–10537.

Vasques MS, Lang CBS, Grindeland RE, Roy R, Daunton N, Bigbee AJ, Wade CE. (1998) Comparison of hyper- and microgravity on rat muscle, organ weights and selected plasma constituents. Aviat Space Environ Med 69: 2–8.

Vibert P, Craig R, Lehman W. (1997) Steric-model for activation of muscle thin filaments. Journal of Molecular Biology 266: 8–14.

Vikstrom KL, Seiler SH, Sohn RL, Strauss M, Weiss A, Welikson RE, Leinwand LA. (1997) The vertebrate myosin heavy chain: genetics and assembly properties. Cell Struct Funct 22: 123–129.

Wada M, Inashima S, Yamada T, Matsunaga S. (2003) Endurance training-induced changes in alkali light chain patterns in type IIB fibers of the rat. J Appl Physiol 94: 923–929.

Wahrmann JP, Winand R, Rieu M. (2001) Plasticity of skeletal myosin in endurance-trained rats (I). A quantitative study. Eur J Appl Physiol 84: 367–372.

Wang J, Jin JP. (1998) Conformational modulation of troponin T by configuration of the NH2-terminal variable region and functional effects. Biochemistry 37: 14519–28.

Wang Y, Kerrick GL. (2002) The off rate of Ca^{2+} from troponin C is regulated by force-generating cross bridges in skeletal muscle. J Appl Physiol 92: 2409–2418.

Watson PA, Stein JP, Booth FW. (1984) Changes in actin synthesis and α-actin-mRNA content in rat muscle during immobilization. Am J Physiol 247: 39–44.

Whalen RG, Butler-Browne GS, Gros F. (1976) Protein synthesis and actin heterogeneity in calf muscle cells in culture. Proc Natl Acad Sci USA 73: 2018–2022.

Widrick JJ, Knuth ST, Norenberg KM, Romatowski JG, Bain JL, Riley DA, Karhanek M, Trappe SW, Trappe TA, Costill DL, Fitts RH. (1999) Effect of a 17 day spaceflight on contractile properties of human soleus muscle fibers. J Physiol 516: 915–30.

Wilkinson JM. (1980) Troponin C from rabbit slow skeletal and cardiac muscle is the product of a single gene. Eur J Biochem 103: 179–188.

Wu, YZ, Baker MJ, Crumley RL, and Caiozzo VJ. (2000) Single-fiber myosin heavy-chain isoform composition of rodent laryngeal muscle: modulation by thyroid hormone. Arch Otolaryngol Head Neck Surg 126: 874–880.

Xu C, Craig R, Tobacman L, Horowitz R, Lehman W. (1999) Tropomyosin positions in regulated thin filaments revealed by cryoelectron microscopy. Biophysical Journal 77: 985–992.

Yki-Järvinen H, Virkamaki A, Daniels MC, McClain D, Gottschalk WR. (1998) Insulin and glucosamine infusions increase O-linked N-acetyl-glucosamine in skeletal muscle proteins in vivo. Metabolism 47: 449–455.

Yonemura I, Hirabayashi T, Miyazaki J. (2000) Heterogeneity of chicken slow skeletal muscle troponin T mRNA. J Exp Zool 286: 149–156.

Yonemura I, Mitani Y, Nakada K,. Akutsu S, Miyazaki J. (2002) Developmental changes of cardiac and slow skeletal muscle troponin T expression in chicken cardiac and skeletal muscles. Zoolog Sci 19: 215–223.

Zacchara NF, O'Donnell N, Cheung WM, Mercer JJ, Marth JD, Hart GW. (2004) Dynamic O-GlcNAc modification of nucleocytoplasmic proteins in response to stress. A survival response of mammalian cells. J Biol Chem 279: 30133–30142.

Zak R, Martin AF, Prior G, Rabinowitz M. (1977) Comparison of turnover of several myofibrillar proteins and critical evaluation of double isotope method. J Biol Chem 252: 3430–3435.

Zhu L, Lyons GE, Juhasz O, Joya JE, Hardeman EC, Wade R. (1995) Developmental regulation of troponin I isoform genes in striated muscles of transgenic mice. Dev Biol 169: 487–503.

Zot AS, Potter JD. (1989) Reciprocal coupling between troponin C and myosin crossbridge attachment. Biochemistry 28: 6751–6756.

Zot, H. G. and J. D. Potter. (1982) A structural role for the Ca^{2+}-Mg^{2+} sites on troponin C in the regulation of muscle contraction. Preparation and properties of troponin C depleted myofibrils. J Biol Chem 257: 7678-7683.

CHAPTER 9

MUSCLE ARCHITECTURE AND ADAPTATIONS
TO FUNCTIONAL REQUIREMENTS

MARCO NARICI AND COSTANTINOS MAGANARIS
Manchester Metropolitan University, Alsager Campus, Alsager, Cheshire, UK

1.   INTRODUCTION

Amongst the main factors that determine the functional properties of a muscle are the type and contractile properties of muscle fibres, their internal structural organisation and macroscopic geometry in relation to the whole muscle-tendon unit. This internal design of a muscle describing the spatial arrangement of muscle fibres with respect to the axis of force development of a muscle is known as *muscle architecture*. In this chapter we specifically deal with the latter parameter and review basic theoretical concepts and recent information from experimental measurements and modelling studies aimed at revealing functional implications and physiological adaptations.

1.1   Definitions and Basic Concepts

Skeletal muscles may be categorised under two main types of architectural design, longitudinal architectures and pennate architectures. In longitudinal-architecture muscles, the muscle fibres run parallel to the action line of the muscle-tendon unit, spanning the whole muscle belly length (Fig. 1a). On the other hand, muscles with fibres arranged at an angle to the muscle-tendon action line are classified as pennate muscles. This specific angle is referred to as the pennation angle and necessitates that the fibres extend to only a part of the whole muscle belly length. If all the fibres attach to the tendon plate at a given pennation angle, the muscle is called unipennate muscle (Figs. 1b, c). Multipennate structures arise when the muscle fibres run at several pennation angles within the muscle, or when there are several

265

R. Bottinelli and C. Reggiani (eds.), Skeletal Muscle Plasticity in Health and Disease, 265–288.
© 2006 *Springer.*

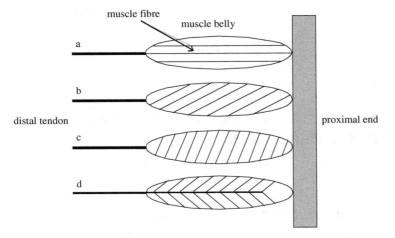

Figure 1. The main muscle architectures. a, longitudinal muscle; b and c, unipennate muscles of different pennation angles; d, bipennate muscle

distinct intramuscular parts with different pennation angles (Fig. 1d). Out of the approximately 650 muscles in the human body, most have pennate architectures with resting pennation angles up to ~30 deg (Wickiewicz et al. 1983; Friederich and Brand, 1990).

From the above definitions and illustrations it becomes readily apparent that pennation angle impacts on muscle fibre length: for a given muscle volume, or area (if volume is simplified by projecting the muscle in the sagittal plane), the larger the pennation angle the shorter the muscle fibre length relative to the whole muscle belly length. Since muscle fibre length is determined by the number of serial sarcomeres in the muscle fibre, the above relation means that pennation angle penalises the speed of muscle fibre shortening and the excursion range of fibres. However, pennation angle also has the positive effect of allowing more muscle fibres to attach along the tendon plate. The existence of more in-parallel sarcomeres therefore means that the muscle can exert greater contractile forces. However, in contrast to the proportionally increasing penalising effect of pennation angle on contractile speed and excursion range, the positive effect of pennation angle on maximum contractile force is not linear, because as pennation angle increases an increasing portion of the extra force gained in the direction of the fibres cannot be transferred through the muscle-tendon action line and effectively reach the skeleton and produce joint moment. The exact amount of this "force loss" is difficult to quantify realistically, but theoretical planar geometric models assuming that the extramuscular tendon and the intramuscular tendon plate (also known as aponeurosis) are in line (Fig. 2), indicate that it is proportional to 1 – cosine of the pennation angle. Thus, despite the trade-off between the simultaneous force gain and loss by pennation angle, it seems that as long as pennation angle does not exceed the value of 45 deg (Alexander and Vernon, 1975) the overall effect on the resultant tendon force remains positive.

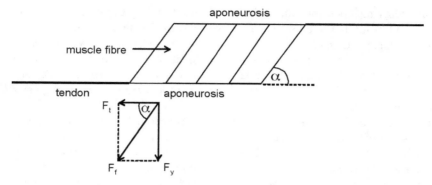

Figure 2. Vectorial analysis of forces based on a simple 2-D muscle model with tendons and aponeuroses lying over straight lines. Ff is the fibre force, Ft is the tendon force and α is the pennation angle. From trigonometry it follows that Ft = Ff̃ cos α

The quantity that best describes this very capacity of muscle to generate maximum contractile force and transmit it through its distal tendon to produce joint moment is the physiological cross-sectional area (PCSA) of muscle. This is because PCSA represents the sum of cross-sectional areas of all the fibres in a muscle (Fig. 3) and it is therefore a measure of the amount of in-parallel sarcomeres in the muscle (Fick, 1911). Diffusion tensor imaging has recently enabled the quantification of human muscle PCSA in vivo (Galban et al. 2004), but many authors have estimated PCSA from the equation $PCSA = V \cdot \cos \varphi \cdot Lo^{-1}$ (e.g., Wickiewicz et al. 1983; Fukunaga et al. 1992; Narici et al. 1996) based on the principles of muscle model in Figs. 2 and 3, where V is the muscle volume, which can be measured from consecutive magnetic resonance imaging (MRI) scans along the muscle's whole length (e.g., Fukunaga et al. 1992; Narici et al., 1992 and 1996; Maganaris et al. 2001; Maganaris, 2004), Lo is the optimal muscle fibre length and φ is the pennation angle

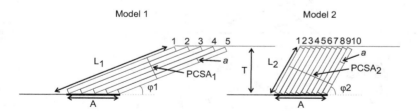

Figure 3. Two muscle models in 2-D (as in Fig. 2) with different architectural characteristics. The muscle fibres are shown as a series of tilted parallelograms between the two aponeuroses (horizontal thick line segments). A, total fibre attachment area on the aponeurosis; L, fibre length, φ, pennation angle; a, fibre cross-sectional area; PCSA, muscle physiological cross-sectional area, T, muscle thickness. The two models have the same a, A and T. However, φ2>φ1 and L2<L1. Note that although the two muscles occupy equal areas (A×T), the number of muscle fibres is 5 in model 1 and 10 and in model 2. Therefore, PCSA2 (10a) = 2PCSA1 (5a)

(measurements methods for quantification of muscle fibre length and pennation angles follow below). The above equation best exemplifies that two muscles with the same volume and anatomical cross-sectional areas (ACSA, the area of a cross section at right angles to the muscle-tendon direction rather than muscle fibre direction) may have different capabilities for maximum force generation (PCSA), and shortening speed and excursion (Lo); these concepts are discussed in greater detail in Section 2.0.

The muscle model in Figs. 2 and 3 is neither conceptually new nor complex. It was first introduced by Nicolaus Steno more than 300 years ago (Steno, 1667) and makes several simplifications. The main characteristic of the model is that all the muscle fibres are identical, i.e., they have the same length, the same CSA and the same number of in-series sarcomeres. Moreover, all the muscle fibres insert at a given pennation angle in the two aponeuroses, which run parallel to each other at a distance that does not change by muscle length changes (e.g., caused by contraction). More importantly, the aponeuroses are in-line with the extramuscular tendons, so the only factor responsible for contractile force loses from the muscle fibres to the tendon is the angle formed between the muscle fibres and the aponeurosis-tendon line of pull. Despite its conceptual and computational simplicity, this model is mechanically unstable when used for describing unipennate structures because it creates a force couple during contraction that tends to rotate and distort the muscle (Otten, 1988; Van Leeuwen and Spoor, 1992). One way to eliminate this shortcoming is to introduce an additional angle θ between the aponeuroses and the tendons, so that the line of pull of the proximal and distal tendons tendons coincide (Fig. 4). In this geometrical arrangement the force transmitted along the tendon is a function of the product $\cos \varphi \cdot \cos \theta$ (rather than $\cos \varphi$ alone in the model in Figs. 2 and 3). However, this additional angle creates another problem: mechanical equilibrium does no longer exist between the tendons and the aponeuroses where they intersect due to the different directions of the forces

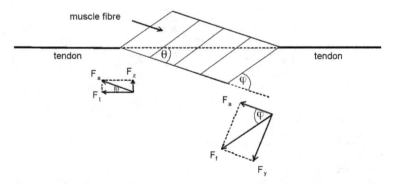

Figure 4. Vectorial analysis of forces based on a 2-D model with tendons and aponeuroses lying at an angle θ. Ff is the fibre force, Fa is the aponeurosis force, Ft is the tendon force and φ is the pennation angle. From trigonometry it follows that $Fa = Ff \cdot \cos \varphi$ and $Ft = Fa \cdot \cos \theta \rightarrow Ft = Ff. \cos \varphi. \cos \theta$

acting along them (Van Leeuwen and Spoor, 1992). Again, if this muscle model is run it will collapse, unless additional geometrical constrains are imposed, e.g., preservation of muscle area (Epstein and Herzog, 2003). Others have achieved mechanical stability in unipennate muscle models by introducing curvatures in the orientation of the muscle fibres and aponeuroses (Van Leeuwen and Spoor, 1992). For bipennate muscles, mechanical instability is not an issue of concern as long as the two unipennate parts of the muscle are symmetrical. From a modelling point of view, it is unfortunate that such muscles are rare in the human body (e.g., the tibialis anterior muscle, see *bottom* Fig. 5). Until further studies are performed to further understand and model realistically the way that myotendinous force transmission occurs, it seems that we may have to rely on computationally efficient models that are conceptually simple, or even oversimplistic. Alternatively, we may use more complex and phenomenologically correct models, whose input parameters

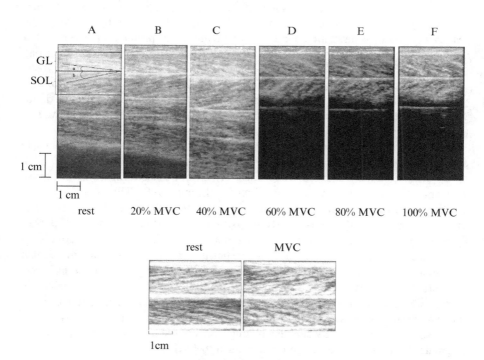

Figure 5. Top, sagittal-plane ultrasound images of the gastrocnemius lateralis (GL) and soleus (SOL) muscles at rest (A), 20% (B), 40% (C), 60% (D), 80% (E) and 100% (F) of plantarflexion maximal voluntary contraction (MVC). The horizontal stripes are ultrasonic waves reflected from the superficial and deep aponeuroses of each muscle and the oblique stripes are echoes derived from fascia septas between muscle fascicles. a is the GL pennation angle and b is the SOL pennation angle. Note the gradual increase of a, b and muscle thickness in GL and SOL from A to F. Bottom, similar sonographs of the symmetric bipennate tibialis anterior muscle at rest and dorsiflexion MVC (Reprinted with permission from Maganaris et al. 1998a, (© The Physiological Society, Blackwell Publishing) and from Maganaris & Baltzopoulos 1999 (© Springer-Verlag))

must often be tuned to match experimental results for each specific muscle studied (for a review see Maganaris 2004).

1.2 Muscle Architecture at Rest vs. Contraction: Theoretical Considerations

Methodologies for experimental characterisation of muscle architecture were first based on dissection of embalmed cadaveric specimens (Wickiewicz et al. 1983; Friederich and Brand, 1990). In such experiments V can be calculated from the mass and density of muscle, Lo can be obtained by measuring the length of dissected muscle fibre fascicles, and φ can be measured as the average value of several angles between muscle fibres and the muscle's action line, on the surface of the specimen. Comparative results of muscle architecture based on anatomical dissection have been very useful in identifying and differentiating the distinct structural characteristics of muscles (Lieber and Friden, 2000). Generally speaking, the antigravity muscles have architectures that favour force production (large PCSA values) while the antagonistic flexors are more appropriate for excursion (long muscle fibres). Based on such criteria, classification of muscles in a standardised and functionally relevant manner became possible (Lieber and Brown, 1992), thus providing guidance to the clinical decision-making in situations where a donor muscle is sought to restore functional loss, as is the case in surgical tendon transfer (Lieber and Friden, 2000).

However, it must be recognized that cadaver-based measurements of muscle architecture are unlikely to reflect accurately the physiological state of a given muscle under in vivo conditions. This is because fixation can cause substantial specimen shrinkage (Friederich and Brand, 1990). This problem can be circumvented by architectural measurements in resting alive muscles or cadaveric muscles, based on MRI scan analysis (Narici et al. 1992; Scott et al. 1993). Clearly, however, these measurements cannot account for contraction effects on muscle architecture. It may be self-evident, but it must always be remembered that muscular function cannot exist without the muscle contacting and generating force, and although resting architectural measurements have often be taken for calculating PCSA, it is the contracting state where these measurements become more relevant, especially when seeking to infer in vivo functional properties. Differences in muscle architecture between the resting and contracting states would mainly originate from the presence and mechanical behaviour the in-series tendon(s). Tendons are collagenous tissues and would therefore elongate on contraction, even if the distal and proximal ends of the muscle-tendon unit were truly fixed. The elongation of the tendon theoretically depends on a) the force produced by activation and b) the shortening potential of muscle fibres. The greater these parameters the longer the tendon should become during isometric contraction, and therefore the greater should also be the differences in muscle architecture between the resting and contracting states. For a given muscle, we would specifically expect that as the force of contraction increases, both muscle length and fibre length decrease.

The geometrical constraint imposed by the isovolumetricity of muscle (Baskin and Paolini, 1967; Epstein and Herzog, 2003) would also mean that the fibre shortening on contraction is accompanied by simultaneous pennation angles increases. Indeed, as discussed in detail below, recent advancements in imaging techniques have verified these notions through architectural measurements in human muscles in vivo, but have also revealed some unexpected findings with important implications for muscle and joint function and mechanics.

1.3   Muscle Architecture at Rest versus Contraction: Experimental Results

The main imaging technique used for characterization of muscle architectural parameters during isometric contraction is B-mode, real-time ultrasonography, the same basic scanning equipment commonly used by radiologists, obstetrics and angiologists for diagnostic purposes. The applicability of ultrasound scanning for muscle architecture measurements relates to the differential penetration of ultrasound waves to muscle fibre and collagenous tissues. Fascicles of muscle fibres are echoabsorptive and in sagittal-plane ultrasound scans appear as oblique black stripes in relation to the axis of the enire pennate muscle, with the white stripes in-between showing the arrangement of the interfascicular echoreflective collagen (Fig. 5). Muscle fascicle length (which is assumed to also represent muscle fibre length) is measured as the length of the fascicular path between the two aponeuroses, usually in more than one sites on the muscle, with or without accounting for any curvature present. If the muscle fascicles are longer than the scan window, a simplification that they extend linearly beyond the boundaries of the window has often been made without introducing large computational errors. Pennation angle is measured as the angle formed between the muscle fascicle trajectory and the aponeuroses visible on the scan, usually in proximity to the attachment points of the fascicle in the aponeuroses if curvature effects are not neglected for simplicity.

The first reports on muscle architecture measurements using ultrasonography appear in early 90's (Henriksson-Larsen et al. 1992; Rutherford and Jones, 1992). Shortly after, this technique was validated through comparisons with direct anatomical measurements of muscle fascicle lengths and pennation angles on human cadaveric muscles (Kawakami et al. 1993; Narici et al. 1996). Since then, ultrasound scanning has been applied in a number of relatively big muscles in the upper and lower-limbs to characterise the effect of contraction on muscle architecture (quadriceps muscle: Ichinose et al. 1997; Fukunaga et al. 1997, biceps femoris: Chleboun et al. 2001, triceps surae muscle: Narici et al. 1996; Kawakami et al. 1998; Maganaris et al. 1998a; tibialis anterior muscle: Ito et al. 1998; Maganaris and Baltzopoulos, 1999; Hodges et al. 2003, brachialis and biceps brachii muscles: Herbert and Gandevia, 1995; Hodges et al. 2003) The results are consistent and confirm the theoretical considerations above: as the intensity of an isometric contraction increases, fascicle length decreased and pennation angle increased in all the studies above. These two effects seem to take place along and across the entire muscle volume (Narici et al. 1996; Maganaris et al. 1998a) and may also occur in other situations where muscle length is altered, e.g., by joint rotation at a given contracting state (Narici et al. 1996; Kawakami et al. 1998; Maganaris et al. 1998a), but the effect of contraction on muscle architecture is more prominent. Contraction-induced increases in resting pennation angle by ~60–250% and reductions in muscle fascicle length by ~30–55% have been reported (Herbert and Gandevia, 1995; Narici et al. 1996; Kawakami et al. 1998; Maganaris et al. 1998a; Maganaris 1999; Hodges et al. 2003), with larger changes typically occurring at joint positions where the muscle-tendon unit is more stretched and smaller changes when the muscle-tendon unit is slacker. In many studies these changes take place in a curvilinear fashion with contraction intensity, with greater relative changes occurring at lower forces and smaller changes at greater forces (Ito et al. 1998; Maganaris et al. 1998a; Herbert and Gandevia, 1995; Hodges et al. 2003). This behaviour is in perfect agreement with the expected increasing stiffness of the in series tendons with tensile load (contractile force), a finding that has also recently been confirmed by direct measurements of in vivo human tendon mechanical properties (for a review see Maganaris, 2002).

A muscular dimension that is also related to the changes in resting fascicle length and pennation angle by contraction and has not been dealt with in detail so far is the thickness of the muscle. Muscle thickness is defined as the distance between the superficial and deep aponeuroses of the muscle and can be measured on both sagittal- and axial-plane ultrasound scans (see Fig. 5 for examples of sagittal-plane thicknesses). In pennate muscles, simultaneous increases in pennation angle and decreases in muscle fascicle length would suffice to ascertain isovolumetricity in the transition from rest to contraction. The model in Fig. 6 is similar to those in Figs. 2 and 3, but it also shows the transition from the resting to the contracting states. Note that the area of the muscle is maintained in the two states by the pivoting of the fibres about their attachment points, which results in both increasing the pennation angle and decreasing the fibre length, without the need for any additional changes in muscle thickness. The 3-D equivalent of this effect is that the ACSA of the muscle

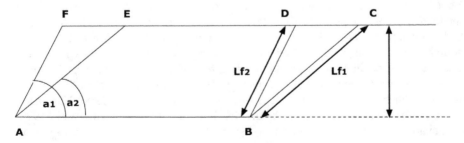

Figure 6. Modelled muscle architecture at rest (ABCE) and during isometric contraction (ABDF). Lf1 and Lf2 are the fibre lengths (Lf1>Lf2), and a1 and a2 are the pennation angles (a1>a2), in the two contracting states. t is the muscle thickness, which remains constant in the transition from rest to contraction (Reprinted with permission from Maganaris & Baltzopoulos, 1999 © Springer-Verlag)

would remain constant during contraction. However, experimental evidence has shown that this is not always the case. Several in vivo muscle architecture studies have shown that the thickness of some muscles increases with contraction intensity, in a similar fashion with the changes occurring in muscle fascicle length and pennation angle (?, ?; Maganaris et al. 1998a; Hodges et al. 2003). For example, in the gastrocemius lateralis and soleus muscles, a thickness increase of ~45% has been reported in the transition from rest to maximum isometric plantraflexion (Maganaris et al. 1998a). A thickness increase of similar magnitude occurs in the abdominal muscles by contraction, while the brachialis and biceps brachii muscle thicknesses increases gradually during contraction by up ~70% and 115% (?, ?; Hodges et al. 2003). Such sizeable muscle thickness increases by contraction can be seen even with naked eye over the skin! Regional differences in the mechanical properties of cytoskeletal and collagen structures within and around some muscles might be responsible for muscle thickening on contraction, but no firm conclusions can be reached at present.

1.4   Implications of Experimental Results

The finding of muscle architecture changes by contraction has important functional and computational implications. Failing to account for the effect of contraction on pennation angle will result in overestimating the force component transmitted from the muscle fibres to the tendon. Forward dynamics models used to simulate in vivo muscle output (e.g., Maganaris, 2004) will therefore overestimate the resultant joint moment, while inverse dynamics models used to quantify in vivo parameters such as the muscle's force-length characteristics (e.g., Ichinose et al. 1997, Maganaris, 2001; Reeves et al. 2004a) force-velocity characteristics (Ichinose et al. 2000; Reeves et al. 2005) and muscle specific tension (force/PCSA, e.g., Narici et al. 1992 and 1996; Maganaris et al. 2001; Reeves et al. 2004b; Morse et al. 2005), will underestimate the forces involved. On the other hand, failing to account for the effect of contraction on fascicle length will mainly result in underestimating PCSA and hence overestimating specific tension. Differences found in fascicle length reduction by contraction across the joint range of movement compared to fascicle lengths at rest (Herbert and Gandevia, 1995) indicate that although the external joint angle may be the same in different muscle contracting conditions, the muscle spindles may not be able to sense this, thus potentially affecting the accuracy of joint position control.

Apart from the significance of contraction-induced changes in pennation angle and muscle fascicle length, there are also several theoretical and practical implications arising from the unexpected muscular thickening by contraction.

Firstly, if the increase in muscle thickness by contraction is an active, primary process that would occur even if the surrounding muscles were absent, this means that the medio-lateral width of the muscle will have to decrease so that the muscular volume is preserved. If it wouldn't, then the intramuscular pressure might drop and the effectiveness of muscle for transmitting the contractile force it generates through its main body might be compromised. If, however, the thickness increase we observe in one muscle is the secondary effect of a medio-lateral width increase in a nearby muscle, this means that there is some mechanical work done in the axial-plane on the "weaker" muscle, which causes its deformation. Whether this is clinically relevant and to any local muscle discomfort and pain after repeated forceful contractions, for example after running – especially when unaccustomed to physical activity, remains to be investigated.

Secondly, muscle thickening combined with the high intramuscular pressure could force the in-series tendon to shift its trajectory away from the joint, thus increasing the tendon's moment arm length. The geometric model in Fig. 7 predicts that a muscle thickness increase of the order measured by ultrasound in the triceps surae muscle would rotate the Achilles tendon around its bony attachment by

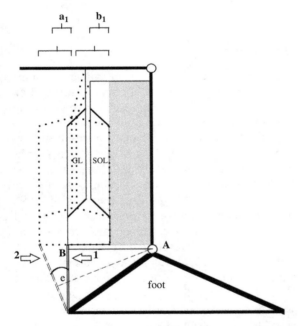

Figure 7. Musculoskeletal model of the lower extremity in 2-D. Continuous lines show the positions of muscles at rest and dotted lines show the positions of muscles during plantarflexion maximal voluntary contraction (MVC). K, knee joint; A, ankle joint; GL, gastrocnemius lateralis muscle; SOL, soleus muscle; a1 and b1, muscle thickness of GL and SOL, respectively, at rest; a2 and b2, muscle thickness of GL and SOL, respectively, during plantarflexion MVC; 1 and 2, orientations of the Achilles tendon at rest and during plantarflexion MVC, respectively; e, angle between orientations of the Achilles tendon at rest and during plantarflexion MVC; (AB) and (AC), Achilles tendon moment arm lengths at rest and during plantarflexion MVC, respectively. Tissue deformation behind the deep aponeurosis of SOL (presented as grey area) from rest to plantarflexion MVC as a result of a thickness increase in the GL and SOL muscles due to high intra-muscular pressure (Reprinted with permission from Maganaris et al. 1998b © The Physiological Society, Blackwell Publishing)

~10 deg. This estimate approximates the actual value measured (~10 deg) from ankle joint MRI analysis and accounts for most of the Achilles tendon moment arm length change caused by contraction (~25% increase, Maganaris et al. 1998b). Clearly, failure to account for this muscle architecture-related effect of contraction will result in unrealistic estimations of musculoskeletal forces and joint moment using modelling.

Finally, a change in muscle thickness means that simple models, such as the one in Fig. 6 which relies on a thickness constancy principle (as an oversimplification of the isovolumetricity principle) for predicting contraction-induced changes in muscle fascicle length and pennation angle, may be severely unrealistic. In fact, when changes in muscle architecture predicted by such a model were compared with actual, in vivo measured changes in the gastrocnemius lateralis and soleus muscles, it was found that the model underestimated the actual muscle architecture changes by ~20–40% (Maganaris et al. 1998a). An additional factor that influenced this difference between theory and experiment was that the trajectory of muscle fascicles was curved rather than straight as assumed by the model. Measurements of pennation angles at both insertion points of the fascicles indicated that the fascicular curvature increased gradually with contraction intensity, which was also directly shown in following studies and further exemplified the role of intramuscular pressure (Muraoka et al. 2002). However, it may be the case that despite the increasing intramuscular pressure, in joint positions where the muscle is very slack the curvature of fascicles decreases by contraction up to a certain intensity level, until the slackness is taken up.

2. INFLUENCES OF MUSCLE ARCHITECTURE ON MUSCLE MECHANICAL PROPERTIES

2.1   Effect on the Length-Force and Force-Velocity Relations

Muscle architecture strongly influences the two fundamental mechanical properties of skeletal muscle, the length-force relation and the force-velocity relation. On purely theoretical grounds, if one considers two muscles with an identical PCSA (same total number of sarcomeres in parallel) but with different fibre lengths (different number of sarcomeres in series), the two muscles will have distinctly different force-length and force-velocity relations (Figs. 8a,b and 9a,b). The muscle with longer fibres will generate the same absolute force as that of the shorter-fibred muscle (since their PCSA will be the same) but will be able to generate force over a greater range of lengths and will display a greater optimum length than that of the shorter-fibred muscle (Fig. 8a). As a consequence, the length-force relation of pennate muscles is narrower in shape than that of paralled-fibred muscles (Woittiez et al. 1984; Gareis et al. 1992). If one considers the force-velocity relation, a muscle with longer fibres will not only display a higher maximum shortening velocity, but will also be able to shorten at a faster velocity at any given submaximal force (Fig. 8b). Conversely, if one considers two muscles with the same fibre length

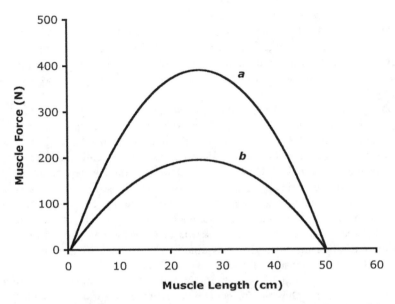

Figure 8a. The length tension relation of two muscles with the same fibre length but with a different PCSA. Note that muscle a with a larger PCSA (greater number of sarcomeres in parallel) generates a much larger force than muscle b with a smaller PCSA; however, the optimum length for maximum force generation is the same for both muscles

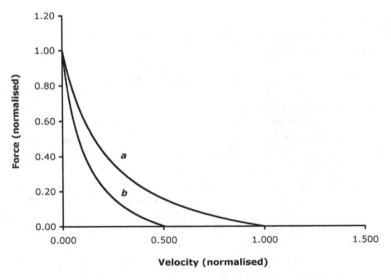

Figure 8b. The force-velocity relation of two muscles with the same PCSA but with a different fibre length. The muscle with longer fibres (a) has a much higher maximum shortening velocity than the muscle with shorter fibres (b). Since the PCSA of a and b are the same, their maximum isometric force does not differ but the force generated at any given submaximal shortening velocity is lower for muscle b

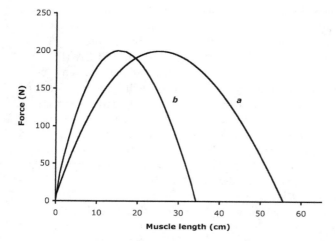

Figure 9a. The length tension relation of two muscles with the same PCSA but with different fibre lengths. Note that muscle a with longer fibres (greater number of sarcomeres in series) is able to generate force over a much larger range of lengths than muscle b with smaller fibre length and has a greater optimum length for maximum force generation than muscle b. Because their PCSA is the same, both muscles generate the same maximum isometric force

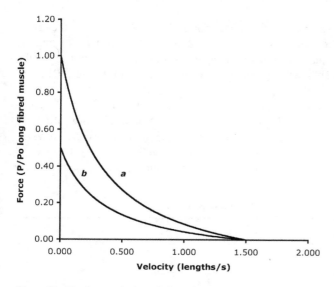

Figure 9b. The force-velocity relation of two muscles with the same fibre length but with a different PCSA. Muscle a with a larger PCSA generates much higher isometric force and dynamic force at any given submaximal shortening velocity than muscle b. However, because fibre length of a and b are the same, their maximum shortening velocity will also be the same

(same number of sarcomeres in series) but a different PCSA (different number of sarcomeres in parallel), the muscle with a greater PCSA will have the same optimum length and length-range of the length-force relation but will be able to generate a greater absolute force than that of the muscle with a smaller PCSA (Fig. 9a). Although no differences will be observed in maximum shortening velocity (since fibre length is the same) the muscle with a larger PCSA will be able to develop greater isometric force and also greater dynamic force at any given submaximal velocity (Fig. 9b). Hence its force-velocity relation will also be different from that of the muscle with a smaller PCSA. The maximum shortening velocity, however, not only depends on fibre length but also on the change in pennation angle during contraction. This is because during shortening, fibres pivot about their origin, pulling the opposite aponeurosis towards proximal end of the muscle, resulting in linear shortening of the muscle (Fig. 10).

Hence the shortening of the muscle will depend on, a) the resting length of the fibre (number of sarcomeres in series), b) the resting pennation angle, c) the pennation angle during contraction. It is thus possible to estimate the shortening velocity by simultaneously taking into account the changes in fibre length and pennation angle:

$$V = [(Lf_1^* \cos \alpha_1) - (Lf_2^* \cos \alpha_2)]/t$$

(where Lf_1 is fibre length at rest, Lf_2 is fibre length in the contracted state, α_1 is pennation angle at rest and α_2 is pennation angle in the contracted state and t is the time of the shortening phase)

All of the above factors depend in turn on tendon stiffness; for instance, the effect of a stiffer tendon will be that of decreasing the amount of shortening of muscle fibres during contraction placing fibres at a longer resting length and at a greater pennation will be associated with a smaller increase in pennation angle (Reeves et al. 2006).

When comparing the force-velocity characteristics of muscles with different fibre composition and muscle architectural characteristics, normalisation of shortening velocities for fibre length (sarcomere number) and forces for physiological cross-sectional area (PCSA) can be useful to estimate the influence of differences in myosin ATPase activity on maximum shortening velocity in the

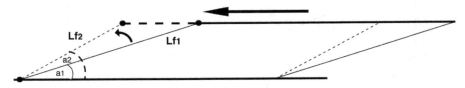

Figure 10. Scheme of a pennate muscle at rest (R) and during contraction (C). As the muscle contracts, fibres pivot about their origin increasing pennation angle from a1 to a2 and shortening from Lf1 to Lf2. In doing so, they pull the opposite aponeurosis towards proximal end of the muscle, resulting in linear shortening of the muscle

absence of difference in architecture. This approach was indeed followed by Spector et al. (1980) to compare the specific force (force/CSA) and the force-velocity characteristics of the cat soleus (SOL) and medial gastrocnelius muscles (GAS). They found that despite a nearly three-fold difference in maximum isotonic shortening velocity (Vmax) between the sarcomeres of the GAS (38.3 μm/s) and those of the SOL (13.4 μm/s), normalisation of shortening velocity to fibre length and pennation angle reduced this difference to 1.5 fold. In other words, despite a 3-fold difference in the intrinsic velocity of the GM and SOL fibres, in toto, because of the different muscle architecture, this difference in Vmax is reduced to 1.5 fold and is probably due to different myosin ATPase activities of these muscles' fibres.

Similarly, despite a 5-fold higher maximum isometric tension of the GAS compared to the SOL, accounting for differences PCSA (i.e. accounting for muscle mass, fibre length and pennation angle) produced very similar values of specific tensions (SOL 23 N·cm^{-2}, GAS 21.2 N·cm^{-2}, Spector et al. 1980).

With the introduction of modern imaging techniques such as ultrasound, it has become possible to study the force-velocity features of muscles in vivo by measuring the shortening of muscle fibres during isokinetic contractions. Using this technique, Ichinose et al. (2000) investigated human vastus lateralis (VL) fascicles behaviour during shortening contractions at 30°/s and 150°/s. Interestingly, they found that despite the imposed contraction was 'isokinetic' the VL fascicles changed their speed of shortening during the movement. Also, reading peak torque at constant joint angle did not correspond to a constant length of the fascicles, the latter was more constant if torque were read at the joint angle corresponding to peak torque. These finding are particularly interesting since they showed that fascicle behaviour can be very different from that imposed mode of contraction and that in order to estimate the F-V characteristics of the VL, it is more correct to use peak torque rather isoangular torque. The authors concluded that the change in fascicle shortening velocity during the isokinetic contractions at the above two angular velocities was mainly due to the elongation of the tendon. These conclusions are similar to those reached by Reeves & Narici (2003) regarding the behaviour of tibialis anterior fascicles during lengthening and shortening 'isokinetic' contractions recorded in vivo using ultrasound. These authors showed that fascicle length increased with contraction velocity, and attributed this to phenomenon a progressively smaller stretch of the series elastic component (Fig. 11). During the eccentric phase, because of the rather constant torque across the different lengthening velocities, fascicles behaved quasi isometrically (Fig. 11).

2.2   Muscle Fascicle Behaviour in Common Locomotory Tasks

Other examples of fibre fascicle behaviour strikingly different from those expected from classical physiology, are found in common locomotory tasks such as walking, jumping and stair negotiation. In a pioneering experiment in which the behaviour of human muscle and tendons during treadmill walking was studied in vivo using ultrasound, Fukunaga et al. (2001) showed that during the active push phase, in

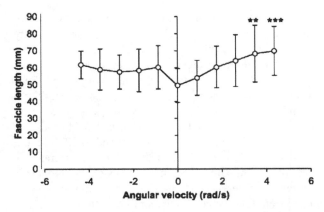

Figure 11. Tibialis anterior muscle fascicle length during concentric and eccentric contractions. Values presented are means±SD. ** and *** respectively denote significant differences of P<0.01 and P<0.001 with respect to 0 rad/s. (Reprinted with permission from Reeves ND, Narici MV (2003) J Appl Physiol 95:1090-6 © American Physiological Society)

which the muscle is expected to shorten, the GAS fascicles actually behave almost isometrically while the gastrocnemius tendon is actually stretched and only at the end of the push phase, the energy stored in the tendon is released as the tendon recoils to its original length while the fascicles shorten (Fig. 12). Similarly, during drop landing from different heights, Ishikawa & Komi (2004) demonstrated that during the breaking phase, contrary to reasonable expectations, lengthening of the VL fascicles decreased and tendinous lengthening increased as the dropping height increased. An even greater paradoxical behaviour of fascicles is observed during stair descent in older people. During this task, Spanjaard et al. (2006) found that the fascicles of the GAS muscle actually shorten, rather than lengthen, during the 'eccentric' muscle breaking phase.

Hence these findings illustrate that, *in vivo*, muscle fibre fascicles can behave very differently from what predicted from isolated muscle experiments; fascicles can actually shorten or behave quasi isometrically while joint angles are increasing and even when the breaking force required is increased, as in drop landing.

2.3 Interaction Between Muscle Architecture and Tendon Mechanical Properties

It is clear from the preceding section that the changes of muscle architecture upon contraction are heavily dependent on tendon extensibility. This is because tendons are both anatomically and functionally placed in series with the muscle. Tendons can be viewed as springs attached in series with the contractile component, the sarcomeres, (Fig. 13). Their extensibility is directly proportional to their length and inversely proportional to their cross-sectional area. Hence long and thin tendons, such as the Achilles tendon, are more easy to stretch than short and thick tendons, such as the patellar tendon.

From Fig. 13 it is clear that the degree of overlap of the actin and myosin filaments depends on the stiffness of the tendon. As pointed out in Section 2.1, the shortening of the entire muscle, however, will also depend on the number of contractile units (sarcomeres) placed in series, that is to say will depend on fibre length. Hence the muscle shortening upon contraction will depend on fibre length (pennation angle, already discussed in the preceding section) and tendon stiffness. Such interaction can now be studied *in vivo* using ultrasonography. By placing an ultrasound probe on the myotendinous juction of the tibialis anterior muscle, Maganaris & Paul (1999) were able to measure *in vivo* the tendon deformation with increasing tendon loads during a series of isometric contractions. By plotting tendon deformation, expressed as percent change from its initial length (strain), against tendon load, expressed as load per cross-sectional area (stress) they obtained the classical stress-strain relation, the slope of which represents tendon material stiffness (Fig. 14). This approach has now been applied by several authors on various tendons of the lower limbs, both in young adults as well as on older individuals. The results obtained on older individuals are of particular interest in this context since they illustrate very clearly the close interaction between muscle and tendons. In old age, fibre fascicle length and pennation angle decrease because of sarcopenia (Narici et al. 2003; Fig. 15). Hence, because of the shorter fibre length, and thus lower number of sarcomeres in series, the fibre shortening upon contraction would be expected to be smaller in old age. However, because tendon stiffness decreases with ageing (Onambele et al., 2006), tendons are more easily to stretch upon contraction, resulting in greater fibre shortening than what would be obtained if tendon stiffness did not decrease. Thus there seems to be a compensation between the changes in muscle architecture and in tendon stiffness with ageing and it has been suggested that this effect may enable muscle fibres to operate closer to the optimum region of force generation of the length-tension relationship (Reeves et al. 2006). Although ageing leads to marked alterations in muscle architecture and in tendon mechanical properties, even in old age the musculoskeletal system displays considerable plasticity to training. These adaptations are another excellent example of the close interaction operating between muscles and tendons.

2.4   Adaptations to Training

Muscle architecture displays considerable malleability to regimes of increased loading. This is clearly shown by the muscular adaptations of body-builders after years of training and also of sedentary individuals after months of strength training. Kawakami et al. (1993) found pennation angle of the triceps brachii to be greater in body builders than in sedentary individuals. Further evidence that muscle hypertrophy is associated with an increase in pennation angle has been provided by studies in which changes in muscle architecture were measured before and after

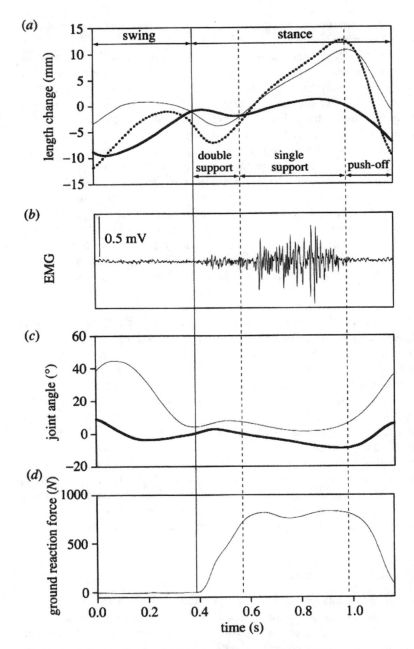

Figure 12. a) Typical changes in GAS fascicular length (thick line), muscle-tendon complex length (dotted line), and tendon length (thin line) in one step cycle. Positive and negative values indicate elongation and shortening, respectively; data are shown relative to heel strike. b) EMG recordings of the GAS muscle, c) ankle joint (thin line) and knee joint (thick line) angles, and d) vertical component of

strength training both in young and elderly people (Aagaard et al. 2001; Kanehisa 2002; Morse et al. 2006). The reason why pennation increases with hypertrophy is simply due to the fact that as fibres undergo hypertrophy, the only way to arrange more contractile tissue along the tendon aponeurosis (of set length) is to increase pennation. In addition to the increase in pennation angle, strength training in older individuals has been shown to lead to an increase in fibre fascicle length. As a matter of fact, after 14 weeks of concentric/eccentric weightlifting exercises of the knee extensors, VL fascicle length increased by 11% (Reeves et al. 2005). This is of particular interest for it demonstrates that some of the changes in muscle architecture associated with ageing may be actually reversed by strength training.

The adaptations to strength training not only concern the muscular but also the tendinous tissue. In older individuals, 14 weeks of resistance training has been shown to increase tendon stiffness by 65% (Reeves et al. 2003). These tendinous adaptations combined with the aforementioned increase in fascicle length are thought to play a positive role in the mechanical behaviour of fibre fascicles upon contraction. Since the fascicles of the VL muscle operate on the left of the length-tension relation, an increase in tendon stiffness would cause a smaller shortening of the actomyosin filaments enabling the fascicles to operate closer to the optimal region of force generation (Reeves et al. 2006).

From these findings it may be concluded that in order to understand the mechanical behaviour of skeletal muscle and its functional adaptations to strength training, the role of tendons cannot be neglected but must be considered together with that of the muscle.

2.5   Adaptations to Disuse

Disuse has long been known to lead to muscle atrophy, a phenomenon that mainly arises from an increase in protein breakdown as well as from a depression, though of minor magnitude, of protein synthesis (Edgerton & Roy, 1996; Edgerton et al. 2002; di Prampero & Narici, 2001). Similar to what observed in senile sarcopenia, disuse atrophy in humans also entails a decrease in fascicle length and in pennation angle (Narici & Cerretelli, 1998; Kawakami et al. 2000; Bleakney & Maffulli 2002; Reeves et al. 2002). After 90 days of strict bed rest in healthy young males (the study of longest duration out of those listed), pennation angle and fascicle length of the GAS muscle were found to decrease by 10% and 13%, respectively, in those participants not performing exercise countermeasures. Always in the same study, those participants performing countermeasures (high-intensity resistive exercise performed every three days), showed only a partial mitigation of muscle atrophy and

---

Figure 12. ground reaction force during the step cycle. For joint angles, zero degree represents the neutral position for the ankle and full extension for the knee. For the ankle, positive and negative angles respectively represent plantarflexion and dorsiflexion angles. (Reprinted with permission from Fukunaga et al. Proc. R Soc Lond B, 229–233, 2001, © Royal Society)

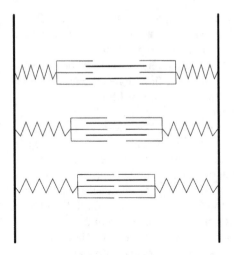

Figure 13. A scheme of a muscle tendon complex. The tendons are represented by springs connected in series with the contractile component represented by a single sarcomere. The overlap of the actin (thin lines) and mysosin filaments (thick lines) depends on the elongation of the springs an thus on tendon stiffness

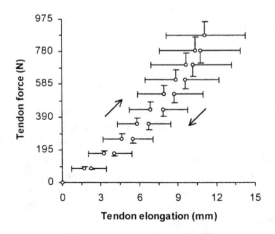

Figure 14. Gastrocnemius tendon force elongation relationship during loading (upward arrow) and unloading (downward arrow). Values are means±SD. (Reprinted with permission from Maganaris & Paul, J Biomech 35, 1639–1646, 2002, © Elsevier B.V.)

changes in muscle architecture were only marginally smaller than those of the no-exercise group (7% decrease in fascicle length and 13% decrease in pennation angle) indicating that a greater volume of exercise would be required to prevent GAS atrophy. These findings suggest that changes in the changes in muscle architecture associated with disuse involve a loss of sarcomeres both in parallel (reduction in

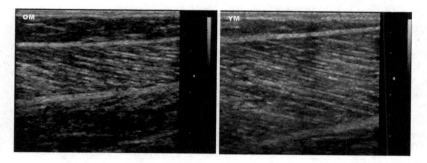

Figure 15. Ultrasound pictures obtained in an old (OM, aged 75 years) and in a young man (YM, aged 27 years). Note that in OM both fascicle length and pennation angle are much smaller than in YM

muscle CSA) and in series (reduction in fascicle length) and as such would be expected to play a significant role in the loss of muscle force and power observed after prolonged disuse.

CONCLUSIONS

Muscle architecture is a main determinant of skeletal muscle's static and dynamic contractile behaviour. However, the study of muscle architecture *in vivo*, mainly by ultrasonography, has highlighted that fascicle behaviour may be very different from that observed in *in vitro* conditions. Imposed contraction modes such as 'isometric', 'isokinetic', 'concentric' and 'eccentric' are often not reflected by the contractile behaviour of the fibre fascicles. This different fascicle behaviour between *in vitro* and *in vivo* conditions is mainly due to the fact that in vivo muscles operate in series with tendons whose extensibility and tensile properties affects the degree and speed of shortening of muscle fibres, range of movement and energy cost of locomotion. Muscle architecture shows remarkable plasticity in response to training, disuse and ageing. The challenge for the future will be that of identifying the molecular mechanisms and signalling pathways involved in the addition or removal of sarcomeres in these conditions and further elucidate the complex interaction between fibre fascicles and tendons *in vivo*.

REFERENCES

Aagaard P, Andersen JL, Dyhre-Poulsen P, Leffers AM, Wagner A, Magnusson SP, Halkjaer-Kristensen J, Simonsen EB (2001). A mechanism for increased contractile strength of human pennate muscle in response to strength training: changes in muscle architecture. J Physiol 534:613–23.

Alexander RMcN and Vernon A (1975). The dimensions of knee and ankle muscles and the forces they exert. J Human Mov Studies 1:115–23.

Baskin RJ and Paolini PJ (1967). Volume change and pressure development in muscle during contraction. Am J Physiol 213:1025–30.

Bleakney R, Maffulli N (2002). Ultrasound changes to intramuscular architecture of the quadriceps following intramedullary nailing. J Sports Med Phys Fitness 42:120–5.

Chleboun GS, France AR, Crill MT, Braddock HK and Howell JN (2001). In vivo measurement of fascicle length and pennation angle of the human biceps femoris muscle. Cells Tissues Organs 169:401–9.

di Prampero PE, Narici MV (2001). Muscles in microgravity: from fibres to human motion. J Biomech 36: 403–12.

Edgerton, VR and Roy, RR (1996). Neuromuscular adaptations to actual and simulated spaceflight. In: Handbook of Physiology. Section 4. Environmental Physiology. III. The Gravitational Environment, Chapt. 32. M.J. Fregly and C. Blatteis (eds.), Oxford University Press, New York, p 721–763.

Edgerton VR, Roy RR, Allen DL, Monti RJ (2002). Adaptations in skeletal muscle disuse or decreased-use atrophy. Am J Phys Med Rehabil 81:S127–47. Review.

Epstein M and Herzog W (2003). Aspects of skeletal muscle modelling. Philos Trans R Soc Lond B Biol Sci 358:1445–52.

Fick R (1911). *Spezielle Gelenk und Muskelmechanik*. Vol. I. Gustav Fischer, Jena.

Friederich JA and Brand RA (1990). Muscle fiber architecture in the human lower limb. J Biomech 23:91–5.

Fukunaga T, Kubo K, Kawakami Y, Fukashiro S, Kanehisa H, Maganaris CN (2001). In vivo behaviour of human muscle tendon during walking. Proc Biol Sci 268: 229–33.

Fukunaga T, Ichinose Y, Ito M, Kawakami Y and Fukashiro S (1997). Determination of fascicle length and pennation in a contracting human muscle in vivo. J Appl Physiol 82:354–8.

Fukunaga T, Roy RR, Shellock FG, Hodgson JA, Day MK, Lee PL, Kwong-Fu H and Edgerton VR (1992). Physiological cross-sectional area of human leg muscles based on magnetic resonance imaging. J Orthop Res 10:928–34.

Galban CJ, Maderwald S, Uffmann K, de Greiff A and Ladd ME. (2004). Diffusive sensitivity to muscle architecture: a magnetic resonance diffusion tensor imaging study of the human calf. Eur J Appl Physiol 93:253–62.

Gareis H, Solomonow M, Baratta R, Best R, D'Ambrosia R (1992). The isometric length-force models of nine different skeletal muscles. J Biomech 25:903–16.

Henriksson-Larsen K, Wretling ML, Lorentzon R and Oberg L (1992). Do muscle fibre size and fibre angulation correlate in pennated human muscles? Eur J Appl Physiol 64:68–72.

Herbert RD, Gandevia SC (1995). Changes in pennation with joint angle and muscle torque: In vivo measurements in human brachialis muscle. J Physiol 484: 523–532.

Hodges PW, Pengel LH, Herbert RD and Gandevia SC (2003). Measurement of muscle contraction with ultrasound imaging. Muscle Nerve 27:682–92.

Ichinose Y, Kawakami Y, Ito M, Kanehisa H and Fukunaga T (2000). In vivo estimation of contraction velocity of human vastus lateralis muscle during "isokinetic" action. J Appl Physiol 88:851–6.

Ichinose Y, Kawakami Y, Ito M and Fukunaga T (1997). Estimation of active force-length characteristics of human vastus lateralis muscle. Acta Anat (Basel) 159:78–83.

Ishikawa M, Komi PV (2004). Effects of different dropping intensities on fascicle and tendinous tissue behavior during stretch-shortening cycle exercise. J Appl Physiol 96:848–52.

Ito M, Kawakami Y, Ichinose Y, Fukashiro S and Fukunaga T (1998). Nonisometric behavior of fascicles during isometric contractions of a human muscle. J Appl Physiol 85:1230–5.

Kanehisa H, Nagareda H, Kawakami Y, Akima H, Masani K, Kouzaki M, Fukunaga T (2002). Effects of equivolume isometric training programs comprising medium or high resistance on muscle size and strength. Eur J Appl Physiol 87:112–9.

Kawakami Y, Ichinose Y and Fukunaga T (1998). Architectural and functional features of human triceps surae muscles during contraction. J Appl Physiol 85:398–404.

Kawakami Y, Abe T and Fukunaga T (1993). Muscle-fiber pennation angles are greater in hypertrophied than in normal muscles. J Appl Physiol 74:2740–4.

Kawakami Y, Muraoka Y, Kubo K, Suzuki Y, Fukunaga T (2000). Changes in muscle size and architecture following 20 days of bed rest. J Gravit Physiol. 2000 Dec;7(3):53–9.

Lieber RL and Friden J (2000). Functional and clinical significance of skeletal muscle architecture. Muscle Nerve 23:1647–66.

Lieber RL and Brown CC (1992). Quantitative method for comparison of skeletal muscle architectural properties. J Biomech 25:557–60.

Maganaris CN (2004). A predictive model of moment-angle characteristics in human skeletal muscle: application and validation in muscles across the ankle joint. J Theor Biol 230:89–98.

Maganaris CN (2002). Tensile properties of in vivo human tendinous tissue. J Biomech 35:1019–27.

Maganaris CN (2001). Force-length characteristics of in vivo human skeletal muscle. Acta Physiol Scand 172:279–85.

Maganaris CN, Baltzopoulos V, Ball D and Sargeant AJ (2001). In vivo specific tension of human skeletal muscle. J Appl Physiol 90:865–72.

Maganaris CN and Baltzopoulos V (1999). Predictability of in vivo changes in pennation angle of human tibialis anterior muscle from rest to maximum isometric dorsiflexion. Eur J Appl Physiol 79:294–7.

Maganaris CN, Baltzopoulos V and Sargeant AJ (1998a). In vivo measurements of the triceps surae complex architecture in man: implications for muscle function. J Physiol 512:603–14.

Maganaris CN, Baltzopoulos V and Sargeant AJ (1998b). Changes in Achilles tendon moment arm from rest to maximum isometric plantarflexion: In vivo observations in man. J Physiol 510: 977–85.

Maganaris CN, Paul JP. In vivo human tendon mechanical properties (1999) J Physiol 52: 307–13.

Morse CI, Thom JM, Reeves ND, Birch KM & Narici MV (2005). In vivo physiological cross-sectional area and specific force are reduced in the gastrocnemius of elderly men. J Appl Physiol 99:1050–5.

Morse CI, Thom JM, Mian SM, Birch KM & Narici MV (2006). Gastrocnemius specific force is increased in elderly males following a twelve month physical training programme. Eur J Appl Physiol (in press).

Muramatsu T, Muraoka T, Kawakami Y, Shibayama A and Fukunaga T (2002). In vivo determination of fascicle curvature in contracting human skeletal muscles. J Appl Physiol 92:129–34.

Narici MV, Binzoni T, Hiltbrand E, Fasel J, Terrier F and Cerretelli P (1996). In vivo human gastrocnemius architecture with changing joint angle at rest and during graded isometric contraction. J Physiol 496:287–97.

Narici M & Cerretelli P (1998). Changes in human muscle architecture in disuse-atrophy evaluated by ultrasound imaging. J Gravit Physiol 5: P73–4.

Narici MV, Landoni L, Minetti AE (1992). Assessment of human knee extensor muscles stress from in vivo physiological cross-sectional area and strength measurements. Eur J Appl Physiol 65:438–44.

Narici MV, Maganaris CN, Reeves ND, Capodaglio P (2003). Effect of aging on human muscle architecture. J Appl Physiol 95: 2229–34.

Onambele GL, Narici MV, Maganaris CN (2006). Calf muscle-tendon properties and postural balance in old age. J Appl Physiol. 100: 2048–2056.

Otten E (1988). Concepts and models of functional architecture in skeletal muscle. Exerc Sport Sci Rev 16:89–137.

Reeves NJ, Maganaris CN, Ferretti G, Narici MV (2002). Influence of simulated microgravity on human skeletal muscle architecture and function. J Gravit Physiol 9:P153–4.

Reeves ND, Narici MV and Maganaris CN (2004a). In vivo human muscle structure and function: adaptations to resistance training in old age. Exp Physiol 89:675–89.

Reeves ND, Narici MV and Maganaris CN (2004b). Effect of resistance training on skeletal muscle-specific force in elderly humans. J Appl Physiol. 96:885–92.

Reeves ND, Maganaris CN and Narici MV (2005). Plasticity of dynamic muscle performance with strength training in elderly humans. Muscle Nerve 31:355–364.

Reeves ND, Narici MV, Maganaris CN (2006). Myotendinous plasticity to ageing and resistance exercise. Exp Physiol [Epub ahead of print]

Reeves ND, Narici MV, Maganaris CN (2003). Strength training alters the viscoelastic properties of tendons in elderly humans. Muscle Nerve 28:74–81.

Reeves ND, Narici MV (2003). Behavior of human muscle fascicles during shortening and lengthening contractions in vivo. J Appl Physiol 95:1090–6.

Rutherford OM and Jones DA (1992). Measurement of fibre pennation using ultrasound in the human quadriceps in vivo. Eur J Appl Physiol 65:433–7.

Scott SH, Engstrom CM and Loeb GE (1993). Morphometry of human thigh muscles. Determination of fascicle architecture by magnetic resonance imaging. J Anat 182:249–57.

Steno N (1667). Elementorum Myologiae Specimen, seu Musculi Descriptio Geometrica. Frolence.

Spanjaard M., Reeves N.D., van Dieën J.H., Baltzopoulos V., and Maganaris C.N. (2006). Human muscle fascicle behaviour during stair negotiation. 5th Word Congress of Biomechanics. Munich, July-August 2006.

Spector SA, Gardiner PF, Zernicke RF, Roy RR, Edgerton VR (1980). Muscle architecture and force-velocity characteristics of cat soleus and medial gastrocnemius: implications for motor control. J Neurophysiol 44: 951–60.

Van Leeuwen JL and Spoor CW. (1992). Modelling mechanically stable muscle architectures. Philos Trans R Soc Lond B Biol Sci 336:275–92.

Wickiewicz TL, Roy RR, Powell PL and Edgerton VR (1983). Muscle architecture of the human lower limb. Clin Orthop Relat Res 179:275–83.

Woittiez RD, Huijing PA, Boom HB, Rozendal RH (1984). A three-dimensional muscle model: a quantified relation between form and function of skeletal muscles. J Morphol 182: 95–113.

CHAPTER 10

RESPONSES AND ADAPTATIONS OF SKELETAL MUSCLE TO HORMONES AND DRUGS

STEPHEN D.R. HARRIDGE

King's College London, London SE1 1UL, UK

1.   INTRODUCTION

The prime, but not only function of muscle, is to serve as a biological machine; producing force, generating power and braking movement. For these purposes muscle converts the energy stored in the chemical bonds of adenosine triphosphate (ATP) into mechanical work and heat. As an example, during cycling we exercise with an efficiency of about 20–25%. The mechanical properties of a muscle, as characterised by the length-tension and force-velocity relationships, are ultimately determined by the behaviour of the molecular motors, the myosin cross bridges which work in concert with other myofibrillar proteins to generate contraction. The maximum isometric force or "strength" of a muscle is determined primarily by its physiological cross-sectional area, or the number of sarcomeres working in parallel with one another. At the other end of the force-velocity relationship, the prime determinants of maximum shortening velocity (V_{max} or V_o), are muscle length (sarcomeres in series), temperature and myosin heavy chain (MHC) isoform composition. At any given instant, the power output of a muscle is a product of speed of movement and force of contraction. For the athlete, the generation of maximum power outputs are prerequisites for high level performance in "explosive" activities such as weight-lifting, sprinting, and in field events such as jumping and throwing. The generation of critical levels of power are also a prerequisite for performing simple tasks of daily living, such as rising from a chair or climbing onto a bus. For a young healthy person these activities can be performed easily, but for the frail elderly person or those suffering from diseases of which muscle loss is a feature these simple tasks become increasingly difficult and sometimes impossible to perform. In these situations, the loss of muscle mass has serious implications for

289

R. Bottinelli and C. Reggiani (eds.), Skeletal Muscle Plasticity in Health and Disease, 289–314.
© 2006 *Springer.*

independent living and quality of life. However, muscle is not only a biological machine. Muscle also has a role as a dynamic metabolic reservoir, as a source of heat and as a form of protective padding, roles which are all compromised when muscle mass is lost such as in the sarcopenia associated with old age (Griffiths et al., 2000). This chapter will concern itself with how hormones and/or drugs may modulate the ability of a muscle to generate force, its speed of movement and its potential to generate power. This will be viewed in the first instance in the context of substances that cause acute effects on muscle force and velocity. In other words, substances that act by regulating function without altering muscle size or composition. A rise in muscle temperature, for example, affects the velocity of shortening and power output without phenotypic changes. This is followed by an examination of some of the key hormones and drugs that may alter function as a result of causing phenotypic adaptations through changes in size and protein isoform composition.

2.   REGULATING MUSCLE FUNCTION WITHOUT CHANGING PHENOTYPE

During a voluntary contraction the amount of force generated by a muscle is the end product of a number of events which involve the recruitment and firing of specific motor units. During a maximal contraction this involves the recruitment of all motor units and their firing at sufficiently high frequencies. At the level of each muscle cell sufficient Ca^{2+} must remain in the sarcoplasm bound to troponin to allow cross-bridge cycling to stretch the series elements such that the external force produced eventually reflects that being generated internally by the cross-bridges. As muscle force is directly proportional to its physiological cross-sectional area, its strength might be increased by obtaining superior activation of the muscle through better neural drive, improving intracellular $[Ca^{2+}]$ or increasing the sensitivity of troponin to Ca^{2+}.

It has long been known that cardiac output can be increased through the administration of inotropes such as adrenaline and noradrenaline to increase force of contraction and thus stroke volume. Such approaches are commonplace in patients in critical care. This increase in force comes as a result of increasing Ca^{2+} flux into cardiac muscle cells, following the phosphorylation of the Ca^{2+} channels of the plasma membrane by protein kinase A. In skeletal muscle there is no contribution from extracellular Ca^{2+} and such potent inotropes do not exist. However there is evidence for inotropic-like effects of at least three substances, namely caffeine, nitric oxide and oestrogen.

2.1   Caffeine

One of the main causes of force decline during repeated fatigue-inducing contractions has been demonstrated to be a reduction in the amount of Ca^{2+} released by the sarcoplasmic reticulum (SR) (Westerblad and Allen, 1991). Simultaneous measurements of both force and intracellular $[Ca^{2+}]$ through Ca^{2+}indicators such as aquaorin and fura-2 AM have shown that caffeine administration can stimulate Ca^{2+} release from the SR. Caffeine does this by acting on the ryanodine receptors which regulate the opening of the Ca^{2+} release channels. In fatigued isolated single mouse muscle fibres, administration of caffeine results in a restoration of Ca^{2+} release and a recovery of force (Westerblad and Allen, 1991). More recently, infusion of caffeine to fatiguing canine muscle *in situ* has also shown a similar effect (Howlett et al., 2005), although the magnitude of recovery is considerably smaller than that reported in isolated single fibre experiments.

2.2   Nitric Oxide

Nitric oxide (NO) is an important signalling substance that affects a wide variety of physiological functions through cGMP-dependent and independent pathways (Moncada et al., 1991). NO is synthesised from L-arginine by NO synthases (NOS) in a Ca^{2+} / calmodulin-dependent manner. Skeletal muscle is one of the few tissues to simultaneously express all the known NOS isoforms although little is known about the mechanisms which control NOS expression. A number of studies have been performed on muscle using both NOS blockers and NO donors to study the effects of NO on contractile function. As regards the use of NOS blockade the results seem to be quite consistent. NOS blockade reduces local production of NO and its physiological effects. For example, inhibition of NOS leads to a reduction in maximal force and Ca^{2+} sensitivity, the latter reflected by a leftward shift in the force-frequency relationship (Maréchel and Guilly, 1999). In contrast, there is less agreement between studies using NO donors, which Maréchel and Guilly (1999) suggest is because in normal muscle the level of NO may be close to maximal, and thus NO donors may have only a relatively small effect. However, administration of NO donors is sufficient to reverse the negative effects of NOS inhibitors.

In contrast to force, the speed of muscle shortening is affected by MHC isoform composition. Fibres containing the MHC-I isoform exhibit V_o and power outputs that are markedly lower than those of fibres containing MHC-IIa isoforms whilst fibres containing the MHC-IIx isoforms have the highest values (Bottinelli et al., 1996). Morrison et al., (1998) showed that bundles of rat diaphragm muscle fibres exhibited a reduction in V_{max} and power output when incubated in Krebs-Ringer solution containing the NO synthase inhibitor N^{ϖ} –nitro-L-aginine (10nm) (Fig. 1). In contrast to V_{max}, V_o (as determined by slack tests) was unchanged. In a mixed bundle of fibres, V_o mainly reflects the behaviour of the fastest fibres and this suggests that NO sensitivity differs among fibre types and plays a more regulatory role in the slower, type I fibres. As all studies were performed with muscles fully active prior to shortening, it would appear that NO plays a role in facilitating the rate of cross-bridge cycling. Maréchel and Guilly (1999) propose that this occurs by a cGMP/PKG-mediated myosin phospho-rylation, which does not alter force, but means that the rate limiting steps of the cross-bridge cycle are accelerated. There thus seems to exist in muscle an NO regulated "acute" plasticity in contraction, which like changes in temperature, allow

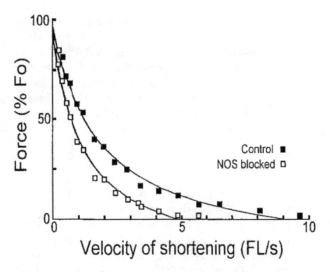

Figure 1. Alteration (slowing) of the force-velocity relationship in rat muscle following NOS blockade
(Reproduced with permission from Maréchel and Guilly, 1999, © Birkhäuser Verlag, Basel, Switzerland)

short term adaptability to challenges placed upon a muscle without the need for
recourse to remodelling of the contractile proteins, a process that may take many
days / weeks.

2.3   Oestrogen

There is some evidence to suggest that the female sex hormone oestrogen plays a
role regulating muscle force production. Sarwar et al. (1996) reported that maximal
voluntary isometric knee extensor strength in young females fluctuated with different
phases of the menstrual cycle. Force was on average 50N greater during the mid-
cycle phase when levels of oestrogen are highest. These data were supported by
Phillips et al. (1996) who studied the force of the adductor pollicis muscle and reported
a 10% increase in maximum voluntary isometric force during the follicular phase
when oestrogen levels are rising. In support of a role played by oestrogen, a cross-
sectional study of the adductor pollicis muscle in women of increasing age showed a
noticeable reduction in force per unit area (specific force) at a time corresponding to
the onset of the menopause, when oestrogen levels are dramatically falling (Phillips
et al., 1993a). In women who were taking oestrogen and progesterone supplements
(hormone replacement therapy, HRT) this fall was not observed. In a subsequent longitudinal study (Skelton et al., 1999), these observations of a positive effect of HRT
(oestrogen/norgestrel) on specific muscle force production were confirmed, with those
women prescribed HRT showing a 12% increase in adductor pollicis force as compared
to a decrease of 3% over the 12 months in the non-treated control group. No change

was observed in the muscle CSA of either group. However, transient increases in circulating oestrogen status, such as the supraphysiological production of oestrogen that occurs in response to pituitary gonadotrophin releasing hormone (GnRH) in women undergoing fertility treatment, do not seem to have an effect on strength (Greeves et al., 1997). The lack of any apparent regulation of muscle strength by oestrogen has also been observed in some other studies in young (e.g. Elliot et al.., 2002) and older women (Taaffe et al., 1995). However, it is noteworthy that in the study of Skelton et al., (1999), no significant effects were observed in the HRT supplement group until after 26 weeks of HRT administration. Thus, whatever the mechanism of the effect of oestrogen on muscle specific force, it would not appear to adapt rapidly to changes in oestrogen status.

A possible mechanism of action is a direct effect of oestrogen on the cross-bridges themselves. Electron paramagnetic studies of single muscle fibres from aged and young rats (Lowe et al., 2002) showed that a 27% lower specific force in the older animals was associated with a lower percentage of cross-bridges in the high-force state (22% versus 32%) Interestingly, in contrast to the decline in isometric force, Philips (1993b) reported no effect on adductor pollicis force during stretch, i.e. stiffness remained unchanged. This suggests no change in the number of cross-bridges with age, but a reduction in the force generated by each cross-bridge. Indeed, a similar phenomenon, (change in isometric, but not stretch forces) has been observed in frog muscle fibres with changes in muscle temperature, which X-ray diffraction studies suggest is due to a greater or lesser axial tilting of the myosin heads at higher and lower temperatures (Linari et al., 2005). Further studies are required to clarify the precise effect of oestrogen and to understand the mechanisms by which muscle force may be affected by oestrogen status and / or old age.

3.   DRUGS AND HORMONES WHICH ALTER MUSCLE PHENOTYPE

In the media, when drugs, hormones and muscle are mentioned together, it is often in the context of the misuse of substances by sportsmen and women aiming to improve athletic performance, or in the case of body building, for the "aesthetic" benefits of hypertrophied muscles and sculptured physiques. However, the loss of muscle mass that occurs as a result of old age (sarcopenia), immobilisation (disuse atrophy), or in disease states such as cancer and HIV-AIDS (cachexia) impacts largely on health and quality of life. It is thus important that we understand the mechanisms which regulate muscle mass and recognise that the legitimate development of therapeutic treatments to increase muscle mass and improve function is crucial.

Muscle makes up approximately 40% of the total protein in the adult body. Proteins are being continually turned over (synthesised and degraded) such, that for most of us, the body is a state of nitrogen balance. In the fasted state the protein turnover of human skeletal muscle is approximately 1.5% per day, which is roughly ten-fold lower than the protein turnover in liver cells. For a muscle to grow there must be a net gain in protein either by increasing the rate of muscle protein synthesis, decreasing the rate of its degradation, or both (Rennie et al.,, 2004). Furthermore, as individual muscle cells grow, the number of nuclei needs to be increased so as to maintain a constant nuclear domain, the area of cytoplasm controlled by each nucleus (Kadi et al., 1999). The source of these new nuclei is satellite cells (the muscle stem cells). These cells are anatomically distinct from muscle fibre nuclei being located between the basal lamina and sarcolema. Satellite cells, when activated, can also differentiate into myoblasts that can fuse with a muscle fibre and produce the contractile proteins for addition to existing filaments.

The complex mechanisms which regulate protein gain and muscle enlargement are being elucidated and involve the interaction of mechanical signals, the action of systemic and local hormones and some metabolic control. It is apparent that there are both positive and negative regulators of protein synthesis; i.e. factors which act to promote the synthesis of new protein, and factors which serve to act as brake on this process. The mechanisms which regulate muscle protein breakdown are also being investigated, allowing the development of target therapies to specific parts of the anabolic process. Thus a drug could potentially be anabolic without actually promoting the synthesis of new muscle proteins, by preventing or slowing its breakdown. The following section focuses on some of the key hormones and drugs which play a role in altering mass and fibre type and thus alter the force potential of a muscle, its shortening characteristics and thereby its power producing ability.

3.1   Growth Hormone & IGF-I Axis

The growth hormone / insulin like growth factor I (GH/IGF-I) axis plays a key role during growth and development. Levels of these hormones reach their peak during adolescence and are maintained at somewhat lower levels during adulthood. With increasing age circulating levels of GH and IGF-I decline (Rudman et al., 1981; Corpas et al., 1993; Lamberts et al., 1997; Morley, 1995). GH secretion by the anterior pituitary gland is mediated by the actions of two hypothalamic hormones (growth hormone releasing factor and growth hormone release inhibiting factor, also known as somatostatin). The "somatomedin" hypothesis originated in the 1950's in early efforts to understand how somatic growth was regulated by factors secreted by the pituitary. It was determined that pituitary-derived GH did not act directly on its target tissues to promote growth, and that there were intermediary substances involved (Daughaday, 1966). The term 'somatomedin' was later adopted to reflect the growth promoting actions of these substances (Daughaday, 1972), and some two decades later, the insulin like growth factors (IGF-I and IGF-II) were characterised. IGF-I was identified as the somatomedin regulated by GH (Rinderknecht and Humble, 1978; Klapper et al., 1983). Both substances were termed 'insulin-like' due to their ability to stimulate glucose uptake into fat cells and muscle. The original hypothesis became the widely accepted model of IGF-I action, that being that the effect of GH on longitudinal body growth was mediated in an endocrine fashion, solely by liver-derived IGF-I. This hypothesis was later revised in the 1980's when D'Ercole and co workers discovered that extra-hepatic tissues also expressed IGF-I, thus suggesting that IGF-I also had an autocrine / paracrine effect (D'Ercole et al., 1984). GH itself has not been shown to affect prenatal development, as GH and GH receptor (GHR) null mice appear no different from their wild type littermates at birth. Contrastingly, *Igf-1* null mice are smaller compared with their wild type counterparts when they are born, and most die early in the neonatal stages. Those that survive are severely retarded and also infertile (Liu et al., 1993).

GH circulates in a complex with its binding protein, growth hormone binding protein (GHBP). The actions of GH are mediated by its binding to the transmembrane receptor GHR. As with most cytokine receptors, the GHR utilises the JAK-STAT signal transduction pathway. Upon binding to the hepatic GHR, GH induces the expression of IGF-I. The liver has the highest expression of the GHR, and accordingly expresses a large fraction of circulatory/serum IGF-I. However, it is also known that other tissues, including the kidney, bone and muscle, also express the GHR and may therefore contribute to the levels of serum IGF-I. Brahm et al., (1997) showed that there was net a release of IGF-I from a working muscle into the circulation by measuring femoral arterial and venous differences during isolated knee extension exercise. Circulating IGF-I itself inhibits GH secretion, providing a negative feedback loop on the actions of GH in peripheral tissues. It is through the action of IGF-I that the effects of GH are primarily mediated.

In both human beings and rats, the IGF-I gene spans more than 70 kilobases and consists of six exons and at least five introns (Stewart and Rotwein, 1996). Two promoters, one adjacent to exon 1 and the other to exon 2 govern gene transcription. The resulting variant mRNA transcripts with different 5' untranslated regions (UTR) and signalling peptides have been classified as Type/Class 1 (exon 1) and Type/Class 2 (exon 2) (for schematic representation of the IGF-I gene see Fig. 2). These 5' mRNA variants are differentially regulated during development in a tissue specific manner (Lowe et al., 1988). In addition to transcription from two promoters, IGF-I is regulated by post-transcriptional events, which yield several mature mRNA transcripts. In all of these variants exons 3 and 4, which encode the mature 70 amino acid peptide, are constant (Bell et al., 1986; Gilmour, 1994), whereas exons 5 and 6 are subject to a complex alternative splicing pattern (fig 2). Splicing is a complex mechanism by which exons are arranged in different combinations from pre-mRNA. The characteristics and potential roles of the IGF-I splice variants are expanded upon in the next section.

It is known that the biological activity of a hormone or growth factor is not simply reflected by its concentration, but also depends on the concentration of the receptor and the affinity of its interaction with the receptor. Over 97% of IGFs are bound to members of a family of six proteins, the IGF binding proteins (IGFBPs 1-6). IGFBPs bind IGF-I and IGF-II with an affinity equal to, or greater than that of the IGF-I receptors (Jones and Clemmons, 1995), and modulate IGF action. In the circulation, over 90% of IGF-I is found as heterotrimeric complex bound to the most abundant binding protein, IGFBP-3, and the acid labile subunit (ALS). One role of ALS is the prevention of the non-specific metabolic effects of the IGFs, such as causing severe hypoglycaemia (reviewed by Boisclair et al., 2001). Other

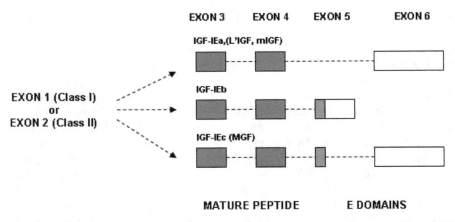

Figure 2. Schematic representation of part of the human IGF-I gene showing the signalling peptide (exons 3 and 4) and E domains (exons 5 and 6). The diagram shows how 3 different C terminus transcripts (along with the different nomenclatures that have been used to describe them) may result from the process of alternative splicing. A color figure is freely accessible via the website of the book: http://www.springer.com/1-4020-5176-x

IGFBPs found in the bloodstream include IGFBP-1, IGFBP-2 and IGFBP-4, but are present at much lower concentrations. IGFBP-4, IGFBP-5 and IGFBP-6 are expressed in many muscle cell lines and have also been reported to be present in muscle tissue *in vivo* (Ewton and Florini, 1995; Bayol et al., 2000). There is some evidence to suggest that alterations in IGFBP expression, namely IGFBP-4 and IGFBP-5 occur in unloaded or overloaded soleus muscle, and that this may play a role in skeletal muscle adaptation to changes in loading.

The anabolic effects of IGF-I have been clearly demonstrated by numerous *in vitro* studies, where it has been shown that IGF-I acts to increase the diameter of myotubes, suppress protein degradation, increase amino acid uptake and stimulate protein synthesis (Florini, 1987; Vandenburgh et al., 1991; Florini et al., 1996; Semsarian et al., 1999; Rommel et al., 2001; Bodine et al., 2001).*In vivo* studies, employing models known to result in muscle growth, have reported that IGF-I expression in the muscle increases early on in the process of hypertrophy and that localized infusion of IGF-I to rat muscle stimulates muscle hypertrophy (Adams and McCue, 1998, Adams, 2002). Studies utilising the model of stretch-induced hypertrophy of the muscle have reported an increased expression of muscle IGF-I mRNA (Schlechter et al., 1986; Czerwinski et al., 1994). A study by DeVol et al., (1990) demonstrated that there was a three-fold increase in IGF-I mRNA levels in the soleus and plantaris muscles in 11-12 week old female rats following tenotomy-induced hypertrophy. This particular study employed hypophysectomized rats where GH production was prevented. This important observation showed that IGF-I mRNA expression could be GH independent in muscle. Later studies utilising a similar model of functional overload in both normal and hypophysectomized rats found that both mRNA and protein levels of IGF-I were increased in muscle, prior to the attainment of significant hypertrophy, and remained elevated for up to 28 days during the hypertrophy process (Adams and Haddad, 1996). Further work using treadmill running in rats showed that GH-suppressed rats were still able increase muscle IGF-I mRNA and IGF-I protein by 55% and 250% respectively (Zanconato et al., 1994).

Until relatively recently, GH could only be obtained from the pituitary glands of cadavers which posed a serious risk of transmitting Creutzfeldt-Jakob Disease (CJD). Recombinant technology has allowed synthetic (rhGH) to be manufactured and used for therapeutic purposes. Such developments have lead to the use of rhGH to treat GH deficient children and adults and have been shown to be effective in promoting longitudinal body growth and increases in muscle cross-sectional area (Cuneo et al., 1991, Roemmich et al. 2001). Low levels of circulating GH and IGF-I are also evident in elderly people. The successful treatment of GH deficient adults with rhGH prompted hopes that rhGH would serve as an effective therapeutic agent to ameliorate the age related loss of muscle mass (sarcopenia). It was also viewed as an illicit means by which some athletes and body builders could increase muscle mass. However, for both groups of individuals the experimental results have been disappointing. In elderly people, administration of rhGH, despite resulting in higher levels of IGF-I in the circulation, has not been shown to increase muscle mass (Rudman et al., 1990, Lange et al., 2002). Even when combined with strength training exercise, the gains in muscle mass and strength are not greater than when exercise is performed alone (Lange et al. 2002, Yarasheski et al., 1995). Likewise, in young subjects, no gains in rates of muscle protein synthesis have been observed in either non-trained males (Yarasheski et al., 1992) or in well trained weightlifting athletes (Yarasheski et al., 1993) following rhGH treatment over and above those induced by high-resistance strength training exercise. GH may, however, cause water retention and consequent weight gain. There is thus little scientific data to support the use of rhGH as an effective anabolic agent even though it significantly increases circulating levels of IGF-I (Rennie 2003).

3.2   IGF-I Splice Variants in Muscle

In humans, alternative splicing of IGF-I pre mRNA leads to the production of three different transcripts at the 3' end, resulting in different C-terminal peptides (E peptides, fig 2). In rodents, only two splice variants have been identified. There is some confusion as to the nomenclature of the splice variants (Hameed et al., 2004a) and the terms IGF-IEa, IGF-IEb and IGF-IEc (MGF) are used here. Evidence is emerging that the splice variants differ in their regulation and in their physiological roles. For example, using a model of stretch to evoke rapid hypertrophy in rabbit tibialis anterior and extensor digitorum longus muscles, Yang et al., (1996) were able to clone and sequence the cDNA of two splice variants of the IGF-I gene. The first isoform was equivalent to the transcript most common in the liver (IGF-IEa). The second, which in this particular study was not detected in the resting control muscle, differed by the inclusion of a 52 base pair insert in exon 5. This observation lead to the conclusion that the IGF-IEb isoform was sensitive to mechanical signals or to the micro-damage caused by the protocol. It was thus re-named mechano-growth factor (MGF). Confusingly, in humans the rodent IGF-IEb slice variant is termed IGF-IEc and the insert from exon 5 is 49 and not 52 base pairs long. In a subsequent study, McKoy et al., (1999) showed that when stretch was combined with electrical stimulation MGF mRNA upregulation was even greater than that caused by stretch alone.

Overexpression of different IGF-I splice variants in mice leads to muscle hyper-trophy. Barton-Davis et al., (1998) using a viral construct and a myosin light chain (LC3f) promoter evoked increased expression of IGF-IEa in the muscles of mice. This lead to muscle hypertrophy and a preservation of muscle mass in the older animals. Subsequently, Musaro et al., (2001) used a transgenic mouse model, which overexpressed IGF-IEa in muscle (not the circulation) and obtained similar results. Gene transfer studies using cDNAs for MGF have also shown this splice variant to have marked anabolic effects (Goldspink, 2001). Effective monoclonal antibodies to the E-domain peptides, i.e. to IGF-IEa and MGF, have proved somewhat elusive (Shavlakadze et al., 2005), and it thus remains unclear whether circulating IGF-I includes these different E-domains, or whether they are cleaved and function independently. The IGF-IEa splice variant is that most commonly expressed in the liver and its transcript expression in human muscle is several-fold higher than MGF (Hameed et al., 2003, 2004).

High resistance strength training is an effective means of increasing muscle mass, strength and power in both young and old people. Analysis of muscle biopsy samples taken 2.5 hours after a single intense bout of exercise, revealed an increase in MGF mRNA, but not in IGF-IEa (Hameed et al., 2003). This observation of different mRNA expression kinetics, suggests different physiological roles for the two splice variants. It is has been proposed that MGF may serve as a "kick starting" mechanism in the repair and adaptation processes by activating satellite cells (Hill and Goldspink 2003). This is supported by studies on C2C12 cells which show different responses following transfection with either MGF or IGF-IEa. With MGF transfection, differentiation was prevented, but the cells increased in number, whereas IGFI-Ea caused myotubes to hypertrophy (Yang & Goldspink 2002). Both IGF-I splice variants possess the same coding region for the mature IGF-I, but as IGF-IEa is expressed at higher concentrations (Hameed et al., 2003) it could contribute more to the mature IGF-I production and serve to stimulate muscle protein synthesis.

Initially, evidence suggested that IGF-I induced hypertrophy via the calcineurin pathway (Musaro et al., 1999). However, pharmacological and transgenic blockade of calcineurin signalling do not prevent overload induced hypertrophy (Bodine et al., 2001). Rommel et al., (2001) provided evidence that IGF-I served to increase protein synthesis by activating the AKT (also known as PKB)-mTOR pathway. It is not the focus of this chapter to probe further into these complex intracellular signalling events. However, an important study by Baar and Esser (1999) clearly demonstrated that a protein kinase (p70S6k) downstream of mTOR (mammalian target of rapamycin) is a key regulatory step. The degree of phosphorylation of p70S6k was found to be closely associated with the subsequent muscle growth in overloaded soleus and gastrocnemius muscles of rats.

In addition to splice variants Ea, Eb and MGF, all IGF-I mRNAs are categorised into either class 1 (containing exon 1 but not exon 2) or class 2 (containing exon 2 but not exon 1) variants, depending on their transcriptional initiation from either promoter 1 or promoter 2 respectively. It has been shown in rats that promoter 2 is GH-sensitive (Bichell et al., 1992). Whilst clear evidence is lacking, it is possible that Class 2 transcripts are GH sensitive whilst Class I transcripts initiating from promoter 1 are sensitive to other signals, including mechanical signals. This would make some sense as Class-2 IGF-I transcripts are the most abundant in the liver, which is the organ chiefly responsible for IGF-I's secretion under endocrine control.

The direct effect of GH and /or overload on muscle IGF-I splice variant activity was recently studied in elderly people. Elderly subjects were separated into three groups. The first was given daily injections of rhGH (0.5 − 0.1 IU/kg/day), the second received daily injections of rhGH and participated in a 12 week high resistance strength training regimen, and the third undertook the strength training regimen without rhGH administration. After 5 weeks, rhGH resulted in significantly increased IGF-IEa mRNA levels, but no change in MGF (although the latter was increased after 12 weeks). In contrast, 5 weeks of strength training had the effect of significantly increasing the expression of MGF mRNA. Interestingly, when strength training and rhGH administration were combined the result was a significantly greater increase in MGF compared with IGF-IEa mRNA. One possibility was that GH facilitated an increase in the primary IGF-I transcript which was then subjected to alternative splicing. When combined with high resistance exercise, the splicing was towards MGF and in the absence of large mechanical signals, towards IGF-IEa. It is noteworthy that despite the larger increases in IGF-IEa and MGF, there was not a significantly greater increase in muscle CSA as determined by magnetic resonance imaging (MRI) in the group of subjects exercising and receiving GH injections when compared to the group that underwent strength training alone (Lange et al., 2002). These data, although not strictly comparable, are in contrast to those reported by Lee et al. (2004). These workers undertook experiments on rats which were subjected to either viral administration of IGF-I (IGF-IEa), resistance training or a combination of both. They observed that the biggest gains in muscle mass and strength were observed in the muscles that underwent resistance training in combination with IGF-I treatment. In the elderly human study, rhGH induced a marked increase in IGF-IEa and MGF mRNA, but not seem to add a further hypertophic stimulus to that of exercise alone. The discrepancy between the two studies may be due to age or species differences. In a separate study in young subjects, in which daily rhGH injections were given for 2 weeks, no effect on IGF-IEa mRNA expression was observed in muscle, despite a three-fold elevation in circulating IGF-I levels (?, ?). These human experiments are in general agreement with other recent animal studies (Iida et al., 2004), where GH- deficient lit/lit mice responded positively to a bolus of GH (increasing both IGF-IEa and MGF mRNA), whilst wild type animals with normal endocrinology did not show a muscle increase in either muscle splice variant. Thus it appears that GH sufficiency is in someway modulating the muscle IGF-I response to exogenous rhGH.

In terms of muscle composition Lange et al., (2002) reported that 12 weeks of rhGH supplementation in elderly people resulted in a significant increase in the proportion of muscle occupied by MHC-IIx isoforms in elderly people.

3.3   Insulin

Like its close relatives (IGF-I) insulin is a protein peptide hormone (around 50 amino acids long) secreted by the pancreas in response to food ingestion. Its role is to transport nutrients in to the cells of the body particularly glucose into skeletal muscle. Prior to its clinical availability, patients with Type I diabetes experienced continuous loss of protein from all tissues. The treatment of patients with insulin dramatically restored muscle mass. Early cell culture experiments and those on immature rodents suggested that insulin served to increase protein synthesis. However, data on humans is less conclusive (see Rennie et al.,, 2004). It seems that the major effect of insulin is, in contrast to that of IGF-I, to prevent protein breakdown. Experiments have shown that protein accretion occurs in the recovery phase after exercise, rather than during the exercise period itself, with rates of protein synthesis and degradation being simultaneously accelerated (Rennie et al., 2004). Physiological hyperinsulineamia studied on the setting of constant amino acid concentrations increased muscle protein synthesis in normal resting volunteers (Biolo et al., 1995). In further studies (Biolo et al., 1999) reported that the injection insulin into the femoral artery did not further increase muscle protein synthesis after a bout of high resistance exercise, but did decrease the rate of muscle breakdown.

3.4   Testosterone

The dramatic increase in the production of the hormone testosterone at the time of puberty stimulates the acceleration in growth and muscularisation in males. Testosterone is a steroid hormone synthesised from cholesterol in the Leydig cells of the testes of male and the ovaries of females. It circulates bound to albumin and globulin, whilst $\sim 4\%$ of testosterone is not bound to plasma proteins and circulates as free testosterone. The bioavailability of testosterone is a function of the total amount of testosterone and the binding capacity of the plasma proteins. Testosterone is the most important circulating androgen in men and has a number of andronergenic effects. In muscle it promotes growth. Of the free testosterone that acts on some tissues, much is converted to dihydrotestosterone (DHT), a more potent androgen, by the enzyme 5-alpha reductase, but some is also converted into oestrogens (estriol, estrone and estradiol) by the aromotase complex which is present mainly in the liver, brain and fat tissue.

Testosterone is the basis for the anabolic steroid family of illegal substances taken by some athletes. It can be modified in various ways to either reduce other unwanted androgenic effects, to make it injectable, or ingestible. For example, the addition of alkyl group to the hydroxyl group at carbon 17 allows the compound to be injected and the addition of an alkyl group to carbon 17 allows them to be taken orally. Modifications of the first three benzene rings allow the steroid compounds to be taken orally and can also increase their potency and reduce their masculinising effects. Nandralone is a steroid gaining notoriety as a steroid of choice as it is believed to maximise the anabolic effects and minimize other unwanted effects. Despite the anecdotal evidence of athletes and coaches, only relatively recently have well controlled, randomized studies been conducted to provide scientific credibility to these claims. For example, Bhasin et al., (1996) performed a randomised double blind study where 43 young eugonadal males were randomised to receive either placebo, placebo plus strength training, 100mg / week testosterone enanthate or 100mg /week testosterone enanthate plus strength training. Compared with placebo alone, the testosterone group increased lean body mass, quadriceps CSA as determined by MRI, and the weight lifted during a single squat. The changes due to testosterone were similar to those evoked by exercise alone. However, when testosterone administration was combined with resistance exercise, the change in all three parameters was greater than the changes in exercise alone.

In vivo binding studies suggest that the androgen receptors in most tissues are either saturated or down regulated at testosterone concentrations at the lower end of the normal range. Athletes may take more than 10 times the doses reported by Bhasin and collaborators. Therefore it is possible that the supraphysiological doses evoke hypertrophy through a non-androgen receptor mediated mechanism. Urban et al., (1995) reported that administration of testosterone to elderly men (so that their circulating levels were equivalent to young males) increased muscle strength and increased rates of muscle protein synthesis. Furthermore, mRNA concentration of IGF-I were increased and IGF-BP4 levels were lower, leading the authors to conclude that the testosterone induced increase in protein synthesis might be mediated by the IGF-I system. Therefore, this is potentially a mechanism for amplifying the androgenergic anabolic signal. Sinha-Hikim et al., (2002) investigated the effects of 20 weeks of testosterone enanthate at either 125, 300 or 600mg / week and reported that fibre size increased in a dose dependent manner (fig. 3). Subsequently, analysis of these biopsy samples showed that both myonuclei and satellite cell number also increased in a dose dependent manner up to a dose of 600mg/week (fig 3). These data are in general agreement with the data reported by Kadi et al., (1999) who showed that the size of both type I and type II muscle fibres of weightlifting athletes regularly taking anabolic steroids were significantly larger than the size of fibres in athletes who did not take drugs. In addition, there was a direct relationship between fibre size and myonuclear number. In contrast, lowering circulating testosterone levels by administration of a gonadatrophin releasing hormone agonist (GnRH) is associated with a reduction in fractional rates of muscle protein synthesis (Mauras et al., 1998).

The mechanisms of action of testosterone on the satellite cells are not clear. It is possible that testosterone might promote entry of the cells in to the cell cycle,

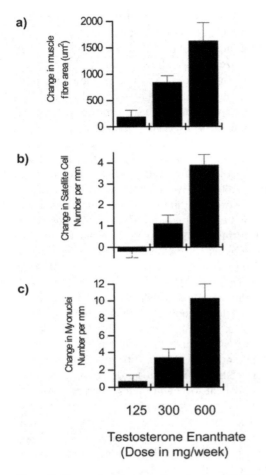

Figure 3. Changes in (a) muscle fibre size (type II), (b) satellite cell number and (c) myonuclei number in response to male subjects being treated with increasing doses of testosterone enanthate. (Data adapted from Sinha-Hikim et al., 2002, 2003)

or that androgens might promote differentiation. These events are those which are also believed to be regulated in part by IGF-I. As with GH/IGF, a progressive decline in circulating testosterone levels (andropause) is associated with ageing. However, testosterone replacement in elderly hypogonadal men has produced rather inconsistent and modest increases in muscle mass and strength and is associated with a number of side effects such as fluid retention, gynaecomastia, polycythaemia and acceleration of benign or malignant prostatic tumours (Borst 2004).

In terms of muscle composition, it appears that endogenous anabolic / androgenic steroids do not play a role in regulating MHC isoform composition in fast or slow muscles of the rat (Noirez & Ferry, 2000) or relative fibre type distribution in humans (Sinha-Hikim et al., 2002).

Dehydroepiandersterone (DHEA) and its sulphated ester, DHEA sulphate (DHEAS) is an androgen synthesized primarily from the adrenal cortex. It has a similar structure to the sex hormones testosterone and oestrogen. Androsterone is an intermediate (precursor) hormone between DHEA and testosterone. Using data from a sample of 558 men (20-95 years) serum DHEAS was found to be an independent predictor of muscle strength and mass, but only for men between 60 and 79 years of age (Valenti et al., 2004). In young men, treatment with DHEA increased serum androstenedione concentrations, but did not enhance serum testosterone concentrations or adaptations associated with resistance training (Brown et al., 1999).

3.5   Myostatin

Muscle is responsive to factors that work as stimulators, or accelerators of muscle protein synthesis such as IGF-I and testosterone, but also factors which act as brakes. Myostatin is such a factor which has come to prominence in recent years as a potential therapeutic target for ameliorating muscle wasting. A member of the TGF family of growth factors, myostatin (or growth differentiation factor 8, GDF-8) gained attention when it became apparent that the highly hypertrophied or double muscled cattle (Belgian Blue or Piedmontese variety) exhibited mutations in the myostatin gene (McPherron and Lee, 1997). Subsequently, myostatin knockout mice were shown to exhibit a similar hypertrophied phenotype (McPherron et al., 1997). The myostatin gene comprises three exons and two introns, and is transcribed as a 3.1 kb mRNA species that encodes a 375 amino acid precursor protein. Myostatin is expressed uniquely in human skeletal muscle as a 26 kDa mature glycoprotein (myostatin-immunoreactive protein) and secreted into the plasma (Gonzalez-Cadavid et al., 1998). Myostatin immunoreactivity is detectable in human skeletal muscle in both type I and II fibers. In humans, evidence for its role as an endogenous negative regulator of muscle mass has come from a number of studies. For example, high resistance exercise in both young and old men women induce post-exercise reductions in myostain mRNA levels (Roth et al., 2003). Furthermore, mutations in the myostatin gene have been reported in a young child with excessive muscle development (Schuelke et al., 2004). By contrast, HIV men with muscle wasting have significantly increased serum and intramuscular concentrations of myostatin-immunoreactive protein (Gonzalez-Cadavid et al., 1998).

How does myostatin work? Myostatin appears to work by negatively regulating satellite cell activity in an autocrine fashion. It seems to inhibit the progression of satellite cells from the G_0 or G_1 cell cycle phase to the S phase by upregulating cyclin-dependent kinase inhibitors. This inhibits withdrawal from the cell cycle that is necessary for differentiation. Furthermore, through decreased expression of the myogenic transcription factors MyoD, myf5 and myogenin, myostain also appears to negatively regulate myoblast differentiation (Roth and Walsh, 2004).

3.6   Beta 2 Agonists

Skeletal muscle contains three major pathways of protein degradation: a lysosomal pathway, a calcium-dependent pathway, and an ATP-dependent ubiquitin-proteasome pathway. Of the three, the ubiquitin-proteasome pathway is responsible for the bulk of muscle proteolysis, including the major contractile proteins actin and myosin. Beta-2 agonists (β_2-agonists, of which clenbuterol is the most commonly prescribed) are bronchodilator medicines that open airways by relaxing the muscles in and around the airways, and are traditionally used at low doses in the treatment of asthma. However, there is evidence that when given at higher doses β_2-agonist have potent anabolic effects in muscle, ameliorating the effects of disuse, dennervation and aging, although the mechanisms by which it may do this are not clear. Ryall et al. (2004). reported that 4 weeks of treatment with the β_2-agonist, fenoterol (1.4 mg kg^{-1}) resulted in 41% and 20% increases in rat extensor digitorum longus (EDL) and soleus muscles respectively. This confirmed earlier work on the anabolic effects of clenbuterol in rat muscle (Zeman et al., 1988). Furthermore, similar increases in muscle and fibre size were observed in a group of older (28 month old) animals (Ryall et al., 2004). Recently, Yimlamai et al. (2005) showed that clenbuterol can inhibit the loss of muscle mass in rates evoked through hindlimb unloading. Clenbuterol was shown to cause muscle specific inhibition of the ubiquitin-proteasome pathway, thus reducing muscle protein degradation. However, this effect appeared to be specific to fast EDL muscles. In general agreement with studies on rodents, Signorile et al., (1995) reported that in spinal cord injured subjects 4 weeks of treatment with metaproterenol (80 mg/day) induced a significant increase in arm strength and forearm size when compared with placebo. In addition to its anabolic properties, chronic administration of β_2-agonists also result in a slow to fast phenotypic switch (Zeman et al. 1988). Thus the combined anabolic and fibre switching properties of β_2-agonists which promote increased power output have thus resulted its joining testosterone, growth hormone and IGF-I, on the list of prohibited doping substances issued by the World Anti Doping Agency (WADA).

3.7   Thyroxine

The thyroid gland, located in the neck just below the larynx, secretes two protein-iodine-bound hormones, thyroxine (T_4) and the active form of thyroid hormone, triiodthyroxine (T_3). It has been reported that T_3 plays a crucial role in the normal development of vertebrate skeletal muscle (Gambke et al., 1983) and an intact thyroid gland is required for the normal development of muscle mass, differentiation of biochemical and contractile properties of skeletal muscle, and plays a role in the transition from the expression of embryonic and fetal isoforms to that of adult MHC isoforms (Finkelstein et al., 1991). Fitts et al., (1984) reported that the soleus muscle of rats made thyrotoxic by L-thyroxine and T_3 supplements to their diets exhibited significantly faster twitch times, a feature not observed in the EDL muscles of thyrotoxic animals. In this regard Li and Larsson, (1997) observed a dose response reduction in MHC-I and increase in MHC-IIa isoforms in rat soleus muscles, without a change in the EDL. Although the molecular mechanism of thyroid hormone action on skeletal muscle and its MHC isoform composition is unclear, it seems that T_3 exert its action by binding to specific nuclear protein receptors (TRα1 α2, β1, β2 β3) which then regulate the expression of myogenic helix-loop-helix transcription factors (the myoD gene family) and subsequently repress or activate genes encoding for different MHC isoforms. In the vastus laterialis muscle of patients with hypothyroidism, a selective loss and atrophy of type II fibres has been reported (McKeran et al., 1975). Treatment with L-thyroxine increased fibre size and also induced an increase in the number of type I fibres, analogous to an increase in MHC-I isoforms. Although all members of the MHC multigene family respond to thyroid hormones (Yu et al., 2000) the mode of response, at least in animal muscles, is muscle and muscle fibre type specific. Thus this hormone may function to remodel a muscle for faster contraction. However, this is unlikely to result in a muscle with an increased power output as T_3 treatment evokes significant loss of muscle mass. For example 8 weeks of T_3 treatment caused a 33% and a 50% reduction in soleus and EDL muscle mass respectively (Li and Larsson, (1997).

3.8   Glucocorticoids

Cortisol is the primary glucocorticoid secreted by the adrenal glands. These hormomes are secreted in emotionally charged situations or in response to the stressful demands of physical activity. Among its other actions, cortisol promotes the breakdown of amino acids delivering them to the liver for synthesis to glucose via gluconeogenesis. Glucocorticoids are thus hormones, which in contrast to those discussed thus far are catabolic agents responsible for breaking down muscle. Patients with adrenocortical dysfunction (Cushing Syndrome) are characterised by muscle weakness and atrophy, as well as a fast-to-slow transition of muscle phenotype (Pellegrino et al., 2004). As already mentioned, a loss of muscle mass occur in a variety of diseases (e.g. AIDS, sepsis cancer, diabetes mellitus), but also occurs with disuse, unloading, such as limb immobilization and microgravity, and also as a result of fasting. The mechanisms by which such proteolysis occurs is now becoming better understood and glucocorticoids have been used to induce muscle atrophy and to study the mechanisms which regulate muscle protein degradation. For example, the synthetic glucocorticoid, dexamethasone induces atrophy and evokes a number of parallel changes in gene expression which have been termed "atrogenes". The gene most dramatically induced is a Ub-ligase (E3), atrogin 1 (MAFbx) which is a muscle-specific F box protein that is induced many fold in fasting, diabetes and cancer (Sacheck et al., 2004). In addition, another muscle specific E3, MuRF1 is also highly induced in atrophying muscle (Bodine et al., 2001). Both of these genes have been shown to be upregulated in human muscle atrophy induced by lower limb immobilisation (Jones et al., 2004).

4.   INTERACTION BETWEEN HORMONES

The previous sections have covered the action of some of the key hormones and drugs involved in muscle plasticity, but each has been treated in isolation. This was done deliberately for purposes of clarity, but it is important to highlight the fact that the body is a complex endocrine environment where different hormones exist together. Under physiological conditions they often work in a co-ordinated manner sometimes to amplify the effects of one another, or to work reciprocally. A good example of this is IGF-I and glucocorticoids. The discussion in Section 3.1 focused on the role IGF-I has as an anabolic agent stimulating protein synthesis, through activating the (PI3)/AKT signalling pathway. By contrast, glucocorticoids evoke muscle protein breakdown though activation of the ubiquitin (Ub)1- proteasome pathway. However, simultaneous administration of IGF-I with glucocorticoids to rats significantly attenuates the glucocorticoid induced muscle atrophy and myofib-rillar protein breakdown (Kanda et al., 1999). Similar studies on cells in culture (Sacheck et al., 2004, Sandri et al. , 2004), have shown that this effect is likely to be mediated by the rapid suppression of atrogin-1 by IGF-I and indeed insulin, such that only when insulin is low (such as during fasting) are the catabolic effects of glucocorticoids on muscle evident (Wing & Goldberg, 1993). The mechanism seems to involve the Forkhead box O (Foxo) class of transcription factors. Foxo gene expression is induced by glucocorticoids. AKT blocks the function of Foxo genes by phosphorylation leading to their sequestration in the cytoplasm away from their target genes. Thus when phosphorylated by the action of IGF-I, AKT impairs Foxo's action. However, without IGF-I induced activation of AKT, dephosphorylation of Foxo factors leads to nuclear entry and protein degradation (see Sandri et al. , 2004).

Another example is the interaction between glucocorticoids and β2-agonists, drugs which are frequently co-prescribed for chronic asthma treatment. Long term treatment with either drug alone causes unwanted effects which go in different directions (glucocorticoids induce atrophy and fast to slow MHC transformation, while β2-agonists induce hypertrophy and shift to a fast MHC isoform expression). Pellegrino et al., (2004) have recently shown in mice that treatment with β2-agonists and glucocorticoids causes a marked attenuation of both effects, although to different extents in different muscles. Furthermore, dexamethasone increases the expression of myostatin, the negative regulator of skeletal muscle mass, in vitro, suggesting that the glucocortcoid induced muscle loss is also mediated, at least in part, by the upregulation of myostatin expression through a glucocorticoid receptor-mediated pathway (Ma et al., 2003).

SUMMARY AND CONCLUSIONS

Muscle is a highly plastic tissue. Its contractile properties can be altered in the short term by increasing temperature or by certain substances which may alter / regulate its force and shortening characteristics. These inotropic-like effects, contrast with those effects exerted by certain drugs and hormones which alter function in the longer term (Fig. 4). The latter do this by remodelling a muscle such that either its size, protein isoform composition, or both is altered. The mechanisms underlying the action of these substances are becoming better understood. However, it is clear that further complex agonist and antagonist interactions exist between these factors which ultimately affect the intracellular signalling events regulating the action of these substances. Whilst this chapter has focused on drugs and hormones in regards to muscle plasticity, they have not been discussed in regard to other key physiological processes impacting on muscle, such as those relating to mechanical signals and to feeding. Both of these have a dramatic effect on rates of muscle protein synthesis (Rennie et al.,, 2004). In addition, the effects of gene polymorphsisms on the efficacy and potency of the substances discussed above is likely to be another confounding factor to be considered. Nevertheless, as our understanding of muscle plasticity continues to increase so will the successful development of treatment strategies to ameliorate the effects muscle loss associated with old age and many disease states.

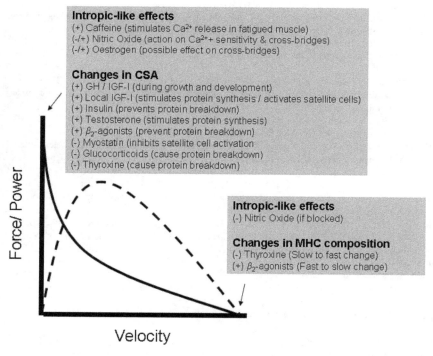

Figure 4. Summary figure showing the force and power-velocity relationship of a hypothetical muscle together with factors which may alter either its force, its velocity of shortening, or both. These hormones and / or drugs may cause acute (inotropic-like) effects without altering muscle size or composition, or result in longer term adaptive responses leading to a change in muscle phenotype

ACKNOWLEDGEMENT

This chapter was written whilst the author was the recipient of a grant from the World Anti Doping Agency (WADA).

REFERENCES

Adams, G. R. (2002) Invited Review: Autocrine/paracrine IGF-I and skeletal muscle adaptation. J Appl Physiol. *93*, 1159–67.

Adams, G. R., and Haddad, F. (1996). The relationships among IGF-1, DNA content, and protein accumulation during skeletal muscle hypertrophy. J.Appl.Physiol *81*, 2509–2516.

Adams, G. R., and McCue, S. A. (1998). Localized infusion of IGF-I results in skeletal muscle hypertrophy in rats. J.Appl Physiol *84*, 1716–1722.

Aperghis, M., Hameed, M., Bradley, L., Bouloux, P, Goldspink, G. & Harridge, S.D.R. (2004) Two weeks of exogenous GH administration does not increase the expression of IGF-I mRNA splice variants in the skeletal muscles of young men. J Physiol *558*, C4

Baar, K., and Esser, K. (1999). Phosphorylation of p70(S6k) correlates with increased skeletal muscle mass following resistance exercise. Am J Physiol *276*, C120–7

Barton-Davis, E. R., Shoturma, D. I., Musaro, A, Rosenthal, N., and Sweeney, H. L. (1998). Viral mediated expression of insulin-like growth factor I blocks the aging-related loss of skeletal muscle function. Proc Natl Acad Sci U.S.A *95*, 15603–15607.

Bayol, S., Loughna, P. T. and Brownson, C. (2000). Phenotypic expression of IGF binding protein transcripts in muscle, in vitro and in vivo. Biochem Biophys Res Commun *273*, 282–286.

Bell, G. I., Stempien, M. M., Fong, N. M., and Rall, L. B. (1986). Sequences of liver cDNAs encoding two different mouse insulin-like growth factor I precursors. Nucleic Acids Res *14*, 7873–7882.

Bhasin, S., Storer, T. W., Berman, N., Callegari, C., Clevenger, B., Phillips, J., Bunnell, T. J., Tricker, R., Shirazi, A., and Casaburi, R. (1996) The effects of supraphysiologic doses of testosterone on muscle size and strength in normal men. N Engl J Med *335*, 1–7

Bichell, D.P., Kikuchi, K. and Rotwein, P (1992) Growth hormone rapidly activates insulin-like growth factor I gene transcription in vivo. Mol Endocrinol. 6, 1899–908.

Biolo, G., Declan Fleming, R.Y. and Wolfe, RR. (1995) Physiologic hyperinsulinemia stimulates protein synthesis and enhances transport of selected amino acids in human skeletal muscle. J Clin Invest. *95*, 811–9.

Biolo, G., Williams, B.D., Fleming, R.Y. and Wolfe, R.R. (1999) Insulin action on muscle protein kinetics and amino acid transport during recovery after resistance exercise. Diabetes. *48*, 949–57.

Bodine, S. C., Stitt, T. N., Gonzalez, M., Kline, W. O., Stover, G. L., Bauerlein, R., Zlotchenko, E., Scrimgeour, A., Lawrence, J. C., Glass, D. J., & Yancopoulos, G. D. (2001). Akt/mTOR pathway is a crucial regulator of skeletal muscle hypertrophy and can prevent muscle atrophy in vivo. *Nat.Cell Biol.* 3, 1014–1019.

Boisclair,Y. R., Rhoads, R. P., Ueki, I., Wang, J., and Ooi, G. T. (2001). The acid labile subunit (ALS) of the 150kDa IGF-binding protein complex: an important but forgotten component of the circulating IGF system. J Endocrinol *170*, 63–70.

Borst, S.E. (2004) Interventions for sarcopenia and muscle weakness in older people. Age Ageing. *33*, 548–55.

Bottinelli. R., Canepari, M., Pellegrino, M.A. and Reggiani, C. (1996) Force-velocity properties of human skeletal muscle fibres: myosin heavy chain isoform and temperature dependence. J Physiol. *495*, 573–86.

Brahm, H., Piehl-Aulin, K., Saltin, B., and Ljunghall S. (1997). Net fluxes over working thigh of hormones, growth factors and biomarkers of bone metabolism during short lasting dynamic exercise. Calcif Tissue Int *60*, 175–180.

Brill, K. T., Weltman, A. L., Gentili, A., Patrie, J. T., Fryburg, D. A., Hanks, J. B., Urban, R. J., and Veldhuis, J. D. (2002). Single and combined effects of growth hormone and testosterone administration on measures of body composition, physical performance, mood, sexual function, bone turnover, and muscle gene expression in healthy older men. J Clin.Endocrinol Metab 87, 5649–5657.

Brown, G.A., Vukovich, M.D., Sharp, R.L., Reifenrath, T.A., Parsons, K.A., King, D.S. (1999) Effect of oral DHEA on serum testosterone and adaptations to resistance training in young men. J Appl Physiol. 87, 2274–83.

Chakravarthy, M. V., Davis, B. S., and Booth, F. W. (2000). IGF-I restores satellite cell proliferative potential in immobilized old skeletal muscle. J Appl Physiol 89, 1365–1379.

Chew, S. L., Lavender, P., Clark, A.J., and Ross, R. J. (1995). An alternatively spliced human insulin-like growth factor-I transcript with hepatic tissue expression that diverts away from the mitogenic IBE1 peptide. Endocrinol 136, 1939–1944.

Coleman, M. E., DeMayo, F., Yin, K. C., Lee, H. M., Geske, R., Montgomery, C., and Schwartz, R. J. (1995). Myogenic vector expression of insulin-like growth factor I stimulates muscle cell differentiation and myofiber hypertrophy in transgenic mice. J Biol Chem 270, 12109–12116.

Corpas, E., Harman, S. M., and Blackman, M. R. (1993). Human growth hormone and human aging. Endocr.Rev. 14, 20–39.

Czerwinski, S. M., Martin, J. M., and Bechtel, P. J. (1994). Modulation of IGF mRNA abundance during stretch-induced skeletal muscle hypertrophy and regression. J Appl Physiol 76, 2026–2030.

Cuneo, R. C., Salomon, F., Wiles, C. M., Hesp, R., and Sonksen, P. H. (1991). Growth hormone treatment in growth hormone-deficient adults. I. Effects on muscle mass and strength. J Appl Physiol 70, 688–694

Daughaday, W. H., & Reeder, C. (1966). Synchronous activation of DNA synthesis in hypophysectomized rat cartilage by growth hormone. J Lab Clin.Med 68, 357–368.

Daughaday, W. H., Hall, K., Raben, M. S., Salmon, W. D., Jr., Van den Brande, J. L., and Van Wyk, J. J. (1972). Somatomedin: proposed designation for sulphation factor. Nature 235, 107.

D'Ercole, A. J., Stiles, A. D., & Underwood, L. E (1984). Tissue concentrations of somatomedin C: further evidence for multiple sites of synthesis and paracrine or autocrine mechanisms of action. Proc Natl Acad Sci. 81, 935–9.

DeVol, D. L., Rotwein, P., Sadow, J. L., Novakofski, J., and Bechtel, P. J. (1990). Activation of insulin-like growth factor gene expression during work-induced skeletal muscle growth. Am J Physiol 259, E89–E95.

Elliot, K.J., Cable, N.T., Reilly, T., Sefton, V., Kingsland, C. and Diver, M. (2005) Effects of supra-physiological changes in human ovarian hormone levels on maximum force production of the first dorsal interosseus muscle. Exp Physiol. 90, 215–23.

Ewton, D. Z., and Florini, J. R. (1995). IGF binding proteins-4, -5 and -6 may play specialized roles during L6 myoblast proliferation and differentiation. J Endocrinol 144, 539–553.

Finkelstein, D. I., Andrianakis, P., Luff, A. R., and Walker, D. (1991). Effects of thyroidectomy on development of skeletal muscle in fetal sheep. Am .J Physiol 261, R1300–R1306.

Fitts, R.H., Brimmer, C.J., Troup, J.P. and Unsworth, B.R. (1984) Contractile and fatigue properties of thyrotoxic rat skeletal muscle. Muscle Nerve. 7, 470–477.

Florini, J. R. (1987). Hormonal control of muscle growth. Muscle Nerve 10, 577–598.

Florini, J. R., Ewton, D. Z., and Coolican, S. A. (1996). Growth hormone and the insulin-like growth factor system in myogenesis. Endocr.Rev. 17, 481–517.

Gambke, B., Lyons, G. E., Haselgrove, J., Kelly, A. and Rubinnstein, N. A. (1983). Thyroidal and neural control of myosin transitions during development of rat fast and slow muscles. FEBS Lett. 156, 335–339.

Gilmour, R. S. (1994). The implications of insulin-like growth factor mRNA heterogeneity. J Endocrinol. 140, 1–3.

Goldspink, G. (2001). Method of treating muscular disorders. United States Patent. No: US 6,221,842 B1

Gonzalez-Cadavid, N. F, Taylor, W. E., Yarasheski, K., Sinha-Hikim, I., Ma K., Ezzat, S., Shen R., Lalani, R., Asa, S., Mamita, M., Nair, G., Arver, S., and Bhasin, S. (1998). Organization of the human myostatin gene and expression in healthy men and HIV-infected men with muscle wasting. Proc Natl Acad Sci U S A. *95*, 14938–43.

Greeves, J. P., Cable, N. T., Luckas, M. J., Reilly T., and Biljan, M. M. (1997) Effects of acute changes in oestrogen on muscle function of the first dorsal interosseus muscle in humans. J Physiol *500*, 265–70

Griffiths, R.D., Newshome, E., and Young, A. (2000). Muscle as a dynamic metabolic store. (Section on "Muscle", Ed: Young, A). Evans, J.G., Williams, T.F, Beattie, B.L., Michael, J.-P, Wilcock,G.K. Eds: Oxford Textbook of Geriatric Medicine 2nd Edition. Oxford University Press, pp 972–979.

Haddad, F., and Adams, G. R. (2002). Selected contribution: acute cellular and molecular responses to resistance exercise. J Appl Physiol *93*, 394–403.

Hameed, M., Orrell, R. W., Cobbold, M., Goldspink, G., and Harridge, S. D. R. (2003). Expression of IGF-I splice variants in young and old human skeletal muscle after high resistance exercise. J.Physiol *547*, 247–254.

Hameed, M., Orrell, R.W., Cobbold, M., Goldspink, G., Harridge, S.D.R. (2004a). Clarification. J. Physiol. *549*, 3.

Hameed, M., Lange, K. H., Andersen, J. L., Schjerling, P., Kjaer, M., Harridge, S. D. R., and Goldspink, G. (2004). The effect of recombinant human growth hormone and resistance training on IGF-I mRNA expression in the muscles of elderly men. J Physiol *555*, 231–240.

Hill, M., and Goldspink, G. (2003) Expression and splicing of the insulin-like growth factor gene in rodent muscle is associated with muscle satellite (stem) cell activation following local tissue damage. J Physiol. *549*, 409–18.

Howlett, R. A., Kelley, K. M., Grassi, B., Gladden, L. B., and Hogan, M. C. (2005). Caffeine administration results in greater tension development in previously fatigued canine muscle in situ. Exp.Physiol. (In Press)

Iida, K., Itoh, E., Kim, D. S., del Rincon, J. P., Coschigano, K. T., Kopchick, J. J., and Thorner, M. O. (2004). Muscle mechano growth factor is preferentially induced by growth hormone in growth hormone-deficient lit/lit mice. J.Physiol *560*, 341–349.

Isaksson, O. G., Jansson, J. O., and Gause, I. A. (1982). Growth hormone stimulates longitudinal bone growth directly. Science *216*, 1237–1239.

Jones, J. I. & Clemmons, D. R. (1995). Insulin-like growth factors and their binding proteins: biological actions. Endocr.Rev. 16, 3–34.

Jones SW, Hill RJ, Krasney PA, O'Conner B, Peirce N, Greenhaff PL. (2004) Disuse atrophy and exercise rehabilitation in humans profoundly affects the expression of genes associated with the regulation of skeletal muscle mass. FASEB J. *18*, 1025–7.

Kadi, F., Eriksson, A., Holmner, S., and Thornell, L. E . (1999) Effects of anabolic steroids on the muscle cells of strength-trained athletes. Med Sci Sports Exerc. *31*, 1528–34.

Kanda, F., Takatani, K., Okuda, S., Matsushita, T. and Chihara, K. (1999). Preventive effects of insulin-like growth factor-I on steroid-induced muscle atrophy. Muscle Nerve *22*, 213–7.

Klapper, D. G., Svoboda, M. E., and Van Wyk, J. J. (1983). Sequence analysis of somatomedin-C: confirmation of identity with insulin-like growth factor I. Endocrinol *112*, 2215–2217.

Lamberts, S. W., van den Beld, A. W., and van der Lely, A. J. (1997). The endocrinology of aging. Science *278*, 419–424.

Lange, K. H., Andersen, J. L., Beyer, N., Isaksson, F., Larsson, B., Rasmussen, M. H., Juul, A., Bulow, J., and Kjaer, M. (2002). GH administration changes myosin heavy chain isoforms in skeletal muscle but does not augment muscle strength or hypertrophy, either alone or combined with resistance exercise training in healthy elderly men. J Clin Endocrinol Metab *87*, 513–523.

Lee, S., Barton, E. R., Sweeney, H. L., and Farrar, R. P. (2004). Viral expression of insulin-like growth factor-I enhances muscle hypertrophy in resistance-trained rats. J Appl Physiol *96*, 1097-104. Erratum in: J Appl Physiol *96*, 2343, 2004

Li, X., and Larsson, L. (1997). Contractility and myosin isoform compositions of skeletal muscles and muscle cells from rats treated with thyroid hormone for 0, 4 and 8 weeks. J Muscle Res Cell Motil *18*, 335–44.

Linari, M., Brunello, E., Reconditi, M., Sun, Y.B., Panine, P., Narayanan, T., Piazzesi, G., Lombardi, V., and Irving, M. (2005). The structural basis of the increase in isometric force production with temperature in frog skeletal muscle. J Physiol *567*, 459–69.

Liu, J. P., Baker, J., Perkins, A. S., Robertson, E. J., and Efstratiadis, A. (1993) Mice carrying null mutations of the genes encoding insulin-like growth factor I (Igf-1) and type 1 IGF receptor (Igf1r). Cell. *175*(1), 59–72.

Liu, J. L. & LeRoith, D. (1999). Insulin-like growth factor I is essential for postnatal growth in response to growth hormone. Endocrinology 140, 5178–5184.

Lowe, D. A., Thomas, D. D., and Thompson, L. V. (2002). Force generation, but not myosin ATPase activity, declines with age in rat muscle fibers. Am J Physiol Cell Physiol *283*, C187–92

Lowe, W. L., Jr., Lasky, S. R., LeRoith, D., and Roberts, C. T., Jr. (1988). Distribution and regulation of rat insulin-like growth factor I messenger ribonucleic acids encoding alternative carboxyterminal E-peptides: evidence for differential processing and regulation in liver. Mol.Endocrinol *2*, 528–535.

Ma, K., Mallidis, C., Bhasin, S., Mahabadi, V., Artaza, J., Gonzalez-Cadavid, N., Arias, J. and Salehian, B. (2003) Glucocorticoid-induced skeletal muscle atrophy is associated with upregulation of myostatin gene expression. Am J Physiol Endocrinol Metab. *285*, E363–71.

Maréchel, G., and Gailly, P. (1999). Effects of nitric oxide on the contraction of skeletal muscle. Cell Mol Life Sci *55*, 1088-102.

Mauras, N., Hayes, V., Welch, S., Rini, A., Helgeson, K., Dokler, M., Veldhuis, J.D. and Urban R.J. (1998) Testosterone deficiency in young men: marked alterations in whole body protein kinetics, strength, and adiposity. J Clin Endocrinol Metab. *83*, 886–92.

McKeran, R. O., Slavin, G., Andrews, T.M., Ward, P., and Mair, W. G. (1975). Muscle fibre type changes in hypothyroid myopathy. J Clin Pathol *28*, 659–63.

McKoy, G., Ashley, W., Mander, J., Yang, S. Y., Williams, N., Russell, B., and Goldspink, G. (1999). Expression of insulin growth factor-1 splice variants and structural genes in rabbit skeletal muscle induced by stretch and stimulation. J Physiol *516*, 583–592.

McPherron, A. C., Lawler, A. M., and Lee, S. J. (1997). Regulation of skeletal muscle mass in mice by a new TGF-beta superfamily member. Nature *387*, 83–90

McPherron, A. C., and Lee, S. J. (1997). Double muscling in cattle due to mutations in the myostatin gene. Proc Natl Acad Sci U S A *94*, 12457–61

Moncada, S., Palmer, R. M., and Higgs, E. A. (1991). Nitric oxide: physiology, pathophysiology, and pharmacology. Pharmacol Rev. 43, 109–42.

Morley, A. A. (1995). The somatic mutation theory of ageing. Mutat.Res. *338*, 19–23.

Morrison, R. J., Miller, C. C. 3rd, and Reid, M. B..(1998). Nitric oxide effects on force-velocity characteristics of the rat diaphragm. Comp Biochem Physiol A Mol Integr Physiol *119*, 203–9.

Morrison, W. L., Gibson, J. N., Jung, R. T., and Rennie, M. J. (1988). Skeletal muscle and whole body protein turnover in thyroid disease. Eur J Clin Invest *18*, 62–8

Murphy, R. J., Hartkopp, A, Gardiner, P. F., Kjaer, M., and Beliveau, L. (1999). Salbutamol effect in spinal cord injured individuals undergoing functional electrical stimulation training. Arch Phys Med Rehabil *80*, 1264–7.

Musaro, A., McCullagh, K.J., Naya, F.J., Olson, E.N. and Rosenthal, N. (1999) IGF-1 induces skeletal myocyte hypertrophy through calcineurin in association with GATA-2 and NF-ATc1. Nature. *400*,:581–5.

Musaro, A., McCullagh, K., Paul, A., Houghton, L., Dobrowolny, G., Molinaro, M., Barton, E. R., Sweeney, H. L., and Rosenthal, N. (2001). Localized Igf-1 transgene expression sustains hypertrophy and regeneration in senescent skeletal muscle. Nat.Genet. *27*, 195–200.

Noirez P and Ferry A. (2000). Effect of anabolic/androgenic steroids on myosin heavy chain expression in hindlimb muscles of male rats. Eur J Appl Physiol *81*, 155–8.

Pellegrino, M. A., D'Antona, G., Bortolotto, S., Boschi, F., Pastoris, O., Bottinelli, R., Polla, B., and Reggiani, C. (2004). Clenbuterol antagonizes glucocorticoid-induced atrophy and fibre type transformation in mice. Exp Physiol *89*, 89–100

Phillips, S. K., Rook, K. M., Siddle, N. C., Bruce, S. A., and Woledge, R. C. (1993a). Muscle weakness in women occurs at an earlier age than in men, but strength is preserved by hormone replacement therapy. Clin Sci *84*, 95–8.

Phillips, S. K., Rowbury, J.L., Bruce, S. A. and Woledge, R. C. (1993b). Muscle force generation and age: the role of sex hormones. IN Sensorimotor Impairment in the Elderly. (Stelmech, G.E. and Hömberg, V. eds) pp 129-141, Kluewr Academic Publishers, The

Phillips, S. K., Sanderson, A. G., Birch, K., Bruce, S. A., and Woledge, R. C. (1996). Changes in maximal voluntary force of human adductor pollicis muscle during the menstrual cycle J Physiol *496*, 551–7.

Rennie, M. J. (2003). Claims for the anabolic effects of growth hormone: a case of the emperor's new clothes? Br J Sports Med *37*, 100–105.

Rennie, M. J., Wackerhage, H., Spangenburg, E. E., and Booth, F. W. (2004). Control of the size of the human muscle mass. Annu Rev Physiol *66*, 799–828.

Rinderknecht, E., and Humbel, R. E. (1978). The amino acid sequence of human insulin-like growth factor I and its structural homology with proinsulin. J Biol Chem *253*, 2769–2776.

Rommel, C., Bodine, S. C., Clarke, B. A., Rossman, R., Nunez, L., Stitt, T. N., Yancopoulos, G. D., and Glass, D. J. (2001) Mediation of IGF-I-induced skeletal myotube hypertrophy by PI(3)K/Akt/mTOR and PI(3)K/Akt/GSK3 pathways. Nat Cell Biol *3*, 1009–1013.

Roemmich, J.N., Huerta, M.G., Sundaresan, S.M. and Rogol, A.D. (2001) Alterations in body composition and fat distribution in growth hormone-deficient prepubertal children during growth hormone therapy. Metabolism. *50*, 537–47.

Roth, S. M., Martel, G. F., Ferrell, R. E., Metter, E. J., Hurley, B. F., and Rogers, M. A. (2003). Myostatin gene expression is reduced in humans with heavy-resistance strength training: a brief communication. Exp Biol Med.*228*, 706–9.

Roth, S. M., and Walsh, S. (2004). Myostatin: a therapeutic target for skeletal muscle wasting. Curr Opin Clin Nutr Metab Care. 259–63.

Rudman, D., Kutner, M. H., Rogers, C. M., Lubin, M. F., Fleming, G. A., and Bain, R. P. (1981). Impaired growth hormone secretion in the adult population: relation to age and adiposity. J.Clin.Invest *67*, 1361–1369.

Rudman, D., Feller, A. G., Nagraj, H. S., Gergans, G. A., Lalitha, P. Y., Goldberg, A. F., Schlenker, R. A., Cohn, L., Rudman, I. W., and Mattson, D. E. (1990). Effects of human growth hormone in men over 60 years old. N.Engl.J.Med. 323, 1–6.

Ryall, J. G., Plant, D. R., Gregorevic, P., Sillence, M.N., and Lynch, G. S. (2004). Beta 2-agonist administration reverses muscle wasting and improves muscle function in aged rats. J Physiol *555*, 175–88.

Sacheck, J.M., Ohtsuka, A., McLary, S.C. and Goldberg, A.L. (2004) IGF-I stimulates muscle growth by suppressing protein breakdown and expression of atrophy-related ubiquitin ligases, atrogin-1 and MuRF1. Am J Physiol Endocrinol Metab. *287*, E591–601.

Sadowski, C. L., Wheeler, T. T., Wang, L. H., and Sadowski, H. B. (2001). GH regulation of IGF-I and suppressor of cytokine signaling gene expression in C2C12 skeletal muscle cells. Endocrinol. *142*, 3890–3900.

Sandri, M., Sandri, C., Gilbert, A., Skurk, C., Calabria, E., Picard, A., Walsh, K., Schiaffino, S., Lecker, S.H. and Goldberg, A.L. (2004). Foxo transcription factors induce the atrophy-related ubiquitin ligase atrogin-1 and cause skeletal muscle atrophy. Cell. *117*, 399–412.

Sarwar, R., Niclos, B. B., and Rutherford, O. M. (1996). Changes in muscle strength, relaxation rate and fatiguability during the human menstrual cycle. J Physiol *493*, 267–72.

Shavlakadze, T., Winn, N., Rosenthal, N., and Grounds, M. D. (2005). Reconciling data from transgenic mice that overexpress IGF-I specifically in skeletal muscle. Growth Horm IGF Res *15*, 4–18.

Schuelke, M., Wagner, K. R., Stolz, L. E., Hubner, C., Riebel, T., Komen, W., Braun, T., Tobin, J. F., and Lee, S. J. (2004). Myostatin mutation associated with gross muscle hypertrophy in a child. N Engl J Med *350*, 2682–8.

Schlechter, N. L., Russell, S. M., Spencer, E. M., and Nicoll, C. S. (1986). Evidence suggesting that the direct growth-promoting effect of growth hormone on cartilage in vivo is mediated by local production of somatomedin. Proc Natl Acad Sci U.S.A *83*, 7932–7934.

Semsarian, C., Sutrave, P., Richmond, D. R., and Graham, R. M. (1999). Insulin-like growth factor (IGF-I) induces myotube hypertrophy associated with an increase in anaerobic glycolysis in a clonal skeletal-muscle cell model. Biochem J 339, 443–451.

Shimatsu, A. & Rotwein, P. (1987). Mosaic evolution of the insulin-like growth factors. Organization, sequence, and expression of the rat insulin-like growth factor I gene. J.Biol.Chem. 262, 7894–7900.

Signorile, J.F., Banovac, K., Gomez, M., Flipse, D., Caruso, J.F. and Lowensteyn, I. (1995) Increased muscle strength in paralyzed patients after spinal cord injury: effect of beta-2 adrenergic agonist. Arch Phys Med Rehabil. 76, 55–8.

Sinha-Hikim, I., Artaza, J., Woodhouse, L., Gonzalez-Cadavid, N., Singh, A. B., Lee, M. I., Storer, T. W., Casaburi, R., Shen, R., and Bhasin S. (2002). Testosterone-induced increase in muscle size in healthy young men is associated with muscle fiber hypertrophy. Am J Physiol Endocrinol Metab 283, E154–64.

Sinha-Hikim, I., Roth, S. M., Lee, M. I., and Bhasin, S. (2003) Testosterone-induced muscle hypertrophy is associated with an increase in satellite cell number in healthy, young men. Am J Physiol Endocrinol Metab 285, E197–205

Singh, M. A., Ding, W., Manfredi, T. J., Solares, G. S., O'Neill, E. F., Clements, K. M., Ryan, N. D., Kehayias, J. J., Fielding, R. A., and Evans, W. J. (1999). Insulin-like growth factor I in skeletal muscle after weight-lifting exercise in frail elders. Am.J.Physiol 277, E135–E143.

Skelton, D. A., Phillips, S. K., Bruce, S. A., Naylor, C. H., and Woledge, R. C. (1999). Hormone replacement therapy increases isometric muscle strength of adductor pollicis in post-menopausal women. Clin Sci. 96, 357–64.

Stewart, C. E., and Rotwein, P. (1996). Growth, differentiation, and survival: multiple physiological functions for insulin-like growth factors. Physiol Rev 76, 1005–1026.

Taaffe, D. R., Pruitt, L., Reim, J., Hintz, R. L., Butterfield, G., Hoffman, A. R., and Marcus, R. (1994). Effect of recombinant human growth hormone on the muscle strength response to resistance exercise in elderly men. J.Clin.Endocrinol.Metab 79, 1361–1366.

Taaffe, D. R., Jin, I. H., Vu, T. H., Hoffman, A. R., and Marcus, R. (1996). Lack of effect of recombinant human growth hormone (GH) on muscle morphology and GH-insulin-like growth factor expression in resistance-trained elderly men. J Clin.Endocrinol.Metab 81, 421–425.

Taaffe, D.R., Luz Villa, M., Delay, R. and Marcus, R. (1995) Maximal muscle strength of elderly women is not influenced by oestrogen status. Age Ageing. 24, 329–33

Thomas, M. J., Kikuchi, K., Bichell, D. P., and Rotwein, P. (1995). Characterization of deoxyribonucleic acid-protein interactions at a growth hormone-inducible nuclease hypersensitive site in the rat insulin-like growth factor-I gene. Endocrinol 136, 562–569.

Turner, J. D., Rotwein, P., Novakofski, J., and Bechtel, P. J. (1988). Induction of mRNA for IGF-I and -II during growth hormone-stimulated muscle hypertrophy. Am J Physiol 255, E513–E517.

Urban, R.J., Bodenburg, Y.H., Gilkison, C., Foxworth, J., Coggan, A.R., Wolfe, R.R. and Ferrando, A. (1995) Testosterone administration to elderly men increases skeletal muscle strength and protein synthesis. Am J Physiol. 269, E820–6.

Valenti, G., Denti, L., Maggio, M., Ceda, G., Volpato, S., Bandinelli, S., Ceresini, G., Cappola, A., Guralnik, J.M. and Ferrucci, L. (2004) Effect of DHEAS on skeletal muscle over the life span: the InCHIANTI study. J Gerontol A Biol Sci Med Sci.59, 466–72.

van Baak, M. A., Mayer, L. H., Kempinski, R. E., and Hartgens F. (2000). Effect of salbutamol on muscle strength and endurance performance in nonasthmatic men. Med Sci Sports Exerc 32, 1300–6.

Vandenburgh, H. H., Karlisch, P., Shansky, J., and Feldstein, R. (1991). Insulin and IGF-I induce pronounced hypertrophy of skeletal myofibers in tissue culture. Am J Physiol 260, C475–C484.

Welle, S., Thornton, C., Statt, M., and McHenry, B. (1996). Growth hormone increases muscle mass and strength but does not rejuvenate myofibrillar protein synthesis in healthy subjects over 60 years old. J Clin Endocrinol.Metab 81, 3239–3243.

Westerblad, H., and Allen, D. G. (1991). Changes of myoplasmic calcium concentration during fatigue in single mouse muscle fibers. J Gen Physiol 98, 615–35.

Wing, S.S. and Goldberg, A.L. (1993) Glucocorticoids activate the ATP-ubiquitin-dependent proteolytic system in skeletal muscle during fasting.Am J Physiol. 264, E668–76.

Yang, S., Alnaqeeb, M., Simpson, H., and Goldspink, G. (1996). Cloning and characterization of an IGF-1 isoform expressed in skeletal muscle subjected to stretch. J.Muscle Res.Cell Motil *17*, 487–495.

Yang, S. Y., and Goldspink, G. (2002). Different roles of the IGF-I Ec peptide (MGF) and mature IGF-I in myoblast proliferation and differentiation. FEBS Lett. *522*, 156–160.

Yarasheski, K. E., Campbell, J. A., Smith, K., Rennie, M. J., Holloszy, J. O., and Bier, D. M. (1992). Effect of growth hormone and resistance exercise on muscle growth in young men. Am J Physiol *262*, E261–E267.

Yarasheski, K. E., Zachweija, J. J., Angelopoulos, T. J. and Bier, D. M. (1993) Short-term growth hormone treatment does not increase muscle protein synthesis in experienced weight lifters. J Appl Physiol. *74*:3073–6.

Yimlamai, T., Dodd, S.L., Borst, S.E. and Park, S. (2005) Clenbuterol induces muscle-specific attenuation of atrophy through effects on the ubiquitin-proteasome pathway. J Appl Physiol. *99*, 71–80.

Yu, F., Gothe, S., Wikstrom, L., Forrest, D., Vennstrom, B., and Larsson, L. (2000). Effects of thyroid hormone receptor gene disruption on myosin isoform expression in mouse skeletal muscles. Am J Physiol Regul Integr Comp Physiol *278*, R1545–54.

Zanconato, S., Moromisato, D. Y., Moromisato, M. Y., Woods, J., Brasel, J. A., LeRoith, D., Roberts, C. T., Jr., and Cooper, D. M. (1994). Effect of training and growth hormone suppression on insulin-like growth factor I mRNA in young rats. Journal of Applied Physiology *76*, 2204–2209.

Zeman RJ, Lineman R, Easton TG, Etlinger JD. (1988) Slow to fast alterations in skeletal muscle fibers caused by clenbuterol, a beta 2-receptor agonist. Am J Physiol. *254*, E726–32.

CHAPTER 11

SKELETAL MUSCLE ADAPTATIONS
TO DISEASE STATES

JOAQUIM GEA, ESTHER BARREIRO AND MAURICIO OROZCO-LEVI
Hospital del Mar, I.M.I.M., Universitat Pompeu Fabra, Barcelona, Spain

1. INTRODUCTION

Human beings have been moving and breathing on our planet for at least the last 1 500 000 years (Carroll, 2003). During this time, skeletal muscles have been essential for both survival and cultural development. The contractile properties of the muscles have provided humans as well as other animals with the ability to move from old lands to new ones, in the hope of finding better living conditions. In addition, skeletal muscles have supplied the appropriate physiological basis for the elaboration of all types of instruments as well as for artistic creation, scientific activity and culture. Thereby, humans were better equipped to defend themselves, multiply and spread out, and also to interact with each other and in this way, build complex societies.

However, when muscles involved in moving from one place to another, manipulating instruments or breathing do not work properly, humans become frail and can even die. Those skeletal muscles located in the lower limbs bestow the ability to move around and socialize. When these muscles become ineffective in doing their job, the mobility of the individual is reduced. They are viewed as being less and less able to play an active role in the community and they also feel incapacitated, which results in a further reduction in their activities. All of this will probably have a deep impact on their quality of life. Skeletal muscles from the upper limbs in turn, are essential to manipulate objects and for self-care. When these muscle groups work inadequately, subjects tend to require assistance even for the simplest of everyday tasks, that is to say they need help from other human beings in order to survive. Respiratory muscles, for their part, are specialized striated muscles whose role is to provide the subject with fresh air, which is essential for its

315

R. Bottinelli and C. Reggiani (eds.), Skeletal Muscle Plasticity in Health and Disease, 315–360.
© 2006 *Springer.*

metabolic functions (including muscle contraction itself). Specifically, respiratory muscles generate changes in intrathoracic pressure, which becomes different from atmospheric pressure, thus allowing the air to enter and exit from the respiratory system. When respiratory muscles fail in their task, oxygen is deficiently supplied to the tissues and carbon dioxide is not removed from them. If this situation persists, aerobic activity and acid-base homeostasis can no longer be sustained, and the subject dies.

Therefore, from a physiological point of view, skeletal muscles can be divided into two main muscle groups: those involved in limb movements (known as *"peripheral" muscles*) and those others responsible for ventilation (*respiratory muscles*). In the following sections we will review most of the nosological situations where these two groups of muscles fail in their task due to intrinsic or extrinsic abnormalities. However, in the first place, because the main cause of death in all these entities is ventilatory failure, the task of respiratory muscles will be more extensively dealt with.

Among the muscles contributing to breathing it is possible to differentiate between *inspiratory* and *expiratory* muscles. Inspiration is the part of breathing that involves the entrance of fresh air into the body. This is an active process that, at rest, accounts for around 5% of the total *oxygen uptake*. Contraction of inspiratory muscles expands the thoracic cage which along with the retractile properties of the lung, increases the negativity of pleural pressure. This "high negativity" is transmitted to lung parenchyma and specifically to alveoli. However, due to the resistance of lung parenchyma, the pressure in the latter is not so negative as it is in the pleural space. Nevertheless, it is still "more negative" than atmospheric pressure (considered zero). This pressure gradient is responsible for the entrance of air into the lungs, where gas exchange occurs. The main inspiratory muscle is the diaphragm, although this is completely true only at rest and in healthy subjects. Other muscles, such as external intercostals, paresternal intercostals and to a lesser degree, scalenes, sternocleidomastoid, latissimus dorsi, serratus and pectoralis help the diaphragm to do its job. This support is especially important when the diaphragmatic function is impaired or an additional effort is required (i.e. during airway obstruction or exercise). The other part of the breathing cycle is expiration. This is generally just a passive process, secondary to inspiratory muscle relaxation and the elastic recoil of the lung. The pressure in the pleural space becomes "less negative" and since the resistance of the lung parenchyma is still present as an obstacle to the transmission of this negativity, the pressure in the alveoli become slightly positive. This creates a new gradient whereby the air is impelled outwards. However, when it is necessary to increase expiratory flow or to overcome increased respiratory loads, other muscles are recruited. The main expiratory muscles are internal intercostals and those located in the abdominal wall. Their contraction results in a further increase in the alveolar positive pressure, thus facilitating expiration.

Both peripheral and respiratory muscles have two main functional properties: *strength* or the ability to develop a maximal effort, and *endurance* or the ability to maintain the task at a submaximal level of intensity. *Muscle dysfunction* would be defined as the situation where skeletal muscles show a reduced strength and/or a reduced endurance, thus being unable to develop their physiological tasks. There are different conditions, and not only diseases, which can result in skeletal muscle dysfunction. Some affect the muscle directly but others involve the nervous structures, the neuromuscular junction (Epstein, 1994), structures contributing to blood flow supply and to metabolic homeostasis, and also morphological changes in the rib cage, the lungs, the osteoarticular system, etc. (Martyn, 1980) (Epstein, 1994) (Appendini, 1996).

However, although skeletal muscles are *very sensitive* to changes occurring in their environment, both at a local and systemic level, they are also known to be *extremely plastic*, being capable of adapting their structure to new conditions (*remodelling*). These changes can facilitate the relative maintenance of their function, but can also result in functional impairment.

2.   NEUROMUSCULAR DISORDERS (*JOAQUIM GEA*)

2.1   Muscular Dystrophy and Other Myopathies

Muscular dystrophies or *dystrophic myopathies* are inherited conditions whose first symptoms generally appear in childhood or adolescence. Muscle dysfunction is progressive and can lead to death. Among these disorders, Duchenne and Steinert dystrophies are the myopathies which most frequently lead to ventilatory failure (Lynn, 1994). The main muscular distrophies along with other inherited muscle disorders and acquired myopathies appear in Table 1.

The **Duchenne muscular dystrophy** also called *Pseudohypertrophic Muscular Dystrophy* is the most prevalent dystrophic myopathy (Figure 1). This condition is recessively inherited and the genetic abnormality is located in the p21 region of the X chromosome, where the gene encoding dystrophin is (Hoffman, 1987). Dystrophin is a protein of 427 kDa, located close to plasmolemma and linked to it by glycoproteins, which in turn are connected to laminin, as laminin is to the basal membrane. Dystrophin strengthens the plasmolemma in such a way that it can better resist the mechanical forces deriving from contraction-relaxation cycles. When dystrophin is absent, injury occurs and although initially the muscle is repaired, later on fibrotic tissues progressively substitute muscle fibers. Only males (with just one X chromosome) are affected by the disease while women are *carriers*. This is due to the absence of descendants from affected men, which means that women always receive at least one *disease-free* X chromosome. The symptoms first appear at around five years of age. Early hip and lower limb muscle involvements lead to the development of walking problems and frequent falls. The disease progresses and extends to those muscles located in the shoulders, abdomen and vertebral column. The appearance of the patient changes since they show lumbar hiperlordosis, with a prominent abdomen. The fat and fibrotic infiltration of muscles gives a false appearance of muscle hypertrophia in the limbs (hence the name of *Pseudohypertrophic Muscular Dystrophy*). Actually there is a muscle atrophy

Table 1. Most prevalent myopathic diseases and neurological-neuromuscular junction disorders

Acquired Myopathies
Polymyositis and Dermatomyositis
Systemic Lupus Erythematosus
Rheumatoid Arthritis
Inflammatoy bowel disease myopathies
Endocrine myopathies
Rhabdomyolysis
Steroid and other drug myopathies
Electrolite disorders
Inherited Myopathies
Muscle Dystrophies
    Duchenne muscular dystrophy
    Steinert myotonic dystrophy
    Facio-scapulo-humeral muscular dystrophy
    Limb-girdle muscular dystrophy
Congenital myopathies
Metabolic myopathies
    Mitochondrial myopathies
    Acid maltase defficiency
Neurological-neuromuscular junction disorders
Amyotrophic lateral sclerosis
Multiple sclerosis
Guillain-Barré syndrome
Parkinson's disease
Myasthenia gravis

which in combination with deformities in the limbs, as a consequence of functional imbalance between flexor and extensor muscles, leads to deformities. At the age of 12–20 years old, the patient would need a wheelchair, and at this stage, respiratory and even cardiac muscles also become affected. These latter two conditions largely determine the survival of the patient, who, without support treatments, die on average at between 20 and 25 years of age.

Steinert Myotonic Dystrophy is less frequent than Duchenne, and the symptoms appear later in life. The inheritance is autosomal dominant and abnormality is located in the region q13.3 of the 19th chromosome, where proteinkynase is encoded (Harley, 1992). The resulting abnormalities in chlorure permeability lead to muscle weakness and delay in relaxation (myotony). First symptoms appear at the age of twenty, when myotony can be observed in the face and neck muscles. This will coexist with the involvement of tongue and pharynx, initiating speech and breath disturbances. Later on, muscles located in the limbs will also be affected. Patients have a characteristic appearance with frontal baldness, ptosis, cataracts, and temporal wasting. In some cases, the involvement of respiratory or heart muscles can become relevant, even when limb weakness is still mild. In many cases, central hypoventilation and sleep-related respiratory disorders add to respiratory muscle involvement (Guilleminault, 1992) (Lynn, 1994). The respiratory

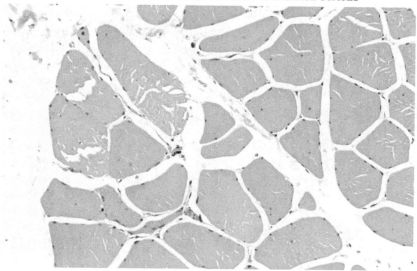

Figure 1. Skeletal muscle from a patient in a relatively early stage of Duchenne muscular dystrophy: Coexistence of normal and atrophic fibers. Note the presence of internalized nuclei. A color figure is freely accessible via the website of the book: http://www.springer.com/1-4020-5176-x

muscle dysfunction can be especially important in patients with relevant comorbidities and in those submitted to mechanical ventilation, as it is very difficult to wean them away from the support device. The overall life expectancy is around 50 years.

The **Facio-scapulo-humeral muscular dystrophy** is an autosomal dominant inherited condition. The molecular substrate of this entity is not very well defined but the genetic abnormality is located in region q35 of chromosome 4. The onset of the symptoms can range from 5 to 20 years, and facial muscles are the first to be affected leading to difficulties in mastication, sucking and eye opening/closing. After a variable time-course, those muscles located in the scapular area and upper limbs are also affected and difficulties in the performance of related movements will appear. The pelvic girdle can also be affected as well as trunk and respiratory muscles. The disease combines periods of progression with others of stability, and generally it does not jeopardize the life of the patient (Lynn, 1994).

Limb-girdle dystrophies (ryzomelic dystrophies) (Figure 2) are a heterogeneous group of inherited autosomal recessive and dominant processes, due to abnormalities in sarcoglicans (DAG), emerin and calpain. Symptoms start between 10 and 40 years, and are predominant in limbs and girdles. Therefore limb movements become difficult and both selfcare and mobility can be affected. In some cases, respiratory involvement is relevant but very rarely does it pose it a threat to the patient's life (Lynn, 1994).

Acid Maltase Deficiency Myopathy is an autosomal recessive disease (chromosome 17, region q23) caused by deficient activity of alpha-1,4-glicosidase,

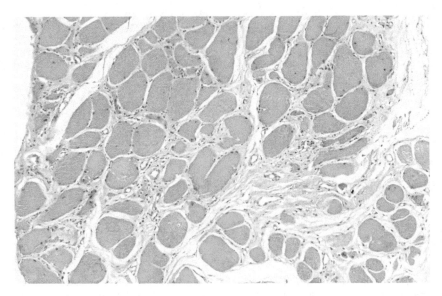

Figure 2. Skeletal muscle from a patient with a Limb-Girdle dystrophy: Atrophic fibers, inflammatory elements, and fibrosis are present. A color figure is freely accessible via the website of the book: http://www.springer.com/1-4020-5176-x

which results in the storage of type II glycogen. Skeletal muscles characteristically show vacuoles filled with glycogen and lisosomal breakdown products. The spectrum of severity of this entity is broad, since it can be fatal in the early childhood (Pompe's disease) or constitute a chronic adult disease. In the adult form, it is the proximal muscles that are particularly affected, though respiratory muscle involvement is also frequent and sometimes this is the first expression of the disorder (Van der Walt, 1987). This is believed to be the consequence of the predominance of oxidative fibers in respiratory muscles, since type I are less efficient than type II fibers for glycogen synthesis and storage (Lynn, 1994).

Mitochondrial myopathies are inherited muscle disorders, with a special peculiarity. They are the result of abnormal mitochondrial DNA (which is independent of nuclear DNA, and has exclusively maternal inheritance). Most of the mitochondrial myopathies show a characteristic accumulation of these organella in muscle fibers (Figure 3) (Tritschler, 1991), a circumstance which gives them a distinctive appearance ("ragged red fibers"). Mitochondria are actually enlarged and distorted, with membrane foldings, crystalline inclusions and concentric whorls. The symptoms secondary to mitochondrial disorders can be heterogeneous depending on the specific myopathy. Thus, MELAS (Mitochondrial myopathy, Encephalopathy, Lactic Acidosis and Stroke) typically results in stroke-like episodes and seizures, while MERRF (Myoclonic Epilepsy and Ragged Red Fibers) causes hypoventilation and myoclonies. Kearns-Sayre syndrome results in ophthalmoplegia, retinal

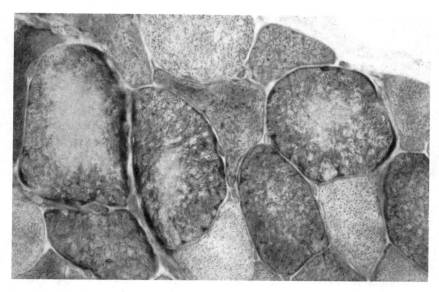

Figure 3. Muscle from a patient with a Mitochondrial Myopathy (NADH histochemical preparation). Note the presence of intense subsarcolemmal staining indicating strong mitochondrial activity in this location A color figure is freely accessible via the website of the book: http://www.springer.com/1-4020-5176-x

degeneration, heart block and ataxia, and Leigh's syndrome in psychomotor regression, brainstem dysfunction, seizures and ventilatory problems.

Inflammatory Myopathies are a very heterogeneous group of disorders, whose main representatives are **Polymyositis, Dermatomyositis** and **Inclusion body myositis**. However, there are grounds for also including **Lupus** and even **Rheumatoid Arthritis** and **inflammatory bowel diseases** in this category. However, since clinical, histological and biological findings are rather different, it is believed that pathogenic mechanisms are also likely to be different (Plotz, 1989).

Dermatomyositis and **Polymyositis** are inflammatory myopathies of unknown origin. Both disorders have an insidious onset, with symmetrical weakness of the shoulders and pelvic girdle muscles. Interestingly, symptoms are more evident in the upper limbs and shoulders than in the lower limbs. Up to 80% of the patients with Polymyositis, Inclusion Body Myositis and Dermatomyositis show diaphragm weakness, which is specially evident in the latter (Teixeira, 2005). However, respiratory failure is uncommon in these disorders.

The main difference between Dermatomyositis and Polymyositis is the involvement of the skin in the former. This involvement characteristically takes the form of erythematous lesions with a subjacent edema in the acute phase followed by residual lesions in the long-term. Muscle biopsies of both disorders show the features which typically characterize myositis, including necrosis, infiltration by macrophages and lymphocytes (Engel, 1984), fibers in different stages of regeneration and a progressive increase in the amount of fibrotic tissue. The mechanism for recruitment of inflammatory cells is believed to be mediated by chemokines. In this regard, the expression of beta chemokines CCL3 and CCL4 has recently been observed to be greater in the muscles of such patients (Civatte, 2005). However, Dermatomyositis and Polymyositis also evidenced differential histological findings. The main ones are the presence of perifascicular atrophy, microvascular injury and endothelial microtubular inclusions in Dermatomyositis (Figure 4). The ways in which blood vessels are affected include arteritis and phlebitis, intimal hyperplasia, and even complete occlusions by thrombi, with subsequent fiber necrosis. Ultrastructural analysis characteristically shows microtubular inclusions in the endothelial cells and multiple signs of capillar degeneration and regeneration (Jerusalem, 1974).

Systemic Lupus Erythematosus is a systemic autoimmune disorder which can affect different organs and systems. This includes the presence of general symptoms as well as those related to the joints, the muscles, the skin, the lungs, the heart and the nervous system. Skeletal muscles are affected in up to 25% of patients (Vilardell, 2004): in particular peripheral muscles, above all those located in proximal territories, and especially lower limb muscles (Foote, 1982). However, the involvement of respiratory muscles is not rare in lupus (Wilcox, 1988) (Wiedemann, 1989). Histological changes are not very different from those typical of other inflammatory myopathies. In this regard, small blood vessels within the muscle can be affected just as they are in many other organs.

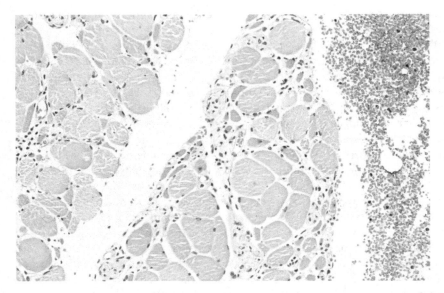

Figure 4. Skeletal Muscle from a patient with Dermatomyositis: Typical perifascicular atrophy and inflammation elements are shown. A color figure is freely accessible via the website of the book: http://www.springer.com/1-4020-5176-x

Rheumatoid Arthritis (RA) is a chronic, systemic and autoimmune disease of unknown aethiology. This disorder is mainly characterized by functional limitation secondary to joint and bone involvements. However, there is increasing evidence that RA also affects skeletal muscles by direct (local and systemic inflammation) or indirect (deconditioning, treatments) mechanisms, reducing their strength (Roubenoff, 1994) (Rall, 1996). Respiratory muscle dysfunction has also been reported in these patients (Gorini, 1990) but generally does not result in severe problems. From a histological point of view, atrophy of type II fibers appears to be common, a finding that supports the role for deconditioning and iatrogenia. In addition, RA myositis is usually associated with changes in blood vessels (Engel, 1994a), which can lead to necrotizing arteriolitis and generation of fibrin thrombi (Schmidt, 1961).

Inflammatory bowel diseases is a generic term which mainly includes ulcerative colitis and Crohn's disease. These disorders share chronicity, recurrence of acute episodes, a central role for inflammation and unknown origin. Musculoskeletal disorders are a non-rare extra-intestinal manifestation of inflammatory bowel diseases (Rankin, 1990) (Salvarani, 2001). Their clinical spectrum is broad and includes fibromialgia (Buskila, 1999) and polymyositis (Voigt, 1999).

2.2   Neurological and Neuromuscular Junction Disorders

Amyotrophic Lateral Sclerosis (ALS) is a progressive, neurodegenerative disorder of the nervous system whose symptoms are the result of degeneration of the upper and lower motor neurones. Therefore, ALS will result in a combination of weakness, fasciculations, spasticity and hyperreflexia. Symptoms can initiate anywhere and progression can eventually involve respiratory muscles (Kaplan, 1994), leading to alveolar hypoventilation. As in many other neuromuscular disorders, this circumstance typically worsens with sleep and in the supine position. In more advanced stages, the ALS patient is confined to a wheelchair and ventilation is critically affected even at rest and during daytime. At this point, the patient dies if ventilatory support is not supplied. From a histological point of view, muscle biopsies of individuals with ALS show signs of neurogenic atrophy (Figure 5).

The **Guillain-Barré Syndrome** is an acute and progressive demyelinating disorder of autoimmune origin. Also called *Acute Inflammatory Demyelinating Polyradiculoneuropathy* (AIDP), it is generally triggered by a viral infection (Arnason, 1984). The AIDP main symptom is weakness, which typically appears symmetrically in the legs and is predominantly proximal. This symptom progresses rapidly to other territories and, in up to a third of the subjects, severely affects respiratory muscles. This is the most serious complication of the disease since it will determine the occurrence of severe hypoventilation and general prognosis (Pascual, 1990) (Teitelbaum, 1994). In addition to respiratory muscle involvement, AIDP can also affect laryngeal muscles and tongue leading to aspiration and/or upper airway obstruction. Finally, the heart is also liable to be affected by AIDP due

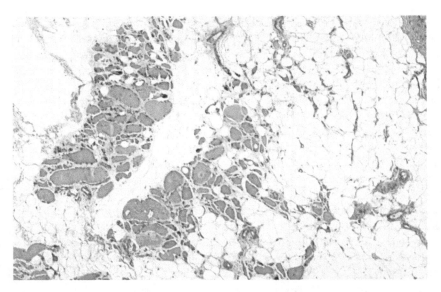

Figure 5. Skeletal muscle from a patient with Amyothrophic Lateral Sclerosis: Signs of neurogenic atrophy, marked adipose infiltration, and coexisting regeneration elements can be observed. A color figure is freely accessible via the website of the book: http://www.springer.com/1-4020-5176-x

to an autonomic dysfunction. Histological findings in muscle are edema, endoneurial mononuclear cell infiltration and segmental demyelinitation (Dick, 1994).

Parkinson's Disease is an entity of unknown origin, although recent studies suggest a role for oxidative stress. Whatever the cause might be, degeneration occurs in the pigmented neurones localized in the *substantia nigra*. As a result, patients with Parkinson's disease develop tremor, rigidity, bradykinesia and postural inability. Upper airway muscles can also be affected (Vincken, 1984) resulting in stridor, respiratory sleep disorders and aspiration. Finally, respiratory muscle dysfunction can appear leading to a mild restrictive abnormality with minor clinical consequences (Brown, 1994) (Cano, 1996).

Myasthenia Gravis is an acquired autoimmune disorder of the neuromuscular junction, caused by antibodies against the acetylcholine receptor (Engel, 1994b). As a result, the number of these receptors is reduced in the postsynaptic membrane, and therefore neuromuscular transmission becomes insufficient. Clinical consequences are fatigue and muscle weakness, which are actually exacerbated by exercise. The most common initial symptoms are ptosis and diplopia, followed by dysfunction of other facial and laryngeal muscles. The course of the disease is variable with periods of exacerbation and remission; and other territories, such as the trunk and the limbs become progressively and asymmetrically affected (Drachman, 1978). Respiratory muscle dysfunction can also appear (Mier-Jedrzejowicz, 1988), and ranges from exertional dyspnea to severe hypoventilation and cough insufficiency

Table 2. Muscle dysfunction in non-primary muscle diseases

Cancer-cachexia

Aging-sarcopenia
Respiratory Disorders
Chronic obstructive respiratory disease
  Bronchial asthma
   Sleep apnea-hypopnea and related syndromes
Deleterious effects of mechanical ventilation
Chronic Heart Failure
Sepsis

(Zulueta, 1994). Muscle biopsy specimens can show IgG and complement components in the neuromuscular junction (Engel, 1977).

Multiple Sclerosis is other idiopathic and inflammatory demyelinating disorder of the central nervous system. Lesions characteristically have the appearance of plaques, with relevant inflammatory activity (lymphocytes, macrophages, rests of myelin) in the acute phase, and extensive gliosis in the chronic phase (Barnes, 1991). Interestingly, signs of active remyelinization can be also observed (Prineas, 1993). Symptoms vary widely depending on the area affected by demyielinization, but muscle weakness is present in around a half of the patients. Respiratory and upper airway muscle involvements can occur as the result of central lesions, and have dramatic consequences in the life of the patients (Howard, 1992).

Skeletal muscles can be also affected in many other conditions which are not directly related to intrinsic abnormalities in neuromuscular structures (Table 2). Among them, disorders involving abnormalities in the rib cage, the heart or the lungs are important due to their prevalence in the general population. In recent years there has been increasing evidence that these entities, in addition to the effects on their primary target organs, also cause skeletal muscle dysfunction. This is the case with cyphoskoliosis, bronchial asthma, chronic obstructive pulmonary disease (COPD) and chronic heart failure (Epstein, 1994) (Hamilton, 1995) (Gea, 2001a). In addition, the deleterious effects of some of the treatments such as mechanical ventilation or steroid therapy should also be briefly considered.

3.   RESPIRATORY DISEASES *(JOAQUIM GEA)*

3.1   COPD

This is an entity of extraordinary relevance due to its high prevalence and elevated health and social costs. The main cause is tobacco smoking, which results in airway and lung inflammation as well as destruction of lung parenchyma. From a functional point of view COPD is mainly characterized by non-reversible airflow obstruction (Global Initiative for COPD, G.O.L.D., 2004). The onset of the symptoms is insidious, with cough and progressive dyspnea, leading with the passing of time to exercise intolerance and even death. However, patients with COPD, in addition to symptoms directly related to changes in their respiratory mechanics, show a variable degree of systemic involvement which includes abnormalities in skeletal muscles, blood, nervous system and even bone metabolism (Montes de Oca, 1996) (Casaburi, 2000) (Noguera, 2004) (Bolton, 2004). Muscle dysfunction occurring in COPD is believed to be of multifactorial origin, with local and systemic factors interacting in each territory to modify in different ways the phenotype of any specific muscle. The following lines include a review of the main findings which have been observed in different skeletal muscles, as well as the causes which have been suggested to explain such changes.

Respiratory muscles. The diaphragm is, as previously mentioned, the main inspiratory muscle. However, its function is deteriorated in COPD (Rochester, 1979) since this disorder both promotes a dramatic increase in lung volume (modifying negatively the length-tension relationships of the muscle) (Goldman, 1978), and also increases resistances within the respiratory system. These two factors increase the work of breathing in individuals whose oxygen delivery can be easily compromised by gas exchange abnormalities occurring simultaneously in the lung. As a result, there is an imbalance between offer and demand in the muscle (metabolic inequality) (Gea, 2001a). In addition to its specificities as a thoracic muscle, the diaphragm of COPD patients is subject to the influence of other local and systemic factors which are common to all other skeletal muscles. These include oxidative stress, inflammation, nutritional depletion and drugs (Di Francia, 1994) (Gosker, 2000) (Barreiro, 2005) (Figure 6). Surprisingly, the final result is not so bad since the diaphragm of COPD patients can develop even greater strength than that of healthy subjects when the latter are forced to breathe at high lung volumes, similar to those which the former are currently facing (Similowski, 1991). This means that the diaphragm undergoes a functional adaptation in COPD. The structural and metabolic basis of this functional adaptation has been recently reported, and includes an increase in the proportion of type I fibers and myosin heavy chain (MyHC) -I, increase in mitochondrial and capillary densities (Figure 7) as well as shortened sarcomerae (Levine, 1997) (Gea, 1998) (Orozco-Levi, 1999) (Gea, 2001a). Experimental models suggest that this is the result of the chronic increase in respiratory loads (Gea, 2000), and can be mediated by the presence of muscle damage (Zhu, 1997) (Orozco-Levi, 2001). Interestingly, these adaptive phenomena can coexist with signs of myopathy (paracristalline inclusions) (Lloreta, 1996) (Figure 8).

Data from other respiratory muscles are scarcer than those from the diaphragm. However, it is clear that structural remodelling and functional adaptation also occur in the external intercostals, paresternal and accessory breathing muscles. External intercostals for instance, show and increase in the size and proportion of type II fibers

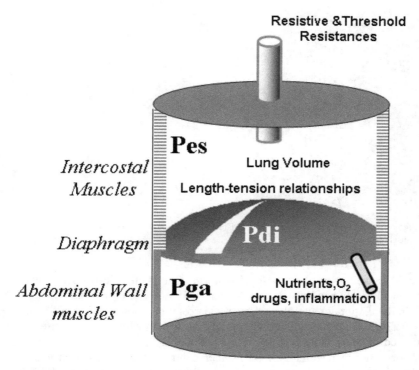

Figure 6. Representation of different factors that can contribute to respiratory muscle dysfunction, as well as that of the main respiratory pressures that can be easily measured (Pes, esophageal-intrathoracic pressure; Pga, gastric-abdominal pressure; Pdi, transdiafragmatic pressure, equivalent to the Pes – Pga difference)

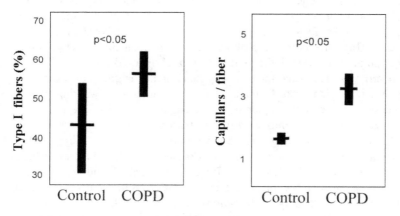

Figure 7. Respiratory muscle adaptation in COPD patients: Increased proportion of both type I fibers (left) and capillaries per fiber (right) occur in the diaphragm

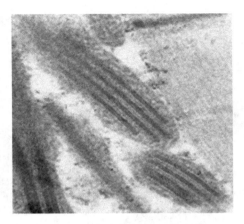

Figure 8. Signs of mitochondrial myopathy (paracristalline inclusions) observed occasionally in the diaphragm muscles of patients with COPD

(fast-twitch), as well as in the MyHC-II isoform (Campbell, 1980) (Gea, 1997), along with increases in capillary density (Jimenez, 1999). Therefore, these changes occurring in the diaphragm and the external intercostal muscles, which can be seen as both opposite and complementary, can supply the system with additional strength and endurance in order more effectively to deal with harmful geometrical conditions and increases in respiratory loads. Other contractile elements occasionally contributing to the respiratory effort, such as *latissimus dorsi* (Orozco-Levi, 1995a), also show signs of remodelling in COPD patients (Orozco-Levi, 1995b).

Although information regarding expiratory muscle adaptation is poor, these muscles are important in COPD patients for both coughing and breathing, as they are chronically activated (Dodd, 1984) (Ninane, 1992). However, their function is deteriorated in this condition (Rochester, 1985) (Ramirez, 2002) and, since in this case the dysfunction cannot be attributed to geometrical changes or deconditioning, it is possible to argue that other factors such as inflammation and oxidative stress could be involved (Gea, 2001a). The fibers of expiratory muscles appear to be smaller in COPD patients, with a slight increase in the proportion of type II fibers (Barreiro, 1999).

Peripheral muscles. This is certainly a very heterogeneous group of contractile elements, with such varied tasks as manipulating instruments or walking, which would not necessarily be equally impaired in severe COPD patients. These patients generally maintain their upper limb activities, at least to some degree, while their overall mobility is reduced. Both upper and lower muscle functions are impaired in such patients (Hamilton, 1995) (Bernard, 1998) (Coronell, 2004), and therefore their exercise capacity is reduced (Killian, 1992). Regarding structural findings, muscles located in the upper limbs and scapular area tend to maintain their histological and biochemical properties, with only slight adaptive changes, while muscles situated in the lower limbs show a clear involution, with atrophy and loss of oxidative capacity. Deltoid muscle for instance exhibits different coexisting fiber size modes (normal, hypertrophic and atrophic) (Hernandez, 2003) and preserved fiber type proportions and oxidative enzyme activity in COPD (Gea, 2001b), whereas brachial biceps show a decrease in the size of the fibers but also maintaining the type proportions (Sato, 1997). Quadriceps muscle, in contrast, clearly shows a reduction in mass (Schols, 1991b), along with smaller fibers, a reduced percentage of oxidative fibers and MyHC-I, less blood vessels, and diminished enzyme oxidative capacity and myoglobin content (Jakobsson, 1990b) (Jakobsson, 1995) (Simard, 1996) (Satta, 1997) (Whittom, 1998). Moreover, the coexisting fact that the oxygen delivery to the muscle is preserved in COPD (Sala, 1999), supports the hypothesis of an inefficient intracellular use of this gas in the quadriceps muscle. In conclusion, it is clear that changes occurring in skeletal muscles of COPD are very heterogeneous, depending on the muscle group which is being analyzed. This suggests that these changes are most probably driven by the complex interaction of different local and systemic factors, the particular interaction being unique for each muscle. In the following paragraphs, the factors involved in the pathogenesis of muscle changes will be briefly reviewed. They include inflammation, oxidative stress, deconditioning and malnutrition among others, and have been evoked not only for COPD muscle dysfunction but for many other chronic diseases.

Inflammation can be considered either a local or a systemic factor since both phenomena have actually been demonstrated in COPD patients (Di Francia, 1994) (Casadevall, 2003). Inflammatory mediators observed in such population are known to be capable of inducing an increase in intracellular protein degradation by either inducing a mismatching in the oxidation-reduction (REDOX) system or activating other proteolytic pathways (Flores, 1989) (Hall-Angeras, 1990). *Oxidative stress* and *nitrosative stress* have also been found in skeletal muscles of COPD patients (Barreiro, 2003) (Barreiro, 2005). It is well known that muscles produce reactive oxidants and nitric oxide (NO) even in normal conditions, since these substances are essential in ensuring contraction and excitation-contraction coupling (Kobzik, 1994). Their overproduction or the deterioration in the antioxidant systems, however, can lead to stress, protein modifications and breakdown, fiber damage and muscle dysfunction (Jackson, 1993) (Barreiro, 2005). *Deconditioning* is the result of the reduction in muscle activity which, as previously mentioned, is present in COPD patients. This factor is especially evident in lower limb muscles. The fact that many of the structural and biochemical changes associated to COPD are similar to those induced by disuse (Diffee, 1991) (Bloomfield, 1997), and that most of them are reversible with training strongly supports the role for deconditioning (Maltais, 1996) in the dysfunction of lower limb muscles. Abnormalities in *nutritional status* are also frequently observed in COPD and have been attributed to a reduction in food intake (due to changes in leptine metabolism) (Schols, 1999), an increase in metabolic costs (Schols, 1991b) and to the presence of systemic inflammation (Di Francia, 1994) (Schols, 1996). Finally, *comorbidity and aging, electrolyte imbalance, hypoxia/hypercapnia* frequently coexist with COPD, which in addition is treated with many different drugs known to modify muscle structure and/or function. All these factors potentially contribute to muscle changes. Taking everything into account, it can be stated that skeletal muscles show structural and functional changes in COPD patients, and that these changes are the result of the interaction of multiple factors, and are particular for each muscle.

3.2   Bronchial Asthma

This is a very prevalent entity characterized by inflammation and edema of the airways, which along with bronchial hyperresponse, leads to reversible episodes of airflow obstruction (Global Initiative for Asthma, G.I.N.A., 2001). Striated muscle dysfunction occurs in chronic asthma and depends mostly on steroid treatment and changes occurring in the thorax shape due to hyperinflation (Martyn, 1980). The latter, as it occurs in COPD, results in variation in the length-tension relationships of respiratory muscles, thus reducing their ability to generate force (Barreiro, 2004a). Hyperinflation in chronic asthma can even be increased during asthma attacks, which further reduces muscle strength. However, both inspiratory and expiratory muscles are actively recruited in chronic asthma and in asthma attacks (Jardim, 2002), suggesting a possible training-like phenomenon. In fact, respiratory muscles maintain or even improve their functional properties in chronic asthma patients (Rochester, 1981) (Mckenzie, 1986) (Peress, 1986) (Bruin, 1997), mostly in those not receiving steroids. This suggests that their muscles, as it occurs also in COPD, undergo functional adaptation and structural and metabolic remodelling. The mass of the diaphragm for instance appears to be greater in patients with chronic asthma (Bruin, 1997). Unfortunately, histological and biochemical findings in asthma are frequently mixed with those deriving from intense steroid and other drug treatments. Therefore, it is difficult to define the "skeletal muscle of asthma". However, it has been observed that type II fibers have a reduced size (Picado, 1990) and that the content of magnesium is reduced in peripheral muscles of asthmatic patients (Gustafson, 1996) independently of the steroid treatment. The physiological consequences of the reduction in magnesium are unclear but probably are related to cell membrane stability or metabolic functions. Finally, steroids, at the doses usually administered in chronic severe asthma, do not seem to cause muscular weakness (Picado, 1990). However, at high doses can induce *steroid myopathy* (see next paragraphs).

A severe asthma attack represents a different situation than that observed in chronic asthma, since the strength of respiratory muscles temporarily deteriorates due to increased hyperinflation and metabolic imbalance (Lavietes, 1988), which would overcome adaptive mechanisms. In severe asthma attacks, rhabdomyolysis appears to occur in as many as one third of the patients (Lovis, 2001) (Tsushima, 2001).

Although there are different drugs with potential effects on muscle, this review does not focus on pharmacological effects. However, since systemic steroids are widely used in asthma and many of the other disorders mentioned in the text, some paragraphs will be devoted to their deleterious effects on skeletal muscles. This is more extensively dealt with in the next chapter.

Systemic steroids can induce both acute and chronic myopathies (Goldberg, 1969). The latter is caused by long-term administration of relatively low doses of steroids (DeCramer, 1996) and is mainly characterized by muscle weakness, which is predominant in proximal muscle groups. Cellular and biochemical substrates of steroid myopathy are fiber atrophy (mostly affecting type II fibers), downregulation of the Insulin Growth Factor I (IGF-I), a decrease in the ratio of protein synthesis and degradation, and abnormalities in carbohydrate metabolism, nitrogen balance, and aminoacid concentration in plasma (Shoji, 1974) (Rannels, 1978) (Lieu, 1993) (DeCramer, 1996). Chronic treatment with higher doses of steroids also appears to induce changes in glycogen metabolism, resulting in glycogen storage on skeletal muscles (Fernandez-Sola, 1993).

Acute myopathy can also develop following few days of administration of high doses of steroids (Kaminski, 1994). This myopathy is characterized by some of the findings observed in chronic myopathy but also by the selective loss of thick filaments, and rhabdomyolysis (Waclawik, 1992) (Ramsay, 1993), resulting in generalized weakness (Nava, 1996).

3.3 Respiratory Disorders Related to Sleep (the Sleep Apnea-Hypopnea Syndrome)

The Sleep Apnea-Hypopnea Syndrome (SAHS) is mainly characterized by the presence of repeated total (apnea) or partial (hypopnea) occlusions of the upper airways during sleep. Its consequences include oxyhemoglobin decrease, loss of the sleep structure, diurnal hypersomnia and arterial hypertension (Hoffmann, 2004) (Parish, 2004) (Yaggi, 2004). Since during apneas-hypopneas respiratory muscles execute repeated and progressive submaximal efforts they can become exhausted (Wilcox, 1990) (Kimoff, 1994). This situation could even be worsened by the fact that these muscles do their task under hypoxia, hypercapnia and reduced blood flow (Garpestad, 1992). In addition, as it occurs in peripheral skeletal muscles, the absence of reparative rest during sleep could further contribute to muscle dysfunction (Chen, 1989). However, repeated efforts could also mimic a training programme, counterbalancing the above- mentioned negative factors. Recent studies suggest that although muscle strength is roughly maintained in SAHS patients, there is an evident decrease in the endurance and the reserve against fatigue (Griggs, 1989) (Aran, 1999). This supports the hypothesis of a central role for sleep deprivation. Interestingly, treatment with Continuous Positive Airway Pressure (CPAP) restores muscle function (Aran, 1999). From a cellular and molecular point of view, the quadriceps muscle of SAHS patients has shown a preservation of the fiber-type distribution along with reduction in the size of type II fibers and protein content, as well as an increase in the activity of different enzymes (such as cytochrome oxidase and phosphofructo-kinase) (Sauleda, 2003). Regarding respiratory muscles, only some preliminary reports are available, and they suggest that SAHS induces an increase in the proportion and size of type-II fibers (Nowinski, 2001).

3.4   Mechanical Ventilation

When the respiratory pump fails in its ventilatory tasks, mechanical devices can help in maintaining pulmonary gas exchange close to the normal range. This therapeutic procedure is called mechanical ventilation (MV), and includes both invasive (endotracheal intubation or tracheostomy) and non-invasive (facial or nasal masks) alternatives. However, since respiratory muscles are relatively inactive with some of the MV modalities, they become deconditioned. Diaphragm for instance has been shown to atrophy following only 48 h of MV (Le Bourdelles, 1994). In a similar way, since this therapeutic procedure results in relative immobilization and drug-induced relaxation, it can also lead to peripheral muscle atrophy. Obviously other factors with deleterious effects on muscle, such as sepsis, multiorganic failure, drugs, malnutrition, malposition and dyselectrolytemia (Deconinck, 1998; Polkey, 2001) can also be present in these critically ill patients. As a result, neuropathies (axonal and demyielinating), as well as neuromuscular junction transmission defects and myopathies can develop. Acute myopathy occurring in critically ill patients has been well defined, and includes loss of thick filaments, fiber athropy, vacuoles, cytoplasmic bodies, and mitochondria with paracrystalline inclusions (Matsubara, 1994) (Lacomis, 1996).

An important issue to be taken into account is that involutive changes occurring in ventilatory or peripheral muscles of ventilated patients will in turn have important consequences. In this regard, weaning patients from MV will be difficult, since in many cases atrophic respiratory muscles would thus be unable to do their job again (Spizer, 1992) (Mehta, 2000), and peripheral muscles will need intense rehabilitation to ensure reintegration into everyday activities.

4.   CHRONIC HEART FAILURE (MAURICIO OROZCO-LEVI)

4.1   Introduction

In the last few years, impaired skeletal muscle function has been acknowledged to be relevant in patients with **chronic hearth failure (CHF)**. Previously, this impairment was ignored, and attention was paid to the relationships between the cardiac limitations and exercise capacity of the patients. However, numerous recent investigations have demonstrated that impairment of skeletal muscle function is also an important predictor of exercise limitation in CHF (Cotes, 1988) (Schols, 1991a) (Hamilton, 1995) (Sullivan, 1995) (Steele, 1997). Regarding muscular involvement which coexists with cardiac impairment in CHF, there are numerous points in common with other chronic diseases (e.g., respiratory and collagen diseases). Typical histological changes within the peripheral muscles of CHF patients include a shift from oxidative (aerobic) to glycolytic (anaerobic) energy metabolism, whereas the opposite is observed in the diaphragm. These findings are in line with the notion that peripheral and respiratory muscles are limited mainly by their reduced endurance and impaired capacity to develop maximal force, respectively (Mainwood and Renaud, 1985; Mancini et al. 1992). The final reason for CHF muscle dysfunction remains unknown, but deterioration in tissular gas exchange, oxidative stress, deconditioning, drugs, nutritional depletion, and low-level, but persistent systemic inflammation might contribute. Each factor carries with it its own potential for developing innovative treatments in the future.

4.2   Evidences of Muscle Impairment in CHF

Patients with CHF show a decreased tolerance to whole-body exercise. By definition, the heart is the primary organ impaired in patients with CHF. However, cardiac dysfunction is unable to entirely explain for the impairment of exercise tolerance. Besides cardiac dysfunction, deterioration of skeletal muscle performance appears as a strong additional predictor of this impairment, which greatly influences the individual's health status. This is coherent with the statement of the World Health Organization (Wood, 1980) which emphasizes that chronic diseases are characterized not only by the primary organ impairments they cause but also by the disabilities and handicaps resulting from them.

Progression of the primary impairment in CHF can be reduced or slowed down with medication and surgery (such as coronary bypass and heart transplantation) (Miller, 1994). These interventions however, aimed at recovering heart function, offer only partial improvement in exercise capacity (Beller, 1997). In addition, a significant percentage of patients have exclusion criteria for surgery/transplantation (e.g., age), and the availability of donor organs is limited. Even more important, surgical interventions do not always confer a survival benefit in CHF and, irrespective of the reversibility of the heart impairment, exercise intolerance can remain. Therefore, it can be deduced that an effective treatment of this disorder would require a detailed insight into its systemic consequences and appropriate whole-body rehabilitation (Stratton, 1994) (Williams, 1997).

4.3   Changes in Muscle Function of Patients with CHF

Muscle function depends, beside other factors, on perfusion system, muscle mass, fiber proportions, and muscle energy metabolism (Armstrong, 1988). It can be inferred that alterations in one or more of these determining factors will play a role in reducing muscle performance. Numerous studies provide evidence that CHF is associated with varied degrees of muscle weakness (Chua, 1995) (McParland, 1995) (Zattara-Hartmann, 1995). An extensive study on the influence of muscle weakness on exercise capacity in cardio-respiratory disorders was carried out by Hamilton et al. (1995). This study compared healthy subjects, patients with heart failure, respiratory failure, and a combination of both, demonstrating that patients have lower strength in both peripheral and respiratory muscles than controls. However, strength and endurance appear to be differently affected in respiratory and peripheral muscles. This notion was reinforced by the poor correlation observed by different authors in the maximal force of both muscle groups in patients with either CHF or respiratory disorders (Chua, 1995) (McParland, 1995) (Gosselink, 1996)

(DeCramer, 1997) (Bernard, 1998). This correlation is much stronger in healthy subjects (Lands, 1993).

In patients with CHF, the exercise limiting symptoms are both central (dyspnea) and peripheral (early perception of limb discomfort) even at low levels of loading (Myers, 1992) (Drexler, 1995) (Hamilton, 1995) (Harrington, 1997). The reduction in muscle tolerance to fatigue appears to be the main peripheral limiting factor. In this regard, early muscle acidosis appears as a critical element contributing to limb muscle fatigue (Mainwood, 1985). In contrast, strength rather than endurance appears to be the main limiting factor in the case of respiratory muscles since only the former correlates with both dyspnea and exercise capacity of the patients. On the whole, it seems that maximal force is the limiting aspect of muscle performance in the respiratory muscles, whereas endurance is the limiting factor in peripheral muscles of CHF patients. However, more detailed studies are required to clarify this point.

4.4   Structural changes in patients with CHF

Marked loss of muscle mass and a decrease in cross-sectional muscle area can be observed in patients with CHF (Wilson, 1990) (Miyagi, 1994) (Shephard, 1997). Moreover, muscle wasting is closely associated with the loss of exercise tolerance in these individuals. Histological and biochemical changes described in peripheral muscles (i.e. quadriceps, biceps and deltoid) of CHF patients include fiber atrophy, endomysial fibrosis, increased lipid deposits, and increased activity of acid phosphatase (a lysosomal enzyme contributing to protein degradation) (Dunnigan, 1987) (Lipkin, 1988). In addition, a normal capillary density has been found (Lipkin, 1988). However, this could be the result of the coexisting fiber atrophy, which would overestimate capillary density, masking an actual reduction in the number of blood vessels. This would agree with the results of two other studies in which both reduced capillary-fiber ratios and atrophy resulted in unchanged capillary densities (Sullivan, 1990) (Schaufelberger, 1995). An unaltered capillary-fiber ratio has also been reported, resulting, however, in a greater capillary density due to fiber atrophy (Mancini, 1989). Finally, other authors have observed a reduced capillary density as well as a decreased capillary-fiber ratio in CHF patients (Drexler, 1992), and even in heart transplant recipients (Lampert, 1996). Finally, the few ultrastructural studies performed using electron microscopy, showed that mitochondrial volume densities in skeletal muscles were lower in CHF patients than in control subjects (Drexler, 1992) (Hambrecht, 1997), and this was still the case even ten months after a heart transplant (Lampert, 1996). All these results suggest that the oxidative capacity of peripheral skeletal muscles may be altered in the disease.

Probably the most remarkable muscle alteration in CHF is the relative shift in fiber type composition, which appears to occur in opposite directions in peripheral and respiratory muscles. Fiber typing is mainly performed histochemically, and is based on differences between fibers in myosin ATPase activities or immunocyto-chemistry (McComas, 1996). A lower percentage of type I fibers and a corresponding higher percentage of type II (mainly type IIb/x) fibers has been reported in limb muscles of CHF patients, compared with those of normal subjects (Lipkin, 1988) (Mancini, 1989) (Sullivan, 1990) (Drexler, 1992) (Schaufelberger, 1995). In contrast, a shift from type IIb/x to type I fibers has been reported in the diaphragms of CHF patients (Lindsay, 1996). In this regard, control subjects present in their diaphragms around 50% of type I, 25% of type IIa, and 25% of type IIb/x fibers (Mizuno, 1991), whereas the diaphragm of CHF patients contains 60% of type I, 35% of type IIa, and only 10% of type IIb/x fibers (Lindsay, 1996). Similarly, a IIb/x to I type shift was observed when the distribution of myosin heavy chain (MyHC) isoforms was analyzed in the diaphragms of CHF patients (Tikunov, 1997). Furthermore, a larger population of type I fibers appears to be present in the diaphragm of CHF patients when compared with sedentary control subjects (Hughes, 1983) (Mainwood, 1985). It is possible that the shifts observed in fiber types (I to IIb/x in peripheral muscles, and IIb/x to I in the diaphragm) have different functional consequences in the affected muscles, since the distinct fiber types have different contractile properties. For instance, Gosker et al. (2003) hypothesized that the shift from type I to type IIb/x along with less oxidative capacity would be responsible for the loss of resistance observed in peripheral muscles of CHF patients. This change might contribute to their observed lower exercise tolerance because, as previously mentioned, peripheral muscle fatigue is the main exercise-limiting factor in these individuals (Harridge, 1996).

4.5   Changes in Muscle Metabolism

Muscle content of high-energy substrates. Measurements of substrate and cofactor concentrations in peripheral skeletal muscles of CHF patients indicate impaired energy metabolism (Gosker, 2003). Most striking are the observed reduced concen-trations of high-energy phosphates at rest. The ^{31}P-nuclear magnetic resonance (^{31}P-NMR) has ensured a direct and non-invasive assessment of tissue concentra-tions of these high-energy phosphates. Increased concentrations of ATP, creatine phosphate (CrP), and nicotinamide adenine dinucleotide in its reduced form (NADH) reflect a high-energy state, whereas elevated concentrations of ADP, AMP, inorganic phosphate (P_i), and oxidized nicotinamide adenine dinucleotide (NAD^+) commonly reflect a low-energy state. Interestingly, Pouw et al. (1998) observed that higher P_i-CrP and ADP-ATP ratios were associated with slightly but significantly high inosine monophosphate (IMP) concentrations in chronic cardio-respiratory patients. The latter phenomenom may be due to an increased degradation of accumu-lating AMP by deamination, which probably reflects a reduced aerobic capacity (Dudley, 1985). The situation becomes even worse during exercise since a greater increase in the P_i-CrP ratio and a faster drop in pH have been found in the calf muscle of CHF patients when compared with healthy individuals (Mancini, 1989) (Kemp, 1996). Similar results were obtained from the forearm muscle (Tada, 1992)

(Stratton, 1994). In addition, a slower recovery of CrP concentrations was observed after exercise in both COPD and CHF patients (Mancini, 1989) (Tada, 1992) (Stratton, 1994) (Kemp, 1996). These results suggest that rephosphorylation of high-energy phosphates is less efficient in these individuals than in healthy subjects, both during and after muscular exercise. In addition, glycogen contents in CHF patients tend to be lower whereas lactate concentrations are higher than in healthy individuals. Thus, it seems that anaerobic energy metabolism is enhanced in CHF patients, and because this process yields far less ATP than does complete oxidative degradation of glucose, this could explain their reduced high-energy phosphate concentrations (Gosker, 2003).

Muscle Enzymes. Although activities of enzymes involved in muscle energy metabolism measured *in vitro* do not reflect completely a physiologic situation (since maximal activities are obtained under optimal, artificial circumstances), they provide an indication of adaptations in the expression of proteins involved in metabolic pathways. Typical oxidative enzymes are citrate synthase, succinate dehydrogenase, and ß-hydroxyacyl-CoA dehydrogenase (HAD) whereas typical glycolytic enzymes are hexokinase, phosphofructokinase, and lactate dehydrogenase (this catalyzing the last step of the anaerobic glycolysis). In adddition, lactate and glycogen concentrations are often measured for studies of muscle metabolism, but note that their low levels may reflect either increased clearance or reduced formation, and vice versa for high concentrations. Analysis of enzyme activities in peripheral muscles of both CHF and COPD suggests an overall increase in glycolytic and an overall decrease in oxidative enzyme activities. Although muscle fiber metabolism can change with no changes in MyHC isoform content and therefore in fiber type (i.e. expressing an uncoupling between metabolic and contractile "phenotype"), both metabolism and fiber type proportions are usually closely related (Essen-Gustavsson and Henriksson, 1984). Whether enzyme activities adapt to the expression of MyHC isoforms and fiber type redistribution or the fiber type – MyHC redistribution adapts to enzyme activities in CHF still remains unclear (Gosker, 2003).

In addition to chronic changes in enzyme activities, as measured *in vitro*, there is probably also an acute effect of exercise in CHF patients. As a consequence of impaired electron transport, regeneration of NAD^+ from NADH is reduced and citrate synthase and HAD are inhibited by the high $NADH\text{-}NAD^+$ ratio (Stryer, 1988). In addition, elevated AMP concentrations, resulting from the inefficient rephosphorylation of ATP, will stimulate glycolysis. However, this acute effect is invisible in the *in vitro* activity measurements. The additional inverse relation of oxidative enzyme activities with arterial lactate concentrations found during exercise in CHF, emphasizes the assumed shift from oxidative toward glycolytic energy generation (Sullivan, 1991). This loss of oxidative capacity probably accounts for the above-mentioned lipid deposits (Tein, 1996), because fatty acid consumption may be reduced while the supply of blood fatty acids continues (Gosker, 2003). Activities of two additional oxidative enzymes, cytochrome-*c* oxidase (COX) and NADH dehydrogenase, have also been found to be elevated in CHF patients. This seems to be paradoxical in the context of the reduction in the activity of the oxidative enzymes mentioned earlier. However, the latter were enzymes involved in either the citric acid cycle or fatty acid oxidation, whereas COX and NADH dehydrogenase are enzymes of the respiratory mitochondrial chain. COX interacts with oxygen and therefore is the main determining factor in mitochondrial oxygen affinity (Gnaiger, 1998). Since a correlation between COX muscle activity and hypoxemia has been found in other chronic cardio-respiratory condition (i.e. COPD), Gosker et al. (2003) suggested that the increased number of COX molecules found in CHF is a mechanism that would enhance the efficiency of residual oxygen extraction and utilization and thus, of the respiratory chain function.

Because of technical difficulties with ^{31}P-NMR and in obtaining biopsies from the diaphragm and other respiratory muscles, little is known about energy metabolism of this muscle group in CHF patients. However, preliminary results suggest that changes observed in enzyme activities are coherent with those reported in muscle structure.

5.   SKELETAL MUSCLE DYSFUNCTION IN SEPSIS (*ESTHER BARREIRO*)

5.1   Introduction

Sepsis is usually defined as the systemic response to serious infection, including several clinical manifestations such as fever, tachycardia, tachypnea and leukocytosis. When hypotension or multiple organ failure occurs as a consequence of this syndrome the condition is called *septic shock*. Respiratory insufficiency is currently considered to be the most frequent cause of death in patients with septic shock. Although respiratory failure has traditionally been attributed to lung injury in sepsis, there is now growing evidence that septic shock is also associated with ventilatory pump failure. In this regard, sepsis-induced respiratory muscle failure has been the focus of many recent studies. Burke et al. (1963) already described in 1963 how hypercapnic respiratory failure occurred in patients with fulminating septic shock in the presence of close to normal PaO_2. Several years later clinical and electromyographic evidence of diaphragmatic contractile failure was provided in patients with severe sepsis who could not be weaned from mechanical ventilation (Cohen, 1982). This observation was based on the concept previously described by Friman (1977), who reported that both maximum force and endurance capacity of various limb muscles significantly declined during acute infections in humans.

Numerous experimental animal studies conducted in the last two decades have also provided considerable evidence of the existence of an association between septic shock and depressed contractile performance of both limb and ventilatory muscles. Sepsis and/or septic shock in these studies have been induced by the administration of either live microorganisms or bacterial endotoxin to experimental animals, resulting in cardiovascular features similar to those observed in the hypodynamic phase of human septic shock. Based on this, Hussain et al. (1985) were the first to report that ventilatory failure occurring in *Escherichia Coli* endotoxic shock in dogs was due to fatigue of the respiratory muscles. Several years later other studies also confirmed these findings. For instance, it was reported (Leon, 1992) (Boczkowski, 1990) (Shindoh, 1992) that acute *Escherichia Coli* endotoxemia reduced diaphragm force and endurance in response to a wide range of stimulation frequencies in rats. Furthermore, peritonitis was also shown to cause diaphragm weakness in rats, leading to the concept that humans with peritonitis may also be predisposed to respiratory muscle dysfunction (Krause, 1998).

5.2   Factors Involved in the Sepsis-induced Ventilatory Muscle Dysfunction

The factors involved in the respiratory muscle dysfunction can be divided into two different categories. On the one hand, sepsis indices increased ventilation, hypoxemia, higher pulmonary resistances, decreased oxygen delivery to the muscle and poor metabolite extraction, all of which can lead to an imbalance between ventilatory muscle supply and demand. On the other, mediators of muscle dysfunction include specific cellular, metabolic, and immune deficiencies that interfere with a number of processes necessary for normal force generation (Hussain, 1998). These defects are mediated by complex interactions between several local and systemic mediators, which contribute to the respiratory muscle dysfunction described in sepsis. These mediators are discussed in the following paragraphs.

Bacterial endotoxins. Although there is no clear evidence showing their direct impact on muscle function, it has been suggested that they act directly on skeletal muscles, leading to a sequence of events that ultimately would precipitate muscle failure.

Arachidonic acid metabolism. The view has also been put forward that endotoxin alters arachidonic acid metabolism, leading to the release of *prostaglandins* such as prostacyclin and thromboxane A_2 (Hussain, 1998). Pre-injection of chronic endotoxemic rats with indomethacin (cyclooxygenase inhibitor) abrogated both the decline in diaphragmatic force generating capacity and prolongation of twitch relaxation time (Boczkowski, 1990). In line with this, Murphy et al. (1995) showed similar results during acute bacteremia in piglets. Interestingly, in this study the use of an analogue of thromboxane A_2 elicited comparable changes in diaphragmatic force similar to those described in acute endotoxemia.

Cytokines. It has been well established that endotoxin can stimulate monocytes, macrophages, and mast cells to produce *tumor necrosis factor alpha* (TNF-alpha), *interleukin-1* (IL-1), and other cytokines. Among them, TNF-alpha is considered to be a central mediator of immune and inflammatory responses, so it has been the focus of several experimental studies. In this regard, Wilcox et al. (1994) reported that 3 hours after systemic infusion of TNF-alpha was associated with both decreased diaphragmatic pressure and shortening in response to artificial phrenic nerve stimulation, compared to control animals. In line with this, in another study TNF-alpha messenger RNA was shown to increase in the rat diaphragms after 3 hours of endotoxin administration. Moreover, they also observed that endotoxin-induced diaphragmatic dysfunction was partially reversed by pre-treating the animals with anti-TNF-alpha antibodies. These findings led to the idea that TNF-alpha must play a major role in sepsis-induced muscle contractile dysfunction, although the mechanisms whereby TNF-alpha acts on skeletal muscle are not yet very clear. It is very likely, however, that this cytokine acts by inducing secondary messenger molecules such as reactive oxygen species (ROS) and nitric oxide (NO).

Nitric Oxide. NO is a multifunctional molecule participating in numerous biological processes in almost all aspects of life, including vasodilatation, bronchodilation, neurotransmission, inhibition of both phagocyte and platelet aggregation, and antimicrobial activity (Moncada, 1991) (Gaston, 1994) (Fang, 1997). NO is an extremely useful intracellular messenger, since it does not react rapidly with most biological molecules at the dilute concentrations produced *in vivo*. NO is synthesized from $_L$-arginine by a group of hemoproteins known as NO synthases (NOS) in the presence of nicotinamide adenine dinucleotide phosphate hydrogen (NADPH) and tetrahydrobiopterin (BH4) (Knowles, 1994). Three isoforms have been identified, two of which are constitutively expressed and were originally purified in the endothelial cells (eNOS, NOS3), and brain cells (neuronal) (nNOS, NOS1). The third one, which is an inducible isoform, was initially purified in macrophages (iNOS, NOS2). The requirements for calcium and calmodulin differ between the constitutive and the inducible isoforms. Since the inducibility of iNOS is highly dependent on the stimulus rather than the gene product (Stamler, 2001), this classification of NOS as constitutive and inducible isoforms is nowadays not very consistent.

The biological chemistry of NO can easily be simplified in 3 main reactions (Beckman, 1996). Firstly, NO acts as a signalling molecule through binding and activation of guanylate cyclase. Secondly, NO may be destroyed by reaction with oxyhemoglobin within a red blood cell to form nitrate. Thirdly, NO can be transformed to peroxynitrite ($ONOO^-$) by reaction with superoxide anion (O_2^-), whose reaction is usually limited by the micromolar concentrations of superoxide dismutases in cells. Nevertheless, excessive NO production as occurs during active inflammatory-immune processes, leads to detrimental effects of this molecule on tissues, which have been mostly attributed to its diffusion-limited reaction with superoxide anion to form the powerful and toxic oxidant peroxynitrite (Beckman, 1996) (Figure 9). This highly reactive species is considered to be mostly responsible for the majority of the damaging effects of excessive NO release (Beckman, 1996). In fact, peroxynitrite formation is the consequence of excessive production of both NO and superoxide, a condition almost invariably occuring at sites of active inflammatory processes (van der Vliet, 1999).

In general, the oxidative modifications induced by peroxynitrite and other reactive nitrogen intermediates include addition or substitution products in which NO is essentially incorporated into the target molecule (nitrosation and nitration reactions). More irreversible NO-induced modifications, however, include nitration of aromatic amino acids, lipids, and DNA bases (van der Vliet, 1999). In this regard, the

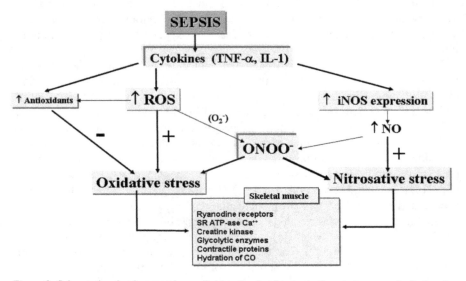

Figure 9. Scheme showing how sepsis may lead to the development of respiratory muscle dysfunction through the induction of oxidative and nitrosative stress. Different molecular and structural targets within the skeletal muscle fibers are also represented. A color figure is freely accessible via the website of the book: http://www.springer.com/1-4020-5176-x

amino acid tyrosine appears to be a primarily susceptible target for nitration. Indeed, the formation of free or protein-associated 3-nitrotyrosine is recently being considered as a potential biomarker for the generation of reactive nitrogen species (RNS) *in vivo* (van der Vliet, 1999). Moreover, there is increasing evidence that nitration of essential tyrosine residues can either inactivate numerous enzymes or prevent phosphorylation of tyrosine kinase substrates, leading to the concept that tyrosine nitration might not only be considered as a marker of nitrosative stress *in vivo*, but as a direct mediator of the damaging effects on tissues observed during active inflammatory-immune processes. Indeed, we have recently demonstrated (Barreiro, 2002a) that sepsis-mediated increase in protein tyrosine nitration is limited to the mitochondria and cell membrane and is highly dependent on the iNOS isoform.

Reactive Oxygen Species. ROS effects on skeletal muscle function have usually been associated with certain pathophysiological muscle conditions, such as muscle fatigue following strenuous exercise, ischemia-reperfusion injury, inflammatory muscle disease, and various myopathies (Jackson, 1990). However, recent evidence has proved that endogenous ROS also regulate contractile function of healthy skeletal muscle. They are produced at a relatively low rate in resting muscle fibers, appear to be essential for normal force production, and their levels progressively increase in response to muscle activation (Reid, 1992). ROS concentration within the muscle fibers is usually kept at relatively low levels by intracellular antioxidants such as superoxide dismutases. For instance, selective reduction of ROS by antioxidant enzymes treatment results in a decline in muscle force that is reversible by enzyme washout (Reid, 1993) (Khawli, 1994). In contrast, exposure to low levels of exogenous ROS increases muscle force (Reid, 1993) (Andrade, 1998). Excessive ROS production, however, as occurs during exogenous ROS exposure, strong muscle contractions or in sepsis, results in a state of oxidative stress leading to a decline in muscle force production.

Several studies have demonstrated that sepsis-induced muscle injury is largely mediated by an increase in ROS levels. In this regard, the group of Supinski (1993) and Shindoh (1992) were the first to report that free radicals contribute to the diaphragmatic dysfunction induced by systemic endotoxin injection. These authors concluded that administration of an antioxidant prevented both malondialdehyde formation (index of free radical-mediated lipid peroxidation) and contractile dysfunction in septic hamsters. In line with this, Peralta et al (1993) also demonstrated increased levels of ROS in septic rat hindlimb muscles, and pre-treatment of the animals with antioxidants attenuated their levels. One year later, the same group of investigators (Llesuy, 1994) reported in another study that oxidative stress occurs early in rats along with both inhibition of active mitochondrial respiration and inactivation of antioxidant enzymes. Besides, Supinski et al (1996) concluded that endotoxin-induced dysfunction in septic hamsters was not limited to ventilatory muscles, but also occurred in limb skeletal muscle, while cardiac muscle appeared to be resistant to it. The same group of investigators (Callahan, 2001a) also showed that free radicals reduce the maximal diaphragmatic mitochondrial oxygen consumption in endotoxemic rats, playing a central role in the alteration of skeletal muscle contractile protein force-generating capacity (Callahan, 2001b). Most recently we have also shown (Barreiro, 2005 b) that different muscle proteins such as glycolytic enzymes, enzymes involved in ATP production and in hydration of carbon dioxide, and contractile proteins are targeted by oxygen free radicals in the diaphragm fibers of endotoxemic animals (Figure 9). Despite the progress, the potential sources of ROS production and the mechanisms whereby oxidative stress contributes to the sepsis-induced muscle dysfunction still remain poorly understood.

Heme Oxygenases. Heme oxygenases (HOs), which were originally identified by Tenhunen et al. (1968) are the rate limiting enzymes of the initial reaction in the degradation of heme to yield equimolar quantities of biliverdin, carbon monoxide (CO) and free iron (Choi, 1996). Biliverdin is subsequently converted to bilirubin through the action of biliverdin reductase, and then free iron is rapidly incorporated into ferritin. Three documented isoforms (HO-1, HO-2, and HO-3) catalyze this reaction (Maines, 1988) (McCoubrey, 1992) (McCoubrey, 1997). Although heme still represents the typical inducer, the 32-kDa HO-1 isoform has been shown to be induced by various non-heme products, such as NO, cytokines, shearstress, heavy metals, endotoxin, hyperoxia, hydrogen peroxide, heat shock and many others

(Tenhunen, 1968). The demonstration of the induction of HO-1 by agents causing oxidative stress led to the notion that HO-1 might have a cytoprotective role against the excess of oxidants. Indeed, HO-1 enzyme is now thought to be part of a more general cellular response to oxidative stress, regulating cellular homeostasis via its three major catabolic by-products: CO, bilirubin, and ferritin. Little is known about the functional significance of HOs in skeletal muscle fibers in sepsis. On the grounds that HOs might play an important protective role in the sepsis-induced diaphragmatic contractile dysfunction, we conducted a study where heme oxygenase enzyme activity was shown to attenuate oxidative stress and promote maximum and submaximum force generation in endotoxemic rats (Barreiro, 2002b).

6.   CANCER INDUCED CACHEXIA *(ESTHER BARREIRO)*

6.1   Introduction

The development of cancer cachexia is the most common manifestation of advanced malignant disease. Indeed, cachexia occurs in the majority of cancer patients before death, and highly contributes to their increased morbidity and mortality (Warren, 1932). The abnormalities associated with cancer cachexia include anorexia, weight loss, muscle loss and atrophy, asthenia, anemia and alterations in carbo-hydrate, lipid and protein metabolism (DeWys, 1985). The degree of cachexia is inversely related to the survival time of the patient and it always implies a poor prognosis (Warren, 1932). Cancer-induced cachexia is always associated with the presence of the tumor, also leading to a malnutrition status owing to anorexia. Furthermore, the competition for nutrients between the host and the tumor implies a hypermetabolic status of the patient resulting in an energetic imbalance.

Muscle proteins are degraded as a source of energy for gluconeogenesis during fasting. Protein breakdown, however, is reduced during prolonged starvation periods since nitrogen and lean body mass have to be preserved. In cancer-induced cachexia this condition is lost and skeletal muscle becomes severely affected since vital structural and functional proteins are depleted and muscle tissue accounts for half of the whole body protein content. What remains a matter of debate is whether the negative protein balance is mainly due to reduced muscle protein synthesis or to increased protein degradation. In the last decade, observations from the group of Argilés et al. (Costelli, 1993) (García-Martínez, 1995) have shown that both protein synthesis and degradation are considerably altered in muscles during tumor growth, and especially a non-lysosomal, ATP and ubiquitin-dependent proteolytic pathway is clearly activated in animals.

6.2   Cytokines

The cytokine TNF-alpha, released by monocytes and macrophages, participates in the muscle loss and nitrogen imbalance of cachectic states. For instance, chronic treatment of rats with recombinant TNF or IL-1 beta led to a decrease in body protein

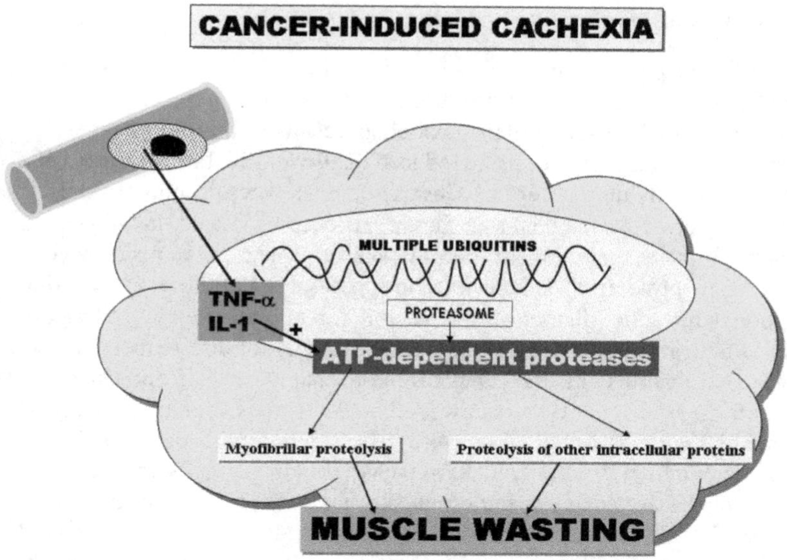

Figure 10. The ubiquitin system is a non-lysosomal ATP-dependent proteolytic pathway in which covalent attachment of ubiquitin is believed to mark proteins destined for degradation. Ubiquitin is also ubiquitinated and this reaction is repeated several times, resulting in the formation of large conjugates that are recognized by the proteasome. In the cachectic state, proteolysis of myofibrillar and other intracellular proteins via the ubiquitin system would be increased leading to muscle wasting. A color figure is freely accessible via the website of the book: http://www.springer.com/1-4020-5176-x

content, associated with reduced expression of myofibrillar proteins, compared with pair-fed control animals (Fong, 1993). Actually, several studies have shown that TNF is clearly involved in the activation of muscle proteolysis, and that this cytokine is mediating the activation of the non-lysosomal ubiquitin-dependent proteases (Fong, 1993) (Llovera, 1994) (Llovera, 1995). Proteins undergoing degradation by an ATP-dependent protease are targeted by ubiquitins (Figure 10), which can be found free or conjugated in an isopeptide linkage to other cellular proteins within the cell. Proteins containing multiple ubiquitins are those targeted for degradation by an ATP-dependent protease. This protein degradation system is integrated in a supramolecular structure, the proteasome, and can also participate in the turnover of long-lived proteins, such as the skeletal muscle ones (García-Martínez, 1994 and 1995). Other cytokines such as IL-1 or interferon gamma have also been shown to activate ubiquitin gene expression (Llovera, 1998). Therefore, it seems that TNF-alpha alone or in concert with other cytokines are the most important mediators of muscle protein catabolism in cancer-induced cachexia. Clearly, muscle proteolysis is the most significant metabolic event occurring in this condition. Future studies should be tailored to abrogate the deleterious effects of such cytokines in cachectic patients.

6.3   Proteolysis of Myofibrillar Proteins

As has already been pointed out, cachexia-induced increased protein degradation mainly implies enhanced breakdown of muscle proteins, especially that of myofib-rillar proteins. Increased muscle protein catabolism results in muscle weakness and fatigue, leading to exercise limitation and reduced quality of life. Furthermore, venti-latory pump failure may also occur in cachectic patients when respiratory muscles undergo enhanced protein catabolism. In fact, it has been estimated that almost one third of deaths in cancer patients are due to muscle protein catabolism. Myosin, the most abundant protein in skeletal muscle, possesses ATPase activity, which in combination with actin filaments, is essential for normal muscle contraction. The activity of myosin creatinine phosphokinase is crucial for differentiated muscle function, provided that this enzyme catalyzes the formation of ATP from phospho-creatine. Kirchberger et al. (1991) showed that the expression of both myosin and myosin creatinine phosphokinase were reduced in cachectic animals. The transcription factor complex myogenin-Jun-D have a binding activity for the myosin creatinine phosphokinase enhancer in normal skeletal muscle. In cachectic muscles the activity of Jun-D is decreased, hence leading to disruption of the binding activity of the transcription factor complex myogenin-Jun-D. Oxidative stress has been shown to decrease *in vitro* the activity of Jun-D, which was prevented by the administration of antioxidants, inhibitors of NOS activity and overexpression of Jun-D (Buck, 1996). It was concluded that oxidative stress and NOS activity, probably through TNF-alpha, are involved in the myofibrillar proteolysis of skeletal muscle.

6.4   Oxidative Stress

Several mechanisms have been proposed to contribute to the development of oxidative stress in cancer patients, such as alterations of both intermediate and energy metabolism, a non-specific chronic activation of the immune system with an excessive production of proinflammatory cytokines, and the use of antineoplastic drugs, particularly alkylating agents and cisplastin, since they have been shown to stimulate ROS production.

As has already been mentioned, oxidative stress also contributes to enhanced muscle protein degradation in cancer-induced cachexia models (Buck et al 1996). Specifically, in the study of Buck et al (1996) the cytokine TNF-alpha was shown to play a central role in mediating the release of both ROS and RNS. Furthermore, in this study malondialdehyde-protein adducts, an index of oxidative stress, were markedly increased in the muscles of the tumour-induced cachectic animals. However, the administration of the antioxidant D-α-tocopherol clearly abrogated such increase, and prevented the muscle wasting of the animals (Buck, 1996). Other authors have also reported an increase in malondialdehyde in the blood of advanced gastric cancer patients (Chevari, 1992) as well as high levels of this type of aldehyde in skeletal muscle during cancer cachexia (Gomes-Marcondes, 2002).

Unpublished observations from our laboratory have also shown that other indices of increased levels of both ROS and RNS in tissues, such as protein carbonylation and 3-nitrotyrosine formation, were greater in the gastrocnemius muscles of rats bearing the Yoshida ascites hepatoma compared with control muscles.

7.   MUSCLE AGING (ESTHER BARREIRO)

7.1   Introduction

Aging or *senescence* is defined as a decline in performance and fitness with advancing age, which is an almost universal phenomenon of multicellular organisms. It has been well established that the detrimental effects of aging are best observed in postmitotic tissues such as brain and muscle provided that damaged or lost cells cannot be replaced by mitosis of intact ones. Furthermore, the deleterious effects of ROS accumulate in those tissues throughout the entire life. Several intrinsic and extrinsic factors have been implicated in the development of muscle dysfunction in aging. Loss of androgens, estrogens, and growth hormone production, reduced glucose and/or fatty acid metabolism are included amongst the intrinsic factors, while diet, exercise, injuries, and sedentary lifestyles are included among the extrinsic ones. Loss of skeletal muscle mass in the elderly is also known as *sarcopenia*, which is precisely defined as a loss of both skeletal muscle mass and function with advancing age. Sarcopenia is a generic name that includes loss of muscle mass, strength and quality, with alterations in muscle structure composition, innervation, capillary density, glucose metabolism, fatigability, and contractility. Sarcopenia has clinical and social implications since elderly individuals are more prompt to falls, leading to greater morbidity and functional autonomy loss. Various cellular and molecular factors have been proposed to be involved in the underlying mechanisms of aging-induced sarcopenia such as changes in muscle fiber phenotype, protein muscle synthesis and degradation, cytokines, mitochondrial function, and oxidative stress among others.

7.2   Changes in Muscle Phenotype

Observations from different studies (Jakobsson, 1990) (Lexell, 1992) indicate that atrophy of type II fibers is the most significant finding of the aging-induced changes in muscle phenotype. In addition, it has been estimated that muscle fiber loss occurs as early as at age 25 and that at age 80 total muscle fiber shows a decrease of almost 40% (Lexell, 1986). Also, in a healthy man, a noticeable decline starts in the fourth decade and the rate of muscle loss is around 10% thereafter (Grimby, 1983). Interestingly, in women a significant decline in muscle mass is noticed in the sixth decade and specifically type II fibers are decreased by about 25% at the age of 70 years old (Sato, 1984). The redistribution of type I and II fibers as well as the age-dependent denervation of type II fibers are both responsible for aging-induced muscle mass loss. In conclusion, skeletal muscles in the elderly are weaker especially because of the loss and reduced number of type II fibers (Kirkendall, 1998).

7.3   Alterations in the Synthesis and Degradation Processes

Activities of specific enzymes are affected in aging muscle. For instance, its respiratory capacity is reduced because activity enzymes of the glycogenolytic and glycolytic pathways are diminished (Kleine, 1976). Glyceraldehyde-3-phosphate dehydrogenase expression and enzyme activity are also decreased in aged muscle fibers (Lowe, 2000). Gluconeogenesis activities are, however, increased in order to compensate for reduced glycogenolysis.

Several observations indicate that aging muscle shows increased catabolic processes, provided that alkaline phosphatase (sarcolemma) and acid phosphatase (lysosomes) activities are reduced and increased, respectively (Reznick, 1988) (Safadi, 1997). Specifically, reduced alkaline phosphatase activity indicates damage to the sarcolemma, whereas acid phosphatase activity points to increased lysosomal degradation. In general, different studies have shown that with advancing aging there is a progressive loss of muscle proteins. Protein synthesis also declines with age in skeletal muscles. The protein synthesis rate of myofibrillar proteins is reduced in muscles of elderly subjects (Welle, 1995) suggesting that muscle ability to synthesize new proteins is reduced. For instance, myosin heavy chain synthesis has been shown to be reduced in aged muscle fibers (Balagopal, 2001). Interestingly, a decrease in the synthesis of mitochondrial proteins has also been demonstrated in other studies (Chretien, 1998). Eventually, different investigators have shown that mainly two proteolytic pathways, the calcium-dependent involving calpains and the intracellular lysosomal pathway show increased activity in aging muscle fibers (Johnson, 1993).

7.4   Cytokines

As mentioned above, certain circulating cytokines are involved in muscle dysfunction and catabolism. In the blood of elderly individuals IL-1, IL-6 and TNF-alpha were shown to be increased (Greiwe, 2001). The best known mechanisms whereby these cytokines lead to muscle dysfunction are those which correspond to the TNF-alpha molecule. Specifically, binding of this cytokine to its sarcolemmal receptors stimulate ROS production by muscle mitochondria (Reid, 2001). The exact pathways of TNF-mediated ROS effects are two-fold: on the one hand, ROS can alter contractile regulation reducing muscle force, and on the other ROS can upregulate the ubiquitin/proteasome pathway, through NF-κβ activation, accelerating protein degradation and muscle mass loss. Current active research is being devoted to inhibiting the deleterious effects of these catabolic cytokines.

7.5   Mitochondrial Function

The mitochondria are the major cellular site for ATP synthesis. The inner membrane of the mitochondria contains all the components of the respiratory chain, which consists of a series of five multi subunit enzyme complexes. Respiratory chain complexes are genetic hybrids: products of both the nucleus and the mitochondrial genomes. The respiratory chain function noticeably declines with advancing age, together with an accumulation of deficient cytochrome c oxidase (Boffoli, 1994). The mechanisms underlying this mitochondrial dysfunction are the result of detrimental mutations of mitochondrial DNA during life. Among these mutations, large deletions, base substitutions, short duplications, and accumulation of 8-hydroxy deoxyguanosine are the ones reported in the literature (Holliday, 2000). These mutations may induce respiratory chain dysfunction along with muscle dysfunction, especially when they interfere with cellular ATP production (Carmeli, 2002). Furthermore, the decline of mitochondrial function occurring with aging is also related to increased electron displacement from the transport chain (Reid, 2002) (Figure 11), leading to the formation of greater levels of superoxide anion, which in turn would also result in the development of oxidative stress.

7.6   Reactive Oxygen and Nitrogen Species in Aging Muscle

The parent molecule in the cascade of ROS formation is the superoxide anion, a free radical generated by addition of one electron to the outer orbital of diatomic oxygen. Other oxygen radical species such as singlet oxygen, hydrogen peroxide, and hydroxyl radicals, are formed by electron exchange reactions from superoxide anion in muscle fibers. Unpaired electrons are lost at different steps in the respiratory chain, and molecular oxygen acts as the electron acceptor molecule (Figure 11). Other enzymes within the muscle fibers such as NAD(P)H oxidases, xanthine oxidase, lipoxygenases, and NOS can also produce superoxide anions. Oxidative stress will take place when either ROS are produced excessively or by a decrease in the antioxidant cellular systems. The imbalance of these two elements in favor of the former will target different cellular components such as DNA, proteins and lipids resulting in the deleterious effects of oxidative stress. Furthermore, heat shock proteins, which act as intracellular chaperones for newly synthesized proteins minimizing protein aggregation, are also reduced in aging muscle (Spiers, 2000) (Figure 11). Levels of both antioxidants and as aforementioned, NO is normally synthesized by NOS in the muscle fibers and its derivatives can also interact with ROS leading to the formation of highly reactive species such as peroxynitrite. As shown by Richmonds et al. 1999 aging seems to have opposite effects on NO synthesis compared to its effects on ROS cascade (Figure 11). Specifically, both NOS expression and enzyme activity were shown to be reduced in aged muscle fibers of rats (Richmonds, 1999).

Reports from the literature with regard to oxidative stress are still controversial and no clear conclusions have been reached yet. For instance, Leeuwenburgh et al. (1998) have shown no accumulation of either 3-nitrotyrosine or o-tyrosine in the heart, skeletal muscle, or liver of rats with advancing age. In another study, antioxidant enzymes were upregulated in aged diaphragm fibers of old mice, and training only improved these systems in the muscles of young mice (Oh-ishi, 1996). Increased levels of lipid peroxidation (index of oxidative stress) did not show any

EFFECTS OF AGING ON REDOX CASCADES

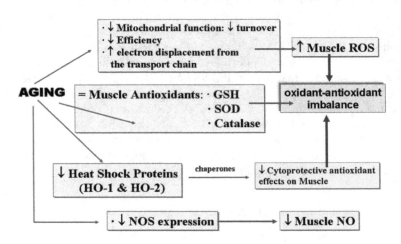

Figure 11. In aging muscle, reactive oxygen species (ROS) formation would be increased because of a reduction in the efficiency of the mitochondrial respiratory chain. Such an increase would not be counteracted by the intracellular antioxidants, since their levels are not upregulated in aging muscle. Furthermore, heat shock proteins, which may act as protein chaperones, have also been shown to be reduced in aging. Eventually, oxidative stress levels would be elevated in aging muscle. The levels of nitric oxide production, however, have been shown to be reduced in this condition. A color figure is freely accessible via the website of the book: http://www.springer.com/1-4020-5176-x. heat shock proteins are increased following exercise, and this response has been shown to be blunted in aged muscle fibers (Spiers, 2000)

relationship with the reduced levels of either NOS expression or activity in the rat limb and respiratory muscles (Richmonds, 1999). Lawler et al. (1997) also demonstrated in their study that the response to oxidant challenge and increased contractile demand was impaired in the aging rat diaphragm. In another interesting report it was shown that the additive effects of aging on mechanical ventilation have dramatic consequences on diaphragm force reserve, increasing weaning difficulties in elderly individuals receiving this type of therapy (Criswell, 2003). Finally, in skeletal muscles of elderly hospitalized patients (more than 65 years of age), the activity of several antioxidant enzymes essentially remained unchanged (Figure 11), except for Mn-superoxide dismutase, while lipid peroxidation and the oxidized fraction of glutathione were increased (Pansarasa, 2000).

7.7   Nutrition and Immobilization

Anorexia and malnutrition are common features of advancing age. On the grounds that oxidative stress contributes to muscle dysfunction in the aging process, some specialists have proposed an increased intake of antioxidant nutrients or even the administration of antioxidant supplements. Specifically, the latter would apply to elderly individuals who either exercise regularly or are chronically ill, provided that both conditions increase oxygen free radical production within the muscle fibers.

Immobilization has shown to be detrimental to aging muscle resulting in the atrophy of this tissue (Carmeli, 2002). The cellular and molecular changes observed in the muscle fibers as a result of disuse are increased protein degradation and reduced protein synthesis, changes in the size and number of sarcomeres, and increased levels of oxidative stress (Carmeli, 2002). In fact, the mechanisms whereby immobilization causes muscle mass loss have been divided into two different stages. For instance, calcium-dependent proteolysis is involved in the early phase, while the lysosomal and ubiquitin-proteasome systems participate in muscle protein degradation in the late phases. Besides, it is thought that white cell infiltration along with secretion of cytokines are most likely involved in the muscle proteolysis of the late stage as well (Zarzhevsky, 2001).

Another important factor to take into consideration are the reduced levels of certain growth factors such as insulin-like growth factor I (IGF-I), which is an essential biological mediator for protein synthesis (Severgnini, 1999). Administration of growth hormone, shown to stimulate IGF-I synthesis, clearly had beneficial effects on muscle atrophy (Zarzhevsky, 2001). Moreover, it has been shown that calcium-activated proteases induces the activity of a xanthine oxidase leading to increased levels of ROS. Eventually, it has been reported that immobilized hindlimb muscles from old animals are more severely affected than those from young ones, and that their capacity for recovery is also significantly reduced (Zarzhevsky, 2001). Treatment with IGF-I restored the capacity for satellite cell proliferation in the gastrocnemius muscles of old rats (Chakravarthy, 2000).

To sum up this part, muscle wasting or loss of muscle mass is the common feature shared by sepsis, cancer and aging. Cytokines, especially TNF-alpha, seem to play a crucial role as mediators of the muscle loss, probably through increasing the levels of ROS production. Increased levels of oxidants would have deleterious effects on contractile regulation on the one hand, and on the other would lead to increased muscle proteolysis, via activation of the ubiquitin/proteasome pathway. Clearly, future research should focus on the design of therapeutic strategies to target either the synthesis or the action of these catabolic cytokines.

ACKNOWLEDGEMENTS

To Josep M. Corominas M.D. for his collaboration in the preparation of the histological figures, and to Roger Marshall for his editing aid.

REFERENCES

Andrade FH, Reid MB, Allen DG, Westerblad H. (1998). Effect of hydrogen peroxide and dithiothreitol on contractile function of single skeletal muscle fibers from mouse. J Phyisiol 509, 565–575.
Appendini L, Purro A, Patessio A, Zanaboni S, Carone M, Spada E, Donner CF, Rossi A. (1996). Partitioning of inspiratory muscle workload and pressure assistance in ventilator-dependent COPD patients. Am J Respir Crit Care Med 154, 1301–1309.

Armstrong RB. (1988). Muscle fiber recruitment patterns and their metabolic correlates. In: Horton HS, Terjung RL, eds. Exercise, nutrition, and energy metabolism. New York: Macmillan Publishing Company, pp 9–26.

Arnason BGW. (1984). Acute inflammatory demyielinating polyradiculoneuropathies. In: Peripheral Neuropathy. Dyck PJ, Thomas PK, Lambert EH and Bunge RP eds. Saunders, pp 2050–2100.

Balagopal P, Schimke JC, Ades P, Adey D, Nair KS. (2001). Age effect on transcript levels and synthesis rate of muscle MHC and response to resistance exercise. Am J Physiol Endocrinol Metabol 2001; 280, E203–E208.

Barnes D, Munro PM, Youl BD, Prineas JW, McDonald WI. (1991). The longstanding MS lesion. A quantitative MRI and electron microscopic study. Brain 114, 1271–1280.

Barreiro E, Gea J, Sanjuás C et al. (1998). Lung volumes and respiratory muscle function before and after exercise in patients with severe asthma. Eur Respir J 1998; 12 (suppl 29), 64s–65s.

Barreiro E, Ferrer A, Hernández-Frutos N, Palacio J, Broquetas J, Gea J. (1999). Expiratory function and cellular properties of the external oblique muscle in patients with extremely severe COPD. Am J Crit Care Med 159 (suppl), A588.

Barreiro E, Comtois AS, Gea J, Laubach VE, Hussain SNA. (2002a). Protein tyrosine nitration in the ventilatory muscles: role of nitric oxide synthases. Am J Respir Cell Mol Biol 26, 438–446.

Barreiro E, Comtois AS, Mohammed S, Lands L, Hussain SNA. (2002b). Role of heme oxygenases in sepsis-induced diaphragmatic contractile dysfunction and oxidative stress. Am J Physiol Lung Cell Mol Physiol 283, L476–L484.

Barreiro E, Gea J, Corominas JM, Hussain SN. (2003). Nitric oxide synthases and protein oxidation in the quadriceps femoris of patients with chronic obstructive pulmonary disease. Am J Respir Cell Mol Biol 29, 771–778.

Barreiro E, Gea J, Sanjuas C, Marcos R, Broquetas J, Milic-Emili J. (2004a). Dyspnoea at rest and at the end of different exercises in patients with near-fatal asthma. Eur Respir J 24, 219–225.

Barreiro E, Gea J, Di Falco M, Kriazhev L, James S, Hussain SN. Protein carbonyl formation in the diaphragm. Am J Respir Mol Cell Biol 2005b; 32, 9–17.

Barreiro E, de la Puente B, Minguella J, Corominas JM, Serrano S, Hussain S, Gea J. Oxidative stress and respiratory muscle dysfunction in severe COPD. Am J Respir Crit Care Med 2005 (in press).

Beckman JS, Koppenol WH. Nitric oxide, superoxide, and peroxynitrite: the good, the bad, and the ugly. Am J Physiol 1996; 271 (40), C1424–C1437.

Beller GA. Selecting patients with ischemic cardiomyopathy for medical treatment, revascularization, or heart transplantation. J Nucl Cardiol 1997; 4, S152–S157.

Bernard S, LeBlanc P, Whittom F, Carrier G, Jobin J, Belleau R, Maltais F. Peripheral muscle weakness in patients with chronic obstructive pulmonary disease. Am J Respir Crit Care Med 1998; 158, 629–634.

Bloomfield SA. Changes in musculoskeletal structure and function with prolonged bed rest. Med Sci Sports Exerc 1997; 29, 197–206.

Boczkowski J, Dureuil B, Pariente R, Aubier M. Preventive effects of indomethacin on diaphragmatic contractile alterations in endotoxemic rats. Am Rev Respir Dis 1990; 142, 193–198.

Boffoli D, Scacco SC, Vergari R, Persio MT, Solarino G, Laforgia R, Papa S. Decline with age of the respiratory chain activity in human skeletal muscle. Biochim Biophys Acta 1994; 1226, 73–82.

Bolton C, Ionescu A, Shiels K et al. Associated loss of fat-free mass and bone mineral density in chronic obstructive pulmonary disease. Am J Respir Crit Care Med 2004; 170, 1286–1293.

Brown LK. Respiratory Dysfunction in Parkinson's disease. In: Respiratory dysfunction in neuromuscular disease. Clin Chest Med. Fanburg BL and Sicilian L eds. WB Saunders Company. Philadelphia 1994, pp 715–727.

Bruin PF, Ueki J, Watson A, Pride NB.. Size and strenght of the respiratory and quadriceps muscles in patients with chronic asthma. Eur Resp J 1997;10, 59–64.

Buck M, Chojkier M. Muscle wasting and dedifferentiation induced by oxidative stress in a murine model of cachexia is prevented by inhibitors of nitric oxide synthesis and antioxidants. EMBO J. 1996; 15 1753–1765.

Burke JF, Pontoppidan H, Welch CE. High output respiratory failure. Ann Surg 1963; 158, 581–594.

Buskila D, Odes L, Neumann L, Odes H. Fibromyalgia in inflammatory bowel disease. J Rheumatol 1999; 26, 1167–1171.

Callahan LA, Stofan D, Szweda LI, Nethery D, Supinski G. Free radicals alter maximal diaphragmatic mitochondrial oxygen consumption in endotoxin-induced sepsis. Free Radic Biol Med 2001a; 30, 129–138.

Callahan LA, Nethery D, Stofan D, DiMarco A, Supinski G. Free radical-induced contractile protein dysfunction in endotoxin-induced sepsis. Am J Respir Cell Mol Biol 2001b; 24, 210–217.

Campbell JA, Hughes RL, Shagal V, Frederiksen J, Shields TW. Alterations in intercostal muscle morphology and biochemistry in patients with chronic obstructive lung disease. Am Rev Respir Dis 1980; 122, 679–686.

Cano A, Gea J, Roquer J et al. Pulmonary function in previously untreated Parkinson Disease (PD) patients. Effects of apomorphine. Am J Respir Crit Care Med 1996; 153 (suppl), A 22.

Carmeli E, Coleman R, Reznick A. The biochemistry of aging muscle. Exp Gerontol 2002; 37, 477–489.

Carroll SB. Genetics and the making of Homo sapiens. Nature 2003; 422, 849–857.

Casaburi R. Skeletal muscle function in COPD. Chest 2000; 117, 267S–271S.

Casadevall C, Coronell C, Ramírez A, Barreiro E, Orozco-Levi M, Gea J. Local expression of the gene encoding TNF-alpha in the external intercostal and quadriceps muscles of severe COPD patients. Am J Respir Crit Care Med 2003; 167 (suppl), A29.

Chakravarthy MV, Davis BS, Booth FW. IGF-I restores satellite cell proliferative potential in immobilized old skeletal muscle. J Appl Physiol 2000; 89, 1365–1379.

Chen H, Tang Y. Sleep loss impairs inspiratory muscle endurance. Am Rev Respir Dis 1989;140, 907–9.

Chevari S, Andial T, Benke K, Shtrenger IA, Free radical reactions and cancer, Vopr Med Khim 1992; 38, 4–5.

Choi AMK, Alam J. Heme oxygenase-1: Function, regulation, and implication of a novel stress-inducible protein in oxidant-induced lung injury. Am J Respir Cell Mol Biol 1996; 15: 9–19.

Chretien D, Gallego J, Barrientos A, Casademont J, Cardellach F, Munnich A, Rotig A, Rustin P. Biochemical parameters for the diagnosis of mitochondrial respiratory chain deficiency in humans, and their lack of age-related changes. Biochem J 1998; 329, 249–254.

Chua TP, Anker SD, Harrington D, Coats AJ. Inspiratory muscle strength is a determinant of maximum oxygen consumption in chronic heart failure. Br Heart J 1995; 74, 381–385.

Civatte M, Bartoli C, Schleinitz N, Chetaille B, Pellissier JF, Figarella-Branger D. Expression of the beta chemokines CCL3, CCL4, CCL5 and their receptors in idiopathic inflammatory myopathies. Neuropathol Appl Neurobiol 2005; 31, 70–79.

Cohen CA, Zagelbaum G, Gross D, Roussos C, Macklem PT. Clinical manifestations of inspiratory muscle fatigue. Am J Med 1982; 73, 308–316.

Coronell C, Orozco-Levi M, Mendez R, Ramirez A, Galdiz JB, Gea J. Relevance of assessing quadriceps endurance in patients with COPD. Eur Respir J 2004; 24, 129–136.

Costelli P, Carbó N, Tessitore L, Bagby GJ, López-Soriano FJ, Argilés JM, Baccino FM. Tumour necrosis factor-alpha mediates changes in muscle protein turnover in a cachectic rat tumour model. J Clin Invest 1993; 92, 2783–2789.

Cotes JE, Zejda J, King B. Lung function impairment as a guide to exercise limitation in work-related lung disorders. Am Rev Respir Dis 1988; 137, 1089–1093.

Criswell DS, Shanely RA, Betters JJ, McKenzie MJ, Sellman JE, Van Gammeren DL, Powers SK. Cumulative effects of aging and mechanical ventilation on in vitro diaphragm function. Chest 2003; 124, 2302–2308.

Decramer M, deBock V, Dom R. Functional and histologic picture of steroid-induced myopathy in chronic obstructive pulmonary disease. Am J Respir Crit Care Med 1996; 153, 1958–1964.

Decramer M, Gosselink R, Troosters T, Verschueren M, Evers G. Muscle weakness is related to utilization of health care resources in COPD patients. Eur Respir J. 1997; 10, 417–423

Deconinck N, Van Parijs V, Bleeckers-Bleukx G et al. Critical illness myopathy unrelated to corticosteroids or neuromuscular blocking agents. Neuromuscul Disord 1998; 8, 186–192.

DeWys W. Management of cancer cachexia. Semin Oncol 1985; 12, 452–460.

Di Francia M, Barbier D, Mege JL, Orehek J. Tumor necrosis factor-alpha levels and weight loss in chronic obstructive pulmonary disease. Am J Respir Crit Care Med 1994; 150, 1453–1455.

Dick PJ. Diseases of peripheral nerves. In: Myology. Engel AG and Franzini-Armstrong C eds. New York 1994, pp 1870–1904.

Diffee GM, Caiozzo VJ, Herrick RE, Baldwin KM. Contractile and biochemical properties of rat soleus and plantaris after hindlimb suspension. Am J Physiol 1991; 260 (Cell Physiol 29), C528–C534.

Dodd DS, Brancatisano T, Engel LA. Chest wall mechanics during exercise in patients with severe chronic airflow obstruction. Am Rev Respir Dis 1984; 129, 33–38.

Drachman DB. Myasthenia gravis. N Engl J Med 1978; 298, 136–142 and 186–193.

Drexler H, Riede U, Münzel T, König H, Funke E, Just H. Alterations of skeletal muscle in chronic heart failure. Circulation 1992; 85, 1751–1759.

Drexler H. Changes in the peripheral circulation in heart failure. Curr Opin Cardiol 1995; 10, 268–273.

Dudley GA, Terjung RL. Influence of aerobic metabolism on IMP accumulation in fast-twitch muscle. Am J Physiol 1985; 248, C37–C42.

Dunnigan A, Staley NA, Smith SA, et al. Cardiac and skeletal muscle abnormalities in cardiomyopathy: comparison of patients with ventricular tachycardia or congestive heart failure. J Am Coll Cardiol 1987; 10, 608–618.

Engel AG, Lambert EH, Howard FM. Immune complexes (IgG and C3) at the motor end-plate in myasthenia gravis: ultrastructural and light microscopic localization and electrophysiologic correlations. Mayo Clin Proc 1977; 52, 267–280.

Engel AG, Arahata K. Monoclonal antibody analysis of mononuclear cells in myopathies: II. Phenotypes of autoinvasive cells on polymyositis and inclusion body myositis. Ann Neurol 1984; 16, 209–215.

Engel AG, Hohlfeld R, Banker BQ. The polymyositis and dermatomyositis syndromes. In: Myology. Engel AG and Franzini-Armstrong C eds. New York 1994a, pp 1335–1383.

Engel AG. Acquired autoimmune Myasthenia Gravis. In: Myology. Engel AG and Franzini-Armstrong C eds. New York 1994b, pp 1769–1797.

Epstein SK. An Overview on Respiratory Muscle Function, Clin Chest Med, 1994: 15, 619–639.

Essen-Gustavsson B, Henriksson J. Enzyme levels in pools of microdissected human muscle fibres of identified type. Adaptive response to exercise. Acta Physiol Scand 1984; 120, 505–515.

Fang FC. Mechanisms of nitric oxide-related antimicrobial activity. J Clin Invest 1997; 100, S43–S50

Fernandez-Sola J, Cusso R, Picado C, Vernet M, Grau JM, Urbano-Marquez A. Patients with chronic glucocorticoid treatment develop changes in muscle glycogen metabolism. J Neurol Sci 1993; 117, 103–106.

Fischman DA, Cerami A, Lowry SF. Cachectin/TNF or IL-1a induces cachexia with redistribution of body proteins. Am J Physiol 1989; 256, R659–R665.

Flores EA, Bristain BR, Pomposelli JJ, Dinarello CA, Blackburn GL, Istfan NW. Infusion of tumor necrosis factor /cachectin promotes muscle metabolism in the rat. J Clin Invest 1989; 83, 1614–1622.

Fong Y, Moldawer LL, Morano M, Wei H, Barber A, Manogue K, Tracey KJ, Kuo G, García-Martínez C, Agell N, Llovera M, López-Soriano FJ, Argilés JM. Tumour necrosis factor-alpha increases the ubiquitinization of rat skeletal muscle proteins. FEBS Lett 1993; 323, 211–214.

Foote RA, Kimbrough SM, Stevens JC. Lupus myositis Muscle Nerve 1982, 5, 65–68.

Friman G. Effects of acute infectious disease on isometric strength. Scand J Clin Lab Invest 1977; 37, 303–308.

García-Martínez C, Llovera M, Agell N, López-Soriano FJ, Argilés JM. Ubiquitin gene expression in skeletal muscles is increased by tumour necrosis factor-alpha. Biochem Biophys Res Commun 1994; 201, 682–686.

García-Martínez C, López-Soriano FJ, Argilés JM. Amino acid uptake in skeletal muscle of rats bearing the Yoshida AH-130 ascites hepatoma. Mol Cell Biochem 1995; 148, 17–23.

Garpestad E, Hatayama H, Parker J et al. Stroke volume and cardiac output decrease at termination of obstructive apneas. J Appl Physiol 1992;73, 1743–8.

Gaston B, Drazen JM, Loscalzo J, Stamler JS. The biology of nitrogen oxides in the airways. Am J Respir Crit Care Med 1994; 149, 538–551.

Gea J. Myosin gene expression in the respiratory muscles. Eur Resp J 1997; 10, 2404–2410.

Gea J, Pastó M, Ennion S, Goldspink G, Broquetas JM. Expression of the genes corresponding to Myosin Heavy Chain isoforms (MyHC I, IIa and IIx) in the diaphragm of patients suffering from COPD. Eur Respir J 1998; 12 (suppl 28), 267 s.

Gea J, Hamid Q, Czaika G, Zhu E, Mohan-Ram V, Goldspink G, Grassino A. Expression of Myosin Heavy Chain isoforms in the respiratory muscles following inspiratory resistive breathing. Am J Respir Crit Care Med 2000; 161, 1274–1278.

Gea J, Orozco-Levi M, Barreiro E, Ferrer A, Broquetas J. Structural and functional changes in the skeletal muscles of COPD patients: The "Compartments" Theory. Mon Arch Chest Dis 2001a, 56, 214–224.

Gea J, Pasto M, Carmona M, Orozco-Levi M, Palomeque J, Broquetas J. Metabolic characteristics of the deltoid muscle in patients with chronic obstructive pulmonary disease. Eur Respir J 2001b; 17, 939–945.

Global Initiative for Asthma. (G.I.N.A) Global Strategy for Asthma Management and Prevention. NHLBI. Bethesda, 2001. www.ginasthma.com.

Global Initiative for Chronic Obstructive Lung Disease. Global Strategy for the Diagnosis, Management and Prevention of Chronic Obstructive Pulmonary Disease. NHLBI/WHO workshop report. Bethesda, 2004. National Heart, Lung and Blood Institute. Update of the Management Sections, GOLD website, www.goldcopd.com.

Gnaiger E, Lassnig B, Kuznetsov A, Rieger G, Margreiter R. Mitochondrial oxygen affinity, respiratory flux control and excess capacity of cytochrome c oxidase. J Exp Biol 1998; 201, 1129–1139.

Goldberg A. Goodman H. Relationship between cortisone and muscle work in determining muscle size. J Physiol 1969; 200, 667–75.

Goldman MD, Grassino A, Mead J, Sears A. Mechanics of the human diaphragm during voluntary contraction: Dynamics. J Appl Physiol 1978; 44, 840–848.

Gomes-Marcondes MC, Tisdale MJ. Induction of protein catabolism and the ubiquitin-proteasome pathway by mild oxidative stress. Cancer Lett. 2002; 180 69–74.

Gorini M, Ginanni R, Spinelli A, Duranti R, Andreotti L, Scano G. Inspiratory muscle strength and respiratory drive in patients with rheumatoid arthritis. Am Rev Respir Dis. 1990; 142, 289–294.

Gosker HR, Wouters EF, van der Vusse GJ, Schols AM. Skeletal muscle dysfunction in chronic obstructive pulmonary disease and chronic heart failure: underlying mechanisms and therapy perspectives. Am J Clin Nutr 2000; 71, 1033–1047.

Gosker HR, Lencer NH, Franssen FM, van der Vusse GJ, Wouters EF, Schols AM. Striking similarities in systemic factors contributing to decreased exercise capacity in patients with severe chronic heart failure or COPD. Chest. 2003; 123, 1416–1424

Gosselink R, Troosters T, Decramer M. Peripheral muscle weakness contributes to exercise limitation in COPD. Am J Respir Crit Care Med 1996; 153, 976–980.

Greiwe JS, Cheng B, Rubin DC, Yarasheski KE, Semenkovich CF. Resistance exercise decreases skeletal muscle tumor necrosis factor alpha in frail elderly humans. FASEB J 2001; 15, 475–482.

Griggs G, Findley L, Suratt P, Esau S, Wilhoit S, Rochester D. Plonged relaxation rate of inspiratory muscles in patients with sleep apnea. Am Rev Respir Dis 1989;140, 706–710.

Grimby G, Saltin B. The ageing muscle. Clin Physiol 1983; 3, 209–218.

Guilleminault C, Stoohs R, Quera-Salva MA. Sleep-related obstructive and non-bstructive apneas and neurologic disorders. Neurology 1992; 42 (suppl 6), 53–60.

Gustafson T, Boman K, Rosenhall L, Sandstrom T, Wester PO. Skeletal muscle magnesium and potassium in asthmatics treated with oral beta 2-agonists. Eur Respir J 1996; 9, 237–240.

Hall-Angeras M, Angeras U, Zamir O, Hasselgren PO, Fischer JE. Interaction between corticosterone and tumor necrosis factor simulated protein breakdown in rat skeletal muscle similar to sepsis. Surgery 1990; 108, 460–466.

Hambrecht R, Fiehn E, Yu J, et al. Effects of endurance training on mitochondrial ultrastructure and fiber type distribution in skeletal muscle of patients with stable chronic heart failure. J Am Coll Cardiol 1997; 29, 1067–1073.

Hamilton AL, Killian KJ, Summers E, Jones NL. Muscle strength, symptom intensity, and exercise capacity in patients with cardiorespiratory disorders. Am J Respir Crit Care Med 1995; 52, 2021–2031.

Harley HG, Brook JD, Rundle SA et al. Expansion of un unstable DNA region and phenotypic variation in myotonic dystrophy. Nature 1992; 355, 547–548.

Harridge SD, Magnusson G, Gordon A. Skeletal muscle contractile characteristics and fatigue resistance in patients with chronic heart failure. Eur Heart J 1996; 17, 896–901.

Harrington D, Anker SD, Chua TP, et al. Skeletal muscle function and its relation to exercise tolerance in chronic heart failure. J Am Coll Cardiol 1997; 30, 1758–1764.

Hernandez N, Orozco-Levi M, Belalcazar V et al. Dual morphometrical changes of the deltoid muscle in patients with COPD. Respir Physiol Neurobiol 2003; 134, 219–229.

Hoffman EP, Brown RH, Kunkel LM. Dystrophin: The product of the Duchennne muscular dystrophy locus. Cell 1987; 51, 919.

Hoffmann M, Bybee K, Accurso V, Somers VK. Sleep apnea and hypertension. Minerva Med 2004; 95, 281–290.

Holliday R. Reflections in mutation research. Somatic mutation and ageing. Mutat Res 2000; 463, 173–178.

Howard RS, Wiles CM, Hirsch NP, Loh L, Spencer GT, Newsom-Davis J. Respiratory involvement in multiple sclerosis. Brain 1992; 115, 479–494.

Hughes RL, Katz H, Sahgal V, Campbell JA, Hartz R, Shields TW. Fiber size and energy metabolites in five separate muscles from patients with chronic obstructive lung diseases. Respiration 1983; 44, 321–328.

Hussain SNA, Simkus G, Roussos C. Respiratory muscle fatigue: A cause of ventilatory failure in septic shock. J Appl Physiol 1985; 58, 2033–2040.

Hussain SNA. Respiratory muscle dysfunction in sepsis. Mol Cell Biochem 1998; 179, 125–134.

Jackson MJ, Edwards RH. Free radicals and trials of antioxidant therapy in muscle diseases. Adv Exp Med Biol 1990; 264, 485–491.

Jackson MJ, O'Farrel S. Free radicals and muscle damage. Br Med Bull 1993; 49, 630–641.

Jakobsson F, Borg K, Edstrom L. Fibre-type composition, structure, and cytoskeletal protein location of fibres in anterior tibialis muscle. Comparison between young adults and physically active aged humans. Acta Neuropathol (Berl) 1990a; 459–468.

Jakobsson P, Jorfeldt L, Brundin A. Skeletal muscle metabolites and fiber types in patients with advanced chronic obstructive pulmonary disease (COPD), with and without chronic respiratory failure. Eur Respir J 1990b; 3, 192–196.

Jakobsson P, Jordfelt L, Henriksson. Metabolic enzyme activity in the quadriceps femoris muscle in patients with severe chronic obstructive pulmonary disease. Am J Respir Crit Care Med 1995; 151, 374–377.

Jardim J, Mayer A, Camelier A. Respiratory muscles and pulmonary rehabilitation in asthmatic patients. Arch Bronconeumol 2002; 38, 181–188.

Jerusalem F, Rakusa M, Engel AG et al. Morphometric analysis of skeletal muscle capillary ultra-estructure in inflammatory myopathies. J Neurol Sci 1974; 23, 391–402.

Jimenez M, Gea J, Aguar MC et al. Capillary density and respiratory function in the external intercostal muscle. Arch Bronconeumol 1999; 35, 471–476.

Johnson P, Hammer JL. Cardiac and skeletal muscle enzyme levels in hypertensive and aging rats. Comp Biochem Physiol 1993; 104, 63–67.

Kaminski HJ, Ruff RL. Endocrine myopathies (hyper- and hypofunction of adrenal, thyroid, pituitary, and parathyroid glands, and iatrogenic corticosteroid myopathy. In: Myology. Engel AG and Franzini-Armstrong C eds. New York 1994, pp 1726–1733.

Kaplan LM, Hollander D. Respiratory dysfuntion in amyothrophic lateral sclerosis. In: Respiratory dysfunction in neuromuscular disease. Clin Chest Med. Fanburg BL and Sicilian L eds. WB Saunders Company. Philadelphia 1994, pp 675–681.

Kemp G, Thompson C, Stratton J et al. Abnormalities in exercising skeletal muscle in congestive heart failure can be explained in terms of decreased mitochondrial ATP synthesis, reduced metabolic efficiency, and increased glycogenolysis. Heart 1996; 76, 35–41.

Khawli F, Reid MB. N-acetylcysteine depresses contractile function and inhibits fatigue of diaphragm in vitro. J Appl Physiol 1994; 77, 317–324.

Killian KJ, Leblanc P, Martin DH, Summers E, Jones NL, Campbell EJM. Exercise capacity and ventilatory, circulatory, and symptom limitation in patients with chronic airflow obstruction. Am Rev Respir Dis 1992; 146, 935–940.

Kimoff J, Cheong T, Olha A, Charbonneau M, Levy R, Cosio M et al.. Mechanisms of apnea termination in obstructive sleep apnea. Am J Respir Crit Care Med 1994; 149, 707–14.

Kirchberger M. Excitation and contraction of skeletal muscle. In West JB (ed) Physiological Basis of Medical Practice. Williams & Wilkins, Baltimore 1991, MD, pp. 62–102.

Kirkendall DT, Garrett WE Jr. The effects of aging and training on skeletal muscle. Am J Sports Med 1998; 26, 598–602.

Kleine TO. Glycolytic and gluconeogenetic enzyme activities in skeletal muscle of differently aged persons. Acta Gerontol 1976; 6, 489–494.

Knowles RG, Moncada S. Nitric oxide synthases in mammals. Biochem J 1994; 298, 249–258.

Kobzik L, Reid MB, Bredt DS, Stamler JS. Nitric oxide in skeletal muscle. Nature 1994; 372, 546–548.

Krause KM, Moody MR, Andrade FH, Addison AT, Miller CC III, Kobzik L, Reid MB. Peritonitis causes diaphragm weakness in rats. Am J Respir Crit Care Med 1998; 157, 1277–1282.

Lacomis D, Giuliani M, Van Cott A et al. Acute myopathy of intensive care: clinical, electromyographic and pathological aspects. Ann Neurol 1996; 40, 645–654.

Lampert E, Mettauer B, Hoppeler H, Charloux A, Charpentier A, Lonsdorfer J. Structure of skeletal muscle in heart transplant recipients. J Am Coll Cardiol 1996; 28:980–984.

Lands LC, Heigenhauser GJ, Jones NL. Respiratory and peripheral muscle function in cystic fibrosis. Am Rev Respir Dis 1993; 147, 865–869.

Lavietes MH, Grocela JA, Maniatis T, Potulski F, Ritter AB, Sunderam G. Inspiratory muscle strenght in asthma. Chest 1988; 93, 1043–1048.

Lawler JM, Cline CC, Hu Z, Coast JR. Effect of oxidant challenge on contractile function of the aging rat diaphragm. Am J Physiol 1997; 272, E201–E207.

Le Bourdelles G, Viires N, Boczkowski J et al. Effects of mechanical ventilation on diaphragmatic contractile properties in rats. Am J Respir Crit Care Med 1994; 149, 1539–1544.

Leeuwenburgh C, Hansen P, Sahish A, Holloszy JO, Heinecke JW. Markers of protein oxidation by hydroxyl radical and reactive nitrogen species in tissues of aging rats. Am J Physiol 1998; 274, R453–R461.

Leon A, Boczkowski J, Dureuil B, Desmonts JM, Aubier M. Effects of endotoxic shock on diaphragmatic function in mechanically ventilated rats. J Appl Physiol 1992; 72, 1466–1472.

Levine S, Kaiser L, Leferovich J, Tikunov B. Cellular adaptations in the diaphragm in chronic obstructive pulmonary disease. N Engl J Med 1997; 337, 1799–1806.

Lexell J, Taylor CC, Sjöström M. What is the cause of the ageing atrophy? Total number, size, and proportion of different fiber types studied in the whole vastus lateralis muscle from 15 to 83 year-old men. J Neurol Sci 1986; 84, 275–294.

Lexell J, Downham D. What is the effect of ageing on type 2 muscle fibres? J Neurol Sci 1992; 107, 250–251.

Lieu F, Powers SK, Herb RA, Criswell D, Martin D, Wood C, Stainsby W, Chen CL. Exercise and glucocorticoid-induced diaphragmatic myopathy. J Appl Physiol 1993; 75, 763–771.

Lindsay DC, Lovegrove CA, Dunn MJ, et al. Histological abnormalities of muscle from limb, thorax and diaphragm in chronic heart failure. Eur Heart J 1996; 17, 1239–1250.

Lipkin DP, Jones DA, Round JM, Poole Wilson PA. Abnormalities of skeletal muscle in patients with chronic heart failure. Int J Cardiol 1988; 18, 187–195.

Llesuy S, Evelson P, Gonzalez B, Peralta JG, Carreras MC, Poderoso JJ, Boveris A. Oxidative stress in muscle and liver of rats with septic syndrome. Free Radic Biol Med 1994; 16, 445–451.

Lloreta J, Orozco-Levi M, Gea J, Broquetas J. Selective diaphragmatic mitochondrial abnormalities in a patient with marked airflow obstruction. Ultrastruct Pathol 1996; 20, 67–71.

Llovera M, García-Martínez C, Agell N, Marzábal M, López-Soriano FJ, Argilés JM. Ubiquitin gene expression is increased in skeletal muscle of tumour-bearing rats. FEBS Lett 1994; 338, 311–318.

Llovera M, García-Martínez C, Agell N, López-Soriano FJ, Argilés JM. Muscle waisting associated with cancer cachexia is linked to an important activation of the ATP-dependent ubiquitin-mediated proteolysis. Int J Cancer 1995; 61, 138–141.

Llovera M, Carbó N, López-Soriano FJ, García-Martínez C, Busquets S, Alvarez B, Agell N, Costelli P, López-Soriano FJ, Celada A, Argilés JM. Different cytokines modulate ubiquitin-gene expression in rat skeletal muscle. Cancer Lett 1998; 133, 83–87.

Lovis C, Mach F, Unger PF, Bouillie M, Chevrolet JC. Elevation of creatine kinase in acute severe asthma is not of cardiac origin. Intensive Care Med 2001; 27, 528–533.

Lowe DA, Degens H, Chen KD, Always SE. Glyceraldehyde-3-phosphate dehydrogenase varies with age in glycolytic muscles of rats. J Gerontol 2000; 55, B160–B164.

Lynn DJ, Woda RP, Mendell JR. Respiratory dysfunction in muscular dystrophy and other myopathies. In: Respiratory dysfunction in neuromuscular disease. Clin Chest Med. Fanburg BL and Sicilian L eds. WB Saunders Company. Philadelphia 1994, pp 661–674.

Maines MD. Heme oxygenase; function, multiplicity, regulatory mechanisms and clinical implications. FASEB J 1988; 2, 2557–2568.

Mainwood GW, Renaud JM. The effect of acid-base balance on fatigue of skeletal muscle. Can J Physiol Pharmacol 1985; 63, 403–416.

Maltais F, Leblanc P, Simard C, Jobin J, Berubé C, Bruneau J, Carrier L, Belleau R. Skeletal muscle adaptation to endurance training in patients with chronic obstructive pulmonary disease. Am J Respir Crit Care Med 1996; 154, 442–447.

Mancini DM, Coyle E, Coggan A, et al. Contribution of intrinsic skeletal muscle changes to 31P NMR skeletal muscle metabolic abnormalities in patients with chronic heart failure. Circulation 1989; 80, 1338–1346.

Mancini DM, Henson D, LaManca J, Levine S. Respiratory muscle function and dyspnea in patients with chronic congestive heart failure. Circulation 1992; 86, 909–918.

Martyn J, Powell E, Shore S, Emrich J, Engel LA. The role of respiratory muscles in the hyperinflation of bronchial asthma. Am Rev Respir Dis 1980; 121, 441–447.

Matsubara S, Okada T, Yoshida M. Mitochondrial changes in acute myopathy after treatment of respiratory failure with mechanical ventilation (acute relaxant-steroid myopathy). Acta Neuropathol 1994; 88, 475–478.

Mckenzie DK, Gandevia SC. Strenght and endurance of inspiratory and limb muscles in asthma. Am Rev Respir Dis 1986; 134, 999–1004.

McComas AJ. Skeletal muscle: form and function. Champaign, IL: Human Kinetics, 1996.

McCoubrey WK, Ewing JF, Maines MD. Human heme oxygenase-2: characterization and expression of a full-length cDNA and evidence suggesting that the two HO-2 transcripts may differ by choice of polyadenylation signal. Arch Biochem Biophys 1992; 295, 13–20.

McCoubrey WK, Huang TJ, Maines MD. Isolation and characterization of a cDNA from the rat brain that encodes hemoprotein heme oxygenase-3. Eur J Biochem 1997; 247, 725–732.

McParland C, Resch EF, Krishnan B, Wang Y, Cujec B, Gallagher CG. Inspiratory muscle weakness in chronic heart failure: role of nutrition and electrolyte status and systemic myopathy. Am J Respir Crit Care Med 1995; 151, 1101–1107.

Mehta S, Nelson DL, Klinger JR, Buczko GB, Levy MM. Prediction of post-extubation work of breathing. Crit Care Med 2000; 28, 1341–1346.

Mier-Jedrzejowicz AK, Brophy C, Green M. Respiratory muscle function in myasthenia gravis. Am Rev Respir Dis 1988; 138, 867–873.

Miller MM. Current trends in the primary care management of chronic congestive heart failure. Nurse Pract 1994; 19, 64–70.

Miyagi K, Asanoi H, Ishizaka S, et al. Importance of total leg muscle mass for exercise intolerance in chronic heart failure. Jpn Heart J 1994; 35, 15–26.

Mizuno M. Human respiratory muscles: fibre morphology and capillary supply. Eur Respir J 1991; 4, 587–601.

Moncada S, Palmer RMJ, Higgs EA. Nitric oxide: physiology, pathophysiology, and pharmacology. Pharmacol Rev 1991; 43, 109–142.

Montes de Oca M, Rassulo J, Celli B. Respiratory muscle and cardiopulmonary function during exercise in very severe COPD. Am J Respir Crit Care Med 1996; 154, 1284–1289.

Murphy TD, Gibson R, Standaert TA, Woodrum DE. Diaphragmatic failure during group streptococcal sepsis in piglets: the role of thromboxane A_2. J Appl Physiol 1995; 78(2), 491–498.

Myers J, Salleh A, Buchanan N, et al. Ventilatory mechanisms of exercise intolerance in chronic heart failure. Am Heart J 1992; 124, 710–719.

Nava S, Bruschi C, Franchia C, Fezicetti G, Callezsri G, Ambrosino N.. Acute treatment with corsticosteroids (CS) alters respiratory (RM) and skeletal muscles (SM) function in humans. Eur Respir J 1996; 9 (suppl), S298.

Ninane V, Rypens F, Yernault J, De Troyer A. Abdominal muscle use during breathing in patients with chronic airflow obstruction.Am Rev Respir Dis 1992; 146, 16–21.

Noguera A, Sala E, Pons AR, Iglesias J, MacNee W, Agusti AG. Expression of adhesion molecules during apoptosis of circulating neutrophils in COPD. Chest 2004; 125, 1837–1842.

Nowinski A, Gea J, Bielew P et al. Structure of the external intercostal muscle in patients with obstructive sleep apnea syndrome (OSAS). Eur Respir J 2001; 18 (suppl 33), 178s.

Oh-ishi S, Toshinai K, Kizaki , Haga S, Fukuda K, Nagata N, Ohno H. Effects of aging and/or training on antioxidant enzyme system in diaphragm of mice. Respir Physiol 1996; 105, 195–202.

Orozco-Levi M, Gea J, Monells J, Arán X, Aguar MC, Broquetas JM. Activity of latissimus dorsi muscle during inspiratory thershold loads. Eur Respir J 1995a; 8, 441–445.

Orozco-Levi M, Gea J, Sauleda J, Cominas JM, Minguella J, Arán X, Broquetas JM. Structure of the latissimus dorsi muscle and respiratory function. J Apply Physiol 1995b; 78, 1132–1139.

Orozco-Levi M, Gea J, Lloreta J et al. Subcellular adaptation of the human diaphragm in Chronic Obstructive Pulmonary Disease. Eur Respir J 1999; 13, 371–378.

Orozco-Levi M, Lloreta J, Minguella J, Serrano S, Broquetas J, Gea J. Injury of the human diaphragm associated with exertion and COPD Am J Respir Crit Care Med 2001; 164, 1734–1739.

Pansarasa O, Castagna L, Colombi B, Vecchiet J, Felzani G, Marzatico F. Age and sex differences in human skeletal muscle: role of reactive oxygen species. Free Rad Res 2000; 33, 287–293.

Parish JM, Somers VK. Obstructive sleep apnea and cardiovascular disease. Mayo Clin Proc 2004; 79, 1036–1046.

Pascual J, Gea J, Aran X et al. Early markers of pulmonary dysfunction in amyotrophic lateral sclerosis. J Neurol Scienc 1990; 98 (suppl), 341.

Peralta JG, Llesuy S, Evelson P, Carreras MC, Gonzalez B, Poderoso JJ. Oxidative stress in skeletal muscle during sepsis in rats. Cir Shock 1993; 39, 153–159.

Peress L, Sybrecht G, Macklem PT. The mechamism of increase in total lung capacity during acute asthma. Am Rev Respir Dis 1986; 134, 994–1004.

Picado C, Fiz JA, Montserrat JM et al. Respiratory and skeletal muscle function in steroid-dependent bronchial asthma. Am Rev Respir Dis 1990; 141, 14–20.

Plotz PH, Dalakas M, Leff RL et al. Current concepts in the idiopathic inflammatory myopathies: Polymyositis, dermatomyositis, and related disorders. Ann Intern Med 1989; 111, 143–157.

Polkey MI, Moxham J. Clinical aspects of respiratory muscle dysfunction in the critically ill. Chest 2001; 119, 926–939.

Pouw EM, Schols AMWJ, Vusse van der GJ, Wouters EFM. Elevated inosine monophosphate levels in resting muscle of patients with stable COPD. Am J Respir Crit Care Med 1998; 157, 453–457.

Prineas JW, Barnard RO, Kwon EE, Sharer LR, Cho ES. Multiple sclerosis: remyelination of nascent lesions. Ann Neurol 1993; 33, 137–151.

Rall LC, Rosen CJ, Dolnikowski G, Hartman WJ, Lundgren N, Abad LW, Dinarello CA, Roubenoff R. Protein metabolism in rheumatoid arthritis and aging. Effects of muscle strength training and tumor necrosis factor alpha. Arthritis Rheum 1996; 39, 1115–1124.

Ramirez A, Orozco-Levi M, Barreiro E et al. Expiratory muscle endurance in chronic obstructive pulmonary disease. Thorax 2002; 57, 132–136.

Ramsay DA, Zochodne DW, Robertson DM, Nag S, Ludwin SK. A syndrome of acute severe muscle necrosis in intensive care unit patients. J Neuropathol Exp Neurol 1993; 52, 387–398.

Rankin GB. Extraintestinal and systemic manifestations of inflammatory bowel disease. Med Clin North Am 1990; 74, 39–50.

Rannels SR, Rannels DE, Pegg AE, Jefferson LS. Glucocorticoid effects on peptide-chain initiation in skeletal muscle and heart. Am J Physiol 1978; 235, E134–E139.

Reid MB, Haack KE, Franchek KM, Valberg PA, Kobzik L, West MS. Reactive oxygen in skeletal muscle: I. Intracellular oxidant kinetics and fatigue in vitro. J Appl Physiol 1992; 73, 1797–1804.

Reid MB, Khawli FA, Moody MR. Reactive oxygen in skeletal muscle.III. Contractility of unfatigued muscles. J Appl Physiol 1993; 75, 1081–1087.

Reid MB, Li YP. Cytokines and oxidative signaling in skeletal muscle. Acta Physiol Scand 2001; 171, 225–232.

Reid MB, Durham WJ. Generation of reactive oxygen and nitrogen species in contracting skeletal muscle. Ann NY Acad Sci 2002; 959, 108–116.

Reznick AZ, Steinhagen-Thiessen E, Silbermann M. Acid and alkaline phosphatase activity in skeletal muscle of aging mice and following prolonged training program. Trends Biomed Gerontol 1988; 1, 91–94.

Richmonds CR, Boonyapisit K, Kusner LL, Kaminski HJ. Nitric oxide synthase in aging rat skeletal muscle. Mech Ageing Dev 1999; 109, 177–189.

Rochester DF, Braun NMT, Arora NS. Respiratory muscle strength in chronic obstructive pulmonary disease. Am Rev Respir Dis 1979; 119, 151–154.

Rochester DF, Arora NS. The respiratory muscles in asthma. In: Lavietes MH, Reichman LB, editors. Diagnostic aspects and management of asthma. Norwalk: Purdue-Frederick CO, 1981.

Rochester DF, Braun NMT. Determinants of maximal inspiratory pressure in chronic obstructive pulmonary disease. Am Rev Respir Dis 1985; 132, 42–47.

Roubenoff R, Roubenoff RA, Cannon JG, Kehayias JJ, Zhuang H, Dawson-Hughes B, Dinarello CA, Rosenberg IH. Rheumatoid cachexia: cytokine-driven hypermetaboolism accompanying reduced body cell mass in chronic inflammation. J Clin Invest 1994; 93, 2379–2386.

Safadi A, Livne E, Reznick AZ. Characterization of alkaline and acid phosphatase from skeletal muscle of young and old rats. Arch Gerontol Geriatr 1997; 24, 183–196.

Sala E, Roca J, Marrades RM, Alonso J, Gonzalez de Suso JM, Moreno A, Barberà JA, Nadal J, de Jover L, Rodríguez-Roisín R, Wagner PD. Effect of endurance training on skeletal muscle bioenergetic in chronic obstructive pulmonary disease. Am J Respir Crit Care Med 1999; 159, 1726–1734.

Salvarani C, Vlachonikolis IG, van der Heijde DM et al. Musculoskeletal manifestations in a population-based cohort of inflammatory bowel disease patients. Scand J Gastroenterol 2001; 36, 1307–1313.

Sato T, Akatsuka K, Kito K, Tokoro Y, Tauchi H, Kato K. Age changes in size and number of muscle fibers in human minor pectoral muscle. Mech Ageing Dev 1984; 28, 99–109.

Sato Y, Asoh T, Honda Y, Fujimatso Y, Higuchi I, Oizumi K. Morphological and histochemical evaluation of biceps muscle in patients with chronic pulmonary emphysema manifesting generalized emaciation. Eur Neurol 1997; 37, 116–121.

Satta A, Migliori GB, Spanevello A et al. Fibre types in skeletal muscles of chronic obstructive pulmonary disease patients related to respiratory function and exercise tolerance. Eur Respir J 1997; 10, 2853–2860.

Sauleda J, Garcia-Palmer FJ, Tarraga S, Maimo A, Palou A, Agusti AG. Skeletal muscle changes in patients with obstructive sleep apnoea syndrome. Respir Med 2003; 97, 804–810.

Schaufelberger M, Eriksson BO, Grimby G, Held P, Swedberg K. Skeletal muscle fiber composition and capillarization in patients with chronic heart failure: relation to exercise capacity and central hemodynamics. J Card Fail 1995; 1, 267–272.

Schols A, Mostert R, Soeters PB, Wouters EF. Body composition and exercise Performance in patients with chronic obstructive pulmonary disease. Thorax 1991a; 46, 695–699.

Schols A, Wousters E, Soeters P, Westertep K. Body composition by bioelectrical impedance analysis compared to deterium dilution and skinfold anthropometry in patients xith chronic obstructive pulmonary disease. Am J Clin Nutr 1991b; 53, 421–424.

Schols A, Buurman W, van den Brekel S et al. Evidence for a relation between metabolic derangements and increased levels of inflammatory mediators in a subgroup of patients with chronic obstructive pulmonary disease. Thorax 1996; 51, 819–824.

Schols A, Creutzberg E, Buurman W, Campfield L, Saris W, Wouters E. Plasma leptin is related to proinflammatory status and dietary intake in patients with chronic obstructive pulmonary disease. Am J Respir Crit Care Med 1999; 160, 1220–1226.

Schmidt FR, Cooper NS, Ziff M et al. Arteritis in rheumatoid arthritis. Am J Med 1961; 30, 56–83.

Severgnini S, Lowenthal DT, Millard WJ, Simmen FA, Pollock BH, Borst SE. Altered IGF-I and IGFBPs in senescent male and female rats. J Gerontol 1999; 54, B111–B115.

Shephard RJ. Exercise for patients with congestive heart failure. Sports Med 1997; 23, 75–92.

Shindoh C, DiMarco A, Nethery D, Supinski G. Effect of PEG-superoxide dismutase on the diaphragmatic response to endotoxin. Am Rev Respir Dis 1992; 145, 1350–1354.

Shoji S, Takagi A, Sugita H, Toyokura Y. Muscle glycogen metabolism in steroid-induced myopathy of rabbits. Exp Neurol 1974; 45, 1–7.

Simard C, Maltais F, Leblanc P, Simard PM, Jobin J. Mitochondrial and capillarity changes in vastus lateralis muscle of COPD patients: electron microscopy study. Med Sci Sports Exerc 1996; 28, S95.

Similowski Th, Yan S, Gaithier AP, Macklem PT. Contractile properties of the human diaphragm during chronic hyperinflation. N Eng J Med 1991; 325, 917–923.

Spiers S, McArdle F, Jackson MJ. Aging-related muscle dysfunction: failure of adaptation to oxidative stress? Ann NY Acad Sci 2000; 908, 341–343.

Spizer AR, Giancarlo T, Mahler L. Nueromuscular causes of prolonged ventilator dependency. Muscle Nerve 1992; 15, 682–686.

Stamler JS, Meissner G. Physiology of nitric oxide in skeletal muscle. Physiol Rev 2001; 81, 209–237.

Steele IC, Moore A, Nugent AM, Riley MS, Campbell NPS, Nicholls DP. Non-invasive measurement of cardiac output and ventricular ejection fractions in chronic cardiac failure: relationship to impaired exercise tolerance. Clin Sci 1997; 93, 195–203.

Stratton JR, Kemp GJ, Daly RC, Yacoub M, Rajagopalan B. Effects of cardiac transplantation on bioenergetic abnormalities of skeletal muscle in congestive heart failure. Circulation 1994; 89, 1624–1631.

Stryer L. Biochemistry. New York: WH Freeman and Company, 1988.

Sullivan MJ, Green HJ, Cobb FR. Skeletal muscle biochemistry and histology in ambulatory patients with long-term heart failure. Circulation 1990; 81, 518–527.

Sullivan MJ, Green HJ, Cobb FR. Altered skeletal muscle metabolic response to exercise in chronic heart failure. Relation to skeletal muscle aerobic enzyme activity. Circulation 1991; 84, 1597–1607.

Sullivan MJ, Hawthorne MH. Exercise intolerance in patients with chronic heart failure. Prog Cardiovasc Dis 1995; 38, 1–22.

Supinski G, Nethery D, DiMarco A. Effect of free radical scavengers on endotoxin-induced respiratory muscle dysfunction. Am Rev Respir Dis 1993; 148, 1318–1324.

Supinski G, Nethery D, Stofan D, DiMarco A. Comparison of the effects of endotoxin on limb, respiratory, and cardiac muscles. J Appl Physiol 1996; 81, 1370–1378.

Tada H, Kato H, Misawa T, et al. 31P-nuclear magnetic resonance evidence of abnormal skeletal muscle metabolism in patients with chronic lung disease and congestive heart failure. Eur Respir J 1992; 5, 163–169.

Tein I. Metabolic myopathies. Semin Pediatr Neurol 1996; 3, 59–98.

Teitelbaum JS, Borel CO. Respiratory dysfunction in Guillain-Barré Syndrome. In: Respiratory dysfunction in neuromuscular disease. Clin Chest Med. Fanburg BL and Sicilian L eds. WB Saunders Company. Philadelphia 1994, pp 705–714.

Teixeira A, Cherin P, Demoule A, Levy-Soussan M, Straus C, Verin E, Zelter M, Derenne JP, Herson S, Similowski T. Diaphragmatic dysfunction in patients with idiopathic inflammatory myopathies. Neuromuscul Disord 2005; 15, 32–39.

Tenhunen R, Marver HS, Schmid R. The enzymatic conversion of heme to bilirubin by microsomal heme oxygenase. Proc Natl Acad Sci USA 1968; 61, 748–755.

Tikunov B, Levine S, Mancini D. Chronic congestive heart failure elicits adaptations of endurance exercise in diaphragmatic muscle. Circulation 1997; 95, 910–916.

Tritschler HJ, Bonilla E, Lombes A et al. Differential diagnosis of fatal and benign cytochorme c oxidase-deficient myopathies of infancy: An immunohistochemical approach. Neurology 1991; 41, 300–305.

Tsushima K, Koyama S, Ueno M, Fujimoto K, Ichiyoshi T, Takei Y, Hanyu N, Kubo K. Rhabdomyolysis triggered by an asthmatic attack in a patient with McArdle disease. Intern Med 2001; 40, 131–134.

Van der Vliet A, Eiserich JP, Shigenaga MK, Cross CE. Reactive nitrogen species and tyrosine nitration in the respiratory tract: Epiphenomena or a pathobiologic mechanism of disease? Am J Respir Crit Care Med 1999; 160, 1–9.

Van der Walt JD, Swash M, Leake J et al. The pattern of involvement of adult-onset acid maltase deficiency at autopsy. Muscle Nerve 1987; 10, 272–281.

Vilardell M, Font J. Systemic Disorders. In: Internal Medicine. Rozman C, Cardellach F eds. Elsevier. Madrid 2004, pp 1091–1140.

Vincken WG, Gauthier SG, Dollfuss RE, Hanson RE, Darauay CM, Cosio MG. Involvement of upper-airway muscles in extrapyramidal disorders. A cause of airflow limitation. N Engl J Med 1984; 311, 438–442.

Voigt E, Griga T, Tromm A, Henschel MG, Vorgerd M, May B. Polymyositis of the skeletal muscles as an extraintestinal complication in quiescent ulcerative colitis. Int J Colorectal Dis 1999; 14, 304–307.

Waclawik AJ, Sufit RL, Beinlich BR, Schutta HS. Acute myopathy with selective degeneration of myosin filaments following status asthmaticus treated with methylprednisolone and vecuronium. Neuromuscul Disord 1992; 2, 19–26.

Warren S. The immediate causes of death in cancer. Am J Med Sci 1932; 184, 610–615.

Welle S, Thornton C, Statt M. Myofibrillar protein synthesis in young and old human subjects after three months of resistance training. Am J Physiol 1995; 268, E422–E427.

Whittom F, Jobin J, Simard PM, LeBlanc P, Simard C, Bernard S, Belleau R, Maltais F. Histochemical and morphological characteristics of the vastus lateralis muscle in COPD patients. Med Sci Sports Exerc 1998; 30, 1467–1474.

Wiedemann HP, Matthway RA. Pulmonary manifestations of the collagen vascular diseases. Clin Chest Med 1989; 10, 677–722.

Wilcox PG, Stein HB, Clarke SD et al. Phrenic nerve function in patients with diaphragmatic weakness and systemic lupus erythematosus. Chest 1988; 93, 352–358.

Wilcox PG, Paré PD, Road JD, Fleetham JA.. Respiratory muscle function during obstructive sleep apnea. Am Rev Respir Dis 1990;142, 533–9.

Wilcox PG, Wakai Y, Walley KR, Cooper DJ, Road J. Tumor necrosis factor α decreases in vivo diaphragm contractility in dogs. Am J Respir Crit Care Med 1994; 150, 1368–1373.

Williams TJ, Snell GI. Early and long-term functional outcomes in unilateral, bilateral, and living-related transplant recipients. Clin Chest Med 1997; 18, 245–257.

Wilson JR, Mancini DM. The mechanism of extertional fatigue in heart failure. Cardioscience 1990; 1, 13–17.

Wood PH. Appreciating the consequences of disease: the international classification of impairments, disabilities, and handicaps. WHO Chron 1980; 34, 376–380.

Yaggi H, Mohsenin V. Obstructive sleep apnoea and stroke. Lancet Neurol 2004; 3, 333–342.

Zarzhevsky N, Carmeli E, Fuchs D, Coleman R, Stein H, Reznick AZ. Recovery of muscles of old rats after hindlimb immobilization by external fixation is impaired compared with those of young rats. Exp Gerontol 2001; 36, 125–140.

Zattara-Hartmann MC, Badier M, Guillot C, Tomei C, Jammes Y. Maximal force and endurance to fatigue of respiratory and skeletal muscles in chronic hypoxemic patients: the effects of oxygen breathing. Muscle Nerve 1995; 18, 495–502.

Zhu E, Petroff B, Gea J, Comtois N, Grassino A. Diaphragm muscle injury after inspiratory resistive breathing. Am J Respir Crit Care Med 1997; 155, 1110–1116.

Zulueta JJ, Fanburg BL. Respiratory dysfunction in Myasthenia Gravis. In: Respiratory dysfunction in neuromuscular disease. Clin Chest Med. Fanburg BL and Sicilian L eds. WB Saunders Company. Philadelphia 1994, pp 683–691.

INDEX

CPSIA information can be obtained
at www.ICGtesting.com
Printed in the USA
LVHW051558030520
654913LV00007B/1452